Birkhäuser

Hermann Graßmann

From Past to Future: Graßmann's Work in Context

Graßmann Bicentennial Conference, September 2009

Hans-Joachim Petsche
Albert C. Lewis
Jörg Liesen
Steve Russ
Editors

Birkhäuser

Editors

Hans-Joachim Petsche
Universität Potsdam
Institut für Philosophie
Am Neuen Palais 10
14469 Potsdam
Germany
petsche@uni-potsdam.de

Jörg Liesen
TU Berlin
Institut für Mathematik
Straße des 17. Juni 136
10623 Berlin
Germany
liesen@math.tu-berlin.de

Albert C. Lewis
Educational Advancement Foundation
Rio Grande St. 2303
Austin, TX 78705
USA
alewis@edu-adv-foundation.org

Steve Russ
University of Warwick
Dept. Computer Science
Coventry CV4 7AL
United Kingdom
sbr@dcs.warwick.ac.uk

ISBN 978-3-0348-0738-8 ISBN 978-3-0346-0405-5 (eBook)
DOI 10.1007/978-3-0346-0405-5

Cover figure: Hermann Günther Graßmann, xylograph after a photograph from 1874 (mirror inverted). Source: Hermann Graßmann. Gesammelte mathematische und physikalische Werke. Bd. 1.1. Herausgeg. von Fr. Engel unter Mitwirkung von E. Study, Leipzig 1894; background: The New Palace campus of the University of Potsdam (Communs of the New Palace at Park Sanssouci, a UNESCO World Heritage Site).

Printed on acid-free paper

Springer Based AG is part of Springer Science+Business Media

www.birkhauser-science.com

Contents

Illustration 1. Conference venue at University of Potsdam

Preface

On the occasion of the 200th birthday of Hermann Günther Grassmann[1] (1809–1877), an international and inter-disciplinary conference took place from the 16th to the 19th September 2009 in Potsdam and Szczecin. For four days more than 75 scientists from four continents – mathematicians, logicians, soft- and hardware specialists, physicists, engineers, historians, philosophers, scholars of religious studies, pedagogues, Indologists, musicologists, and many more – participated in an inter-disciplinary dialogue on the work of Grassmann and his significance for current research.

Shortly before this conference, the first comprehensive English-language biography of Grassmann appeared, *Hermann Grassmann – Biography*. In addition, the newly published volume *Hermann Grassmann – Roots and Traces* presents many previously unpublished documents on the life and work of Grassmann and his family.

The present book, which includes the Proceedings of the Bicentennial Conference as well as additional documents related to Grassmann, is the third and concluding volume of this Grassmann-Trilogy.[2]

Point of departure

The last major conference concerned with Grassmann was devoted to the 150th anniversary of the first publication of his *Lineale Ausdehnungslehre*. It took place in 1994 on the island of Rügen. Two publications with first-class contributions resulted from this conference [Schreiber 1995; Schubring 1996a]. In the following 15 years

[1] Like his father Justus Günther, Hermann Graßmann used to spell his name with the German letter "ß". Hermann's brother Robert and his fosterfather Friedrich Heinrich Gotthilf Grassmann preferred "ss". In this book, we will write Grassmann and always use "ss", except in some titles and quotations.

[2] The conference proceedings will be complemented by a video-recording of the conference on four DVDs, published simultaneously by Universitätsverlag Potsdam [Petsche and Lenke 2010].

the Rügen conference has stimulated numerous new publications. Some of them have a mathematical approach, some deal with philosophical aspects, and some reflect pedagogical contexts of Grassmann's works.

A Spanish translation of Grassmann's *Lineale Ausdehnungslehre* of 1844 by Emilio Oscar Roxin appeared in 1947. A French translation of the same work by Dominique Flament and Bernd Bekemeier followed in 1994. In the most ambitious translation project of Grassmann's works yet, Lloyd Kannenberg published in 1995 and 2000 English translations of the 1844 and 1862 versions of the *Ausdehnungslehre* and several other important papers of Grassmann. Finally, in 2008 the Russian translations of selected papers of Hermann and Robert Grassmann – thoroughly introduced and with extensive commentary – were published by Boris Biryukov, Lyubov Gavrilovna Biryukova, and Zinaida Kuzicheva.

Conference topics

The idea of the conference was to review and present a multi-faceted picture of Hermann Grassmann. An initial goal was to uncover the complexity of factors that had made Grassmann's creativity possible. In addition, we wanted to analyze the variety of the influential roles that his ideas have played and continue to play up to the present. The Conference Program organized the range of subjects in four focal themes:

1 Historical contexts of Grassmann's creativity.
2 The diversity of Grassmann's influence.
3 Present and future of Grassmann's ideas.
4 Grassmann and his hometown: from past to future.

These focal themes evolved into the several sections of the *Contents* of the current volume.

While at the Rügen conference the mathematical and natural scientific works of Grassmann stood at the center of the discussions, the Bicentennial Conference strived for a more comprehensive access to Grassmann's works. In this conference we investigated the significance of his work for philosophical as well as for scientific and methodological questions, for comparative philology in general and for Indology in particular, for psychology, physiology, religious studies, musicology, and didactics. Special attention was given to the question which influences of Grassmann's work are noteworthy for current scientific research. Accordingly, thanks to the breadth of the contributions received, the subdivision of the focal points of this book arrived at a certain arrangement that permits the relevance of Grassmann to emerge more clearly. Readers interested in the natural sciences or mathematics and those who are more humanities oriented will both find that this makes Grassmann more accessible to them.

The 37 contributions of 40 authors from 15 countries presented in this volume convincingly demonstrate that Grassmann's legacy is more important than ever.

Structure

The main part of this volume is formed by papers presented at the Bicentennial Conference. In order to introduce the reader to Hermann and Robert Grassmann, and to achieve a certain independence within the framework of the Grassmann-Trilogy, we start this volume with translations of two biographies: the previously unpublished life history of Hermann Grassmann (written by his son Justus Grassmann), and the life history of Robert Grassmann (written by himself and concealed in a footnote of his – now scarcely available – *Gebäude des Wissens*).

An English translation of the paper *Ueber den Begriff und Umfang der reinen Zahlenlehre* by Justus Grassmann (Robert and Hermann's father) is added as an Appendix. This paper, in a translation prepared by Lloyd Kannenberg is published here for the first time and offers a deep insight into the world of ideas of the father and permits the extent of his influence on his son Hermann to be traced out.

To the book is appended an index of the references of all contributions. This offers an impression of the variety of publications about Grassmann, and simplifies further investigations. The name and citation indexes aid the classification of the individuals mentioned in the contributions. Moreover, from the citation directory the references of the individual contributions to the different sources can be conveniently found.

Acknowledgements

Neither the arrangements for the conference, nor the publication of this book, would have been possible without the active support of the following institutions:

- German Research Foundation (DFG)
- German Federal Ministry of Education and Research (BMBF)
- University of Potsdam
- University of Szczecin

- Universitätsgesellschaft Potsdam e. V.
- Birkhäuser Publishing Ltd.

We have been able to present in this volume several previously unpublished documents on the life of Grassmann thanks to the friendly support of librarians and archivists at the following institutions:

- Archiv Walter de Gruyter of the Staatsbibliothek Berlin PK
- Greifswald University Archive
- Archiwum Państwowe in Szczecin
- Książnica Pomorska im. Stanisława Staszica in Szczecin

Translations of the original documents on the lives of the Grassmann brothers as well as the paper of Justus Grassmann on the pure theory of number were generously contributed by Prof. Lloyd Kannenberg. In addition, we thank Prof. Kannenberg for reviewing several of the manuscripts.

Further thanks are owed to Peter Lenke, Mark Minnes, and John Ohala, who stood by our side with help in checking through the manuscripts, Gottfried Kessler, who was helpful with accessing the archival sources, as well as Matthes Uhlmann, who contributed to the editing of the visual materials. We are also pleased to thank everybody who supported our authors and who contributed to the entire project by checking facts, voicing criticism, or being involved in some other way.

Finally, our warm thanks are due to Dr. Karin Neidhart, Birkhäuser's editor of the entire project of the three Grassmann volumes:

Hermann Graßmann – Biography

Hermann Graßmann – Roots and Traces

Hermann Graßmann – From Past to Future

She has facilitated, guided, supported, and helpfully promoted these volumes at every stage.

For the editors, Hans-Joachim Petsche

On 16 November 2010, the Graßmann-Conference received the Potsdam Conference-Award from the mayor of Potsdam, the capital of the federal state of Brandenburg, making it the first-time winner in the category "Innovative and/or exceptional events".

Illustration 2. Opening session of the Grassmann Bicentennial Conference in Potsdam

Illustration 3. Hermann Grassmann's great-grandson Dietrich Grassmann and the main organizer of the conference Hans-Joachim Petsche in front of a picture showing the Stettin Masonic lodge garden in the nineteenth century

Illustration 4. Inauguration of a Grassmann exhibition unveiling recently discovered documents in the Pomeranian Library (during the conference's stay in Szczecin)

Illustration 5. Participants of the Grassmann Conference in front of the former Marienstiftsgymnasium in Szczecin

Abbreviations for works of Hermann Grassmann

A1 Grassmann, Hermann. 1844. *Die Wissenschaft der extensiven Grösse oder die Ausdehnungslehre, eine neue mathematische Disciplin.* 1. Aufl. Leipzig: Wigand (2. Aufl. Leipzig 1878). [Reprinted in: GW11, 4–312].

$A1_K$ Grassmann, Hermann. 1995b. *Linear extension theory* (Die Lineale Ausdehnungslehre, 1844), transl. Lloyd C. Kannenberg. In [H. Grassmann 1995a, 1–312].

A2 Grassmann, Hermann. 1862. *Die Ausdehnungslehre. Vollständig und in strenger Form begründet.* Berlin: Enslin. [Reprinted in: GW12, 1–383].

$A2_K$ Grassmann, Hermann. 2000. *Extension theory* (Die Ausdehnungslehre, 1862), transl. Lloyd C. Kannenberg. American Mathematical Society, London Mathematical Society.

LA Grassmann, Hermann. 1861: *Lehrbuch der Mathematik für höhere Lehranstalten.* Teil 1: Arithmetik, Berlin: Enslin.

RV1 Grassmann, Hermann. 1876. *Rig-Veda. Übersetzt und mit kritischen und erläuternden Anmerkungen versehen von Hermann Grassmann. Erster Theil. Die Familien-Bücher des Rig-Veda: (zweites bis achtes Buch).* Leipzig: F. A. Brockhaus.

RV2 Grassmann, Hermann. 1877a. *Rig-Veda. Übersetzt und mit kritischen und erläuternden Anmerkungen versehen von Hermann Grassmann. Zweiter Theil. Sammelbücher des Rig-Veda: (erstes, neuntes, zehntes Buch).* Leipzig: Brockhaus.

HGW11 Grassmann, Hermann. 1894. *Gesammelte mathematische und physikalische Werke.* Vol. 1.1. Ed. Friedrich Engel assisted by Eduard Study. Leipzig: Teubner.

HGW12 Grassmann, Hermann. 1896. *Gesammelte mathematische und physikalische Werke*. Vol. 1.2. Ed. Friedrich Engel, assisted by Hermann Grassmann (Jun.). Leipzig: Teubner.

HGW21 Grassmann, Hermann. 1904. *Gesammelte mathematische und physikalische Werke*. Vol. 2.1. Eds. Eduard Study, Georg Scheffers and Friedrich Engel. Leipzig: Teubner.

HGW22 Grassmann, Hermann. 1902. *Gesammelte mathematische und physikalische Werke*. Vol. 2.2. Eds. Jakob Lüroth and Friedrich Engel. Leipzig: Teubner.

HGW31 Grassmann, Hermann. 1911. *Gesammelte mathematische und physikalische Werke*. Vol. 3.1. Eds. Justus Grassmann and Friedrich Engel. Leipzig: Teubner.

HGW32 Grassmann, Hermann. 1911. *Gesammelte mathematische und physikalische Werke*. Vol. 3.2. Ed. Friedrich Engel. Leipzig: Teubner.

Illustration 6. Pestalozzi in Stanz (lithography, 1846). An artistic representation of the many followers of the pedagogue J. H. Pestalozzi (1746–1827). Among them, Hermann's uncle, Friedrich Heinrich Gotthilf Grassmann (1784–1866). The picture documents the great influence of progressive pedagogy on Hermann Grassmann's uncle and on his father.
Further information on p. 490

On the lives of the Grassmann brothers

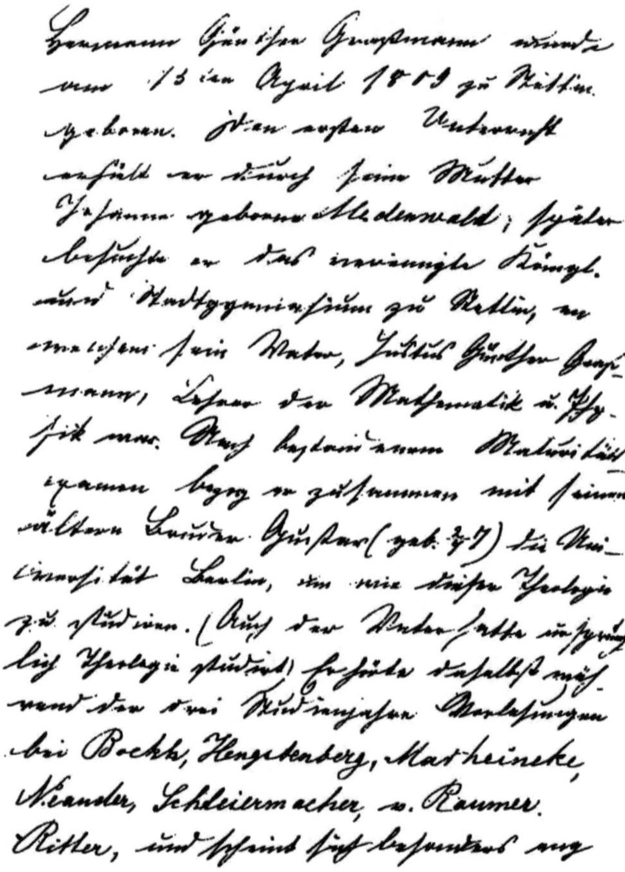

Hermann Günther Grassmann wurde
am 15ten April 1809 zu Stettin
geboren. Den ersten Unterricht
erhielt er durch seine Mutter
Johanne geborene Medenwald; später
besuchte er das vereinigte Königl.
und Stadtgymnasium zu Stettin, an
welchem sein Vater, Justus Günther Grass-
mann, Lehrer der Mathematik u. Phy-
sik war. Nach bestandenem Maturitäts
examen bezog er zusammen mit seinem
älteren Bruder Gustav (geb. 07) die Uni-
versität Berlin, um wie dieser Theologie
zu studiren. (Auch der Vater hatte ursprüng-
lich Theologie studirt) Er hörte daselbst wäh-
rend der drei Studienjahren Vorlesungen
bei Boeckh, Hengstenberg, Marheineke,
Neander, Schleiermacher, v. Raumer,
Ritter, und scheint sich besonders an

Illustration 7. First page of the description of the life of Hermann Grassmann by his son Justus. See also p. 491
[Książnica Pomorska im. Stanisława Staszica w Szczecinie, Oddział Rękopisów]

Description of the life of Hermann Grassmann by his son Justus Grassmann, probably written shortly after the death of his father, 1877

(Published here for the first time since the original was found in the family Nachlass of the Grassmanns, Pomeranian Library in Szczecin, Manuscripts and Old Prints Department)

Hermann Günther Graßmann was born on the 13th of April 1809 in Stettin. He received his first education from his mother, Johanne née Medenwald; later he attended the Unified Roy[al] and City Gymnasium in Stettin, at which his father, Justus Günther Graßmann, was Instructor of Mathematics and Physics. After completing the maturation examination he attended, together with his elder brother Gustav (born 2/7 07), the University of Berlin, in order to study theology. (Their father had also originally studied theology.) There, during his three years of study, he attended lectures with Boekh, Hengstenberg, Marheineke, Neander, Schleiermacher, v. Raumer, Ritter, and seems to have been drawn especially close to Schleiermacher, on whose lectures and works many commentaries were found among the papers he left behind. Of mathematical lectures he heard none; but he did deal with mathematics, in which, even at the Gymnasium, he did not fail in outstanding achievement, and especially *when* his father's most important work, "On Physical Crystallonomy and Geometric Combination Theory" of 1829 appeared, he studied and worked through it completely. He also

H.-J. Petsche et al. (eds.), *From Past to Future: Graßmann's Work in Context*,
DOI 10.1007/978-3-0346-0405-5_1,

concerned himself deeply with the other works of his father, especially his article "On the pure theory of number" (Program of the Stettin Gymnasium, 1827), // his Theory of Space and Trigonometry, and thus the peculiar mathematical viewpoint of these works were of no little influence on his later math[ematical] works. In October 1831 he entered the Royal Seminar for Scholarly and Higher Training, and in November of the same year passed his state philological examination (esp[ecially] for the subjects Lat[in], Gr[eek], and mathematics). Nevertheless, he had not yet given up his favorite wish of becoming a minister, and worked diligently at theology and on 1834 22/5 stood for his first, 1839 12/7 his second theological examination – the former with the grade "good," the latter with the grade "very good." Finally, on 1840 1/5, he com//pleted, with a re-examination, his scientific certificate of Nov[ember] [18]31 and earned his unconditional qualification to teach in math[ematics], physics, mineralogy, and chemistry, for which subjects the certificate recommended him as fully qualified. (Examination paper: On the Theory of Ebb and Flow.) In the mean time, in 1834 he went to the Trade School in Berlin as Instructor, where particularly in the company of Jacob Steiner he received new stimulation *toward math[ematical] works*. On 1836 1/1 he became Instructor at the Otto School (dealing with chemistry), Mich[aelmas] of 1842 at the Gymnasium in Stettin. In the latter year took place his lectures on his most important math[ematical] work, the Ausdehnungslehre, held prior to its publication before a private circle of his brother Robert Gr[assmann], von Kamecke (now Minister of War), Jungklass (now Schools Inspector in Bernberg), and Scheibert (now Privy Government and Schools Inspector, Retired, in Schleswig). At Easter 1843 he became Instructor (on 1847 5/5 Senior Instructor) at the Fried[rich]–Wilh[elms] School (Secondary [School]), 1st Order, in Stettin; (thanks to his instructional work at this school there appeared several small textbooks, thus: Manual for the First Lessons in the Latin Language (1843), Manual for the German Language, 1848 (together with his brother Robert), German Reader (together with W. Langbein) – The year 1848 mainly laid claim to his activity in the political field. In faithful devotion to the royal house he publicly raised his voice, one of the first, against // the revolutionary strife in Berlin (e.g., the news article in the Vossischen Zeitung "The Fruits of the Barricades Battle in Berlin"), and together with his brother Robert founded a newspaper (Deutsche Wochenschrift) in which the most burning questions, principally the newly-to-be-established Constitutions of the German Empire, the Prussian State and the Church in Prussia, would be discussed. In the same year (10/8) he became engaged to Marie Therese Knappe, daughter of the widowed lady of the manor at Alt-Storkow near Nörenberg in Pomerania. On the 12th of April of the following year the wedding took place, and soon there gathered around the parents a happy throng of children (in all 11, still living 8), on whose great and small adventures their father, despite // his professional and scientific activities, always kept an eye both watchful and loving. In the year 1852 on 1/7 he obtained, after his father died on 9/9, the position of First Mathematician at the Unified Roy[al] and City Gymnasium (the later Marienstift-Gymnasium) and soon thereafter the title of Professor. In this position he remained active until the end of his life. *He died as a consequence of heart failure, which had been unknown to him earlier and first made itself felt 2 years before his death and under a gradual decline of strength (dropsy finally appearing in consequence of thinning of the blood) laid him peacefully to sleep.*

In the year 1844 appeared the first part of the Ausdehnungslehre from Wigand in Leipzig (in consequence of a correspondence with Professor Möbius), in 1847 his "Geometric Analysis," a treatment of the prize problem of the Princely Jablonowskian Society for the reconstruction and further development of the geometric calculus invented by Leibniz, or the establishment of a calculus similar to his (prize awarded 1/7 1846), and then followed // numerous articles, most of geometric content, in Crelle's Journal; and a few on physics in Poggendorf's Annalen. Meanwhile, on the whole these works did not produce the desired result. Individual recognition was indeed not lacking (Moebius, Bellavitis, Cremona) and especially many of the smaller geometrical articles gave rise to applied investigations. But the peculiar core of the works and especially the Ausdehnungslehre remained strange, and nowhere did a mention of this work occur. Then he decided, since he believed that this failure must be ascribed to its form of presentation, to undertake simultaneously with the publication of the second part a complete reworking of the first part as well. But after the new work *(appearing 1862)* also received no more attention // than the first and at the same time all attempts to gain an academic chair and thus ~~to~~ about him to gather a circle of students which he could win over to his ideas, came to nothing, with a heavy heart he turned from mathematics to linguistic studies, on which he had already long worked as a relaxation from his math[ematical] works. Already in the year 1852 he had occupied himself with Sanskrit (in 1857 together with Dr. Bressler), later he worked on Greek, Lithuanian, Zend, etc., and when his first ling[uistic] works in Kuhn's Zeitschrift found a favorable reception, he decided to publish a major work, first a Rig-Veda dictionary, then a Rig-Veda // translation, after both of which would follow a grammar. At this time also appears his small work on "German Plant Names" (1870) through which he sought to introduce a German nomenclature for plants into botany. Finally in 1872, after a lapse of more than ten years, he again took up his math[ematical] works, especially after Herm[ann] Hankel, Alfred Clebsch, and Victor Schlegel had called attention to the significance of his Ausdehnungslehre.

The extremely abundant recognitions of him that were in part for his linguistic works and also the joy of seeing his mathematical labors no longer lying idle, but instead seeing them recommended by impor- // tant mathematicians to intermesh in the most varied disciplines, in great measure beautified for him the evening of his life. Only a few days before his death, Prof. Burmeister in Dresden informed him that he was thinking of giving lec[tures] on the Ausdehnungslehre in the coming winter semester; and with joy he could now leave to others that which an unfavorable fate had denied to himself. Signs ~~of the~~ of *external* recognition were the appointment as a Corresponding Member of the *Roy[al]* Association of Sci[ences] at Goettingen (2/2 71), the appointment to an Honorary Doctorate of the Philos[ophical] Faculty of the University of Tübingen (1876), the appointment as a member of the American Oriental // Society (1876) etc.

His activities were also directed to other areas. Thus in the years 61–72 he had had collected many folk songs, written out their melodies by ear and set them for three-part singing in his family (By chance, four of them were published in the Festgruß for Ludwig Erk, Heilbronn 1876) – Further, he led a men's *choral* society of Stettin students. He was a member of the Lodge, into which he entered at Christmas 1835. From the year 1856 he became its Speaker. From 1857 he was a member of the board of the Chinese Missionary Association in Pomerania, from 1859 Secretary and later Chairman. From 1852 he led the Physical Society founded in Stettin by his father. – etc. // Still to be mentioned is a small theological paper "On the loss of faith," which he wrote not long before his death, and which will shortly appear in print.

Editorial Notes:

- Page breaks in the notes of Justus Grassmann are indicated by "//".
- Text struck out is taken over as text struck out.
- Insertions in the margin or above the lines are written in *italics.*
- Proper completions are inserted between brackets "[,]".

(Transcribed from the original by G. Keßler and H.-J. Petsche, August 2009.
Translation by Lloyd Kannenberg, February 2010.)

Königl. Wissenschaftliche
Prüfungs-Commission
zu Greifswald

Acta

betr:

die Prüfung des Schulamts-
Candidaten
Robert Siegfried Ludolph Grass-
mann aus Stettin
de 1840.

Illustration 8. Cover sheet of Robert Grassmann's teacher's examination file from 1840. See also p. 494
[Universitätsarchiv Greifswald, Prüfungsakten]

Life history of Robert Grassmann, written by himself (1890)

(Grassmann, Robert: Footnote to the Foreword of the "Gebäude des Wissens")

To the author it has always been a necessity, on studying a book, to imagine the author of the book alive before him. He therefore obtained a picture of him and strove to learn enough of the course of his life to be able to judge the course of his development. The author assumes a like necessity with many of his readers; he has therefore, in this first volume, included his picture[1] and in the following offers the reader a brief survey of the course of his development.

The author was born on 8 March 1815 and at first raised in the house of his father, Justus Grassmann, Professor of Mathematics and Physics at the Gymnasium in Stettin, then from his 3rd year in the house of his uncle, Privy Government and Schools Inspector and Seminary Director Heinrich Gotthülf Grassmann of Stettin. He obtained his school education from 1825 to 1834 at the Marienstift Gymnasium in Stettin, his academic education from 1834 to 1838 at the Universities of Bonn and Berlin, where he studied natural science (with Treviranus, Dove), philosophy (with Brandis, Fichte), and Protestant theology (with Nitzsch, Bleek, Neander), undertook long journeys through Germany, Switzerland, Italy, and Austria, as well as through Belgium, Holland, and France, and finally at Berlin passed his first theological examination with the certificate "good". After his return to Stettin he was appointed, at first in 1838 for three months as

[1] This announcement seems not to have been realized! Copies of this book in Berlin (Staatsbibliothek and Humboldt University Library), London (British Library), and Strasbourg (Bibliotheque Nationale) do not contain a picture of Robert Grassmann (H.-J. Petsche & J. Liesen).

H.-J. Petsche et al. (eds.), *From Past to Future: Graßmann's Work in Context*,
DOI 10.1007/978-3-0346-0405-5_2, © Springer Basel AG 2011

first instructor at the then School-Teacher Seminar, and then in 1839 for three months to stand in for its Director. At this time he began the study of higher mathematics and physics, mostly from French works, while at the same time serving from 1838 to 1839 as a Pioneer. As such he received the request to produce galvanic-activated mine detonators, which at that time were activated by so-called detonation-sausages (tubes loaded with gunpowder); he succeeded in building an apparatus that ignited three mines simultaneously at up to 600 paces, and this apparatus was soon thereafter introduced into the Prussian army. The author was released from the Service in 1839, after he had passed the examination for Pioneer-Officer with the certificate "good". In the year 1840, before the scientific examination commission at Greifswald, he then achieved the certificate for unrestricted facultas docendi with the authorization to teach in mathematics, physics, philosophy, and theology for Prima, in all other subjects for the middle classes of the Gymnasium.

The author now acted from 1841 to 1852 as Instructor at the higher schools of Stettin, in 1845 became Senior Instructor and in the same year was married to Miss Charlotte Jeanneret from Switzerland, with whom to this day he lives in wedded bliss. From this marriage seven children were born to him, of whom three: a son Gustav and two daughters, Elise and Helene, are still living. In the year 1846 the author established a scientific society in Stettin and delivered scientific lectures in it from 1846 to 1848. The author had at this time undertaken major scientific studies. Already as a Gymnasium student in the upper classes he had devoted 2–3 h daily to his private studies; he still has six quarto volumes of work of mathematical and physical content from that time. Then, in preparation for the examination, he had made independent extracts from numerous works. As an Instructor he continued this activity. For the language classes he studied the works of K.F. Becker, Herling, and Jacob Grimm, and, together with his brother Hermann Grassmann wrote a German grammar and a short Latin grammar. For the physics class he wrote a physics exercise booklet, for the geography class a manual in geography and an atlas. His principal studies however he devoted to the philosophers. He studied Aristotle and Plato among the ancients, and mainly Hegel and Schleiermacher among the moderns. With these studies he soon recognized that philosophy in its present form, despite the many often sparkling intellects in it, was without a scientific basis and without scientific method, and set as his life's problem to create for philosophy a scientific basis valid for all time, similar to that possessed by natural science. He has for 45 years since that time devoted 4–8 h nearly every day to the most strenuous efforts on this problem and submits in the present work the result of those efforts.

But in the year 1848 there entered into the life of the author a great reversal. The great governmental movements and radical changes of the year 1848 aroused the writer and his brother, the later Professor Hermann Grassmann, from their studies.

They recognized that it was necessary to lead the movement in an orderly, steady way. Both at once stood in their conviction with all their powers. They laid down their program in the *Deutsche Wochenschrift*, which appeared from the end of March until June 1848. On 1 July 1848 the two of them then established their own printing works together with a publishing house, at the same time publishing a large political newspaper under the title *Norddeutsche Zeitung* and a trade paper. His brother, the later Professor Hermann Grassmann, withdrew from this business in 1851; however the author continued alone on its account. From 1 July 1848 until the year 1855 he was Editor-in-Chief of the *Norddeutsche Zeitung* and wrote numerous lead articles of political and national-economic content for the paper. The orientation of the newspaper was free conservative with a German national traditional stance. The newspaper advocated with warmth and vigor the unification of the German fatherland under the scepter of the Hohenzollerns, the development of transport and trade and the development of the school system, and supported the government insofar as it agreed with the convictions of the author. It strongly opposed the immoral degeneration and irreligious practices of many circles and pursued the vain and dictatorial decrees of certain party leaders with apt and severe ridicule, while it most willingly published the columns and granted support to every tendency that would serve the country and public weal and would further freedom and civilized behavior.

To the author it was at that time repeatedly given to intervene supportively in political life. When in 1848 the National Assembly turned to a dangerous course and in November 1848 the majority of it, despite its dissolution, continued in session, the author hastened to Berlin in order to gain a reliable judgment of the situation. On 11 November he had in the apartments of the Professors' Group in Berlin a long meeting with several members of the Zentrumspartei and was there convinced that the majority of the National Assembly steered toward a republic, and were also no longer to be dissuaded from this course of action. On the following day he then hastened to former president of the National Assembly Grabow and to the Minister of Interior von Manteuffel, as well as to Potsdam to the Lords of the Court and to Representative Bassermann from Mannheim, who was sent by the German Parliament in Frankfurt a. M. to Berlin, in order to inform himself about this occurrence and to orientate himself to the situation. All these men were agreed that the activities of the National Assembly had to be brought to a halt. Herr von Manteuffel would have imposed the Belgian Constitution. The author pointed out how much more correct it would be to publish the Constitution agreed to by the National Assembly provisionally as the fundamental law of the State and then to improve it further; Herr von Manteuffel recognized this and proceeded accordingly. Meanwhile, in the absence of the author, the officials of Stettin, under the terrorism of the so-called Progress Party, declared against the Ministry and for the National Assembly; a deputation was sent to Berlin to express the assent of the

Stettiners to the National Assembly, and met the author on his return trip from Berlin to Stettin. The problem of the author was now to lead the Stettiners back to the correct path again; the author and his friends succeeded in solving this problem. The City Council of Stettin on 16 November withdrew the resolution earlier agreed upon and on the basis of the report of its Commission declared itself for the government and against the National Assembly. A detailed presentation of this resolution was even sent the same evening from the author to all magistrates of Prussian cities and presented to the Frankfurt Parliament. Numerous letters of thanks were the response. The danger was eliminated.

In the year 1849 the author was called to Berlin in order to undertake the editing of the German Reform. In a short time he succeeded in doubling the number of subscriptions to the paper; but in consequence of differences with the publisher, which would have decisively interfered with the editing, the author returned to Stettin after a few months.

Now in Stettin, in 1850 the author received an order from the Protestant Consistory for Pomerania to undertake a critical edition of the text of the Bollhagen and Stargard hymnals. The Hymnological Section of the Royal Library in Berlin was placed at his disposal for this purpose; the author succeeded in restoring the original texts of almost the entirety of the hymns of the hymnals mentioned. In the year 1852 the author established for this his own type foundry and had the two hymnals set in the form of stereotype plates and issued them to bookshops. In the year 1853 he then built his own printing plant, but a steam boiler that was installed in the plant, and served for the operation of a Fournier cutting machine, exploded in August 1854, while the author was living in the country, destroying the greater part of his print works and forcing the author to sell the *Norddeutsche Zeitung* and resign the editorship.

The author now found time again to turn his attention exclusively to his literary studies. In 1855 he worked, together with his brother, the later Professor Hermann Grassmann, on the theory of number and extension theory, and in 1856 logic and combination theory as well. Then in 1856 the author alone published the Biblical History for country schools, for city schools and for higher schools, as well as the Life of Jesus and the Chronology in the Life of Jesus, and worked on individual branches of his great work, at least in their first outlines. Also in the following years 1857–1865 he occupied himself almost exclusively with these works, but found besides the inspiration to prepare three drafts for a new appendix to the Bollhagen and Stargard Hymnals in the years 1857, 1858, and 1859. The new hymns in the new edition of the Stargard hymnal and likewise those in the new edition of the Bollhagen hymnal are taken from these drafts in Hessenland.

In the year 1861 the author founded a Hebrew print works, which in 4 1/2 years produced from stereotype plates a great number of Hebraic-Talmudic works (Talmud, eight volumes, Mishnah in two editions of six volumes, Shulkhan Arukh in two editions of five volumes, Siddur four editions, Chumash in two editions of four volumes) and sent many thousand copies to Poland; the author established his own warehouse in Warsaw, which managed these sales.

In the year 1862 the author let the first book of his great work appear on a trial basis in bookshops under the title *Atomenlehre.*

In the year 1865 he again founded a large newspaper, the *Stettiner Zeitung*, which was followed in 1867 by a second paper, the *Pommersche Zeitung*, both under the editorship of the author. Unfortunately soon thereafter the business in Warsaw took a very unfortunate turn. The ruble sank permanently to 2/3 its value; all wares stored in Warsaw were lost, so that the author lost what was then his entire fortune and in 1869 was forced to declare bankruptcy in Stettin.

With the support of his friends, however, he quickly succeeded in overcoming this misfortune; already on 18 January 1871 he could present to the Court proof that all the creditors were fully satisfied by him for all their demands in the bankruptcy as to capital, interest, and costs. In the same year the son of the author, Dr. Gustav Grassmann, entered the business and relieved the author from a non-trivial part of his business work; but the business from that time achieved a significant upswing. At present the business of the author publishes three large daily newspapers in the format of the *Kölnischen Zeitung*, with 48,000 subscriptions, prints the Bibles of the Main Prussian Bible Society and numerous other works, operates 12 high-speed presses and a rotary press continually, and his own type foundry and galvano-plastic institution with electric machines, as well as a large book bindery and mint with numerous machines that are combined in the business. All these works are driven by two steam engines and employ well over 100 men. The author has now again devoted the greater part of his energies to his {literary} works, while at the same time he is active as City Marshal and leader of the Citizens Party for the welfare of his native city.

In the year 1872 the author again published the *Atomistik* in a completely revised form. In the same year then followed the first edition of the *Formenlehre oder Mathematik*, proceeding purely in formulas. Logic is here for the first time derived strictly scientifically in formulas and has found approval with the mathematicians of Europe and America. In the year 1873 followed the *Erdgeschichte oder Geologie* and from 1875 to 1876 the *Wissenschaftslehre* in four books. The latter was to present an introduction to the *Gebäude des Wissens*; it introduces philosophy in the first book, natural science in the second, political science in the third, and the science of God in the fourth, and each book presents a synopsis of those individual sciences. The Work appeared in 1882 under the title *Einleitung in das Gebäude des Wissens* in a new edition.

The author recognizes that the published Works are not yet satisfactory in form; rather, they present the subjects as if they were proven. The author now seeks to remedy these shortcomings in his *Gebäude des Wissens*; he added rigorously scientific proofs for the more important theorems. In this form the first eight volumes of his *Gebäude des Wissens* have appeared from 1881 to 1890 [...].

(Grassmann, Robert: Vorwort zum Gebäude des Wissens. In: Das Gebäude des Wissens. Bd. 1: Die Wissenslehre oder die Philosophie. 1. Teil: Das Verstandeswissen oder das formale Wissen, umfassend die auf die Philosophie vorbereitenden Wissenschaften. Stettin: R. Grassmann 1890, footnote, pages XIX–XXIV. Translation by Lloyd Kannenberg, February 2010.)

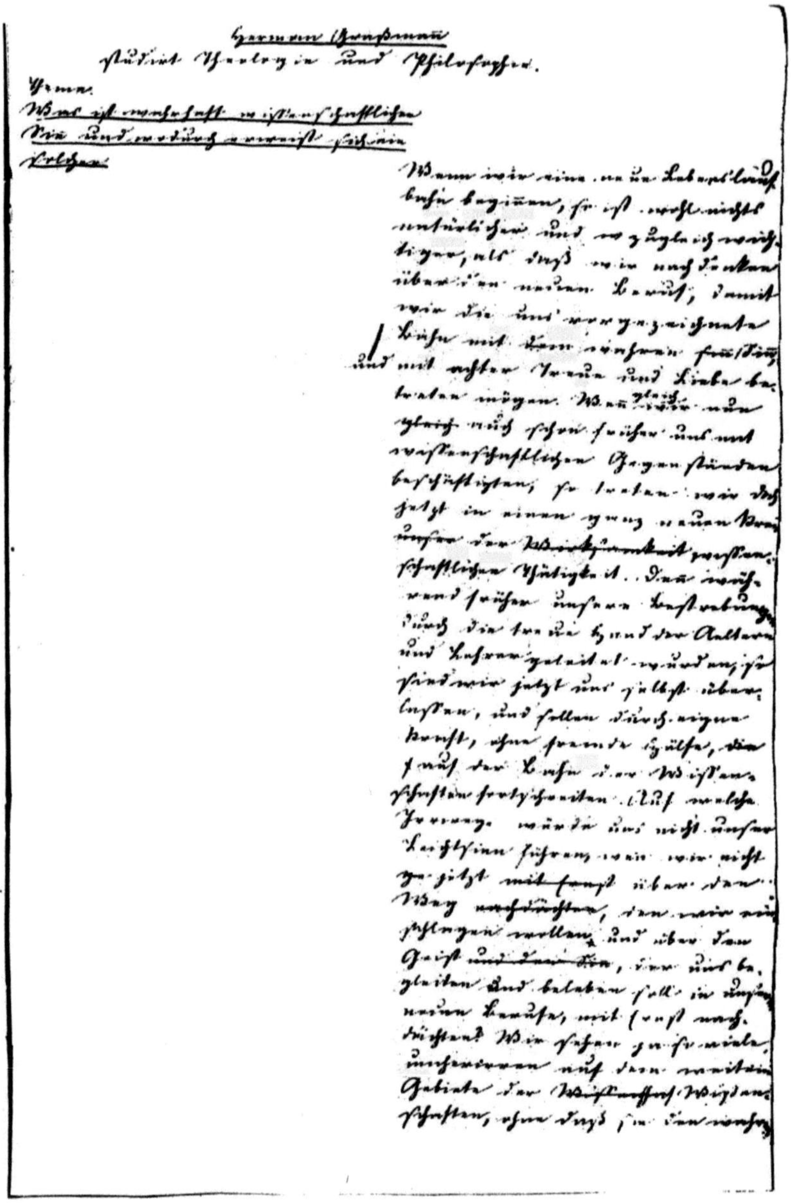

Illustration 9. First page of Hermann Grassmann's final examination essay from secondary school, written in 1827 on the topic: "What is genuine scientific spirit and how does it reveal itself as such?" The transcription and translation are given on p. 495
[Archiwum Państwowe w Szczecine, Gimnazjum Mariackie w Szczecinie]

Historical contexts of Hermann Grassmann's creativity

Discovering Robert Grassmann (1815–1901)

Ivor Grattan-Guinness

An overlooked prolific polymath

In appreciating Hermann Grassmann it is unusually significant to consider his family background, for he belonged to the middle of a trio of generations of professional mathematicians and mathematical educators. In particular, it is known that his relationship with his younger brother Robert (1815–1901) was close, at times active even on a daily basis, and on occasion a professional one between author and publisher. In contrast to the remarkable silence over Robert in the biography [Schlegel 1878] of Hermann, recent research on Hermann has given us new insights into Robert: the proceedings [Schubring 1996a] of a conference on the occasion of the sesquicentenary of Hermann's first *Ausdehnungslehre* included four articles that discussed the brothers' work and collaborations over arithmetic and the foundations of mathematics and on educational issues, while the biography [Petsche 2006, 2009] of Hermann has further information. The recent volume of family "traces" [Petsche et al. 2009, 59–116] provides transcriptions of several interesting documents on Robert's life and that of his foster-father, Justus's brother Friedrich, a school inspector, and in the preceding pages we have seen Robert's own autobiographical summary. Yet still he seems fugitive: the lack of a *Nachlass* is especially unfortunate. Who was this unusual author/publisher who published

H.-J. Petsche et al. (eds.), *From Past to Future: Graßmann's Work in Context*,
DOI 10.1007/978-3-0346-0405-5_3,

so much and yet is remembered, if at all, only as the brother of his famous brother? In particular, what books and pamphlets did he publish?

Plan of the paper

In an effort to answer especially the latter question, this paper is mainly bibliographical in character. The total number of books certainly exceeds 40, but maybe not 50. He was somewhat late into the practice, for his first book appeared in 1851; but he made up for lost time by publishing books in groups, such as at least 20 books on various topics (many as reprints) in 1900! Death in the following year did not prohibit the posthumous (re)appearance of several books until the 1910s. Most of those were religious ones, which brought him some (local) renown, during life as well as afterwards. By contrast, his many academic books appear to have sold badly; for example, copies are difficult to find even in library catalogues never mind losses from stacks. Only the first two parts of *Die Formenlehre oder Mathematik* [1872b, c] seem to have made any notable national or international impact (and for which an English translation [R. Grassmann 2005] is now available).

Section 3 lists Robert's books in philosophy, science, mathematics and logic, and religion. Then in Sect. 4 I catalogue and consider in a little detail his *Die Wissenschaftslehre oder Philosophie* (1872), and in Sects. 5–8 its extension into his "principal work", a philosophical tome of maybe 20 numbered volumes entitled *Das Gebäude des Wissens.* It appeared between 1881 and 1900, including nine volumes in 1890, his 76th year; I shall cite volume *n* below in the form "GW*n*". I conclude in Sect. 9 with some general remarks about the financial circumstances that allowed him to study and self-publish to such an extraordinary extent, and the means by which he acquired such a vast range of knowledge in this city of presumably limited access to information.

Books by the score

After leaving school in the early 1850s Robert seems to have been in a financial position that allowed him to reduce his teaching to some private work and to devote considerable time to study and publish his scores of books. According to the catalogues that I have consulted, the main holders of these books seem to be the Berlin *Staatsbibliothek* (where however several copies are now missing, presumably due to War circumstances), Greifswald and Göttingen *Universitäts-Bibliotheken*, the British Library in London, and the Harvard University Libraries. The national libraries of European countries are most easily accessed via the website http://theeuropeanlibrary.org/portal/en/index.html. Also worth using is the website http://used.addall.com of a very large collective of antiquarian booksellers; they usually offer dozens of copies of Robert's books, mostly the religious ones. The website http://www.emis.de/MATH/JFM of the abstracting journal *Jahrbuch über die Fortschritte der Mathematik* provides some information, including reviews of a few of the mathematical books. Some of his books are now available on the website http://www.uni-potsdam.de/u/philosophie/grassmann/Downloads.htm.

Drawing principally on these sources, I have compiled in Tables 1–6 lists of his other books and pamphlets outside GW, classified by topic and including dates of editions when known. Posthumous editions, usually of religious works, are included. The list is surely incomplete; it might also contain items that he announced but in the end did not publish. Several of his books, including some in GW, contain well-drawn diagrams and/or lithographs; he did not explain whether he prepared these pictures himself, or employed an artist, or obtained them commercially, or followed some combination of these options. Again, while he almost always published with "Druck und Verlag R. Grassmann" (with "ss"), he did not indicate whether he did any of the typesetting.

Table 1. Philosophy

<table>
<tr><td colspan="6">Die Erweiterungslehre der höhere Zweig der Synthese, [1895?]
Die Menschenlehre, oder das geistige Leben der Menschen, 1900
Die Rechtschreibung oder Orthographie der deutschen Sprache, 1897
Wissenschaftliche Erkenntiss der Ausenwelt. Die Basis jeder Philosophie, 1900, 254 pp.
Die Wissenschaftslehre oder Philosophie, 1875–1876 [See Table 2]</td></tr>
<tr><td>Theil</td><td>Title</td><td>Pages</td><td>Theil</td><td>Title</td><td>Pages</td></tr>
<tr><td>1(1875)</td><td>Die Denklehre</td><td>176</td><td>3(1876)</td><td>Die Erkenntnisslehre</td><td>264</td></tr>
<tr><td>2(1875)</td><td>Die Wissenslehre</td><td>176</td><td>4(1876)</td><td>Die Weisheitslehre</td><td>244</td></tr>
<tr><td colspan="6">Die Wissenschaftslehre oder Philosophie, 2ter Ergänzungstheil.
Die Formenlehre oder Mathematik, 1872. [Repr. 1966.]</td></tr>
<tr><td>Buch</td><td colspan="4">Title</td><td>Pages</td></tr>
<tr><td></td><td colspan="4">Einleitung</td><td>14</td></tr>
<tr><td>1</td><td colspan="4">Die Grösenlehre</td><td>52</td></tr>
<tr><td>2</td><td colspan="4">Die Begriffslehre oder Logik</td><td>43</td></tr>
<tr><td>3</td><td colspan="4">Die Bindelehre oder Combinationslehre</td><td>24</td></tr>
<tr><td>4</td><td colspan="4">Die Zahlenlehre oder Arithmetik</td><td>62</td></tr>
<tr><td>5</td><td colspan="4">Die Ausenlehre oder Ausdehnungslehre</td><td>26</td></tr>
</table>

Table 2. Science

<table>
<tr><td colspan="5">Das natürliche System der Maase, Gewichte und Münzen, 1865</td></tr>
<tr><td colspan="5">Die Physik oder die Lehre von den Gesetzen der Natur…, 1900, 111 pp.</td></tr>
<tr><td colspan="5">Die Physik, die Chemie und Krystallonomie oder die Gesetze der unorganischen Wesen, 1900, 350 pp.</td></tr>
<tr><td colspan="5">Leitfaden der Geographie fuer hoehere Lehranstalten, 1861, 8th ed. 1873, 15th ed. 1896</td></tr>
<tr><td colspan="5">Die Weltwissenschaft oder die Physik
Theil 1, Die Lebenslehre oder die Biologie</td></tr>
<tr><td>Buch</td><td>Title</td><td>Date</td><td>Pages</td><td>GW</td></tr>
<tr><td>1</td><td>Das Weltleben oder die Metaphysik</td><td>1862, 1882</td><td>xii + 350</td><td>3,1</td></tr>
<tr><td>2</td><td>Die Pflanzenlehre oder die Physiologie der Pflanzen</td><td>1882</td><td>xiii + 301</td><td>3,2</td></tr>
<tr><td>3–4</td><td>Das Tierleben</td><td>1883</td><td>xii + 633</td><td>4</td></tr>
<tr><td colspan="5">Theil 1,2</td></tr>
<tr><td>1</td><td>Die Körperlehre oder die Atomistik</td><td colspan="3">2nd ed. 1872</td></tr>
<tr><td>2</td><td>Die Erdgeschichte oder Geologie</td><td colspan="3">1873, 273 pp.</td></tr>
</table>

Table 3. Mathematics and logic

Aufgaben zu den Gleichungen ersten Grades mit einer und mehreren Unbekannten, 1898
Die Differential- und Integralrechnung bei Vermeidung der Trugschlüsse eine höchst leichte Wissenschaft, 1900, 44 pp.
Die Differential- und Integralrechnung streng wissenschaftlich dargestellt, 1900, ii + 44 pp. [Same as previous?]
Die Funktionenlehre namentlich die Differential- und Integralrechnung in strenger Formelentwickelung, 1900, xii + 189 pp. [Reissue of next.]
Die Logik und die anderen logischen Wissenschaften, 1890, 189 pp. [See also GW25.]
Formelbuch der Ausdehnungslehre, 1891, 14 pp.
Formelbuch der Folgelehre oder Funktionenlehre, 1895, 27 pp.
Formelbuch der Formenlehre oder Mathematik, 1895, 24 + 27 + 14 + 5 pp.
Formelbuch der logischen Wissenschaften, 1890
Fünfstellige logarithmische und trigonometrische Tafeln, 1895

Before GW: Robert's *Wissenschaftslehre*

Robert's most ambitious philosophical project before GW was called *Die Wissenschaftslehre oder Philosophie* (hereafter, "WP"). The first portion to appear was published in 1872 as five short "books" forming a "second supplementary part" of the work; the first two books were the *Formenlehre* and *Begriffslehre* mentioned above as his relatively well received writings. The heart of the work was an 800-page book published in 1875 and

Table 4. Social matters

Der Krieg von 1870–1871 zwischen Frankreich und Deutschland, 2nd ed. 1873
Die landwirthschaftliche Statik und die Methode landwirthschaftliche Beobachtungen, 1851, 24 pp.
Die thönernen Unterdrains: Nutzen, Ausführung und Kosten derselben im nördlichen Deutschland, Nach der besten Quellen, 1851, 39 pp.
Die Menschenlehre, oder das geistige Leben der Menschen, 1900
Das Wahlrecht der neuen Städteordnung, 1876, 23 pp.
Fürst Bismarck: Ein Lebensbild, 1876, 336 pp.
Grassmannsches Familienbuch, aus alten Urkunden zusammengetragen, 1876
Die Verkehrslehre oder Haus und Gewerbe, 1900, 396 pp.
Die Rechtschreibung oder Orthographie der deutschen Sprache, 1897, 15 pp.
Schutzmassregeln gegen die asiatische Cholera, 1892, 16 pp.
Über die Resorption der Nahrung bei Herzkrankheiten, Berlin, 1888, 34 pp.
Wie läßt sich die Wehrkraft des Preußischen Volkes vermehren?, 1860, 35 pp.

Table 5. Biblical history

Biblische Geschichte des Alten Testamentes zum Gebrauch für Schule, 1856, 1885
Biblische Geschichte des Alten Testaments für höhere Schulen, 1897, 1903
Biblische Geschichte des Neuen Testamentes zum Gebrauch für Schulen, 1856
Biblische Geschichte des Neuen Testaments für höhere Schulen, 1903
Biblische Geschichte für Landschulen, 1883, 1903
Biblische Geschichte für Stadtschulen, 12th ed. 1885, 15th ed. 1903
Das Leben Jesus: Ein wörtlicher Abdruck aus den vier Evangelien nach der Zeitfolge geordnet, 1855
Das Leben Jesus nach wissenschaftlicher Feststellung von Ort und Zeit, 1897, 1898
Die biblische Geschichte des Neuen Testaments nach wissenschaftlicher Feststellung von Ort und Zeit, 1898, 2nd ed. 1900
Die wissenschaftliche Feststellung von Ort und Zeit für die Biblische Geschichte des Alten Testamentes, 2nd ed.1900
Die Zeitfolge im Leben Jesu nach wissenschaftlichen Prinzipien dargestellt, 1858, 2nd ed. 1860, 52 pp.

1876, divided into four separately paginated Parts, each of which was divided into four Books. The names of all the Parts and the first three of each quartet of Books were of some kind of "lehre", while each final Book expounded some kind of "denken" or "wissen". Table 7 gives the summary details, including the initial page numbers of each Book and the page numbers of the index and table of contents (marked "SI") of each Part. Many of the Books were divided further into "Sections" ("Abschnitte").

Table 6. Christian doctrine

Auszüge aus den theologischen Moralen der römisch katholischen Kirche oder Welche Fragen soll der Priester in der Beichte den Frauen und Mädchen auf Befehl des Papstes vorlegen..,: als Ms. gedruckt [1893]
Auszüge aus der von den Päpsten Gregor XVI., Pius IX. und Leo XIII. als Norm für die römische Kirche sanktionierten Moraltheologie des Heiligen D. Alphonsus Maria de Liguori und die furchtbare Gefahr dieser Moraltheologie für die Sittlichkeit der Völker, 8th ed., 1899. 9th ed. 1901. "Umgearbeitete Ausgabe als Ms. gedruckt" 1903. 1904, 1909, 1910, 1911 (new ed.), 1912, 1913, 1916. [Around 94 pp. Dutch translation 1903.]
Briefe an seine Heiligkeit den Pabst über die Fragen, wo ist die Pestbeule der christlichen Kirche, und die antichristliche Partei in der christlichen Kirche zu finden?, 1893, iv + 124 pp. 8th ed. 1904
Die Christen und die Antichristen in der christlichen Kirche namentlich in der römisch katholischen Kirche, 1901
Die christliche Gotteslehre oder streng wissenschaftliche Theologie, gegründet auf die Gesetze Gottes in der Natur und auf Gottes Lehre in der Bibel, 2 vols., 1900, 627 pp.
Die Gotteslehre oder die Theologie, 2 Theile, 1900, 92 pp.
Die Verfluchungen und Beschimpfungen des Herrn Christus und der Christen durch die Päpste, Bischöfe und Priester der römischen Kirche und der Pflicht jedes Christen diesen Verfluchungen gegenüber, 1895, 106 pp. 3rd ed. 1900

Table 7. Structure of *Die Wissenschaftslehre oder Philosophie*

Part, Book	Part, Book
1 (1875) Die Denklehre SI 165–176	3 (1876) Die Erkennntnislehre SI 251–264
1,1; 17 Die Wahrnehmungslehre	3,1; 9 Die Bewusstseinlehre
1,2; 33 Die Gestaltungslehre	3,2; 33 Die Bildungslehre
1,3; 73 Die Sprachlehre	3,3; 75 Die Geisteslehre
1,4; 89 Das Formdenken	3,4; 139 Das Staatswissen
2 (1875) Die Wissenslehre SI 168–176	4 (1876) Die Weisheitslehre SI 225–244
2,1; 7 Die Beobachtungslehre	4,1; 7 Die Vertiefungslehre
2,2; 33 Die Ableitungslehre oder Philosophie	4,2; 43 Die Offenbarungslehre
2,3; 51 Die Wesenlehre oder Ontologie	4,3; 105 Die Heilslehre
2,4; 79 Das Weltwissen oder die Naturphilosophie	4,4; 173 Das Gotteswissen

The course of the book is well reflected in its preface, with sections successively devoted to "the branches of human thought", physics, politics, theology, philosophy, "the edifice of human thought", "the parts of human thought", and "the books of '*Denklehre*'"; he added that the last two Parts constituted "*Gedankenlehre*". Clearly he was a philosopher keen on classifying knowledge into its various named departments. Starting out in WP1,1 from sense data and other standard philosophical categories such as types of thought, he recorded information over a very wide range, perhaps unevenly; for example, a lot on plants in WP2,4 but little on other parts of botany. WP1,2 is striking for containing quite a lot of both mathematics and logic between the same covers; the *Ausdehnungslehre* formed part of the glue. He affirmed the importance of mathematics for school education in WP2,3. He treated at length both ethnology and ethnography in WP3,2, and gave a long list of short descriptions of countries in WP3,2. He ended up with "human belief and wisdom" in WP4,2 in discussing Christianity, which he described as "the essence of all essences" in the preface to WP4.

In a long review in an educational journal [Ballauff 1878] recognised that he was reading "an overview of all philosophy", and liked the empirical psychology and the advocacy of teaching language and the discussion of memory in WP1,3. But he queried somewhat the network of spirits, essences and atoms with which Robert entwined his science and his religion, mainly in WP2,4.

The first planned version of GW

Robert indicated that GW was the successor of WP in 1882 when he launched its publication by reprinting WP complete, but with new, and very similar, titles given to the book as a whole and to each part. According to the catalogue of the *Staatsbibliothek Berlin*, the four volumes were given the GW numbers 21,1–22,2.

In the new foreword Robert stated that GW would follow its predecessor but with newly "strong scientific proofs" of its various assertions. Table 8 contains the details of "gw", as I shall call it; the numbers in square brackets mark the places in GW (listed in Tables 9–11 below) where the topics were eventually treated, if at all. Each volume number marked "B" was called "Band" by Robert, and was presumably that of the title of a gw *Band,* with the subsequent unlettered rows indicating the planned volumes themselves.

The house that Robert Grassmann built: the structure and chronology of GW

GW was Robert's main production; in that autobiography of 1890, written when he was 75 years old, he stated of himself that he "has for 45 years since that time devoted 4–8 h nearly every day to the most strenuous efforts on this problem and submits in the present work the result of those efforts" [p.10]. Tables 9–11 list its 18 known volumes by his organization into *Band, Teil* and when necessary *Hälfte.* The first table covers the preliminary matter up to GW2,2, the second continues up to GW10 in 1899,

Table 8. GW as originally conceived (1882)

Band	Title
B1 [1]	*Die Wissenslehre oder die Philosophie*, Band 1
[1,1,2]	*Die Sprachlehre oder philosophische Grammatik*
[1,2?]	*Die Formenlehre oder Mathematik*
B2 [2]	*Die Wissenslehre oder die Philosophie*, Band 2
[2,1]	*Die Denklehre oder die logischen Wissenschaften*
[2,2]	*Die Wesenslehre oder Ontologie (die Metaphysik der Philosophen)*
B3 [3]	*Die Lebenslehre oder die Biologie*, Band 1
[3,1]	*Das Weltleben oder die Metaphysik*
[3,2]	*Das Pflanzenleben oder die Physiologie der Pflanzen*
B4 [4]	*Die Lebenslehre oder die Biologie*, Band 2
[4]	*Das Tierleben oder die Physiologie der Wirbeltiere und des Menschen*
B5 [5]	*Die Sittenlehre oder die Ethik*, Band 1
[5]	*Die Menschenlehre oder die Anthropologie*
[6]	*Die Verkehrslehre, und zwar Haus und Gewerbe*
B6 [6]	*Die Sittenlehre oder die Ethik*, Band 2
	Die Kunstlehre oder die Aesthetik
	Die Wissenschaftslehre oder die Encyclopädie der Wissenschaften
B7 [7]	*Die Staatslehre oder die Politik*, Band 1, *Die Rechtslehre*
[7]	*Die Staatspflege oder das Staatsministerium*
[7]	*Die Schatzpflege oder das Finanzministerium*
[7]	*Die Schutzpflege oder das Ministerium des Innern*
[7]	*Die Rechtspflege oder das Justizministerium*
B8 [8]	*Die Staatslehre*, Band 2, *Die Reichslehre*
[8]	*Die Bildungspflege oder das Kultusministerium*
[8]	*Die Wirthschaftspflege oder das Ministerium für Gewerbe*
[8]	*Die Wehrpflege oder das Kriegsministerium*
[8]	*Die Reichspflege oder das Ministerium des Auswärtigen*
B9 [9]	*Die Gotteswissenschaft*, Band 1 *Die Weisenlehre*
[9]	*Die Gotteslehre oder die Theologie*
	Die Schöpfungslehre
B10 [10]	*Die Gotteswissenschaft*, Band 2, *Die Kirchenlehre*
[10?]	*Die Heilslehre*
[10?]	*Die Himmelslehre*
11	*Die Sprachgeschichte*, Band 1, *Die Geschichte der arischen Worte und Formen*
12	*Die Sprachgeschichte*, Band 2, *Die Geschichte der deutschen Worte und Formen*
13	*Die Erdgeschichte*
14	*Das Landschaftsgemälde der Erde*

Table 9. Bibliographical details of GW, WP and Bände 1–2 (1882–1890)

	Prospekt zum Gebäude des Wissens (1890)	38 pages
Die Einleitung in das Gebäude des Wissens oder die wissenschaftliche Propädeutik (1882) "Vorwort" pages iii–xv		
WP/GW	Title	Pages
1/21,1	*Die Einleitung in die Denkwissenschaften oder die philosophische Propädeutik*	176
2/21,2	*Die Einleitung in die Weltwissenschaften oder Naturwissenschaften oder die physische Propädeutik*	176
3/22,1	*Die Einleitung in die Staatswissenschaften oder die ethische Propädeutik*	264
4/22,2	*Die Einleitung in die Gotteswissenschaften oder die theologische Propädeutik*	244
Erster Band, *Die Wissenslehre oder die Philosophie* Erster Teil, *Das Verstandeswissen oder das formale Wissen, umfassend die auf die Philosophie vorbereitenden Wissenschaften* (1890) "Vorwort zum Gebäude des Wissens" pp. v–xxix		
Hälfte	Hälfte title	Pages
[1,1,1]	*Die Geschichte der Philosophie und erste Einleitung in der Philosophie*	120
[1,1,2]	*Die Sprachlehre, d.h. die Lehre von den Arten von Laut-, Wort und Satzbildungen welche dem Menschen möglich sind, von ihren Formen und Gesetzen* [Repr. 1900.]	xii+216
[1,1,3]	*Formelbuch der Denklehre* [Repr. 1900.]	50
1,2	*Die Denklehre, oder die Lehre der wissenschaftliche Denkakte und Verknüpfungen, welche dem Geiste des Menschen möglich sind, von ihren Formen und Gesetzen, in strenger Formelentwicklung* [Also called] *Die Grösenlehre, die Grundstamm der Denklehre*	xxii+531
Zweiter Teil, *Das erkennende Wissen oder das wesentliche Wissen umfassend die der Philosophie im engeren Sinne angehörenden Wissenschaften* (1890)		
Hälfte	Hälfte title	Pages
2,1	*Die Erkenntnislehre*	x+284
2,2	*Die Wesenslehre oder Ontologie, d.h. die Lehre von den Wesen des Alls, ihren Bewegungen, Kräften und Einwirkungen, ihren Formen u. Gesetzen* [Repr. 1900.]	ix+628

and the third covers the later numbered volumes. The numberings are surprisingly inconsistent in places, especially early on; the numbers in square brackets in Table 9 are my guesses.

Table 10. Bibliographical details of GW, Bände 3–10 (1881–1899)

Band	Date	Gebiet title	Teil or Band title	Pages
3,1	1882, 1881	*Die Lebenslehre oder die Biologie*	*Das Weltleben oder die Metaphysik*	xiii+350
3,2	1881		*Die Pflanzenlehre oder die Physiologie der Pflanzen*	xii+301
4	1883		*Das Tierleben oder die Physiologie der Wirbeltiere und namentlich des Menschen*	xii+633
5	1883, 1884	*Die Sittenlehre oder die Ethik*	*Die Menschenlehre und die Verkehrslehre oder die Anthropologie und die Agatholie*	xvi+432
6	1890		*Die Verkehrslehre oder die Güterlehre*	ix+395
7	1890, repr. 1900	*Die Staatslehre oder die Politik*	*Die Rechtslehre, enthaltend die Staatspflege, die Schatzpflege, die Schutzpflege und die Rechstpflege*	xii+493 no index
8	1890, repr. 1900		*Die Reichslehre, enthaltend die Bildungspflege, die Wirthschaftspflege, die Wehrpflege und die Reichspflege*	vii+646
9	1890	*Die Gotteslehre oder die Theologie*	*Die Gottesweisheit oder die Lehre vom Gottesreiche*	xiv + 257
10	1899		*Die Heilsweisheit oder die Heilsreich auf Erden und das Himmelsreich*	[vi]+373

For years of publication I rely upon Robert's title pages, which appear in some volumes in even more profusion than in WP; in particular, GW3,1 and GW5 are given two different years of publication because they contain consecutive title pages with those dates! Each volume is well indexed, apart from GW7 for some reason; but none contains a bibliography or many references.

The main bulk of the book, GW1,1,1–GW10, comprises about 6,000 pages in total. While it follows the same ambition and scope of the original gw1–gw10, it fulfilled that aim in far fewer volumes and with less emphasis upon governmental administration. He summarised the contents in some detail in an undated *Prospekt* of 38 pages. No trace remained in GW of gw11–gw14 on the history of languages and geology, although the subjects were not entirely ignored; indeed, he published a volume on the latter topic in 1873 as a supplementary volume of WP (Table 2).

The numbering of the volumes of GW and the chronology of their publication suggest a highly non-chronological execution of the project; the five preparatory volumes

Table 11. Bibliographical details of GW, Apparent Bände 20–25 (1891–1900)

Band	Date	Gebiet title	Teil or Band title	Pages
20,1	1900	*Die Geschichte des Gottesreiches nach streng wissenschaft- licher Forschung*	*Die Geschichte der vorchristlichen Zeit*	886 in total
20,2	1900		*Die Geschichte der christlichen Zeit*	
23	1891	*Formenlehre oder Mathematik in strenger Formel-entwicklung*	*Die Formenlehre oder Mathematik in strenger Formelentwicklung*	ix + 132
	1891		*Die Zahlenlehre oder Arithmetik streng wissenschaftlich in strenger Formelentwicklung*	xii + 242
	1895		*Die Folgelehre oder Funktionenlehre streng wissenschaftlich in strenger Formelentwicklung*	xii + 189
	1891		*Die Ausdehnungslehre oder die Wissenschaft von den extensiven Grösen in strenger Formelentwicklung*	ix + 132
	??		*Die Erweiterungslehre der höhere Zweig der Analyse*	38
25	1900		*Die Logik* [Repr. of *Die Logik*, 1890?]	xiii + 188

and four of the later ones did not appear until 1890, while some of the later ones had started to appear from 1881 onwards. Then there are the late extra dollops of religion in GW21 (where I rely on the catalogue of the *Staatsbibliothek Berlin* for the volume numbers), of mathematics in GW23 and of logic in GW25, adding nearly 2,000 more pages. I have not found any trace of GW11–19 or of GW24, especially not analogues to gw11–14, unless a copy of any of them is or was held in some library that I have missed. If in fact they were never published, then they will not be the only gap, for in his autobiographical note in GW1,1 [p. xxviii], and at gw6, he promised GW6,2 on aesthetics that I am sure never appeared.

Some characteristics of GW

What is going on in this massive enterprise? As in WP on a smaller scale, Robert's main concern was in classifying and naming theories of all kinds, often using "-lehre" as suffix;

my favourite is "Polizeiwissenschaft" in GW7. While WP carried the word "knowledge" in its title, Robert was also much concerned there with knowing ("Wissen"), and this aspect came to the philosophical fore in GW, as its title suggests. Unlike other large-scale classifications of theories [Fisher 1990], this one is a study of people (and animals) knowing things, not just of knowledge itself. The word has been mistranslated as "knowledge" in [Petsche 2009]; in [Petsche et al. 2009] it is left untranslated.

This emphasis on knowing as well as knowledge gives GW an attractive and unusual character among large-scale philosophical writings. For example, GW1,1,1 is taken up with a review of the history of the various philosophies, where system builder Aristotle is a hero; and it is followed immediately by GW1,1,2 on phonetics, placed this early because it is essential for our means of communicating what we know. Again, Robert's elaborate account of ontology includes over 40 pages on the physiology of the brain, since that organ plays the crucial role in cognitive processes; moreover, he stressed strongly the two halves of the brain, and furnished his account with appropriate illustrations [GW2,2, 327–361].

Throughout GW Robert presented pairs or trios of numerically adjacent volumes as "Teile" or "Hälfte" under a common title, which itself was a logical disjunction. Each common title covered a "Region" ("Gebiet"). GW1,1,1–GW2,2 started off with types of knowing, followed up to GW10 by the sciences, ethics, politics and theology. GW5–GW6 offered the striking pairing of anthropology with aspects of economics under the common general heading of morals and ethics. To us today the association of economics with any kind of ethical activity is not even funny; but in the late nineteenth century a strong element of ethics attended economics for example, so that when trading was being conducted the prices or bartering arrangements were fair, and also in the distribution of income across society (see, for example, [Walras 1896]). Robert shows the same sentiment, though as usual he does not cite contemporaries. However, the title of GW6 was mistaken; the habitual disjunction between transport and goods should have been replaced by a conjunction.

The quantity and variety of information both factual and theoretical is staggering; for example among many, statistics on the provision of railways in some African countries and Turkmenistan in GW6, and on the national incomes of many countries in GW7. The glorification of Protestantism and alleged details of the life of Jesus fill GW9–GW10 (and also GW20,1–GW20,2 and many other books listed in Table 1, including WP4). These parts constituted the climax of GW, of course, with the aether given much credit; GW10 ends with praise "of the wisdom and knowledge of God. Amen". Nobody seems to have emulated Ballauff with WP and reviewed GW in total.

Robert Grassmann on the calculus and logic

Space, and indeed competence, prevents any systematic or detailed survey of the many pages of GW; the couple of features discussed here serve as examples.

Firstly, as part of his account of several parts of mathematics, including Hermann's *Ausdehnungslehre*, Robert presented his preferred version of differential and integral calculus [GW1,2, 252–256]. He chose to follow [Lagrange 1797] in believing that, apart from exceptional values of x, any mathematical function f(x) could be expanded in a Taylor series, despite the fact that [Cauchy 1822] had refuted this position with counterexamples of infinitely flat functions such as exp $(-1/x)$ at $x = 0$. Interestingly, in a rare example from him of actually citing another text, Robert mentioned [Moigno 1840], a textbook based upon Cauchy's teaching; but he seems to have misunderstood the issue as involving the convergence of the series rather than its nonexistence, although Moigno dealt with both issues in lectures 8–9.

Secondly, the last (?) volume, GW25, provides a very welcome account of logic, expanding on the treatment given in GW1,2 and in other books. This book, or one very similar, appeared in 1890, for on p. 174 he alluded to that year in a reminiscence; it is listed for 1890 in Alonzo Church's bibliography of symbolic logic [1936, 131], under the (odd) title *Die Logik und die anderen logischen Wissenschaften,* with the same number of pages (188) but no numbering of front matter. In line with his stress upon knowing, he admitted roles for (a) logic connected to the creation of knowledge and acquaintance with it, and not just its deductive layout. Table 12 outlines its contents, following the layout of Table 7 for WP including initial page numbers. As so often with Robert, fours are much to the fore, in the numbers of "branches" and "sections", though 2,4 and 4,4 were not developed; the third branch is a very unusual exception.

In his foreword Robert noted parts of the history of logic, especially in recent decades. He preferred symbolic to prosodic logic, specifically "the derivation of the laws of thought without words in equations of single-valued quantities and single-valued connections" [GW25 vii]; like his contemporaries of similar inclination he developed a version of what we recognise as algebraic logic. Unusually for him, he discussed other authors at length, especially George Boole, Augustus De Morgan, Ernst Schröder and Wilhelm Wundt; he also recalled his own *Begriffslehre* (1872) and his collaboration with Hermann. The key word in the quotation above is "quantities", my rendering of his favourite word "Grösen"; he took them to be "objects of thought" but seemed to have intended only propositions [pp. vii–viii]. In the lengthy introduction he developed a propositional calculus, followed in the first section by "concepts" ("Begriffe"), a predicate calculus to us, though with no logic of relations or explicit quantification theory. Both presentations included logic diagrams.

Table 12. Structure of GW25, *Die Logik* (1900)

Zweig, Abschnitt	Zweig, Abschnitt
1; 1 Die Bestimmungslehre oder Logik, die niedere Analytik	3; 133 Die Bindelehre oder Kombinationslehre, die niedere Synthetik
-,-; 1 Einleitung in die Logik 1,1; 50 Die Begriffslehre 1,2; 72 Die Urteilslehre 1,3; 95 Die Schlusslehre 1,4; 112 Die Beweislehre	3/1; 136 Die Gebinde oder Kombinationen 3/2; 145 Die Klassen der Gebinde
2; 123 Die Wandlungslehre oder Tropik, die höhere Analytik	4; 167 Die Erspähungslehre oder Spekulationslehre, die höhere Synthetik
2/1; 125 Die Aenderungslehre 2/2; 127 Die Artungslehre 2/3; 129 Die Sippungslehre 2/4; 130 Die erweiternde Wandlungslehre	4/1; 171 Die Erspähungslehre 4/2; 179 Die Gliederungslehre 4/3; 187 Die Gesellungslehre 4/4; 188 Die erweiternde Erspähungslehre

Robert's liking for knowing and not just knowledge fitted in well with the prevailing philosophy of his time, which saw logic as normatively linked with thinking. Of his various branches "Erspähungslehre" is particularly congenial; an English translation might be "calculus of insights", or his own synonym about "speculations", or guided heuristics. This branch was characterised by the property that when two notions are connected "in an internal manner" to produce a synthesis, they are both affected. The Hegelian triad comes to mind, but in fact Robert criticised Hegel for "breaking down the boundary wall" between concepts and thus the "spaces" surrounding their "essences", thereby making "impossible any strictly scientific thought" [pp. 174, 176]. The predecessor branch handled "the lower synthetic" where the original notions remain unchanged after connection, such as the disjunction of two unrelated propositions.

As in several aspects of his (and Hermann's) philosophy, the dominant influence on Robert here was the dialectics of Friedrich Schleiermacher (1768–1834), who advocated an (anti-Hegelian) method of "dialectics" in which he promoted the consideration and maintenance of pairs of contrasting notions as a means of both developing theories (that is, the category of knowing) and assessing them epistemologically, especially as candidates for certainty. Hermann adapted the same methodology to develop his theories in a framework of pairs of pairs of contrasted notions, such as intension/extension and addition/multiplication (not necessarily arithmetical):

$$(a + \neg a)(u + \neg u), \text{ that is, } au + a\neg u + \neg au + \neg a\neg u \tag{1}$$

[Petsche 2009, 153–155]. Robert often used the same method of grouping topics, and maybe also sub- and sub-sub-topics, in fours. On occasion a = u; for example, = and ¬ = twice over in the *Formenlehre* of 1872.

This quartet scheme was extolled in Robert's calculus of speculations, which he characterised early on in GW as "the highest branch of logical knowledge" and praised Schleiermacher's advocacy of it [GW1,1,1, 89–90]. Robert had envisioned the (meta-)logical self-reference of the scheme to GW itself in his *Denklehre* [GW1,2, 590], and (as Sebastian Berchtold pointed out to me) he fulfilled it in this book on logic at the end of his account of the predicate calculus [GW25, 70–71, shown in Fig. 1]. Take the quartet of concepts (1) and conjoin them with each of four (sic) "main regions" A, B, C and D. A hexadectet of compound concepts arises, which may be laid out in four quartets, as shown in the first display in Fig. 1 (where Robert used overbars to represent negations). "The Gebäude des Wissens gives us a beautiful example of this division", and he gave three examples of how the regions of the work were designed; the first quartet related to the entire structure GW1–GW2, GW3–GW4, GW5–GW8 and GW9–GW10 more or less as summarised in the 1890 prospectus, while the other two divided up respectively the theories of "world living" and animal life of GW3,2 and GW4, and of morals of GW5–GW8, in both cases in order. Reporting this excellent example of conscious self-reference furnishes a fine finale to this brief survey of an extraordinary work. Let us end with four questions, though not quite in a quartet scheme: a pair on practical matters followed by a pair on intellectual ones.

Four final queries

Firstly, apart from income from some private teaching, how did Robert maintain himself and his family after resigning from the school in 1852? Maybe some of the religious and educational books listed in Table 1 were a main source of income, for certain print runs were considerable. In particular, the title pages of editions of his book of extracts from papal decrees proclaims it to be selling in its 330,000s in 1903, in the 350,000s in 1909, and in the 370,000s in 1916, and so must have done well financially in his lifetime. The production of editions of Talmudic texts in Hebrew in the 1860s was presumably also profitable. But he does not seem to have given *Verkaufslehre* great attention in his many academic books. Further, very little is known about his print works and publishing house, such as other authors, apart from Hermann, whom he published, though information is available from electronic catalogues when one searches for "Grassmann" under "publisher". His activity as a newspaper editor is also poorly studied, and there electronic search tools are much less powerful. Finally in this part, the brothers were

160. Logik. 70

160. **Satz. Bei jeder Teilungsregel werden nur die Begriffe in Betracht gezogen, welche innerhalb der zu teilenden Hauptbegriffe liegen oder in jedem Hauptbegriffe H_a werden beide Seiten des Gegenſatzes g und $\overline{g}$ mit dem Hauptbegriffe H_a verwebt (multiplizirt).**

Beweis. Unmittelbar aus 155 und 156.

Beispiele für den kreuzenden Gegenſatz $(a+\overline{a})(u+\overline{u})$ und die vier Hauptgebiete A, B, C, D.

Einteilung der vier Hauptgebiete.

Teilbegriffe	A	B	C	D
$a\overline{u}$	$a\overline{u}A$	$a\overline{u}B$	$a\overline{u}C$	$a\overline{u}D$
au	auA	auB	auC	auD
$\overline{a}u$	$\overline{a}uA$	$\overline{a}uB$	$\overline{a}uC$	$\overline{a}uD$
$\overline{a}\overline{u}$	$\overline{a}\overline{u}A$	$\overline{a}\overline{u}B$	$\overline{a}\overline{u}C$	$\overline{a}\overline{u}D$

Das Gebäude des Wissens giebt uns ein schönes Beispiel dieſer Teilung durch dieſelbe Teilungsregel. Hier bilden die beiden Begriffe Einzelleben und Zuſammenleben, bez. die vier Begriffe: Lehre, Beschreibung, Geschichte und Gemälde die vier Teilbegriffe der Teilungsregel (des principium divisionis). Die vier Begriffe: Denkwissenschaften, Weltwissenschaften, Staatswissenschaften und Gotteswissenschaften bilden die vier zu teilenden Hauptbegriffe und ergiebt ſich darnach folgende Einteilung.

Einteilung des Gebäudes des Wissens.

Teilbegriffe	Denkwissenschaften.	Weltwissenschaften.	Staatswissenschaften.	Gotteswissenschaften.
Lehre.	Wissenslehre	Lebenslehre	Sittenlehre	Weiſenlehre
Beschreibung.	Wissensbeschreibung	Lebensbeschreibung	Sittenbeschreibung	Weiſenbeschreibung
Geschichte.	Geistesgeschichte	Weltgeschichte	Völkergeschichte	Heilsgeschichte
Gemälde.	Geistesgemälde	Weltgemälde	Völkergemälde	Heilsgemälde

Die Lebenslehre bietet uns ein zweites Beispiel dieſer Teilung durch eine und dieſelbe Teilungsregel. Hier bilden die beiden Begriffe: Erhaltung und Fortpflanzung, bezüglich die vier Begriffe: Nahrungsleben, Arbeitsleben, Blüteleben und Fruchtleben die vier Teilbegriffe der Teilungsregel (das principium divisionis). Die vier Begriffe: Körperleben, Pflanzenleben, Geweideleben und Leibesleben bilden die vier Hauptbegriffe und ergiebt ſich darnach folgende Einteilung.

71 Teilung der Begriffe. 161.

Einteilung der Lebenslehre.

Teilbegriffe.	Körperleben.	Pflanzenleben.	Geweideleben.	Leibesleben.
Nahrungsleben.	Lichtleben	Dammleben	Darmleben	Genussleben
Arbeitsleben.	Wärmeleben	Saftleben	Blutleben	Muskelleben
Blüteleben.	Mischleben	Blüteleben	Geschlechtsleben	Vorstellungsleben
Fruchtleben	Füllleben	Fruchtleben	Keimleben	Seelenleben

Die Sittenlehre zeigt uns ein drittes Beispiel dieſer Teilung durch eine und dieſelbe Teilungsregel. Hier bilden die beiden Begriffe: leibliche Pflege und geistige Pflege, bezüglich die vier Begriffe: Nahrungspflege, Arbeitspflege, Gestaltungspflege und Wissenspflege die Teilbegriffe der Teilungsregel (des principium divisionis). Die vier Begriffe: Menschenlehre, Verkehrslehre, Rechtslehre und Reichslehre bilden die vier Hauptbegriffe und ergiebt ſich darnach folgende Einteilung.

Einteilung der Sittenlehre.

Teilbegriffe.	Menschenlehre.	Verkehrslehre.	Rechtslehre.	Reichslehre.
Nahrungspflege.	Genusspflege	Hauspflege	Staatspflege	Bildungspflege
Arbeitspflege.	Arbeitspflege	Gewerbepflege	Schatzpflege	Wirtschaftspflege
Gestaltungspflege.	Gestaltungspflege	Kunstpflege	Schutzpflege	Kriegspflege
Wissenspflege.	Bewusstſeinspflege	Wissensch.-pflege	Rechtspflege	Reichspflege

Jeder Zweig der Wissenschaft bietet uns weitere Beispiele dieſer Teilungsregel.

Satz. Jeder Gegenſatz umfasst alle Begriffe, welche dem Hauptbegriffe angehören und teilt, wenn er den Hauptbegriff schneidet, den Hauptbegriff in zwei ſich ausschliessende Unterbegriffe ein. 161

Beweis. Unmittelbar aus 89.

Fig. 1. Robert Grassmann's self-analysis

Freemasons [Schubring 2009]; how important (or not) was this membership upon their work?

Secondly, by what means did Robert collect so much information in a medium-sized town with no university and possibly modest libraries and archives. Was he wealthy enough to buy books and journals and employ researchers? Did he find useful contacts through his newspapers? If so, then where is all the resulting material today?

Thirdly, how do we appraise the similarities between the works of the brothers, and the direction of influence? Parallels include several papers by Hermann on problems in physical sciences and Robert's review of these sciences especially in GW2,2 and in some other books; Hermann's book *Deutsche Pflanzennamen* of 1870 (published by Robert) and Robert's GW3,2 (1881) on plants; and Hermann's Sanskrit dictionary (1873–1875) and papers on German "Sprachlehre" (1842, 1877), and Robert's GW1,1,2 (1890) on phonetics. Hermann is the first in chronology of similar writings: Robert shows an even wider overall range of interests than does Hermann, but surely he exhibits the less penetrating mind; for example, his *Ausdehnungslehre* (1891) is a cookbook version of results in Hermann's books of 1844 and 1862. As was mentioned in Sect. 1 above, their collaborations have been studied already, but our understanding would be enhanced by greater knowledge of Robert's concerns. Some information on Hermann's attitude to GW would be welcome; at present there seems to be none.

In this context it is suitable to ponder upon the influence of Schleiermacher on the brothers, especially Robert. We have noted his advocacy of differences leading to their quartet scheme, the common interest in knowing and not just knowledge, and the enthusiasm for Christianity; among other candidate categories is Schleiermacher's role in the rise of German nationalism in the early nineteenth century [Dawson 1966, Chaps. 4, 8], which is consonant with their preference for words with exclusively German etymology.

Finally, what drove this man to such extremes of production, especially as he seems not to have supported it with any strong desire for sales or publicity? The quantity of output is partly explained by repetitions – I have noticed even a few straight transcriptions of text from one book into another (for example on Cauchy and the calculus), and doubtless there were more – but the extent of the effort is astonishing. Who else published nine books at the age of 75, and around 20 a decade later? Clearly being the younger brother, especially with Hermann as an elder bother and also Justus as father, must have provided stimuli in all sorts of ways, as maybe did his further upbringing by his unremarkable uncle Friedrich. The third son in the family, he does not show the penchant for rebellion that [Sulloway 1997] claims to be noticeable among non-first borns; neither does second-born Hermann.

As far as understanding Robert Grassmann is concerned, our "Gebäude des Wissens" is but a modest dog kennel. His attainment was remarkable, at least at the level of collecting, classifying and communicating information; but his influence was usually slight, with the possible exception of effects upon Hermann. Maybe we can dig deeper into his life and work in time for his bicentenary in 2015 and sort out some of these questions.

Acknowledgements

For help and advice I thank Utz-Uwe Haus, Hans-Joachim Petsche, Sebastian Berchtold and Karin Reich.

Hermann Grassmann's theory of religion and faith

Johann Ev. Hafner

I

Whoever comes to Potsdam will inevitably find traces of the Orientalist movement, of which Hermann Grassmann was a part. Frederick the Great furnished Potsdam with different *chinoiseries*, his nephew Frederick William II built some seemingly Turkish buildings, and his grandson and king during the time of Grassmann, Frederick William IV, was driven by an admiration for India, which even led him to learn Sanskrit [Strohmeier, Knauer and Stás 2007, 5870]. At the age of 21, he wrote the novel *Die Königin von Borneo* [Kroll 1997].[1] It is a story about a prince (according to letters of his, this was the king himself) who travels to a far-away kingdom in Asia with the help of *rocs* (the eagles from *Thousand and One Nights*). His romantic longing was the peak of a European Wanderlust which was interested in cultures beyond the Christian Occident. Grassmann lived in the aftermath of this movement. According to Andrea Polaschegg, we can call this turning point a linguistic turn in German Orientalism [Polaschegg 2005].[2] It was triggered by three factors: first, by the establishment of historical and comparative philology, second, by the all-encompassing historization of cultures and

[1] For his architectural activities cf. [Johannsen 2007].

[2] I thank Dr. Catherina Wenzel for this valuable suggestion.

H.-J. Petsche et al. (eds.), *From Past to Future: Graßmann's Work in Context*,
DOI 10.1007/978-3-0346-0405-5_4, © Springer Basel AG 2011

religions, and third, by the rediscovery of mythical texts as material for the humanities (Creuzer, Schelling). At this time, the Protestant theologians who dealt with Semitic literature because of their knowledge of Hebrew practiced a kind of division of labour with scholars of Antiquity, who did research on Sanskrit and Persian texts.[3] Hermann Grassmann was the first to write a Sanskrit dictionary for the Rig-Veda, while being a staunch Protestant Christian at the same time. Therefore, he stood for a kind of bridge between those two approaches. His commitment to the Protestant Church on the one hand, and his interest in the ancient Indian religion on the other, are like two pylons on two riversides, so as to support a bridge which was still in the making.

II

This paper will present and analyze the text *Über den Abfall vom Glauben*,[4] which has a unique and isolated position in Grassmann's opus. It was published only after his death and picks up a thread dating back to the beginning of Grassmann's intellectual life. In 1827, he began his studies in Protestant theology at what is today the Humboldt University in Berlin ("Friedrich Wilhelm University" at the time), where he was impressed and influenced especially by Neander and Schleiermacher. In 1834, he passed his first theological exam and five years later, the second. Only one year later, in 1840, he received permission to teach (*facultas*) mathematics, since he had pursued intense philosophical and mathematical studies during his studies of theology. After forty years of research in mathematics (except for a small review of two books for religious instruction in schools [H. Grassmann 1851]), he returned to theology in the last year of his life, 1877. In this little text, which he – with some degree of understatement – used to call *leaflet (Schriftchen)*, he pronounced a general sentence against superstitious Catholics, materialistic scientists, unorganized Pietists, but especially against his own Church, while also showing signs of very conservative Protestantism. *Über den Abfall vom Glauben* [H. Grassmann 1878] is a polemical text written with a certain amount of aggression. The subtitle *Mahnungen an die wissenschaftlich Gebildeten der Neuzeit*[5] alludes to Schleiermacher's *Über die Religion: An die Gebildeten unter ihren Verächtern* from 1799 [Schleiermacher 1913].

His starting point was the observation that educated people were leaving Christianity after making a short stop in Christian ethics, and that the uneducated were

[3] *Indogermanistik* was established only in the nineteenth century as a consequence of this new enthusiasm for India and Sanskrit.

[4] *On the loss of faith* [H. Grassmann 1878].

[5] *Appeals to the scientifically educated in modern times.*

prone to social-democratic ideas. Grassmann diagnosed two forms of apostasy: first, the deconstruction of revelation as a historical and man-made text (among the educated), second, the massive overestimation of mankind and its needs as the central fulcrum of the world (mostly among the uneducated).[6] A symptom of this apostasy could also be found in libertinism within the Church, where each community aimed to define its own creed instead of holding onto the fixed doctrines of the Gospel. Grassmann criticized Catholicism and its practices (pilgrimages, works of goodwill, veneration of saints, indulgence) and Pietism, which places the subjective experience of faith over the acceptance of the Gospel. Grassmann situated true faith in between three negative poles: the materialism of science, the externalized belief of Catholics and orthodox Lutherans, and the internalization by Pietism and liberal Protestants. Grassmann's arguments can be reconstructed around five questions.

Why have people stopped believing in miracles?

Ever since empirical science explained the world according to Newtonian laws, God's intervention in history was no longer considered to be possible [H. Grassmann 1878, 15]. Unfortunately, this denial of miracles also intruded into theology, which now was attempting to demythologize the reports on miracles in the Gospels. As a consequence, the Gospel as a whole lost much of its authority. Grassmann saw this as a hypocritical way out for theologians who opted to deny the reality of miracles by making the theological argument that God could not override laws which He Himself had instituted. But – Grassmann asked rhetorically – do we really know the laws of nature? Can we explain the history of mankind, the sensitive life of animals, the growth of plants, gravitation, or electricity? Does not all progress in the natural sciences entail a revision of the laws which we assumed to be true? And do we not have to concede that we *refer to the whole by taking one of its tiny aspects as a point of departure*?[7] Does not every newly discovered law appear to us like a miracle, in comparison to earlier laws? Especially the rise of Christianity in late Antiquity seems like a historical miracle, if one takes into account how unlikely it actually was. Judging by Christian texts, rituals and historical figures from the year 200 A.D., nobody would have been able to conclude that this small group of people would develop into a world religion which would provide the basis for the Prussian state in the nineteenth century.

---

[6] Cf. [H. Grassmann 1878, 9].

[7] "von diesem Bruchteile [, den wir wissen,] auf das Ganze schließen" [H. Grassmann 1878, 16].

Where does the knowledge of mankind come from?

In the main chapter of his book, Grassmann resorts to a Kantian distinction. First, our knowledge arises from experience, *thanks to which we are able to perceive external events.*[8] Second, it comes from analytical conclusions, which combine our experiences in logical ways. But if conclusions already require that something is known to us and if experiences are determined by human needs, these operations will not guarantee true knowledge. Grassmann takes mathematics as an example for a science which does not produce anything new, but only unfolds something knowable in an analytic way. But unlike animals, man does not only have a perceptive and logical capacity, but also something for which he uses the German term *Geist*. *Geist* is where our experiences and conclusions meet in unity. Usually we do this by stating hypotheses, which we try to verify or falsify with new experiences. In the latter case, we replace a false hypothesis with a better one. Moreover, *Geist* has the ability to reflect on itself. It does not only produce relative, hypothetical certainty, but it achieves complete certainty [H. Grassmann 1878, 23, 24].[9] Our *Geist* only stands on safe ground if it focuses its thinking on itself, and not on objects. This is obviously a new formulation of the Cartesian *fundamentum inconcussum*. Grassmann enriches this thought about self-certainty with some arguments from German Idealism: self-consciousness is not only a solipsistic self-experience in thinking but needs stimuli from and to the outside. The more one divests *(entäußert)* oneself and becomes aware of one's own weaknesses, the more one expands one's "personal consciousness to an objective consciousness".[10] In the first place, this is a kind of inter-subjective consciousness of mankind, but only if we recognize our dependency on the exterior world can it expand to a divine consciousness or a consciousness of the divine (*Gottesbewusstsein*). Self-consciousness can become aware of the fact that it is not reflecting on its own, but that it is a reflection, the mirror of a higher personal being. This knowledge is rather a kind of acknowledgment*(Anerkenntnis)*, than some kind of conclusion.

Now Grassmann suddenly jumps into ethics. This higher being is not merely the role model and starting point for finite subjectivity. Rather, it demands that its imperfect avatar accepts the claim: "Ye shall be holy for I am holy." (cf. *Lev.* 11, 44 par) It is easy to detect the influence of Schleiermacher, who defined religion as *Gefühl*, a feeling of absolute dependence in the subject's self-consciousness. In Schleiermacher as well, two results arise from the basic act of reasoning: first, the acknowledgment of the infinite, and second, the acceptance of ones own finiteness. In Christian language: God

---

[8] "durch welche wir äußere Vorgänge wahrnehmen" [H. Grassmann 1878, 20].

[9] "Der Geist des einzelnen Menschen tritt seinem denkenden Geiste zugleich als Gegenstand der Beobachtung gegenüber, es entsteht das Selbstbewußtsein." [H. Grassmann 1878, 23].

[10] "persönliches Bewußtsein zu einem objectiven Bewußtsein" [ibid., 25].

is true and unique in His holiness, while I am a sinner. This is the formal principle of Schleiermacher's main work *Glaubenslehre* [Schleiermacher 1984], and it separates this book into theology and anthropology: the split within objective consciousness that humans are the representation of the absolute *and* the negative "shadow" of the absolute. This thought is grounded in a dialectic of its own necessity, in Christian language one might say that human reason turns towards itself and can only break through its limitations by reaching divine consciousness, which enables human reason to observe itself.

Where do we find absolute knowledge?

Grassmann reads the history of Christian revelation as an action of God who wants to help mankind to resolve its contradictions. Man may achieve a level of thought in which he realizes that there is a God, but man cannot find a path which reconciles the two consciousnesses. Therefore, the God of the Bible spoke to Abraham, Isaac and Jacob, to Moses, David and the prophets. He revealed to Abraham and Isaac that their houses were chosen ones; He promised Jacob to make his house the foundation of a great people; He gave Moses the law; He bestowed an inner experience of God upon David, the alleged author of the Psalms, and He commanded the prophets to proclaim the fear of God [H. Grassmann 1878, 28]. According to Grassmann, all this was said in dark allusions. It was only in Christ that the Revelation came to "bright clarity"[11] This is – and we hear the faithful Protestant speaking – written down in the Bible, with highest authority. To accept the Bible, it is not enough to believe that it provides important information, but one must have faith in the one who authored the Bible. We do obey the sayings of Christ because Christ spoke *as* God, not *about* God, and we do not merely accept them as the opinion of somebody with powerful philosophical wisdom. To hear the word of the Bible is to encounter Christ, the word, the *logos* itself. This is not just one experience among others, nor is it a conclusion of reason, because the latter would only lead to subjective certainty. Christ awakens in the listener – this is the language of Lutheran theology – the *verbum internum*, the inner word. According to Grassmann Christ reveals himself to each faithful person from within. The New Testament reached the disciples and the Church directly and was handed down to the present generation without any alteration. God gave this document to the disciples and then to the Church, whose duty it was to keep it alive and interpret it in a way that people could understand

[11] "lichte Klarheit" [ibid., 29].

its meaning, not only its textual content. Had it been a verbal decree of God, it would only foster human submissiveness.

Grassmann avoids addressing the tricky question concerning the authority of the books of the Bible. Are they all written with the same degree of authority? Since the deists of the eighteenth century, there were several proposals on how to distinguish between the important sayings of Jesus and the less important stories of miracles. And how does he answer the question arising from the fact that the canon of the scriptures of the New Testament was fixed by decisions made by members of the early Church, which is why the Apocalypse of John is canonical and the Apocalypse of Peter considered apocryphal. The Protestant counterargument to these questions usually was that the scripture establishes its own authority through its life-changing effects and its ongoing use in the Christian communities. "Therefore I believe in the truths that are revealed in the Bible, not because they are part of the Bible, but because I have experienced their sanctifying power, their eternal divine truth in my consciousness."[12] This is of course a circular argument if communities using other scriptures are seen as nonorthodox.

Is the Bible the absolute word?

Grassmann disapproves of Catholics whose dogmas go beyond the Bible, regarding these dogmas as mere human legislation *(Menschensatzung)* and their creators as lightheaded visionaries *(Schwärmer)* who expand the Bible with new revelations in the spirit. In contrast, the task of the true Church is simply to repeat the text of the Bible and remind the faithful of their duties. This is because the Bible is an objective and perfect document of divine revelation.

What does Grassmann have to say about the scholarly investigation of the New Testament? Here, his embarrassment becomes very clear: on the one hand, he clings to the immediate revelation, and yet he rejects verbal inspiration in which the authors of the New Testament would be mechanical instruments of the divine spirit. Then it would not be the meaning, but the letters on the paper which carry this inspiration. Christianity always opted for the concept of real inspiration because the New Testament was not written in the language of the Lord, but was always already a Greek translation. However, Grassmann does not deduce such arguments from the history of religion; instead he quotes the Bible itself: Paul never said that he was possessed by the Spirit while writing down his letters, and the authors of the Gospels do not mention any

---

[12] "Ich glaube also den in der Bibel geoffenbarten Wahrheiten nicht darum, weil sie in der Bibel stehen, sondern weil ich ihre seligmachende Kraft, ihre ewige, göttliche Wahrheit in meinem Bewußtsein erfahren habe" [ibid., 31].

experience of ecstasy during their writing process, but quote the testimonies of others who saw and heard Jesus. This is another classical *petitio principii*: Grassmann explains his concept of inspiration by relying on arguments from the Bible, which is, in turn, the text whose inspiration is up for discussion.

Grassmann goes into detail on this topic because he observes how modern philology erodes faith in the Bible. Faith is threatened by scholars who discover dissenting manuscripts of the Gospels and direct parallels in other religious traditions. Grassmann fears that these theologians could become the new teachers of the Church and that they will be the ones to determine which passages in the Bible are to be regarded either as original, or as the product of later editing processes, which passages can in fact be traced back to the historical Jesus, and which are only legends about the mythological Christ. Grassmann's conviction is the following: where the credibility of the Bible is weakened, apostasy is inevitable. He projects a model of a slippery slope. Whoever breaks the unity of the scripture by distinguishing early Jesuanic and later communitarian sayings will soon come to a distinction between authentic and unauthentic, then to a division of important and unimportant passages, and end up bringing forward the opinion that everything in the Bible could be considered unimportant. In some respect, Grassmann anticipated modern historical research on the New Testament scriptures in which Jesus is described as a charismatic Palestinian preacher of a baptismal movement. All sentences about his divine qualities, about revelation and so forth, will collapse under this verdict. The slippery slope has seven stages:

a) The word of God loses its absolute authority if one doubts the authority of Christ and rejects his law as too rigorous. *This person has then created his personal interpretation of the Saviour.*[13]
b) Others will follow in this apostasy and mocking of the faithful, declaring them unscientific, naive people.
c) Christ will be criticized, called a dreamer, too radical or too harmless a teacher of wisdom.
d) The divine creator will be substituted by the God of the deists, a kind of higher intelligence which does not interfere with history.
e) The immortality of the soul will be questioned and man will shrink to a mere natural being.
f) Truth will be reconstructed as mere subjective certainty and brain functions.
g) Morals will lose their power because there is no good and bad, but only biological and psychological pleasure or distress.

[13] "[...] hat sich einen Heiland zurecht gemacht nach seinem eignen fleischlichen Sinn" [Grassmann 1878, 38].

At the end of Grassmann's slippery slope we find the pure materialists, whom he considers to be *worse than pagans and cannibals* [H. Grassmann 1878, 41].

Who interprets scripture?

By rejecting the concept of verbal inspiration and yet holding on to the theory of the absolute text, Grassmann runs into the following problem: if we do not have to take the holy text word-by-word and if it still requires an interpretation, who then has the authority to decide between contradictory interpretations? Religion needs institutions outside the scripture because the text is not unambiguous.

This calls – according to Grassmann – for "steadfast and uncompromising commitment to belief".[14] Commitment to belief *(Bekenntniß)* does not refer to the subjective conviction of individuals, but to the beliefs as they are written down in The Book of Concord *(Bekenntnisschriften der lutherischen Kirche). Bekenntniß* has to be disseminated and deepened through means of print media and communities for "inner mission" (according to the nineteenth-century concept of instruction by catechism of churchgoers). Those who still oppose infant baptism (for instance reformed Protestants), who react polemically to Christianity (for instance scientific critics), who live in sin (for instance unmarried couples), who refuse the *Bekenntnisschriften* (for instance liberal theologians) should be banned from the lists of candidates and voters in Christian parishes. Grassmann wants to collect the holy remnant in small communities which will form the "true Christian Church"*(Wahrhaft christliche Kirche)* within the official Protestant State Churches *(Evangelische Landeskirchen).* If a *Landeskirche* shows tendencies of apostasy, Grassmann threatened, a new Church had to be established.[15]

Grassmann's *On the loss of faith* is a document against the hard-line conservatives in Frederick William IV's Prussia. In contrast to the liberals and the moderate conservatives, these conservatives defended the idea of a Christian state, with a king as God's official representative *(Amtsmann)* at the top of the state hierarchy. The Synod of the Church of Prussia *(Preußischer Kirchenrat)* should preside below him. Both the king as *summus episcopus* and the *Kirchenrat* as the highest synod are obliged to enforce the law

[14] "[Das] feste und rücksichtslose Bekenntniß." Grassmann continues: "[...] jede Nachgiebigkeit gegen die oft mit großer Kunst vorgetragenen Einwürfe und Meinungen der Unglücklichen, jeder Versuch der Vermittelung ist ein Verrath an der Wahrheit" [ibid., 42].

[15] Grassmann refers to the so-called *Gustav-Adolph-Vereine.* These societies helped Protestant brothers and sisters in the diaspora and considered themselves the protagonists in a cultural battle (*Kulturkampforganisationen*), considering Catholics enemies of the German Reich (*Reichsfeinde*) and trying to expel them from public life and education.

as it is revealed in the Bible. The Church is seen as the second power beside the state, but its authority is reduced to the realm of faith and belief. In the long run, Churches would have become a mere source of religious positions *(Weltanschauungsorgan)* within the state. A hypothetical Ministry for Faith and Culture, the Church could never represent a kind of ecclesiastical parliament of small parishes and synods, but would inevitably turn into an instrument of state administration. It seems that Grassmann feared that the cooperation of throne and altar would monopolize all religious authority and that religious convictions would become instruments of the state.

III

After having looked at this polemic book, which sets out in a philosophical manner and ends with a quite staunch Lutheran position which rejects all ecumenical dialogue, it seems rather strange that Grassmann focussed his research on Hinduism. Between 1872 and 1875 he wrote his Rig-Veda dictionary, *Wörterbuch zum Rig-Veda*, and published the two volumes of his translation of the Rig-Veda in 1876/77 [H. Grassmann 1873–1875, 1876, 1877a]. In a paper called *Die älteste Religion der Inder* [H. Grassmann 1874][16] – only three years before his book *On the loss of faith* – he expressed his admiration for the early Vedic religion: "that there was no people among the pagans whose fervour and depth of religious feeling could be compared to that of the Indians of the Vedic period"[17] He sees parallels between Christianity and Hinduism in their "amenity and simple clarity"[18] in the connection of spirituality and sensuality, in the fatherly relation of God to the faithful, in the faith in immortality, in the belief in the presence of God during sacrifice, and so forth. One should know that the texts of the Rig-Veda are mostly hymns to different gods (the God of the sun, storm or fire) and that these gods are identified with the sun, storm, or fire. Grassmann explicitly reads these texts not as mythology but as a poetic means of expressing original and pure faith in a benevolent God (singular!). My assumption is that Grassmann was influenced by Creuzer's book *Symbolik und Mythologie der alten Völker, besonders der Griechen* [Creuzer 1812–1821]. According to Creuzer, the ancient religions were based on a pure and primal monotheism which deteriorated into polytheistic forms due to priestly teachers, their

[16] According to present research these hymns were written around 1000 B.C. and probably compiled to form the Rig-Veda only around 500 B.C. The earliest commentaries are from the third century B.C., cf. [Witzel and Gotō 2007, 467–478].

[17] "daß es kein Volk unter den Heiden gegeben hat, welches an Innigkeit und Tiefe der religiösen Empfindung mit den Indern der vedischen Zeit verglichen werden kann" [H. Grassmann 1874, 480].

[18] "Anmut und einfachen Klarheit" [H. Grassmann 1874, 473].

symbols and narratives*(Symbole und Mythen)*, folk legends, poetic transformation and the experience of a spirit-filled nature. Creuzer aimed to find the kernel of this pure religion by comparing it to Oriental religions and their Greek transformations.

Grassmann assumes that his spiritualist interpretation of the Vedic texts was already the intention of the authors of the Rig-Veda: "Every time the myth [of Indra killing the snake Vrta] occurs, it is only a metaphor, and the poets are well aware of the symbolism of this story."[19] In order to prove this assumption, he quotes hymn 203 (II, 12) *To Indra (An Indra)*. The first seven stanzas are about Indra, King of Gods, the creator, dragon-slayer and tamer of the castes, lord over the other tribes, helper of the obedient and lord over animals, sun and water. Grassmann stops his quotation here. Perhaps the quote was too long, or perhaps he wanted to suppress the following stanzas in which Indra is depicted as an angry warrior and partaker of the sacrifice. Similarly Grassmann avoids those hymns to Varuna in which Varuna is a member of the sacrificial community, which would weaken his idea that the early Vedic religion shows "a clear reminiscence of original monotheism"[20]

If one shares Grassmann's clear distinction between gods of nature and high gods, and if one reads the hymns as selectively as he does, one can certainly find elements of monotheism there. But his determination to find such monotheism is quite startling if we keep in mind how Grassmann envisioned a history of religion in which only Christianity represented the full presence of self-consciousness as a consciousness of God. Since only three years separate these two texts, we cannot explain the dichotomy in his conclusions – on the one hand, Lutheran orthodoxy, on the other, religious openness – by a biographical change of opinion. It is much more likely that Grassmann's thinking represents the rift within Protestant self-consciousness in the nineteenth century: interest in old religions and documents of a pure monotheistic faith, clashing with a feeling of dissatisfaction with one's own religion, so closely connected to the Prussian state. The identity of Protestantism was based on the fact that it considered itself an alternative, not merely an amendment to dogmatic Catholicism. But Protestants were also aware of the fact that their alternative would not necessarily lead to a compact faith, but to a broad spectrum of divergent denominations (especially between Reformed and Lutheran Protestants, who merged in the Prussian union of 1817). The rift was even deepened by the upcoming historical approach to religion. It forced Christians to view their own religion as one among many and showed them that Christianity had only gained its present form by undergoing a long development throughout the centuries. There were two possible reactions to the pluralization and historization of

[19] "So oft dieser Mythus [Indra erschlägt die Urschlange Vrta] vorkommt, so ist der doch nur ein Bild, und die Dichter sind sich des Bildlichen in dieser Darstellung wohl bewußt." [H. Grassmann 1874, 478].

[20] "deutliche Anklänge an einen ursprünglichen Monotheismus" [ibid., 479].

Christianity: one could interpret it as a development brought about by the Spirit, or as a historically contingent process, so that the Church and dogma are subsequent disciplinary functions. Grassmann, of course opts, for the second possibility. He wants to save the core of religion – which in his view is the Holy Scripture – from the force of historical interpretation. He is not motivated by stubborn conservatism. Rather, it is the result of a philosophical insight: consciousness is unable to ground itself or criticize itself. Whenever consciousness turns itself into an object of reflection, it is thrown back upon its own means and remains confined to its own perspective. In order to break out of transcendental, circular reasoning about the ego there must be something that comes from outside and from above, something that is not posited by consciousness itself *(vom Bewusstsein gesetzt)*. This support from above is – Grassmann's view – the Bible. Therefore, this is the point at which his religious openness reaches its limitations.

Grassmann thus avoided the fallacy of many Orientalists who saw the Bible as a bad copy of pre-biblical texts. He insisted on the revelatory character of the Bible, motivated by philosophical reasons. But how can his high esteem for the Rig-Veda and his faith in the Bible be reconciled? Let me add one last assumption.

Perhaps Grassmann relied on Max Müller's insight, who declared the study of the Rig-Veda the most important source for understanding mankind. The ancient Indian hymns (Müller dated them 1500–1000 B.C.) originated centuries before the biblical and Buddhist texts came into existence and therefore were closer to the beginnings of human self-reflection. Thus, the Rig-Veda could be the first source for anthropology, the Bible the first source for theology. This division of labour would have fitted the Protestant self-perception very well. According to Paul, even the pagans, who had never heard of God, understood the reality of God in the works of creation (cf. *Romans* 1,20). The Vedic hymns could then be viewed as an expression of universal human reflection, which would lead to the paradox, unresolved dichotomy *(unversöhnter Widerstreit)* that man is capable of understanding his own finiteness, but simultaneously reaches a state of consciousness in which he realizes that his reflection on finiteness *and* infinity is finite! The biblical revelation would then function as a philosophically inspired liberation from this dilemma, a liberation which man himself can never accomplish.

The Significance of *Naturphilosophie* for Justus and Hermann Grassmann

Marie-Luise Heuser

The tradition of dynamist *Naturphilosophie* undoubtedly influenced the work of Justus Günther Grassmann and his son Hermann. The founder of dynamist crystallographic theory, Christian Samuel Weiss, had a great impact on the formulation of the early vectorial symbolism [Scholz 1996]. Weiss was able to transform central hypotheses of dynamist *Naturphilosophie* into a scientific research program which then played an important part in J. G. Grassmann's mathematics. Weiss' dynamist theory of crystallization formed the basis for J. G. Grassmann's "geometric theory of combination", which can be interpreted as a mathematical antecedent of H. Grassmann's 1844 *Ausdehnungslehre* [Lewis 1981]. H. Grassmann's crystallographic works also show influences of the dynamist theory of crystal formation. In its application to crystal figures H. Grassmann saw "the most beautiful application" of extension theory in accordance with the goal of *Naturphilosophie*, which consisted of finding a mathematical theory for the self-constructing (*selbstconstruierenden*) processes of nature.

Naturphilosophie not only influenced J. G. and H. Grassmann in creating extension theory, but it was also important as a philosophical background and motivation for Bernhard Riemann's non-Euclidean theory of manifolds, for Carl Gustav Jacob Jacobi's critical position to analytical mechanics and, last but not least, Georg Cantor's transfinite set theory. The emergence of extension theory has to be located in a broader intellectual movement in which *Naturphilosophie* was the underlying stimulant for new mathematical concepts going far beyond human perception and imagination.

H.-J. Petsche et al. (eds.), *From Past to Future: Graßmann's Work in Context*,
DOI 10.1007/978-3-0346-0405-5_5,

One central motivation for this new approach to mathematics was to decipher the so-called "inner side of nature", which was thought to be invisible but nonetheless of a physical nature. The inner side of nature was not understood in idealistic terms, only as an idea, but in physical terms as inherent to the organizing force creating the visible world.

One has to start with Weiss because he played a main role in transforming central concepts of *Naturphilosophie* to a scientific program, which then went on to form the basic feature of early expressions of vectorial ideas in the work of J. G. Grassmann. Who was Weiss? Recently, some new insights into the theoretical development of Weiss appeared, ever since his lectures on philosophy were discovered and edited.[1] He held the following lectures at the University of Leipzig in 1803 and the University of Berlin from 1810 to 1817:

Lectures in philosophy at the University of Leipzig

SS[2] 1803	"Kants Metaphysische Anfangsgründe der Naturwissenschaft"

Lectures in philosophy at the University of Berlin

SS 1811	"Die philosophischen Principien der gesammten Naturwissenschaft"
SS 1812	"Ueber das philosophische Studium der Naturwissenschaft"
WS 1813/14	"Die Anfangsgründe der philosophischen Naturlehre"
SS 1814	"Philosophische Naturlehre"
SS 1815	"Die Grundsätze der philosophischen Naturlehre"
SS 1817	"Die philosophischen Grundsätze der allgemeinen Naturlehre"

These lectures show that Weiss followed mainly the *Naturphilosophie* of Friedrich Wilhelm Joseph Schelling in his attempt to advance a theory of the emergence of matter and – in his terms – a theory of "self-organizing" and "self-structuring" processes of nature. Around 1800, crystal formation was a central example for the building forces of inorganic nature because it was assumed to show that additional forces of life did not necessarily have to enter the equation. Many scientists who were part of the *Naturphilosophie*-movement studied mineralogy at the Academy of Freiberg and were disciples of the famous mineralogist Abraham Gottlob Werner: Franz von Baader, Friedrich von Hardenberg (Novalis) and Alexander von Humboldt, among others. Crystal formation was understood to be a crucial model for the self-structuring processes of physical nature and an epistemic bridge to a biological theory of self-organization.

[1] I found the manuscripts of this lectures hidden between some crystallographic notes and edited them. The publication of these texts is now underway.

[2] SS = Summer semester; WS = Winter semester.

The philosophy of Christian Samuel Weiss

Weiss is not very well known in the history of mathematics. For this reason it may be useful to begin with some biographical notes. He was born in 1780 in Leipzig, where he studied medicine and the natural sciences from 1796 to 1800. After completing his dissertation he went to Berlin, where he participated in meetings of the *Gesellschaft naturforschender Freunde*. Then he set off for Freiberg to study mineralogy with Werner. After finishing his studies he returned to Leipzig, where he gave a course of lectures in the field of natural sciences and on Immanuel Kant's "Metaphysical foundations of natural science". In 1810 he became a professor of mineralogy at the newly opened University of Berlin. This is where he built his crystallographic theory. He died in 1856. As a young man Weiss was already well acquainted with the main exponents of dynamist *Naturphilosophie*. He was friendly with Achim von Arnim, Johann Wilhelm Ritter and Hans Christian Oersted. He led discussions with Friedrich Wilhelm Joseph Schelling, Franz Xaver Baader, Friedrich von Hardenberg (Novalis), Lorenz Oken and Carl Gustav Carus. At the University of Berlin, the famous Johann Gottlob Fichte and equally famous Friedrich Schleiermacher were among his colleagues.

Weiss elaborated his main philosophical ideas in the years between 1800 and 1817 (then Georg Wilhelm Friedrich Hegel continued Weiss' lectures on the philosophy of nature). His dynamist conception of matter was inspired by Kant and Schelling. Weiss preferred Schelling's philosophy to Kant's: "Of course, with him [Schelling] a new deeper and truer *philosophical spirit* has spread in the *studies of nature*; [...]" [Weiss 1810–1817, 299]. In his view, the main difference between the scientific works of Kant and Schelling – concerning their aims, not their achievements – lay in this: Kant only reflects on the notion of single and particular matter and on the general and abstract nature of any kind of matter and its predicates; Schelling reflects on nature as a *whole* and as a *system*, from which singularities or particularities emerge. Kant regards matter as *given*, Schelling views it as *emerging* [Weiss 1803, 2].

The central aims of Weiss' research work were the following:

- He searched for a theory of the ultimate beginning of the universe and the history of nature.
- He intended to establish a theory of the origin of matter as a "*creatio ex zero*".
- He developed a concept of crystal formation with respect to a universal physical theory of self-formation ("Selbst-Gestaltung"), considering the physical theory of self-formation (in the Romantic tradition) a starting point for a biological theory of self-organization.

According to Weiss, nature has an external and an internal side. The outer side of nature is visible, perceptible appearance, limited in space, separated into individual entities *("in Einzelnes zerfallen")*, limited to specific localities *("auf jeweilige Orte eingeschränkt")*. It is characterized by the fact that it takes up space *("Raumerfüllung")*, "finite existence in space". The internal side of nature is invisible, imperceptible. Every entity is everywhere at the same time, chaotically merged into the whole *("im Ganzen aufgegangen")*, one with the whole, one with the entire world *("Eins mit ihm, Eins mit Allem")*. This is how Weiss put it: "Things are separated only as outward phenomena, in truth they are one with all." [Weiss 1810–1817, 355]. This means that on the inside of nature nothing is localized: "The nature of matter is not at all limited to the place in which it first reaches the senses. The entire filling of space, its finite existence there in space, is but one side of its nature and its existence. – Therefore its invisible side, the true supplement of its nature, is everywhere, merged into the whole, in it, one with it, and one with all and everything."[2]

With respect to Kant's concept of space, Weiss stated: "Space shows some similarities with this foundational, primordial unity *(Realgrund)* because in space any part must be determined only in relation to the whole. Thus the possibility of the whole is a prerequisite for the possibilities of the parts." [Weiss 1810–1817, 244].

The main result of this piece of speculative thinking is that the emergence of matter is thought of as a dynamist process which constitutes matter by processes of separation and combination. Another result of this holistic view of the inner side of nature is a criticism of the mechanistic theory of matter and motion, which – seen from this perspective – is nothing but a theory concerning the outer side of nature. He was of the opinion that a truly dynamical treatment of the theory of motion "is yet to come and must be elaborated in a completely new approach." [Weiss 1810–1817, 350].

Emergence of matter

Weiss criticized the atomistic conception of crystallization brought forward by René Just Haüy, the French mineralogist. He rejected the idea that ontologically the last things remaining are little blocks of eternal atoms of matter. According to Weiss, matter is not given as an eternal ontological substance, neither as atoms nor as a continuum. Weiss proposed that matter should emerge as a product, as a result of "separations in space". He did not think of these "separations" as separations of existing matter, but

[2] "Das Wesen der Materie ist keineswegs auf *den Ort* eingeschränkt, wo sie zuförderst in die Sinne fällt. Die *ganze Raumerfüllung*, ihr endliches Daseyn im Raume *ist nur die eine Seite* ihres Wesens u[nd] ihrer Existenz. Zufolge ihrer *unsichtbaren Seite*, der wahren *ergänzenden* ihres Wesens ist sie *überall*, ist sie im *Ganzen*, aufgegangen in ihm, Eins mit ihm, Eins mit Allem." [Weiss 1810–1817, 355].

very abstractly as a "*creatio ex zero*". In order to understand this better, I give you a longer quotation from his commentary on Haüy's "Minéralogie" of 1804:

"The whole of nature can be thought of as being – in a purely mathematical sense – a development of opposing quantities of nought or zero. One must not attribute an absolute existence to any single entity of matter; each one is nothing but a quantity, one pole separated from its opposite pole, solely existing in isolate separation. At this point I would refer to the mathematicians and state that, were it to be united with its opposite pole, the two poles would fuse and become nought; like positive and negative electricity, like the south and the north poles of magnetism, any and all opposing quantities result in nought, and it was a notion that Kant had already conceived of, that the sum total of all substances, of the entire universe would be zero equals nought." [Weiss 1804a, 367f].

Thus, the essence of Weiss' conception of matter is the following.

- Matter is the product of symmetry-breaking processes, which separate a primordial zero into positive and negative entities. He calls this "*creatio ex zero*".
- Vice versa, he holds that if all opposites were unified, matter would perish and become zero.
- This means that the existence of matter is a secondary, derived product of more fundamental processes. Matter is an epiphenomenon of deeper processes.
- In Weiss' conception the opposites are not only given as repulsive and attractive forces – as in Kant's dynamist theory of matter in his *Metaphysische Anfangsgründe der Naturwissenschaft* [Kant 1786], a work which Weiss criticized as being too mechanistic.
- Weiss also did not assume only magnetic and electric forces as being constitutive for matter. This was the philosophical conception of his Romantic colleagues Achim von Arnim, Ritter, Schubert, and others.
- His astonishingly modern hypothesis was that the most fundamental opposites are matter and its counterpart, anti-matter. He assumed that in nature anti-matter opposes matter, which means that when they meet, they will both become zero. Our "globe" for instance "would be reduced to zero", if it met its counterpart of anti-matter.

In his lectures in Berlin he stated: "One could think of matter that would reduce other matter, for example the earthly sphere, to *zero*, if they were to unite. Does it really exist? Can it be found in nature? This question is not yet answered! Until now we have only a speculative investigation, nothing but a logical development. Such an idea is not at all contradictory, but something which is based directly on the nature of positing the finite." [Weiss 1810–1817, 352]. In this quotation, Weiss refers to Kant's *Versuch, den Begriff der negativen Größen in die Weltweisheit einzuführen* [Kant 1763]. This philosophical work of Kant was published in 1763. Weiss preferred it to Kant's *Metaphysical foundations of natural science.* Weiss held the view that "Every property of matter has its *opponent*. The whole magnitude of matter, in which it is expressed, is opposed by a shadow, an idea of a magnitude in which all that is posited would be negated." [Weiss 1810–1817, 352]. The idea that matter is made up of *interior* opposites is embedded in a philosophical framework which is known as German *Naturphilosophie* [see Heuser 1992].

Concept of extension

As a result of the research done on the philosophical writings and unpublished manuscripts of Weiss, I found that he conceptualized a notion of extension which has some similarities with the concept of extension from the 1844 *Ausdehnungslehre.* Perhaps Weiss' conception can be seen as one major source for Hermann Grassmann's *Ausdehnungslehre* through the mathematical theory of his father Justus.

Where are the similarities? Hermann Grassmann defined the extended magnitude or *Ausdehnung* as the "magnitude created by the generation of difference".[3] What is essential here is the fact that elements separate and become fixed as something apart.[4]

Weiss also defined extension as something generated by difference. The originally un-extended unit becomes extended by unfolding what used to be one.[5] Extension is the same as this process of unfolding. In his lectures at the University of Berlin, which he held there from 1810 to 1817, he claimed: "Extension itself has to be recognized as a development of the interior [i.e. the inner, un-extended side of nature, M. H.] into an infinite manifold, which has to be grasped as a continuous multiplicity."[6] Weiss had a non-linear concept of the continuum. The continuum is produced by omnipresent opposites. Matter is the product of symmetry-breaking processes. Weiss also was aware of an extension in time (not only in space). Organic life can be viewed as a continuous

[3] "Ausdehnung ist das durch Erzeugung des Verschiedenen gewordene" [A1, 26].

[4] "ein Auseinandertreten der Elemente und ein Festhalten derselben als aus einander seiender" [A1, 28].

[5] "das Außer-einander-treten dessen, was vorher nicht außer einander war" [Weiss 1804c, 645f].

[6] "Ausdehnung selbst muß zuförderst als Entwickelung des Inneren zu einer unendlichen Mannichfaltigkeit, als eine stetige Vielheit im räumlichen Daseyn der Einheit setzender Act erkannt werden" [Weiss 1810–1817, 308].

multiplicity extended in time.[7] Time and its duration are qualitatively differentiated. Weiss demanded a new theory of dynamics with respect to the emergence of matter and the internal laws of change.[8] This new theory required a new mathematical theory. In his words: "The mathematician understands movement as a simple dimension, homogeneous in itself. He never understands it as being internally organized. Which however it is; it is an infinite number of varied and heterogeneous internal states, of which external motion is the last (and visible) final success." [Weiss 1810–1817, 350].

Weiss demanded not only a new theory of dynamics, but also new mathematical concepts which had to go beyond mathematics as a pure instrument of quantification. He was unsatisfied by the state of mathematics at the time because it remained confined to being a theory of quantity *(Größenlehre)*. The notion of quantity as the foundation of mathematics was not acceptable to him and had to be overcome.[9] Mathematics had to consider not only the quantitative development of given magnitudes, but also the qualitative change within the infinity of heterogeneous internal states. "This requires a higher kind of mathematics than what we have today, a type of mathematics capable of dealing with the infinite in an original manner, not only as something derived from the finite or an approximation to the infinite." [Weiss 1810–1817, 351]. This required a mathematical concept of actual infinity, and not merely of potential infinity, "mathematics for which infinity is original and the finite is derived from infinity, as in nature, and not vice versa." [Weiss 1810–1817, 351].

Weiss identified the method of Pierre Simon Laplace and his French school as the type of mathematics relying on quantity which he wanted to criticize as being insufficient. We know that H. Grassmann criticized the French school in a similar way in his *Theory of Tides*. From his point of view, the development of formulas was carried out in a "mechanic" way, offering no clues on how the material developed: in his words, a mind-numbing procedure *(Geist tödtende Formelentwicklung)*. He was confident that his own mathematical approach would be more appropriate when it came to dealing with the "idea" of nature than the method relying on coordinates, which allowed only

[7] "Leben ist das nach einem Einheitsprinzip stetig und ins unendliche als verschiedenartig gesetzte Daseyn in der Zeit, oder die Ausdehnung des Daseyns in die Zeit als stetige Verschiedenartigkeit" [Weiss 1810–1817, 343].

[8] Lawvere interestingly outlined "the internal law of change" in "Grassmann's dialectics and category theory" [Lawvere 1996].

[9] "Theorie wurde = Betrachtung der Dinge als *endlicher Größen*, der *Quantität* allein nach, und als *endlich*, nach den Grundsätzen der allg. Größenlehre. *Mathematik* wurde alle theoretische Naturlehre; sie selbst war *Größenlehre*; der Begriff der Größe ein *letzter*, *einfacher*, für die Wissenschaft einziger. Alles gieng daraus hervor durch *Gleichung*, reine Verstandessache, Nöthigung, das Gleiche vermöge der Gleichung in der *andern* Form wiederzuerkennen. – Nie ein *Höher aufsteigen*, nie ein *Begründen*, nie ein urspr. neues *Erkennen*, nie wahres *Entwickeln*" [Weiss 1810–1817, 302].

a "symbolic representation of the idea". It is worth mentioning that H. Grassmann emphasized that it was this adequateness of his mathematics to nature which motivated him to elaborate his new approach at a later point in time. Obviously, the main goal of his life-task was to find a better relation between mathematics and nature.[10] His mathematical thinking was accompanied by reflections about nature and seems to have been driven by the aim of obtaining a better knowledge of nature.

H. Grassmann created an abstract mathematical theory of extension which he separated from the theory of space and geometrical interpretation [Hestenes 1996]. Nevertheless, he stated that his extension theory was linked to the fact that algebra had no notion of dimensionality. Extension theory was therefore grounded in the fact that "a notion of differences" had to be taken into account, "corresponding to the dimensions of space" [HGW11, 297]. He created a theory of extension which he separated from perception in space, but not completely from the theory of space. In the short essay mentioned above, in which Grassmann explained the essence of his new theory, he emphasized that his extension theory was also a foundation for the theory of space.

Extension is generated by continuous heterogeneity. In Weiss' writings we find a similar conception of extension. Extension is not a homogeneous empty stage, allowing bodies to move from point *a* to point *b*, but extension is originally generated by continuous processes of separation which produce irreducible heterogeneity.

Weiss found the physical model of continuous multiplicity in phase transitions, especially in the emergence of crystalline matter, as well as in the emergence of organic matter. In fluids we have an isotropic distribution of opposites, in solid bodies this changes to an anisotropic multiplicity of inner opposites "which are not uniform to all sides and directions, but specifically different, in ways which are different to others, i.e., specific directions which are not the same." [Weiss 1810–1817, 311]. In crystal states the interior opposites of forces are given in number, position, direction and relative strength. Weiss was not only interested in the transition of fluid to solid matter, but also in the transition of inorganic to organic matter. Many of the discussions in Berlin were focussed on the differences and similarities of inorganic *self-formation* and organic *self-organization*. Trying to clarify the differences between crystalline and organic matter, Weiss argued as follows: "The organic state is yet by nature a necessary *combination* of fluidity and solidity, and therefore cannot be subordinated to one or the other. It can only be put to both contraries as a combining element. The

[10] "Durch diesen Erfolg nun hielt ich mich zu der Hoffnung berechtigt, in dieser neuen Analyse die einzig naturgemässe Methode gefunden zu haben, nach welcher jede Anwendung der Mathematik auf die Natur fortschreiten müsse, und nach welcher gleichfalls die Geometrie zu behandeln sei, wenn sie zu allgemeinen und fruchtreichen Ergebnissen führen solle. Es reifte daher [!, M. H.] in mir der Entschluss, aus der Darstellung, Erweiterung und Anwendung dieser Analyse eine Aufgabe meines Lebens zu machen" [A1, 9].

organic state contains within itself a *nonhomogeneity*, namely a *constant* nonhomogeneity, which hence includes infinite nonhomogeneities or heterogeneities; it consequently cannot be a state of inner homogeneity; it necessarily implies different structural elements in spatial separation, spatially apart from one another, yet not separated, but continuously integrated as a unity, in itself, and this is why we may call it a spatial unity in this respect, organized as a continuous nonhomogeneity. That is why the relationship of the contraries, which perhaps rule and really dominate it, changes everywhere and continuously, supposing you are looking at the *solid* mass of the organic state. In organic structures, internal contraries are originally indeed given in number, direction and relative strength; but these oppositions are not *constant* in a given finite continuum, however small it may be, but display a continuously changing state of in-homogeneity, which creates the continuous inner heterogeneity of the whole. And therefore a confusion of the organic with the crystalline state, or a subordination of one to the other, which has been tried before, now is no longer possible."[11]

Weiss is saying that crystalline matter shows a homogeneous anisotropy of the continuum and grasps it in linear terms, while organic matter is an inhomogeneous continuum and must be grasped in curvilinear terms.

By introducing the concept of "axis" Weiss was able to describe all seven crystal systems: cubic, tetragonal, orthorhombic, monoclinic, triclinic, hexagonal and trigonal [Scholz 1989, 1994]. This was his main contribution to the transformation of speculative *Naturphilosophie* to scientific theory.

[11] "Der organische Zustand ist schon seiner Natur nach eine nothwendige *Verbindung* von Flüßigem und Festem, und läßt sich eben deshalb weder dem einen noch dem andern unterordnen, sondern nur beiden Entgegengesetzten als verbindendes Glied gegenüberzustellen. Der organische Zustand enthält eben so nothwendig in sich eine *Ungleichartigkeit*, und zwar eine *stetige* Ungleichartigkeit, die also Ungleichartigkeiten oder Heterogenitäten ins unendliche in sich schließt; er kann also nie der Zustand eines in sich Homogenen seyn; er faßt nothwendig qualitativ verschiedene Glieder räumlich getrennt, oder wirklich räumlich außer einander, wiewohl auch diese nicht abgetrennt von einander, sondern von Stelle zu Stelle zur Einheit stetig ineinander greifend, in sich, und kann also in diesem Betracht eine zur stetigen Ungleichartigkeit räumlich gegliederte Einheit heißen. Eben deshalb ändert sich auch, wenn man auf die *feste* Masse des Organischen allein sieht, für dieselbe das Verhältniß der Gegensätze, das auch in ihr herrschen mag und gewiß herrscht, überall und stetig. Allerdings sind gewiß auch in den organischen Bildungen {innre} Gegensätze ursprünglich sowohl der Zahl, als der Richtung und der relativen Stärke unter einander nach gegeben; aber auch durch kein noch so kleines endliches Continuum hindurch *gleichförmig*, sondern mit stetig sich verändernder Ungleichförmigkeit, die es zur stetigen innern Ungleichartigkeit des Ganzen bringt. Und somit kann von einer Verwechslung des organischen mit dem krystallinischen Zustande, oder von einer Unterordnung des einen unter den andern, wie wohl sonst versucht worden ist, nicht mehr die Rede seyn." [Weiss 1810–1817, 311].

The question of influence

In his 1829 *Zur physischen Krystallonomie und geomtrischen Combinationslehre* J. G. Grassmann wrote that he could not refer to Weiss' publications in the *Abhandlungen der Königlich Preussischen Akademie*. But he also said that he was aware of Haüy's *Mineralogie* [J. Grassmann 1829, 71, 73, 177] and of Haüy's *Physik* [ibid., XI]. Both works had been translated by Christian Samuel Weiss, and in both works he enclosed his own concepts of nature as supplements [Weiss 1804a,b,c,d,e]. J. G. Grassmann must have been aware of these five contributions, in which Weiss outlined his dynamist philosophy and his dynamist conception of crystal formation, especially in [Weiss 1804b, d]. Moreover, Grassmann mentioned the work of Carl Michael Marx, *Geschichte der Crystallkunde* [J. Grassmann 1829, 168], in which Marx gave a detailed description of Weiss's crystallographic theory [Marx 1825, 225–242]. So it was not surprising that he followed Weiss' theory and not the morphological approach of Carl Friedrich Christian Mohs. (Scholz overlooked these connections in [Scholz 1996, 41] but brought forward the right interpretation that Grassmann's approach relied on Weiss and not on Mohs.).

J. G. Grassman and Weiss shared the main goal of dynamist science. They wished to find new mathematical foundations for a science of natural history. Grassmann intended a "fluid geometry" to help him gain better insights into the inner synthesis of patterns in nature. He wanted to understand the inner processes of nature. Like Weiss, he followed the epistemological idea that pure mathematics with its constructive power should be able to decipher the internal, dynamic structure of nature. Not a mimetic, but a theoretical relation to nature was the key to the inner synthesis of its beings. Because of the principal similarities and analogies between cognitive and natural processes, the products of the mind (that is, theory) are linked to products of nature (for example crystals). J. G. Grassmann followed Weiss in his model of a dynamist, morphogenetic mathematics, distancing himself from a structural type of static mathematics. Like Weiss, Grassmann searched for a theory that accounted for the emergence of patterns and that did not remain confined to the theory of algebraic structures. It was decisive for Weiss and for Grassmann that mathematics should not remain limited to a theory of quantity and that a theory independent of all relations of quantity had to be built. The main point was that mathematics should not merely quantify the extrinsic (mechanical) behaviour of nature but recognize the intrinsic possibilities for structuring and organizing nature. This methodological aim was also constitutive for the mathematical works of Bernhard Riemann, Johann Benedict Listing, Carl Gustav Jacob Jacobi and Georg Cantor [Heuser 2007].

Inspired by dynamist *Naturphilosophie* and looking for a mathematics applying to *natura naturans*, the founders of modern mathematics obtained deep insights into the realm beyond perceptible space *(Anschauungsraum)*. We can summarize these contributions as follows:

- A loss of interest in the morphological point of view (Weiss).
- Therefore also a loss of interest in Kant's restriction to finite figures of geometry (triangles, circles etc.).
- A turn towards a structural view of mathematics (Weiss's "Structur", "Structur-Gegensätze").
- A shift from coordinate systems to a topological theory of mathematics, "independent of all relationships of quantity" (Listing: "unabhängig von allen Größenverhältnissen").
- A turn away from the mechanistic view of nature (Weiss, Jacobi, Riemann, Cantor).
- Turn away from the Euclidean view of space (Riemann).
- Turn away from the finite view of mathematics (Cantor).
- A shift towards the structuring and organizing processes of nature, while the original, primordial source of nature is thought of as lacking a representational substratum (Weiss, J. G. Grassmann, Riemann, Cantor).

Some of the founders of modern mathematics were interested in the writings of dynamist *Naturphilosophie*: Jacobi appreciated Novalis, Riemann read the early writings of Gustav Theodor Fechner and Cantor saw his antecedents in Nicholas of Kues and Giordano Bruno, see [Heuser 1991, 1992, 2007].

Justus and Hermann Grassmann: philosophy and mathematics

Michael Otte

During the eighteenth century in particular the empirical sciences grew enormously. Increasing breadth, complexity and unclearness of human experience brought an end to the "Classical Age" (Foucault) and its traditional forms of classification and representation. "More than ever before, the sciences were faced with the inevitability of experience in the eighteenth century. Even though quantitative extensions of knowledge had always led to changes in scientific methods, techniques and theories, this increase in knowledge accelerated to such a degree that the capacity of the traditional information processing technologies, based on the spatial organization of the stock of knowledge, seemed exhausted." [Lepenies 1978, 16 f].

Therefore, evolutionary conceptions of epistemology, being concerned with the circular connection between conditions and conditioned and its development in time, gained ground during the nineteenth century. Schelling's *Naturphilosophie* of 1799 provides a pertinent example. Its first maxim is to conceive of *Nature* as a self-productive system and to explain everything on the basis of natural powers alone. Schelling applied the guiding principles of *Naturphilosophie* two or three years later in his criticism of Kant's conception of mathematics [Schelling 1802/03].

We may in fact read off the beginnings of these developments already from Kant's own philosophical development. In his *Enquiry concerning the clarity of the principles of natural theology and ethics* [Kant 1764], which was written as a reply to the question – posed by the Royal Academy at Berlin – whether metaphysical truth could be

H.-J. Petsche et al. (eds.), *From Past to Future: Graßmann's Work in Context*,
DOI 10.1007/978-3-0346-0405-5_6, © Springer Basel AG 2011

equated with mathematical truth, Kant draws his well-known distinction between analytic and synthetic truths. He classifies mathematics as based on arbitrary definitions and thus as synthetic and affirms that it is, in contrast to philosophy, much too early for metaphysics and natural philosophy to proceed according to the synthetic method. The empirical method should be nothing but a variant of the analytical one, simply confined to those characteristics which sound and secure experience detects about things. Its principles are not given, but have to be inferred by analysis of given experiences. This means, therefore, that philosophy finds itself jointly with natural science in one camp and both stand in opposition to mathematics, which is synthetic knowledge. Kant writes:

> "The genuine method of metaphysics is basically identical with that which Newton introduced into natural science and which was there of such useful consequences. One should, it is said there, seek the rules according to which given appearances of nature proceed through secure experience, at best *(allenfalls)* with the help of geometry. [...] Equally so in metaphysics." [Kant 1764, 286].

Only after having understood from Hume that metaphysics must be something different from what it had been so far, because even causal relations are not analytical, in fact, all relations must be considered external, Kant changed his conceptions of philosophy, natural science and mathematics. In his *Critique of Pure Reason* [Kant 1781] mathematics is no more merely a language or a formal science based on the law of contradiction alone, but is objective and the nature of its objects is responsible for the fact that mathematics was able to develop its particular synthetic method, which enabled it to make real cognitive progress, differently from logics. Kant writes:

> "Mathematics affords us a brilliant example, how far, independently of all experience, we may carry our *a priori* knowledge. It is true that mathematics occupies itself with objects and cognitions only in so far as they can be represented in intuition.
>
> But this circumstance is easily overlooked, because the said intuition can itself be given *a priori* and therefore is hardly to be distinguished from a mere pure conception." [Kant 1781, B 8].

Kant still distinguishes between the method of philosophy and mathematics, stating that philosophy explains objects from concepts analytically, whereas mathematics and natural science explain them from the construction of concepts in intuition synthetically [Kant 1781, A 713, B 741]. Schelling abandoned these divisions when he stated the principles of Kant's constructive science and mathematics as principles of *Naturphilosophie* [Schelling 1799, Sect. 1].

But nineteenth-century mathematics and philosophy reacted to Kant in different ways. On the one hand, a strictly anti-Kantian rigorous movement of arithmetization of pure mathematics was established, which reduced philosophical considerations more or less to logical ones and denied any foundational role at all to intuition. There arose, however, also a very different view of mathematical epistemology, – represented by Romantic *Naturphilosophie*, for example – which conceived of Kantian *construction* in symbolic terms, rather than in geometrical or substantial ones and which accordingly tried to transform Kantian intuition into an *inner* or intellectual intuition, which was meant to make all things and cognitions dynamic [Jahnke 1990, 53].

This kind of intuition is based on what has been called "hypostatic abstraction" by Peirce or "reflective abstraction" by Piaget. Hypostatic abstraction, that is, idealization and compressed reification, is by its nature a kind of observation with respect to one's own mathematical activity and it transforms mathematics into meta-mathematics. An important example is provided by modern axiomatics in the sense of Hilbert. Geometrical axioms are nothing but hypostatization of constructions or perceived invariants, and Peano's axioms of number – which were in fact anticipated by Hermann Grassmann's *Lehrbuch der Arithmetik* of 1861 [LA] – are nothing but crystallized operations.

Justus Grassmann had in this sense written in 1824 already that by the term "intuition" (Anschauung) we should understand "inner intuition, intuition of the activity of the constructing mind or intuition of the inner constructions." [J. Grassmann 1824, VIII]. And his son Hermann wrote in 1844:

> "The essence of the philosophical method is that it proceeds by means of contrasts to arrive at the particular from the general; the mathematical method on the other hand, proceeds from the simplest concepts to the more complex, and thus, through the connecting of the particular, attains new and more general concepts." [A1, 30]

In this sense philosophy is analytic and mathematics is synthetic, in accordance with the Kantian view. Grassmann continues however: "Since both mathematics and philosophy are sciences in the strictest sense, so must the methods in both have something in common which makes them thus scientific. Now, we add the scientific quality to a method of treatment when the reader is, on the one hand, led by it necessarily to the recognition of each individual truth, and is at each point of the development put in the position of seeing the direction of further progress, on the other. [...] Thus the scientific presentation in essence is an interlocking of two series of developments of which one consistently leads from one truth to another and makes up the essential content, while the other governs the process itself and determines the form." [A1, 30–32].

The opening sentences are almost a repetition of Kant's famous distinction between philosophy and mathematics. A distinction that is qualified in the sentences that follow in a manner which reminds one of the criticism of Kant's notion of "constructing a concept in intuition" as it had been put forward by Schelling [1802/03]. Thus, some denied the relevance of intuition for mathematics and wanted to come back to the rationalist programs of Descartes and to complete the arithmetization of mathematics. Others wanted to stress the theoretical character of science and mathematics and emphasized the necessity to generalize. It is only the process of generalization which distinguishes mathematics, for example from a mere mechanical doing.

Also, in order to generalize one must employ some kind of abductive or hypothetical reasoning based on an intuited idea or perceived analogy [A1, 31]. Kant's notion of *construction in intuition* implied a limitation to mathematical and scientific knowledge, because mathematical concepts should have realizations in (pure) intuition. But this pure intuition should be capable of development and must thus be based on activity, rather than on sensible objects [Jahnke 1990, Chap. II.1].

What is important here is to emphasize that mathematics must necessarily no longer be satisfied with constructing triangles and the like, but has to construct more fundamental concepts, like the notion of equality or the concept of space – Hermann Grassmann was the first to conceive geometrical spaces of arbitrary dimensions – and even the notion of a mathematical theory as a whole. For that it needs a philosophical overview in addition to calculation and empirical construction. The view that the notion of (mental) construction has to be objectified, along with the idea that the subject matter of mathematics must be extended, and based on the unity of the general and the particular realized through the construction, became important to both the Grassmanns.

Hermann Grassmann subsequently accounts for this analytical momentum of "clarity" with the observation that he who does not acquire that overview for himself, remains totally dependent on the particular manner in which the truth was found [A1, 30; see also J. Grassmann 1824, XVII]. This observation, namely that mathematical knowledge has to have a quality of general clarity, that in mathematics we do not primarily gain truths, but reasons for those truths and that mathematical proofs had accordingly to be constructed from necessary conditions, became a general trend during the nineteenth century.

Bolzano's *Wissenschaftslehre* also, for example, contains a distinction between proofs that verify, being intended to create conviction or certainty, and others, which "derive the truth to be demonstrated from its objective grounds. Proofs of this kind could be called justifications (*Begründungen*) as distinguished from the others which merely aim at conviction (*Gewissheit*)." [Bolzano 1837, Sect. 525]. Proofs should, in order to serve as objective justifications, be *pure*. Bolzano criticized, for example, Gauss's proof of the fundamental theorem of algebra of 1799, because Gauss had on that occasion employed geometrical considerations to prove an algebraic theorem. Bolzano did not, as is often claimed, doubt the validity of Gauss's arguments and he did not question the certainty of our geometrical knowledge, but criticized the *impurity* of Gauss's proof of 1799.

The general feeling at the end of the eighteenth century was, as mentioned above, that the traditional means of justification and organization of knowledge became insufficient as the given methods showed themselves as incapable of producing new results. Contemporaries in the eighteenth century even believed that mathematics had come to an end of its growth. "A great upheaval in the sciences is imminent. In view of the present aspiration of the great minds, [...] I should almost like to claim that there will not be three great mathematicians in Europe within a century. This science will suddenly remain fixed to the spot where the Bernoullis, Euler, Maupertuis, Clairaut, Fontaine, d'Alembert, and Lagrange have left it." [Diderot 1754, 421]. Similar statements have come down to us from Lagrange [Stuloff 1968].

Enlightenment philosophy and science had freed itself from the paternalism of theology and now wanted do get rid of the new tyranny of mathematical rationalism – mathematics is "a kind of metaphysics" [Diderot 1754, 420] – in order to develop a descriptive science, based on experience and an elaborated and sound theory of experimentation. Kant had shared these concerns, as the introduction to the second edition of his *Critique of Pure Reason* clearly shows, but he gave them a twist that did not combine well with the new empirical experiences and he had in fact transformed mathematics into a kind of metaphysics. The problematic nature of Kant's entire epistemology of the empirical sciences is nicely reflected in a controversy between Kant and Georg Forster (1754–1794), Alexander von Humboldt's teacher, which took place in 1785. Whereas Kant emphasized that "by mere stumbling about, without a guiding principle defining that which is sought after, nothing useful would ever be found" [Kant 1788, 161], Forster points out the continuous and systemic character of Nature means that, due to the fact that we are only able to think within fixed differences, all our theories must remain one-sided and preliminary, see in more detail [Otte 1997, 338ff]. *Naturphilosophie* shared these views. Nature is not a kind of geometry, Schelling said, but geometry is a productive force of Nature [Schelling 1799, Sect. 1].

Thus, nineteenth-century epistemology will have to be discussed in terms of principles of *complementarity* (contemporaries used the term *dialectic*) and we shall see soon that the contrasts of *distinct and equal* as well as that of *the continuous and discrete* play a fundamental role in the works of both Grassmanns, father and son.

Now it is a well-known fact that Diderot's and Lagrange's fears have not come true; rather, the "great upheaval in the sciences" predicted by Diderot seized mathematics as well, and led to new developments of method and objects not anticipated. The new style of mathematics, which began to emerge at the turn of the nineteenth century, is seen, as most historians of mathematics agree, first of all in the tendency towards rigorous proof and in a more careful elaboration of the foundations and definitions of mathematics. Mathematics should follow the synthetic ideal of antiquity, but without being based on intuition in a Kantian sense. Analysis sees a foundation of its methods, the nucleus of which is described as arithmetization. In retrospect, Felix Klein wrote in 1895:

> "The spirit in which modern mathematics was born, [...] has placed topmost a philosophical principle, that of continuity. This applies to the great pioneers, to Newton and Leibniz, it applies to the whole of the eighteenth century, which, for the development of mathematics, has really been a century of discoveries. It is only gradually that rigorous criticism emerges, which enquires after the consistency of these bold developments [...]. This is the age of Gauss and Abel, of Cauchy and Dirichlet [...] hence the demand for exclusively arithmetical proof." [Klein 1895, 143/144].

This summary by Klein represents a view of the development of mathematics in the nineteenth century current far and wide to this day. It entails, however, some difficulties and even contradictions. The first problem concerns the opposition established by Klein between mathematical discovery in the eighteenth century, and the foundation or codification of mathematics in the nineteenth century. Does this separation between development and foundation, discovery and justification, generalization and proof, really apply to the mathematics of the nineteenth century, which did show the marks of a historically unprecedented productivity?

In fact, as was stated already, the nineteenth century witnessed an internal opposition to developments in logic [Heijenoort 1967], as well as, in mathematics. As well as the rigorous movement of arithmetization, a quite different approach, dedicated to more or less radical conceptual generalization and based on a way of analytical thinking in terms of structural analogies presented itself which had its roots in Grassmann's work and culminated in Hilbert's new axiomatics as presented in his *Foundations of Geometry* [Hilbert 1899].[1]

Hermann Grassmann was very much influenced by his father Justus Grassmann (1779–1852), who was a much more philosophically minded person and both, father and son, searched, like Schelling, for connections between a synthesis of nature and a synthesis of the mind. If Hermann Grassmann is sometimes called a *visionary* such notions could be attributed with much more justification to his father, who had formed his *philosophy of science* under the influence of romantic *Naturphilosophie* [Otte 1989; Heuser 1996]. Hermann Grassmann himself excels more through his immense analytical powers of mind.

[1] See also [Israel 1981].

Justus Grassmann taught at the Stettin Gymnasium from 1806 to 1852 when his son, Hermann, succeeded him as head of mathematics. For Justus Grassmann, the educational context was not just significant because of his position, but he was also convinced that no educational "work is of solid value if it is not at the same time [...] able to claim scientific recognition." [J. Grassmann 1829, XIV]. Hence he published a series of small books between 1817 and 1835 in which he wanted to expound a new philosophical and educational conception of mathematics while at the same time developing mathematics itself. J. Grassmann defines mathematics in the spirit of Schelling, not Kant, as pure constructivity, as a construction which does not start from any content or empirical intuition, but solely considers things according to the principle of noticing their equality or difference. Logic, J. Grassmann says, abstracts from content, thereby presupposing there is one. Mathematics is strictly formal construction without any question of content [J. Grassmann 1827, 3]. Intuition is the result of mathematical activity not its starting point. This reverses the order of the relationship between theory and experience as compared to the inductivist ideal of empiricism.

As Justus Grassmann regards the contrast of equal and unequal to be relative, he gains a division of mathematics according to "which of the two elements serve as the determination that underlies the synthesis, while both are continually present" [J. Grassmann 1827, 5] in the theories of quantities and combination. "The synthesis of equals gives us the quantity", while "the joining together as different results in the theories of combinatorics" [J. Grassmann 1827, 4]. Justus Grassmann is particularly interested in this combinatory element of mathematics or in the "construction from differences". All these considerations reappear in Hermann Grassmann's writings.

As it is rather common among historians to consider Grassmann's work and especially his *Geometrische Analyse* [H. Grassmann 1847] as an extension of Leibniz's ideas, it may be appropriate to emphasize the differences. Leibniz is, like Grassmann [A1, 9], not satisfied with either the Euclidean or Cartesian approach to geometry, and this is because neither of them carries "analysis to its conclusion". This concept of analysis is closely linked to Leibniz's constitutive *principle of the identity of indiscernibles.* The latter consists of the thesis that there are no two substances which resemble each other entirely but only differ numerically because then their complete concepts would coincide. The identity of a substance stems from its properties which make up the complete concept of this substance. The complete concept enables one logically to differentiate a substance from all others. Leibniz interprets a proposition like "all congruent triangles are similar" to mean the concept of similarity is contained in the concept of congruence, for congruent triangles contain all the properties of similar triangles and also others.

Leibniz's conception was determined by the classical ontology of substance and property and his propositions followed the classical subject-predicate form. He tried therefore to found the basic definitions of his geometrical characteristics on congruence and remained thus within the classical conceptual framework. Congruence becomes the most general geometrical relation in Leibniz's project. If we are comparing things according to quantity and to quality, respectively, it follows that congruence expressing equality in both these respects is the most general or concrete geometrical equality and Leibniz sets congruence as the absolute geometric identity, as it represents an equality according to both external and internal definitions, according to quantity as well as to quality.

Grassmann's thinking in contrast was relational through and through. In such a view one does not try and derive or explain the relationship from analyzing the nature of its *relata*, but concentrates on the relations and their overall structure from the very beginning. Grassmann accordingly begins his *Ausdehnungslehre* of 1844 with a "general doctrine of forms", which is but a meta-theoretical framework for all of mathematics [Otte 1989, 29/30]. Grassmann also pointed out that the congruence relation does not lend itself to constitute an extensional ontology. The conjunction of congruent parts does not necessarily result in congruent figures, that is, $A = B$ does not necessarily yield $A + C = B + C$ [HGW11, 331].

According to Leibniz's fundamental "principle of the identity of indiscernibles" there is no distinction without a motive (Leibniz–Clarke Correspondence). Grassmann's rather constructivist approach is based on precisely the opposite view, that is, on a distinction without motive. "It is", writes Grassmann, "irrelevant in what respect one element differs from another, for it is specified simply as being different, without content to the difference. Our science shares this notion of combinatorics." [A1, 47; see also J. Grassmann's Zahlenlehre 1827].

Leibniz's theory of identity derives from the traditional Aristotelian theory of substances. Substances are the subjects of predication. They are in this sense the prerequisites of properties and relations. A is B or $A = B$ means that A and B are appearances of the same substance. In Fregean terminology this is stated saying that A and B are different intensions of the same extension or that they are representations with a shared referent but different meanings. The other, Grassmannian view interprets "A is B" in terms of the idea of shared qualities of different things as they seem relevant from the perspective of a certain theory. The relevant context is always the theory (and all mathematical theories are to be arranged within an overall transcendental framework, as is specified by H. Grassmann's "general doctrine of forms"). The context and unity of knowledge is neither provided by objective reality nor by Kant's "synthetic unity of consciousness" [Kant, B 138], but by the theory as a structured whole and as a reality *sui generis*. In 1824 Justus Grassmann had already documented these views. He writes: "It is of utmost importance, without any doubt, to exercise intuition and the capacity to construct. It should, however, not be denied that the logical aspects, the organized reflections about the product of construction, are also of great value. It might even be admitted that the first can never exist without the latter." [J. Grassmann 1824, ix].

Grassmann begins his "derivation of the concept of the theory of extension" ("*Ableitung des Begriffs der Ausdehnungslehre*") by presenting two contrasts, that of discrete and continuous and that of equal and different and he then continues:

> "From the combination of these two contrasts, of which the first refers to the mode of production and the second to the elements of that production, derive four species of forms and the corresponding disciplines of the doctrine of forms (or mathematics)." [A1, 26f].

This endeavor to establish a general framework is typical for the time and its most famous product is Bolzano's monumental *Wissenschaftslehre* (doctrine of science). Justus Grassmann strives like Bolzano toward a totality of the sciences "from the theories of quantities, combination, and natural history in the wider sense", that can serve as a "conscious presentation of the agreement between nature and mind", and "have the reconciliation and unification of the two separated spheres, the process of freeing the human mind from the torture of empty abstractions and raw empiricism as its object." [J. Grassmann 1829, 172]. Furthermore, he directly addresses the opposition between foundation and development which characterized the epoch when he says in 1827: "The more science gains in breadth, the more necessary it is to organize the mass of material thus presented, not only in order to facilitate the entry of the beginner, but rather, and more importantly, thus to elevate the raw information to a genuinely ordered knowledge, in which the position, the connections, the function of every part is distinctly perceived relative to the whole, and so that this latter can appear as an organism, as a manifestation of an infinite intellect as it becomes clear to us in a particular sphere. These remarks find their application not just in the experiential and particularly in the natural sciences, in which the mass of material mainly accumulates these days; rather, they apply especially to pure science, and particularly to mathematics, by which the laws of outer nature are given to us in inner development as a rich endowment of the human intellect. The extent of mathematics has already increased so significantly, and expanded in the most varied directions to such a degree, that the intellect sinks under the weight of the presentations, and they themselves run the danger of becoming blind instruments that one can only apply, as it were coincidentally, to the appearances of nature, rather than their attending to the explanation of nature in an orderly way, or rather, since they are independent of it, their being a model and standard for its presentation.

This lack of organization obtains not only in the mutual relations of the mathematical disciplines however, but also in each of them individually; indeed it extends to the arrangement of the individual theorems in them, and it is a much too one-sided attitude, tenaciously maintained by mathematicians above all, that it is *only* necessary to verify whether a theorem can be completely demonstrated from what has preceded it. Without a doubt it is the requirements of communication and of instruction that have secured the victory of this attitude over all others; but it can safely be predicted that the methods for science must coincide with the methods for instruction as soon as the inner economy of a discipline has assumed a genuinely organic form, resulting from its inner essence, since that which in every sense belongs together must also be most fully understandable in its unity." [J. Grassmann 1827, 1–2].

Similar views can be identified, as was said, in Bolzano and many other contemporaries. Only Bolzano was more concerned with the foundations and with the context of justification and hence his program of arithmetization, whereas the Grassmanns took a rather genetical or dynamical perspective on mathematics and science and from this they came near in their methodological and theoretical views to a new conception of axiomatics.

One should remind oneself here once more that Hermann Grassmann's *Lehrbuch der Arithmetik* [LA] is, as Wang remarks, probably the first "serious and rather successful attempt to create an axiomatic foundation for numbers." [Wang 1957, 147]. H. Helmholtz (1871), Robert Grassmann (1872), E. Schröder (1873), or O. Stolz (1885), all of them followed more or less explicitly H. Grassmann's approach. Wang also shows in detail that Grassmann's system of characterizing the whole numbers essentially corresponds to the characterization "which is customary in present day abstract algebra. According to the latter, integers form an ordered integral domain in which a set of positive integers has a least element." [Wang 1970, 70]. In particular, Grassmann was probably the first to introduce recursive relations into the axiomatic of arithmetic, and this is why Helmholtz called the recursive characterization of addition (and multiplication as well) "Grassmann's axiom".

Institutional development of science in Stettin in the first half of the nineteenth century in the time of Hermann Grassmann[1]

Maciej Szukała

Pomerania at the turn of the nineteenth century

Hermann Günther Grassmann was associated almost all his life with Stettin (now Szczecin). At his birth in 1809 the city had been under French occupation for three years and had barely 20,000 inhabitants. At the time of his death in 1877 in Stettin, the city had removed its encirclement of blockading walls and had approximately 80,000 citizens. Grassmann's life spanned that part of the history of the Prussian state during which Stettin was located within its borders and was the capital of the Pomeranian province. This was the period from the Napoleonic wars through to the unification of Germany and the proclamation of the Empire, an important time for Central Europe. The period was characterized by important socio-political changes which can be seen from the turn of the eighteenth century through the nineteenth century – from the

[1] This work was completed with the support of the Faculty of Mathematics and Physics of the University of Szczecin.

H.-J. Petsche et al. (eds.), *From Past to Future: Graßmann's Work in Context*,
DOI 10.1007/978-3-0346-0405-5_7,

fading Enlightenment to the rising Romantic era. The consequences of these processes were especially visible in Prussia and other German states. In the years after its humiliating military defeat in 1806 the Prussian state was a good example of the introduction of large-scale reforms. These changes, introduced from above by an absolutist state, included the renewal of many aspects of society. To various degrees the administration was rebuilt, a higher education and middle school level system that was modern for its time was created, the army was modernized, and social change took place in the country. In this way, the foundation stone of a modern state and modern society was laid. The successes and failures of these reforms determined the history of Prussia, especially through the mid-nineteenth century. These new trends – particularly in the field of culture and science – gave expression to the increasingly important educated middle class [Salmonowicz 1987, 424]. The cultural emancipation of this social class was, in turn, the precursor of its political emancipation [Salmonowicz 1987, 236].

The reforms of the Prussian state were introduced into the different provinces, including the province of Pomerania. This latter area was at one time the territory of the Duchy of Pomerania whose local dynasty died out in the seventeenth century. With its integration into Prussia in 1815 it was finally united with the so-called New Pomerania, which had been under Swedish control, with its main cities of Greifswald and Stralsund. The reforms were carried out from above, since the stimulus for change came from the enlightened segment of the bureaucracy. In these years of upheaval, 1806–1815, an effective bureaucratic apparatus made structural changes with deliberation. It is true that even earlier, under the government of Frederick the Great, the civil service, drawn largely from the aristocracy, became a group that was characterized by a high level of professional knowledge. However, not until the reign of Frederick William II did such officials attain a more serious and politically important significance. During the time of his successor, Frederick William III, they played a crucial role in the reforms which were led by Karl Baron vom und zum Stein and by Duke Karl August Hardenberg. This social class was not ideologically homogeneous: its inner circle had different political tendencies fighting each other and there were constant clashes between conservatives and reformers of the liberal strand. Because of its highly developed organization, access to information, and cliquishness, this group was also the only power in the state that was in a position to come to agreement on the determination of political objectives. Some historians take these qualities as ground for the opinion that the conservative and liberal part of the bureaucracy already formed the foundation for the Prussian parliamentary system in 1808–1810 and not – as is generally held – only after 1848 [Vogel 1983].

The time of the Bourgeois reformers

The years 1806–1848 were undoubtedly the period of greatest significance for the Prussian bureaucratic establishment. This class exerted its influence in society throughout all socio-political affairs, although formally it participated neither in the official circles nor local government structures. The high level of professional knowledge among the executive administration, and also the rise of the conviction that an official had a calling as a public servant, were linked with the development of the typical Prussian philosophy of the State, whose synthesis had been carried out by Hegel. The characteristics of this civil service, such as incorruptibility, high level of education, and administrative skill, led the Liberal Otto von Camphausen[2] to refer to the group as an "aristocracy of talent" ([Salmonowicz 1987, 363] see also [Branig 1979; Hartung 1981]). Among the talented were some provincial governors, unchallenged in their positions, who maintained their office beyond the 1815 changes. The system of government (of 1825) granted them significant powers. The governors, as the highest officials in the provincial administration, were in a position to carry out reforms in social, economic, and cultural fields. Within a province, the governor personally represented the Prussian King and exercised on his behalf a number of rights such as the declaration of laws or pardons. On the one hand he was a state representative in internal matters, embodying the authority of the monarch with respect to the provincial legislature, and, on the other hand, he represented the interests of the province to the King and the central government. The governor directed and oversaw such sectors as medical colleges, consistories, General committees, annuity banks, and credit and insurance institutions. He also managed universities and led scientific societies. He thus had wide powers [Salmonowicz 1993, 99; Wielopolski 1965, 61]. One must not forget, however, that this power was given to him by the King and that the governor was limited on the one hand by the existing provincial social structure and, on the other hand, by the centralized, absolute power of the Prussian state. With a strong enough personality however, whoever occupied this office could exert a great influence upon affairs of the province. One of the foremost examples is Johann August Sack, the governor of Pomerania in the years 1815–1831. Another influential individual was the very talented Theodor von Schön, governor of the province of East Prussia, 1816–1820, and of the province of Prussia, 1820–1840. Others should also be mentioned: Friedrich Theodor von Merkel, governor of Silesia, 1816–1820 and 1825–1828; Friedrich Ludwig von Vincke, governor of Westphalia, 1816–1844; and Eduard Heinrich Flottwell, governor of the Posen Province, 1830–1840.

---

[2] Otto von Camphausen (1812–1896), brother of the Prussian prime minister Ludolf von Camphausen, was finance minister in the Bismarck era.

Johann August Sack: governor and reformer in Pomerania

Since the 1830s, we can see a trend to appoint nobles as representatives of the higher offices. In contrast, since the time of Frederick the Great, a large part of the central bureaucracy had been filled by professionals from the middle class [Koselleck 1967, 681–90]. Johann August Sack (1764–1831) was symbolic of this group. He was one of the most interesting personalities of the first half of the nineteenth century in the world of Prussian provincial bureaucracy.[3]

At the beginning of his career Sack came to know the head of the War and Treasury departments in Cleve, the previously mentioned Baron vom und zum Stein. This acquaintance proved relevant in shaping Sack's political career. He rose through the ranks of the state administration and in 1813 attained the post of governor of the Rhine province. In 1816, because of his liberal tendencies, Sack was – despite the support of Stein – put in the equivalent post of governor in far away Stettin [Wehrmann 1921, 279; Eggert 1965, 23]. When Sack arrived in the fortress city, Stettin was – as mentioned above – the capital of the united province of Pomerania.

For some years the province had put into effect self-governing powers that were granted on the basis of the Municipalities Ordinance (Städteordnung). This ordinance was introduced for the whole of Prussia in November 1808 on the initiative of Baron vom und zum Stein. However, one must note that Stettin in the years 1806–1813 was under French occupation. The French left the city only in December 1813, leaving it and its surroundings devastated. The population drastically declined. In 1813, the population was 18,000. It had reached the number which the city of Stettin alone had at the end of the eighteenth century [Stępiński 1994, 19]. The implementing of plans for reconstruction and expansion of the city, and for its economic development, owed a great deal to Governor Sack: for example, through his initiative in modernizing the waterway to Swinemünde (Swinoujscie) and in the development of the local port, which acted as an outer harbor for Stettin [Gwiazdowska 2001, 243–4]. Sack also took note of the weakness of the city in the scientific realm compared with cities in the western provinces of Prussia.

[3] Cf. [Petrich 1887, 1890; Steffens 1931a, b, 1932, 1938a; Branig 1959]; of the Polish literature dedicated to the governor we can mention [Wielopolski 1965] – the author emphasized Sack's role in the consolidation of state influence in the internal affairs of Pomerania; [Turek-Kwiatkowska 1986] – the author focused on Sack's importance for the creation of scientific and cultural structures; [Stępiński 1983, 1994], on the other hand, dealt with Sack's role in the reform of vesting of ownership.

Stettin and its Marienstift Gymnasium

Unlike Breslau, Berlin, and Königsberg, to name some examples, Stettin was not a university town. When the universities in Berlin, Bonn, and Breslau were founded at the beginning of the nineteenth century, Stettin was passed over. This is explained by the fact that in the Prussian province of Pomerania – where the University of Greifswald was founded in 1456 – Stettin remained in the part that was under Swedish rule 1648–1815. The Marienstift Gymnasium formed a substitute for an institution of higher learning for Stettin. It was founded in 1544 and counted among its graduates many people with great achievements in Pomeranian culture and science.

In Stettin, there was also another school – the Council School. In 1805 the schools were brought together under a cabinet decree of King Frederick William III and thereby formed the United Royal and City Gymnasium (Vereinigte Königliche und Stadt-Gymnasium), with 178 students. At the beginning of the nineteenth century this school, under the influence of the reforms initiated by Wilhelm von Humboldt, incorporated the Neo-Humanist educational program, like many other schools of its kind in Prussia.

The highpoint of Neo-humanism occurs just at the turn of the nineteenth century. The humanism of the Renaissance was based on ancient culture and – for formal reasons – focused on literature and art. In the case of Neo-Humanism, this return to antiquity had a more practical and broader objective – the education of man. Neo-Humanism was a specifically German movement which arose from Protestantism and was infused with a zealous spirit [Spranger 1928, 100]. The term "Neo-humanism" embodied the ideal model of education wherein the main rule according to Humboldt was the ethical value of science: the way to the formation of character was through the formation of the mind [von Aster 1969, 356]. Thus, the classical Stettin Gymnasium ascribed great importance to the study of ancient (Greek, Latin, Hebrew) and modern (French, English) languages, history and geography, drawing, calligraphy, and mathematics. The school was highly valued by the bourgeoisie that eagerly sent their own sons there. In the years 1830–1840 the pupils numbered 450 while by 1853 there were 750. Over different periods of the nineteenth century many teachers taught at the Gymnasium who proved to be a great benefit for Pomeranian and German culture and science. Among these were: historians – Wilhelm Böhmer, Ludwig Giesebrecht, Hermann Hering, Martin Wehrmann; philologists – Ferdinand Calo, Friedrich Wellmann; the composer Carl Loewe, known beyond Pomerania; the painter Ludwig Most; and the mathematician and physicist, as well as Sanskrit expert, Hermann Grassmann, who continued the scientific tradition of his father, Justus Günther Grassmann.

Justus Grassmann, a longtime professor at the Stettin school, started there in 1806 and was a founder in 1835 of the Physics Society. Hermann Grassmann was a graduate of the Gymnasium in 1827. After studying theology at Berlin University, he worked from 1831 as a teacher at the Stettin Gymnasium, and taught in the local seminary until 1834. Then he taught briefly in other schools in Berlin and, from 1836, again in Stettin in the Ottoschule. In 1843 he began teaching at the Friedrich-Wilhelm School and in 1853 he was placed back in the Marienstift Gymnasium as professor of mathematics and physics in place of his deceased father [Graumann 2000, 45–7; Wehrmann 1894, 167, 170, 174]. Another student at the Stettin Gymnasium was Hermann's younger brother, Robert Siegfried (1815–1901). In 1848 the two brothers founded the *Norddeutsche Zeitung*, in 1866 the *Pommersche Zeitung* and in 1875 the *Stettiner Tageblatt*. Robert also opened a printing shop.

The *Pommersche Provinzial: Blätter für Stadt und Land* 1820–1825

At the beginning of the nineteenth century the Gymnasium in Stettin was the only source of scientific education in the city and, as a result, was inadequate for the capital of the province. Governor Sack sought to change this. He was aware that the economic backwardness of Pomerania, in comparison with the western provinces of Prussia, for the most part resulted from its social and cultural backwardness, which in turn was due to the fact that science and culture were not accessible to the upper classes of society. Opportunities for emphasizing the integrity of Pomeranian society arose at public festivities, which the governor undertook with great pomp. The first of these was the 300th anniversary in 1817 of the Reformation and another was the 100th anniversary in 1821 of the joining of the port of Stettin, and part of Pomerania (Altvorpommern), to Prussia. The second of these events brought Frederick William III to Stettin along with his whole family. In [Koch 1821] the Director of the Marienstift Gymnasium, Friedrich Koch, extolled the importance of the consequences of the Reformation, as well as of the union of Pomerania with Prussia, for the development of education. The affiliation of another part of Pomerania to Sweden was regarded as a "national disaster" and made this one of the saddest periods in the history of the former territories of the Duchy of Pomerania [von Bohlen 1865, VII].

The third occasion was the 700th anniversary of the Christianization of Pomerania, celebrated as a Church and state event in 1824. It provided another occasion – in addition to those already mentioned – to emphasize the special ties to the dynasty and the state through the provincial government and the humanistic circle.[4]

[4] For more information see [Stępiński 1998]. The problem of national integration at the beginning of the nineteenth century in the most easterly Prussian province, East Prussia, is addressed in [Niedzielska 1997].

Governor Sack was of the opinion that the first step towards eliminating – at least in part – the cultural backwardness of Pomerania should be the founding of a press of a scientific nature. He also had the idea of organizing a historical society in cooperation with the Gymnasium teachers. For Sack history was an essential discipline for current times and for the political actions aimed at state integration. One of the means for achieving these plans was the aforementioned intention of establishing a scientific periodical. The governor took as a model for the new journal the monthly *Schlesische Provinzialblätter* of Upper Silesia, established in 1785. The press as it existed up to then in Pomerania did not tie in with Sack's ambition, although he supported the *Stettinische Zeitung* established in 1765 and the *Greifswalder Wöchentlicher Anzeiger*. The Silesian periodical, on the other hand, met the governor's expectations.

The journal was completely apolitical and published a chronicle of events, literary prose and articles about provincial history [Steffens 1938b, 235]. It was important for Sack next to find a person who would serve as publisher and, at the same time, chief editor. The governor had become aware of the Superintendent of Treptow an der Rega, the pastor Johann Christian Haken (1767–1835). Haken was an author whose works included historical and literary writings. Noteworthy among the historical works are his articles about figures such as Schill and Nettelbeck [Kelchner 1879]. Sacks choice was no accident. After a meeting with Haken the governor received his proposal for a new journal intended to solidify the binding of these regions to the Prussian State by picturing the story of Pomerania's past.[5] The periodical should be suitable, not only for the educated bourgeoisie (teachers, civil servants, clergymen), but also, Haken emphasized, for the middle class – merchants and craftsmen. The journal should combine rationalist Enlightenment ideas with new trends. Thus, the *Pommerschen Provinzialblätter für Stadt und Land* arose, whose first issue was published early in 1820.

In January 1820 the official Stettin governmental gazette contained Governor Sack's notice of the creation of the journal.[6] In it were the words that have been cited many times since: "*that such a country and such a people were worth every cultural resource.*" Sack emphasized how important the existence of such a publication is for the intellectual integrity of the society of both parts of Pomerania. The document ended with other equally important words: "*that in Pomerania a Second and Third Pomerania should be created with respect to culture and people.*" The governor's announcement also appeared in the official journals of the Köslin and Stralsund administrative districts. The periodical was a private enterprise of Haken and the Berlin bookstore Flittner. The

[5] See Haken's letter to Sack of 1 Jan. 1820. In: Archiwum Państwowe w Szczecinie [State Archive in Szczecin, hereafter: AP Szczecin], Naczelny Prezydent Prowincji Pomorskiej [Governor of Pommerania, hereafter: NPPP], Nr. 3051, K. 1–2.

[6] *Amtsblatt der Königlichen Regierung zu Stettin* Nr. 1, 7 Jan. 1820.

journal was printed in Stettin by Effenbarts Publishing; four issues made up a volume of about 500 pages. After an initial surge the number of subscribers decreased: in 1823 there were 600 and in 1824 only 400 subscribers. A part of these figures were derived from merchants in Stettin. In the countryside, tenants, land managers and land owners subscribed. The largest group of readers, however, was made up of civil servants, teachers, and pastors. In Greifswald, the publication was ordered by seven university professors [Steffens 1938b, 238–44]. It became the forum of educated people, printing historical sketches, literary texts, reviews of new books, and a chronicle of the province. Its columns were a channel of expression for well-known persons such as Leopold von Ledebur, director of the Royal Kunstkammer in Berlin, Konrad von Levezow, professor at the art academy (Kunstakademie) in Berlin, and, most notably, Ludwig Giesebrecht (1792–1873).[7]

In exploring the cultural and scientific history of Stettin one should not ignore this unquestionably most outstanding historian of Pomerania (before Martin Wehrmann) who was also a famous poet and friend and collaborator of the composer Carl Loewe, for whose oratorios he wrote lyrics. Giesebrecht was born in the village of Mirow in Mecklenburg and took up the study of history in 1812. However, in the war against Napoleon in the spring of 1813, he – like many other young Germans – volunteered for a Prussian hussar regiment. From 1814 he continued his studies at the University of Greifswald. In 1816 he arrived in Stettin and began working at the local Gymnasium, where he remained active for nearly 50 years. At the time of the founding of the *Pommersche Provinzialblätter* Giesebrecht was already known, but more as a poet than a historian. Giesebrecht's poetic works appeared in individual issues of the periodical, but it also published his historical works. Despite Giesebrecht's literary activity, the *Pommersche Provinzialblätter* showed signs from the start of a certain structural weakness. The journal did not have a definite principle: it was a literary and scientific almanac in the style of the Enlightenment, as represented by the chief editor Haken. He tried to change the profile of the periodical by increasing the range of historical writings, but the number of subscribers continued to decline. Although Haken wanted to avoid social issues, articles were published on the farmer question, which referred to the start of reforms in the vesting of property in Pomerania. These publications created an angry reaction from the land-owning Junkers who were hostile to Sack's social policy.[8] The inclusion of such social issues had a negative impact and the *Pommersche Provinzialblätter* went

---

[7] Among the works on Giesebrecht are: [Kern 1875, 1879, 1914; Runze 1894; Lemcke 1892; Tacke 1923; Altenburg 1942, 1966], as well as [Turek-Kwiatkowska 1997] whose work also includes sketches of Brüggemann and Sack.

[8] Archiwum Państwowe Szczecin, NPPP, Nr. 3051, K. 94. See further [Stępiński 1993].

noticeably downhill. In 1824 Haken, in his letters to Sack, expressed skepticism regarding the future of the periodical.[9] In his replies Sack appealed to Haken to continue the journal, but Haken decided to cease publication. In the last, sixth *Band* in 1825 he took leave of his readers stressing that the production of the periodical in Pomerania was a difficult task given that there was only a small group of educated people in the area. Haken, however, also expressed the hope that a new scientific quarterly would appear in 1826, this time under the editorship of Ludwig Giesebrecht.

The founding of the "Society for Pomeranian History and Classical Studies"

The *Pommersche Provinzialblätter* fell through because its content was not compatible with the intellectual circles of Pomerania. Nevertheless, this unsuccessful attempt fulfilled its goal, since, by exposing the historical difficulty, it prepared the way for the formation of a scientific periodical. This periodical, entitled the *Neue Pommersche Provinzialblätter* was published as a journal of the newly founded historical society in Stettin – the Society of Pomeranian History and Classical Studies (Gesellschaft für Pommersche Geschichte und Altertumskunde). The emergence of the historical society in Stettin was the result of a general trend in Germany towards a fascination for history that emerged after the Napoleonic wars. The direct incentive for the formation of the society in Pomerania was undoubtedly the earlier founding in Berlin of the Society for Earlier German History (Gesellschaft für Deutschlands ältere Geschichtskunde) by Baron vom und zum Stein. From the beginning of his administration Sack showed a great interest in the history of the region. Pastor Ludwig Wilhelm Brüggemann, the author of a multi-volume, statistical-historical-topographical description of Pomerania, died in 1817 whereupon Sack ordered the purchase of Brüggemann's scientific Nachlass [Müller 1934].

Sack's efforts to establish the historical society would not have succeeded if the governor had not won the support of the representatives of the intellectual elite of Stettin and Greifswald. Of special note, besides Ludwig Giesebrecht mentioned above, were: Heinrich Wilhelm Böhmer, Karl Friedrich Hasselbach, and Johann Ludwig Gottfried Kosegarten. Böhmer was interested primarily in questions concerning the medieval history of Stettin. Hasselbach, who spent several years as director of the Stettin Gymnasium, also showed interest in the medieval history of Pomerania. Together with

[9] Archiwum Państwowe Szczecin, NPPP, Nr. 3051, K. 56–57.

Kosegarten, an eminent linguist and historian, and professor at the University of Greifswald, he edited the *Pomeranian Diplomatic Codex* in 1843. All of these belonged – like Sack – to the educated bourgeoisie.

Thanks to his winning over these four outstanding intellectuals to the idea of founding a historical society, and thanks to the simultaneous support of the central government in Berlin, the governor was able to announce the formation of the Society of Pomeranian History and Classical Studies. He used the occasion of the 700th anniversary celebration of the Christianization of Pomerania in June 1824 for the announcement. The historical society brought together not only people with a humanistic background, but also from mathematics and science education, such as Justus Grassmann, a member until his death [Homeyer 1854, 7]. Membership in the Society, whose sponsor was the Prussian Crown Prince (later King Frederick William IV), was a matter of prestige. The society began with the systematic study of the history and archaeology of Pomerania and these two directions dominated through its 120-year existence. Historical interest concentrated mainly on the editing of sources and on scientific publications. Over time, this historical trend culminated in the founding of the Historical Commission for Pomerania. The results of archaeological research, which initially appeared in the form of collections, eventually found its expression in the institution of a museum. The historical society in Stettin developed a variety of public channels to achieve its purposes, such as the publication of articles in the previously mentioned journal *Neue Pommersche Provinzialblätter*, the successor to the *Pommersche Provinzialblätter*. When the *Provinzialblätter* ceased publication in 1832, the society published *Baltische Studien* as its own journal and it still serves this function today.

The establishment of the Stettin Provincial Archives

Another undertaking important for the development of science in Stettin was the establishment of a Provincial Archives. This idea arose as a result of system reforms in Prussia which prepared the way for the formation of archives as public state institutions. Its founding was influenced to a large extent by the ideas of Romanticism, but these roots were downplayed in the interest of the country's past and of the development of the historical sciences.

Duke Karl August von Hardenberg took the initiative for the reorganization of the archives, whose most important function was to be the collecting and making available of source texts for scientific purposes. The idea was proposed of opening an archive in every Prussian province. This process proceeded in stages since it relied on enactment at each change of power at the highest offices in the years 1810, 1814, and 1819. In the Prussian state by the mid-nineteenth century there were eight archives: a central one, the Secret State Archives in Berlin, and seven provincial ones: in Düsseldorf, Koblenz, Magdeburg, Meiningen, Münster, Sigmaringen, Stettin, and Breslau (Wroclaw) [Gollmert 1856, 113].

The Stettin archive was established in 1827, when Governor Sack charged Baron Friedrich Ludwig von Medem with the task of its organization. Medem proved in the next few years to be a gifted practitioner and theorist of archival work. In his published reflections he defined an "Archive" as an institution that has archival resources of a public nature that are useful for the areas of learning and government [von Medem 1835]. These were innovative views for the time, which met with little understanding among Sack's successors. Under the decree of January 1831, both the archive in Stettin, as well as archives in the other Prussian provinces, were subordinated to the archive directorate in Berlin and reported directly to the respective provincial governors.

Medem's work in the archives began with classifying the ducal documents which formed the root of the collection. Together with Hasselbach and Kosegarten he was an editor of the first *Pomeranian Diplomatic Codex*, mentioned above, published in 1843. His work was innovatively carried on by the next head of the Stettin Archives, Robert Klempin, a historian and archivist [von Bülow 1882; Randt 1934]. Leopold von Ranke regarded Klempin as one of the most talented students at the University of Berlin. In 1843 Klempin defended his dissertation and three years later qualified as professor at the University of Greifswald. He had to leave the Berlin Hochschule, however, because of a progressive illness, but attained a place in the Stettin archive thanks to Ranke's support before the former director of the Prussian State Archives, Professor Karl Wilhelm de Deleuze Lancizolle. As a historian he devoted himself to studies of Pomerania in the Middle Ages. The history of the edition of the *Pomeranian Book of Deeds* (Urkundenbuch) began with Klempin.

The flowering of scientific life in Stettin

Over the years the historical society and the provincial archives in Stettin thrived and, thanks to the initiative of its educated citizens, a number of other societies were founded. In 1828, the Geographical Society was created, in 1835 the Art Association of Pomerania and the Physical Association (whose initial chairman was Justus and then, from 1852, Hermann Grassmann [Engel 1936, 80]), in 1837 the Etymological Association, in 1858 the Technical Association, and in 1862 the Polytechnic Association.

With all of these initiatives for the institutional formation of the scientific foundations of science itself, the first half of the nineteenth century proved to be very fruitful for Stettin. Positive effects could be observed for years following. These achievements were the work of the representatives of a generation that died in the 1860s and 1870s. In 1864 Hasselbach died, in 1872 Calo, in 1873 Giesebrecht, in 1874 Klempin, and in 1877 one of the most outstanding German mathematicians, Hermann Grassmann.

Illustration [illegible]: Hermann Grassmann's secondary school certificate from 18[illegible]. The translation [illegible]

[illegible]

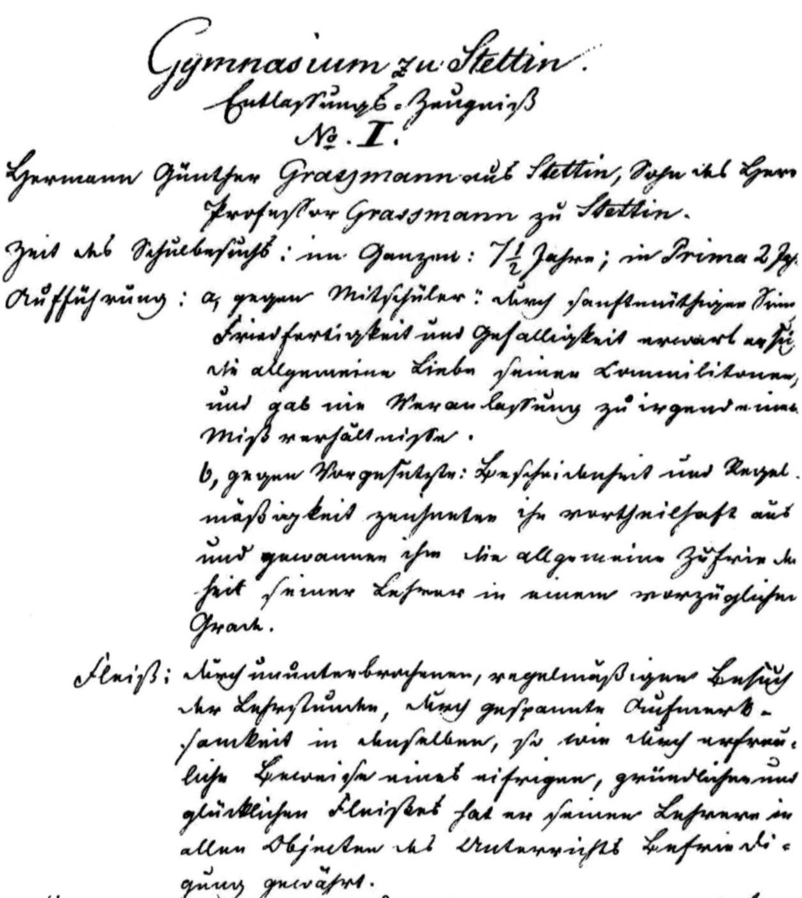

Gymnasium zu Stettin.
Entlassungs-Zeugniß
No. I.

Hermann Günther Grassmann aus Stettin, Sohn des Herrn Professor Grassmann zu Stettin.

Zeit des Schulbesuchs: im Ganzen: 7½ Jahr; in Prima 2 Jahr.

Aufführung: a, gegen Mitschüler: durch Sanftmüthigkeit, Friedfertigkeit und Gefälligkeit erwarb er sich die allgemeine Liebe seiner Commilitonen, und gab nie Veranlassung zu irgend einem Mißverhältnisse.

b, gegen Vorgesetzte: Bescheidenheit und Regelmäßigkeit zeichneten ihn vortheilhaft aus und gewannen ihm die allgemeine Zufriedenheit seiner Lehrer in einem vorzüglichen Grade.

Fleiß: Durch ununterbrochenen, regelmäßigen Besuch der Lehrstunden, durch gespannte Aufmerksamkeit in denselben, so wie durch erfreuliche Beweise eines eifrigen, gründlichen und glücklichen Fleißes hat er seinen Lehrern in allen Objecten des Unterrichts Befriedigung gewährt.

Kenntnisse: in Sprachen und Wissenschaften hat er so befriedigende Fortschritte gemacht, daß ihm das Zeugniß der unbedingten Tüchtigkeit zur Universität nicht hat versagt werden können.

Stettin d. 17ten September 1827.

Illustration 10. Hermann Grassmann's secondary school certificate from 1827. The translation is given on p. 497
[*Archiwum Państwowe w Szczecine, Gimnazjum Mariackie w Szczecinie*]

Philosophical and methodological aspects of the work of the Grassmann brothers

Brief outline of a history of the genetic method in the development of the deductive sciences

Lyubov Gavrilovna Biryukova

I

As is well known, the axiomatic method is in general predominant in deductive knowledge. It is less known, however, that apart from the axiomatic method another appeared quite long ago called the "genetic method" in the methodology of the deductive sciences. The genetic approach accompanied the use and development of the deductive-axiomatic method for centuries. It did not compete with the axiomatic method, which can be explained by the epistemological fruitfulness of axiomatization always containing a certain heuristic potential. At the end of the nineteenth century the axiomatic method, which had thoroughly penetrated into mathematics, especially after the development of non-Euclidean geometries, was strengthened by the development of the interpretation of axiomatic systems and supplemented by means of logical formalization. Along with the set-theoretical method, axiomatization had for decades been one of the main means of ordering deductive knowledge and obtaining new results.

H.-J. Petsche et al. (eds.), *From Past to Future: Graßmann's Work in Context*,
DOI 10.1007/978-3-0346-0405-5_8, © Springer Basel AG 2011

The situation changed in the middle of the twentieth century due to "machine" mathematics, cybernetics and informatics, which "strengthened" human thought with technical data processing devices. Before the emergence of this situation, the "universality" of the axiomatic method, at least in the form built up in classical logic and mathematics, as well as the set theoretical method had been seriously called into question. The point is that the development of the axiomatization method as such directly resulted in the genetic approach in the construction of deductive theories.

II

It is possible to single out the properties of the genetic method and compare it with the axiomatic method at the semantic level – the level of informal theories and interpretations of formalized theories. First of all this concerns the subject domain; if in the case of axiomatization the subject domain is considered to have been set before the introduction of axioms, then in the case of the genetic approach this domain is built up in terms of some elementary objects using a system of generating rules or operations. In axiomatic theories the emphasis in proofs is placed on axioms; in genetically constructed theories inferences on the basis of equality relations and the principle of "complete induction" and its generalizations as well as recursive procedures come to the fore. We can point out that the genetic viewpoint is often identified with the intuitionist-constructive position in the logical substantiation of mathematics and logic. However, it is possible to understand constructivity in a more general sense as, for example, [Smirnov 1962] where constructivity is extended to "mental construction" in the sense of reducibility algorithms (see for example [Yanovskaya 1958]).

It is natural to understand the "geneticity" of deductive theory even more broadly as a construction in which the world of the objects in question is considered to appear as a result of the application of some generating rules that do not necessarily have a finite (effective) nature. The development of the genetic method, the distinctions (from the standpoint of constructivity) that refer to the understanding of the subject domain, and the character of the generating operations and proof procedures make it possible to single out three meanings of genetic theory. The first is a presupposition that is identified with constructivism in logic and in the philosophical foundations of mathematics. The second meaning involves the extension of this presupposition at the expense of including reducibility algorithms (and not only the usual algorithms) into the means used within it. The third meaning involves the connection of the genetic method with some generating method, not necessarily algorithmic in the broader sense.

III

Analyzing the history of the development of deductive theories such as mathematics and logic, we can note the main stages in the formation of the genetic approach (method). The first stage is associated with Euclid's *Elements*. Euclid's work, which has played the principal role in the development of the axiomatic method, is at the same time historically a realization of the first form of the geneticity of a deductive theory (in the second meaning): the rules of constructing geometric figures in compass-and-ruler geometry represent some reducibility algorithms. Further development proceeds in two ways: on the one hand, there is the formation of constructive geneticity, on the other hand, there is a transition to geneticity in a broader, non-constructive sense.

A study of the history of deductive knowledge and its logic-mathematical interpretation finds everywhere the unity of axiomatic and genetic ideas. Thus, in Leibniz's works and the underlying idea of a "universal language" and a corresponding logical calculus (calculus rationator) we can see distinctly the unity of axiomatic and genetic ideas. The emergence of mathematical analysis was an important step in the formation of the genetic method.

Starting from the seventeenth century genetic presuppositions appear explicitly in the form of ideas concerning a fundamental role for complete mathematical induction. The advent of distinct constructive ideas is usually associated with the ideas of Leopold Kronecker and Henry Poincaré. Thoralf Skolem's recursive arithmetic of positive integers published in 1923 is considered to be the first logico-mathematical theory developed in terms of realized constructivist positions. However, a study of Hermann and Robert Grassmann's mathematical-logical heritage shows that they should be considered the originators of the genetic-constructive approach in the deductive sciences.

IV

The analysis of the introductory parts of H. Grassmann's work [Al] allows one to bring out those aspects of his scientific activity that concern the interpretation of the foundation of deductive knowledge: the opposition of the genetic method to the axiomatic one, a recursive-inductive methodology, the desire to pass from elementary abstract structures to complex ones; these gave him the opportunity to pass from the general "theory of forms" to the genetic (in the second meaning) construction of the *Ausdehnungslehre* and then to the inductive-recursive (constructive) theory of positive integers in [LA]; it was a general methodological achievement that is made clear from the fact that R. Grassmann used this method in the theory of forms (magnitudes in a broad sense) preceding, from his point of view, any precise knowledge including mathematics and logic.

V

In R. Grassmann's works [R. Grassmann 1872a, 1890b] we find – in the form of his "general theory of magnitudes" *(Grösenlehre)* a detailed finitary-constructivist concept of constructing the principles of deductive knowledge, naturally within the limits of the development of mathematics and logic of the last century. R. Grassmann also creates a proof theory based on the properties of an equality relation and the recursive setting of operations. In contrast to the works of other logicians of the last century, R. Grassmann's logical theory [R. Grassmann 1872c, 1890c] was constructed in a constructivist style representing a theory of concepts (Begriffe) of finite extent and being an extension of the theory of magnitudes constructed with the same methods.

VI

In the nineteenth century the genetic method was developed in both constructivist and non-constructivist directions. An examination of Georg Cantor's set theory shows that the hierarchy of infinite sets is developed genetically in a non-constructive way – on the basis of generating rules (principles); the resulting method of proof by way of transfinite induction is a non-constructive generalization of constructive complete induction.

VII

Despite the availability of genetic-nonconstructive concepts, their application to effectivist (finitary, constructivist) logico-methodological presuppositions proved to be principally in genetic methods. A well-known characterization of the genetic method made by David Hilbert acknowledges the fact that the constructivist inductive-recursive meaning became dominant in the concept of theory geneticity. T. Skolem's work [Skolem 1923] shows that the concept of a "recurrent way of thinking" was in principle a further extension of the inductive-recursive approach founded by the Grassmanns.

Grassmann's epistemology: multiplication and constructivism[1]

Paola Cantù

Introduction

Grassmann's epistemological insights have attracted less attention than his mathematical results. Mathematicians have considered him as a precursor of modern mathematical theories – vector algebra, exterior algebra, Clifford algebra, and geometric algebra – paying attention to the development of mathematical results rather than to the philosophical differences between Grassmann's own project and modern mathematical reconstructions [Cartan 1907; Hestenes 1985]. Historians of mathematics have analyzed the reception and influence of Grassmann (see for example the papers collected in [Schubring 1996a]), and philosophers' attention has been driven especially to the *Introduction* to the 1844 edition of *Ausdehnungslehre*, or to a comparison with other philosophers.[2]

[1] This work was completed with the support of ANR Chaire d'Excellence "Ideals of Proof," Université de Nancy, Université de Paris 7 Diderot (France), but had its origin in my dissertation discussed at the University of Genova (Italy) in 2003.

[2] See for example the articles by A. C. Lewis on Schleiermacher [Lewis 1977] and M. L. Heuser on Schelling [Heuser 1996], or recent papers by S. Russ on Bolzano, A. C. Lewis on Cassirer, J. Riche on Whitehead, and M. Hartimo on Husserl in this volume.

H.-J. Petsche et al. (eds.), *From Past to Future: Graßmann's Work in Context*,
DOI 10.1007/978-3-0346-0405-5_9,

This paper will follow a different path, namely an inquiry into the reciprocal influences of Grassmann's mathematical and philosophical insights, approaching Grassmann's epistemology from the perspective of his mathematical work. The starting point will be an analysis of the notion of product between extensive magnitudes, to which the philosophical literature has paid insufficient attention. Several works of Grassmann will be examined, including the two editions of *Ausdehnungslehre* [A1; A2], *Theorie der Ebbe und Flut* [H. Grassmann 1840], *Kurze Uebersicht über das Wesen der Ausdehnungslehre* [H. Grassmann 1845], *Geometrische Analyse* [H. Grassmann 1847], *Sur les différents genres de multiplication* (SD) [H. Grassmann 1855], and *Der Ort der Hamilton'schen Quaternionen in der Ausdehnungslehre* (HQ) [H. Grassmann 1877c].[3]

The second section of the paper will be devoted to a comparison with three vector-based systems: vector analysis, exterior algebra, and geometric algebra. Considering similarities and differences with respect to these mathematical theories, the attention will be directed to some philosophical issues, such as the question of the homogeneity of the elements, the consideration of the domain as having a fixed or a variable number of dimensions, and the general properties of multiplicative operations. I will claim that there is a certain continuity in Grassmann's epistemological program – contrary to the common tendency in the literature to emphasize the elements of discontinuity between A1 and A2.

In the last section of the paper some aspects of Grassmann's epistemology – the difference between numbers and magnitudes, the relation between geometry and extension theory, and the development of a constructivist approach to mathematics – will be related to the role played by the operation of multiplication in Grassmann's mathematical theory.

The product between extensive magnitudes

To understand Grassmann's notion of product between extensive magnitudes, one has first to introduce the concept of an extensive magnitude itself, which is not an easy task, especially if one wants to take into account the various definitions and names introduced by Grassmann in different works. Since I am interested here in the philosophical implications of the choice of characterizing extensional magnitudes by means

[3] The 1844 and 1862 editions of the *Ausdehungslehre* and the latter two articles have been translated into English by L. C. Kannenberg [$A1_K$; $A2_K$]. References are to Grassmann's *Gesammelte Werke* [HGW11–HGW32]; English quotations are taken (with some modifications) from the English translation.

of their product, I will only briefly summarize some relevant features of the definition of extensive magnitudes.[4]

Extensive magnitudes

In A1 Grassmann distinguished between a form (or thought form), which constitutes the general object of mathematics, and a concept of extensive magnitude which constitutes the object of extension theory, i.e., the mathematical theory that studies continuous forms generated by different elements. A form, or thought form, is defined as a "particular existent that has come to be by an act of thought" [$A1_K$, 24]. Each form is determined by its generating elements, which might be equal or different, and by its generating act, either continuous or discrete. Forms are thus classified according to opposite concepts: equal/different, discrete/continuous. On the basis of this partition of forms into four kinds, which is dependent on their laws of generation, Grassmann classified mathematics in four branches – Number Theory, Theory of Intensive Magnitudes, Combinatorial Theory, and Theory of Extensive Magnitudes – and claimed that each one should be grounded independently.

Grassmann introduced an extensive formation as the totality of elements obtained by variation of the generating element [HGW11, 28]. He then defined an extensive magnitude as the class of extensive forms that are generated according to the same law by means of equal variations [$A1_K$, 47], in other words, the class of all extensive formations that have the same direction, the same orientation, and the same size. In modern parlance, an extensive magnitude of first grade is a vector, while an extensive formation of first grade is a bound vector. Continuous numbers are numerical magnitudes, obtained as the quotient of magnitudes of the same grade [HGW11, 130].

In A2 Grassmann seems to drop his former plan to build extension theory independently from all other mathematical branches, as he assumes from the beginning a continuous system of numbers, and he defines extensive magnitudes by means of them.[5] The use of numbers to define magnitudes seems to contradict A1, where numbers were introduced as quotients of magnitudes, but it is to be noted that Grassmann added that

[4] I developed a more detailed analysis in the second part of my PhD thesis [Cantù 2003, 153–345], where the accent was put on the discontinuities between the two editions, and on Grassmann's criticism of the "traditional" definition of mathematics as a theory of magnitudes.

[5] "I define as an *extensive magnitude* any expression that is derived from a system of units (none of which need to be the absolute unit) by numbers, and I call the numbers that belong to the units the *derivation numbers* of that magnitude; for example the polynomial $a_1e_1 + a_2e_2 + \ldots$ or $\sum ae$ or $\sum a_re_r$, where $a_1, a_2, \ldots\ a_1, a_2, \ldots$ are real numbers and $e_1, e_2, \ldots$ form a system of units, is an extensive magnitude, specifically the one derived from the units $e_1, e_2, \ldots$ by the numbers $a_1, a_2, \ldots$ belonging to them" [$A2_K$, 4].

"the presentation chosen here very closely follows arithmetic, but in the sense that it assumes the numerical magnitude as a continuous magnitude" [$A2_K$ xiv], i.e., as one of the extensive magnitudes introduced in A1. Besides, the following passage – omitted in the English translation – shows that Grassmann still considered numerical magnitudes as derivable from the definition of extensive magnitudes: "If the system [of units, i.e., an extensive magnitude] consists only of the absolute unity (1), then the derived magnitude is not an extensive but a numerical magnitude." [HGW12, 12].

The product between extensive magnitudes

Magnitudes of second, third, ... and n-th grade were obtained by means of the introduction of a multiplication of the generating elements or units. In A2 for example $v = a_1e_1 + a_2e_2$ is a quantity of first grade, while $w = a_1e_1 + a_2e_1e_2$ is a quantity of second grade.

In *Theorie der Ebbe und Flut*, the inner (linear) product of two segments was defined as the algebraic product of a segment by the orthogonal projection of the second on the first [HGW31, 40, 212]. A similar definition was maintained in A1, although the product was only briefly mentioned in the Introduction [HGW11, 11]. In A2 the inner product of two arbitrary magnitudes was defined by means of the outer product: it is equal to the outer product of the first by the supplement of the second, that is, $[A|B]$ is the inner product of the magnitudes A and B. If the grade of A is a and the number of dimensions of the system is n, the grade of the supplement of A is $n - a$ [HGW12, 93–94]. This definition thus includes in some sense the previously defined notion of inner product.

In *Theorie der Ebbe und Flut* the "geometric" product of two segments was the oriented surface of the parallelogram thereby delimited [HGW31, 30]. The definition of the "outer" product in A1 was analogous [HGW11, 80–81]. In SD [HGW21, 214–15] and in A2 [HGW12, 37–38] the definition was based on two axiomatic conditions: the outer product is anticommutative: $e_re_s = -e_se_r$, and it is equal to 0 when the magnitudes are linearly dependent: $e_re_r = 0$.

In A1 Grassmann introduced another notion of product, which he called regressive product or "eingewandt" (a sort of counter-product) [HGW11, 206]. The regressive product is relative to the system that the two extensive magnitudes have in common (e.g., the intersection of the systems they belong to), and the grade of the resulting magnitude depends on the number of dimensions of such a system. This system might vary, and the grade of the result varies according to the system. But the product of two magnitudes can also be considered as relative to a fixed system, for example to a system of dimension 3, as in the case of geometry [HGW11, 243], where the product can be considered as applied.[6] If the grade of the result of the product is $> n$, $< n$, or $= n$, the product increases, decreases, or either increases or decreases the grade of the magnitudes. Geometry is an application of extension theory, exactly because of the three-dimensionality of space: the product is always relative to it. In A2 the regressive product was not abandoned, as a general notion of a product relative to a system of given dimension was developed: the grade of the result of the product of two magnitudes A and B of grade q and r respectively in a domain of dimension n is defined as $q + r(\bmod n)$[HGW12, 66]. Progressive and regressive products correspond to the outer and to the inner product. The latter is generalized by means of the notion of the supplement of a given magnitude in a domain of fixed dimension, so as to decrease the grade of magnitudes, but still expresses orthogonality. It is precisely in A2 that the idea of introducing products as a means to increase but also decrease the dimension of the magnitudes themselves becomes fully explicit. It is exactly because this fact went generally unnoticed that a huge discontinuity between the two editions has been defended in the philosophical and mathematical literature. On the contrary, the general notion of product considered in A1 as relative to a variable domain is restricted in A2 to the case of what we have called an applied product, i.e., a product that is relative to a system of fixed dimension. This seems to be a consequence of Grassmann's aim of unifying the two products under the name of a "product relative to a principal domain."

Finally, in HQ, a late writing that Grassmann published in order to prove the originality of his own theory with respect to Hamilton's quaternions, a third kind of operation is considered: the median product, which is defined as the sum of the inner and the outer product [HGW21, 268]: $ab = \wedge[a|b] + \mu[ab]$. This notion of product was later developed by Clifford, and by the defenders of geometric algebra [Hestenes 1985].

A comparative philosophical analysis

In order to appreciate the philosophical interest of Grassmann's ideas, I will briefly compare it to three other vector systems based on the notion of product: vector analysis, multi-vector algebra, and geometric algebra.

---

[6] My use of the term "applied" should not be confused with Grassmann's use of the term "eingewandt" to denote the regressive product as opposed to the outer or progressive product. An applied product, as I intend it, is a regressive product whose result is univocally determined, since it is relative to a fixed system.

The product between vectors and multivectors

The first approach was developed by Gibbs and Heaviside in the case of a three-dimensional space. They considered the case $V = R^3$ and defined two products. The dot product of two vectors $V \times V \rightarrow R$ is the product of the moduli of the vectors by the cosine of the angle between them: since the resultant is a scalar and not itself a vector, it is also called a scalar product. The cross product of two vectors $V \times V \rightarrow V$ is defined as the vector that is perpendicular to both vectors, so directed that the triple of vectors might be a right triple, and whose modulus is equal to the product of the moduli by the sine of the angle between them. The cross product is thus itself a vector. The algebra generated by the dot product is different from the algebra generated by the cross product: the former is commutative, and not associative; the latter is anticommutative and not associative.

The second approach, exterior algebra, is based on an operation of multiplication – the wedge product – defined as $\wedge V^n \rightarrow \wedge(V)$, which generates multivectors of different grade. Scalars are multivectors of grade $K = \wedge^0(V)$. The domain of this algebra, which is associative and anticommutative, is the direct sum of the subspaces containing the entities of a given grade k. The space V is thus itself a subspace of $\wedge(V)$.

Geometric algebra, first presented by D. Hestenes as a foundation for classical mechanics [Hestenes and Sobczyk 1984], was inspired by Clifford's and Grassmann's works. The main difference with respect to exterior algebra consists in the definition of a single notion of multiplication that allows the construction of a unique algebra for inner and outer product. The basic idea is to introduce a general product to explain – like Grassmann's median product – both the increase and the decrease of a magnitude's grade. The inner product $a \wedge b$ of two vectors a and b is considered as a product that decreases the grade, whereas the outer product $a \cdot b$ is conceived as a product that increases the grade. As in multi-vector algebra, there are magnitudes of various grades, which are called scalars if $K = 0$, and k-blades if $k > 0$, to emphasize the fact that unlike scalars they have directional features.

The geometric product is defined as the sum of the inner and outer product: $ab = a \cdot b + a \wedge b$, where a and b are magnitudes of any grade. If a and b are 1-blades, i.e., vectors, the definition implies for example the addition of a scalar and a 2-blade, i.e., a bivector: geometrically interpreted, this means that one should add a number and an oriented plane segment.

Other geometrical properties such as coplanarity or perpendicularity are expressed by means of the commutative or anti-commutative property of the product: vectors are collinear if and only if their geometrical product is commutative, and they are orthogonal if and only if the geometrical product is anticommutative. Apart from the noncommutativity of multiplication, which of course has some consequences for the definition of division, the algebra thus obtained is very similar to the algebra of scalar quantities (numbers), and "facilitates the transfer of skills in manipulations with scalar algebra to manipulations with geometric algebra." [Hestenes 1985, 36].

Domain and homogeneity

A comparison of the different "vector algebras" raises some questions concerning the nature of the domain of extensive magnitudes. Is it closed under multiplication? Is it possible to define addition between non-homogeneous elements?

In the Gibbs–Heaviside approach, the domain is closed under the additive and the multiplicative operation: the cross product of two vectors is again a vector, i.e., a magnitude of the same dimension. Besides, only homogeneous magnitudes are added or multiplied.

Grassmann's product on the contrary might produce magnitudes of higher or smaller grade; besides, it can be defined between magnitudes of different grade. Addition on the contrary is generally defined between homogeneous elements in such a way that the result is again an element of the same dimension [HGW12, 49]. Yet homogeneity properly applies only to "real" addition, and not to "formal" addition.[7] A kind of addition can be defined even between non-homogeneous magnitudes, as is the case in A1, where the addition of a segment and a point gives a point. But here the addition is just "formal," and the symbol of addition has to be interpreted as a movement from one point to another point rather than as a concatenation of magnitudes [HGW11, 166]. The "formal" addition is not a sum of extensive magnitudes, but an operation that shares the same algebraic properties.

Exterior algebra similarly admits that the product might generate elements of different grade, but does not introduce a geometrical interpretation of addition of entities of different grade. Besides, the multi-vector space is generated by the wedge product from an *n*-tuple of one-dimensional vectors, so the multiplication is primarily introduced between homogeneous elements.

The geometric product defined by Hestenes as the sum of an inner and an outer product is closed under the domain of *k*-blades. In other words the product is defined between any two graded magnitudes and is again a graded magnitude. Grassmann's notion of a unique, applied product (i.e., relative to a principal domain) can be nicely described, yet the idea of a product of magnitudes relative to a variable domain can hardly be explained from this perspective, given the fact that the domain is closed under

[7] I thank Dominique Flament for directing my attention to this point.

the product. Besides, geometric algebra explicitly refuses the law of homogeneity, arguing that it is not only possible but also useful to abandon this mathematical taboo [Hestenes 1985, 30].

This brief comparison between vector analysis, Grassmann's extension theory, exterior algebra, and geometric algebra has already shown that there are some differences, both from a technical and from a philosophical point of view with respect to Grassmann's own theory. On the one hand I have shown that modern algebras are not a complete description of Grassmann's notion of extensive magnitude but rather capture only some aspects of a more general philosophical project. On the other hand I have directed attention to some philosophical questions that rise from the texts of Grassmann, as he distinguishes between the generation of a magnitude of grade n by means of a n-tuple of one-dimensional generators and an applied product, or between a non-homogeneous formal addition and a homogeneous real addition.

Conclusion

As a conclusion, I will now claim that the product between extensive magnitudes is related to Grassmann's non-reductionist interpretation of the relation between numbers and magnitudes, to a new way of introducing the distinction between abstract and applied mathematics, and to the development of a constructivist approach in mathematics.

In A1 continuous numbers are defined in the Newtonian way as the quotient of extensive magnitudes; in A2 they are used to define the notion of extensive magnitude, and they are introduced themselves as numerical magnitudes. In SD, the article written after Cauchy's presentation at the French Academy of Sciences of his work on the "clefs algébriques," the analysis of multiplication allows us to better distinguish the main algebraic difference between numbers and magnitudes, i.e.. the commutativity or anticommutativity of the product [HGW11, 214–215]. So, continuous numbers are defined by means of magnitudes, and not vice versa. Natural numbers are independently defined in the *Lehrbuch der Arithmetik*, as the result of a discrete generation from equal elements [LA, 2–3]. The independent foundation of arithmetic and extension theory has the aim of bringing attention to the structural similarities based on the analogy in the construction process. The product between extensive magnitudes is analogous to the operation of addition in arithmetic, because they are both generating laws. The product is used in the axiomatic definition of numbers and magnitudes: both are characterized by linearity, but the product of the former commutes, thus making the operations simpler, while the product of the latter does not, which makes it easier to "grasp the gradually emerging magnitudes in their simplest concepts." [$A2_K$, 27].

Geometric algebra best explains Grassmann's idea that numerical magnitudes can be indifferently defined as the quotient of two magnitudes of the same grade, or as the only magnitudes whose product commutes, as it defines scalars as vectors of grade 0 that satisfy commutativity. The notion of product is thus essential in all works of Grassmann to grasp the difference between numbers and magnitudes, and also the reason why he refuses a reductionist foundation of the former on the latter, or vice versa.

Such a reductionism was quite widespread in analysis and in the development of algebra as a universal mathesis, but Grassmann developed a new notion of extensive magnitudes, based on abstract constraints concerning their multiplicative generation rather than on a definition of magnitude based on addition, order, and homogeneity conditions (as in the generalized theory of proportions). Besides, he classified the branches of mathematics according to their different operational rules, rather than on empirical criteria of abstraction or on the distinction of different species belonging to a common genus [Cantù 2010]. In the case of extensive magnitudes the fundamental operational rule is exactly the product. Finally, considering geometry as an application of extension theory, and geometrical figures as multivectors generated by vectors rather than sets of points generated by a point, he suggested a radical alternative to analytical geometry. Grassmann's epistemological shift in nineteenth century geometry is due to the previously mentioned factors, which are all connected to the introduction of a new notion of product between extended magnitudes: that is the reason why the originality of Grassmann's philosophical project cannot be appreciated without driving attention to the notion of product.

In particular, the product is related to Grassmann's constructivism, which is based on a different notion of generalization, and on the fact that the knowledge of mathematical forms relies on the understanding of the rules of generation of the forms themselves. Parting from the traditional definition of mathematics as a science of magnitudes, Grassmann considers mathematical forms as particulars rather than universals. Generalization is not conceived as an enlargement of a given domain by means of the addition of new elements; it is rather obtained by modifying the defining conditions of the fundamental operations. In particular, the ambivalent role played by geometry in A1 and often reproached to Grassmann as an incapacity of achieving a truly abstract perspective, should be considered in the light of the distinction between the "general" notion of product defined in A1 for two magnitudes with respect to the domain generated by the magnitudes themselves (regressive product in its general formulation), and the "particular," regressive, applied product defined both in A1 and in A2 with respect to a system of n dimensions. Geometry is an application of extension theory, essentially because multiplication in space is relative to the fixed number of dimensions of the space itself. This is the primary sense in which geometrical magnitudes are "embodied" extensive forms: the operations on geometrical figures are relative to a fixed system, the three-dimensional space.

The idea of considering as more "general" the product relative to a variable domain – a domain that is not closed under the operation but rather a result of our carrying out the operation itself – is one of Grassmann's most interesting philosophical ideas that lacks an adequate representation in the mathematical theories mentioned. One could claim that the general notion of product (relative to a variable domain) was substituted in A2 by a general notion of product (relative to a principal domain) because of technical mathematical difficulties. The problem with this answer is that it does not take into account the fact that Grassmann never really abandoned the idea of considering operations as determined independently from the domain they are applied to. Firstly, both in SD and A2 Grassmann developed an axiomatic definition of the multiplication that is not relative to a principal domain. Besides, the refusal to admit a domain of elements given prior to, or independently from the generation of the elements themselves, is an idea that Grassmann never abandons, and a basic assumption of his epistemological "constructivism." The latter is grounded on the distinction between formal sciences – where no constraint on the domain is taken as granted, and the forms are one and the same with their construction – and real sciences, where some constraints are accepted from the outset, and forms are thus "embodied" in a fixed domain. Grassmann's constructivism is based on the idea that there are some fixed fundamental operations rather than a fixed domain; besides, a general notion is particularized when further conditions are fixed, as in the case of the regressive product which is less general, if considered as relative to a unique domain.

Axiomatics and self-reference

Reflections about Hermann Grassmann's contribution to axiomatics

Mircea Radu

The (never ending?) debate

Hermann and Robert Grassmann's writings published between 1844 and 1891 led to a lively debate among scientists, philosophers, and historians about the contribution they bring to the advancement of axiomatics.

Gottfried Martin, Michael Otte, Hans-Niels Jahnke, Judson Webb, and Arno Zaddach[1] regard some of the general fundamental methodological tenets and/or some of the mathematical techniques developed by the Grassmann brothers as major contributions to the emergence of the new axiomatic spirit which, according to common wisdom, is said to have found its accomplished expression in David Hilbert's work.

Jean Cavaillès, Hao Wang, Jean Dieudonné, Michel Guillaume, Albert Lewis, Jean-Luc Dorier, and Hans-Joachim Petsche hold the opposite view. According to these scholars, even though many of the mathematical concepts and techniques developed by the Grassmann brothers were subsequently incorporated, for instance, in the axiomatic

[1] To these names one could also add those of Hermann von Helmholtz, Giuseppe Veronese, Hermann Hankel, Gottlob Frege, and Otto Hölder. All of them link the contributions of the Grassmann brothers (as a rule those of Hermann) to the axiomatic tradition, compare [Radu 2000a, b, 2005] and the literature quoted therein for details.

H.-J. Petsche et al. (eds.), *From Past to Future: Graßmann's Work in Context*,
DOI 10.1007/978-3-0346-0405-5_10, © Springer Basel AG 2011

treatment of arithmetic and in the axiomatic presentation of the abstract vector-space concept, Hermann Grassmann's approach to the foundations of mathematics is, on the whole, antithetical to axiomatics.

Some of these scholars also argue that the philosophical ideas that shaped Hermann Grassmann's views on the foundations of mathematics, most notably the ideas of Schleiermacher's *Dialectic*, are incompatible with modern axiomatics. The mere fact that modern axiomatic treatments of arithmetic and algebra rely heavily on mathematical techniques developed by Hermann Grassmann or the direct acknowledgment of Richard Dedekind and of Giuseppe Peano of their debt to Grassmann, are not seen as proving the presence of the modern axiomatic spirit in Grassmann's work.[2] Any resemblance with existing modern axiomatic theories, so it seems, is purely accidental, and in any case unintentional.

Last but not least, there is a third group of scholars who are led to consider the impact of the general methodological ideas and of the mathematical techniques (for instance, recurrent definitions and the treatment of abstract algebraic structures) developed by the Grassmann brothers on axiomatics indirectly, while discussing other subjects. Among the scholars belonging to the third group I would mention Abraham Robinson, Michael Crowe, Donald Gillies, Hans Wussing, Paul Shields, and Volker Peckhaus.[3]

On the whole, the ideas put forward by these scholars either support more or less directly the view that the Grassmann brothers did not bring any major contributions to the emergence of the new axiomatic spirit or they contain general statements that seem compatible with an interpretation of the contributions of the Grassmann brothers as belonging to the axiomatic tradition.[4]

A word of warning must be added to the classification attempted here. The approaches of the previously mentioned authors are very different in scope, depth, philosophical perspective, general account of axiomatics underpinning them, and so on. Grouping them together is, therefore, to a certain extent misleading. By doing this, I do not wish to suggest the existence of any wider agreement between the positions defended by the various scholars. The only agreement involved concerns the general assessment of the place of axiomatics in the work of the Grassmann brothers or the

---

[2] This view is held, for instance by Hao Wang in his influential paper from 1957 [Wang 1957] and by Hans-Joachim Petsche 2006 in his biography of Hermann Grassmann [Petsche 2006].

[3] Some of these authors provide substantial space to the discussion of the history of axiomatics without, however, even mentioning Hermann or Robert Grassmann (see, for instance [Radu 2000b, Sect. I.2] for details). This type of approach also contributes to the general tendency to place the work of the Grassmann brothers outside the axiomatic tradition. Almost all authors mentioned in this paper, however, do discuss some of the contributions of the Grassmann brothers, even if not all explicitly address axiomatics.

[4] Compare [Radu 2000b; 2005] for details.

implications of the positions developed for the assessment of the place of axiomatics in the work of the Grassmanns. Despite its obvious limitations, the previous classification does provide a useful overview of some of the most influential writings that shaped and continue to shape the debate dealt with in this paper.

The place of axiomatics in the *Lehrbuch der Arithmetik (1861)*: the positions of Gottlob Frege, Judson Webb, and Hao Wang

One obvious source of the debate outlined above is constituted by the philosophical ideas put forward by Hermann Grassmann in the introductory chapters of A1 and their evolution in subsequent work. Another is the mixture of mathematical and philosophical ideas spread throughout Grassmann's mathematical presentation of the theory of n-dimensional vector spaces given in A1. Grassmann offers, it would seem, too much and too little at the same time. Too much for the mathematician interested mainly in the mathematical ideas, and too little for the philosopher and the historian, looking for clear philosophical lines of thought.

In the midst of this uncertainty, the historian finds himself forced to interpolate Grassmann's scattered methodological ideas by fitting them to broader philosophical traditions. Interpolation problems are not so easy, particularly in philosophy. The stunning diversity of interpretations of Hermann Grassmann's statements concerning the foundations of mathematics proves it. This difficulty is increased by the fact that different historians have different philosophical agendas. As a result of that, the various interpretations are biased in favor of one or another of several competing philosophical conceptions. Particularly the debate over the contribution to axiomatics has turned Grassmann's work into a battleground for philosophical theory.

I will now provide a closer examination of two of the most interesting classical contributions that have shaped the debate about Hermann Grassmann's conception of the foundations of mathematics. I explore some of the implications of the points of view examined, in an attempt to clarify the sources of disagreement. I also briefly explore the possibility of finding a common ground that may contribute to a solution to the questions raised. My solution is based on an examination of the account of the distinction between symbolic mathematical synthesis and conceptual analysis in Hermann Grassmann's work. I argue that Grassmann combines two complementary accounts of this distinction, one emphasizing perfect harmony, the other emphasizing, on the contrary, complete separation. Mathematics advances by means of a dialectic, self-referential, sign-mediated, cognitive activity of the subject during which harmony is permanently created and dissolved. I argue that this game is characteristic not just for

A1 but also for the 1861 treatment of arithmetic. I then propose a definition of axiomatics based on the previous distinction concerning the relationship between synthesis and analysis in mathematics. I argue that my definition makes it possible to bridge the gap between those positions which place the contributions of the Grassmann brothers within the modern axiomatic tradition and those that reject such a view.

In his examination of the history of mathematics, Judson Webb raises the issue of the place of axiomatics in Hermann Grassmann's work. He focuses mainly on the place of recurrent definitions in the 1861 *Lehrbuch der Arithmetik*. Up to a point, Webb repeats arguments already used by Gottlob Frege.

Frege pointed out that, in a recurrent definition, for instance, of the addition of natural numbers or of integers, the operation defined (addition) is used both as *definiens* as well as *definiendum*. Recurrent definitions (of the operations) are, therefore, described as circular. Grassmann's 1861 definition of addition is $a + (b + e) = (a + b) + e$, where a and b stand for arbitrary integers, whereas e stands for the fundamental unit (1 in Peano's treatment) used as a starting point in generating the sequence of the integers:[5]

> "7. Definition. Use a magnitude e to construct a sequence of magnitudes according to the following procedure. Posit e as an element of the sequence. Following this posit $e + e$ (read e plus e) as the next element of the sequence. Continue to apply this procedure by adding $+e$ to the last element of the sequence. Similarly posit $e + -e$ (read e plus minus e) as the element of the sequence that immediately precedes e. Continue to apply this procedure by adding $+ - e$ to the first element of the sequence. In this way one obtains a sequence of magnitudes which is infinite in both directions
>
> $$\ldots e + -e + -e + -e; e + -e + -e; e + -e; e; e + e; e + e + e; \ldots$$
>
> If one takes each element of this sequence to be different from all the others, the resulting sequence is called the fundamental sequence, e the positive unit, and $-e$ the negative unit."[LA, 3; my translation M.R.][6]

[5] Peano constructs the arithmetic of the natural numbers, whereas Grassmann constructs the ring of the integers. This, however, is of no consequence for the following investigation.

[6] "7. Erklärung. Man bilde aus seiner Größe e eine Reihe von Größen durch folgendes Verfahren: Man setze e als ein Glied der Reihe, setze $e + e$ (gelesen e plus e) als das nächste Glied der Reihe, und so fahre man fort, aus dem jedes Mal letzten Gliede das nächstfolgende dadurch abzuleiten, dass man zu jenem $+e$ hinzufügt. Ebenso setze man $e + -e$ (gelesen e plus minus e) als dem e zunächst vorhergehende Glied der Reihe, und so fahre man fort, aus dem jedes Mal ersten Gliede der Reihe das nächst vorhergehende dadurch abzuleiten,

Interestingly, Frege had come to deal with this matter, stimulated by the same 1861 *Lehrbuch der Arithmetik* of the Grassmann brothers. Of course, Frege regards such a definition as a vicious circle, and argues, based on this, that the Grassmann approach to arithmetic is untenable.

Webb, on the contrary, argues that this circle is a virtuous one, and that this type of circularity marks the shift from a strictly genetic to a relational approach to the foundations of mathematics, one based on implicit rather than on explicit definitions of the mathematical objects of the mathematical theory treated. Given that recurrent definitions are indeed the backbone of the Grassmann approach to arithmetic, Webb argues that this proves that a new relational and therefore axiomatic approach to the foundations of mathematics is clearly present in the work of the Grassmann brothers [Webb 1980, 44].

It is not uninteresting to emphasize that Frege's criticism of the Grassmann approach to arithmetic and the reiteration of this approach by Webb, remind us of the famous debate between Frege and Hilbert concerning Hilbert's axiomatics. It almost seems that Webb has managed to identify a firm link between modern axiomatics and the 1861 treatment of arithmetic provided by the Grassmanns.

A major advantage of Webb's analysis is that it is independent of a too sophisticated philosophical apparatus. All that is needed is the view that the main difference between the classical non-axiomatic treatment of arithmetic and the new axiomatic treatment is the shift from a genetic approach, which subordinates structure to ontology, to a relational approach in which ontology is subordinated to structure.[7] Recurrent definitions appear as the backbone of the latter.

There is, however, a poisoned apple here, which is played down by Webb. In the 1861 treatment of arithmetic given by the Grassmann brothers recurrent definitions are used only *after* the sequence of the integers has been genetically introduced. Thus even if, following Frege and Webb, we concede that the treatment of the arithmetical operations given in 1861 is axiomatic, the overall treatment of the arithmetic of the integers seems genetic, for the sequence of the integers is introduced prior to and independently

dass man zu jenem Gliede $+-e$ hinzufügt, so erhält man eine nach beiden Seiten unendliche Reihe

$$\ldots e+-e+-e+-e; e+-e+-e; e+-e; e; e+e; e+e+e; \ldots$$

Wenn man in dieser reihe jedes Glied von allen übrigen ... als verschieden annimmt, so nennt man diese reihe die Grundreihe, e die positive Einheit, $-e$ die negative Einheit."[LA, 3]

[7] Below, I argue that Luhmann's conception provides a more adequate framework for interpreting this shift.

of any relational structure (see above). The approach to the foundations of arithmetic given by the Grassmann brothers remains, it would seem, fundamentally a genetic one.[8]

Another perhaps even more important doubt is raised by the fact that in his arithmetic treatise Grassmann only speaks of "definitions" and that he constantly emphasizes the perfect match between symbolic expression and symbolic proof on the one hand, and intellectual content, on the other. This may even be regarded as the expression of Leibniz' influence on Grassmann. According to Leibniz, mathematics must be built on definitions and axioms. Axioms, however, are identical propositions and definitions are no more than the expression of clear ideas and, most importantly, must be constructive [Otte 1989, 15]. If this is true, and if Grassmann's emphasis on definitions is an expression of such a position, it then becomes difficult to place Grassmann's work within the modern axiomatic tradition.[9]

Unlike Webb who played these aspects down, Hao Wang regards them as firm evidence that the 1861 treatment of arithmetic proposed by the Grassmann brothers had not been intended to be axiomatic, is not axiomatic, and, if examined from the perspective of axiomatics, it is unsatisfactory. Even though Wang writes that the Grassmann approach to arithmetic was "the first serious and rather successful attempt to put numbers on a more or less axiomatic basis" [Wang 1957, 147] he concludes that "Grassmann did not present his development in axiomatic form." [Wang 1957, 147].

Wang's position concerning this remains, as far as I can see, ambiguous.[10] Petsche also briefly considers Wang's views on the Grassmann approach to arithmetic [Petsche 2006, 226 f.]. There is just one notable difference between Petsche's interpretation of Wang's position and my own. Petsche interprets Wang as placing the 1861 approach to arithmetic inside the axiomatic tradition, whereas according to my own interpretation, Wang only points out that Grassmann's work contains most of the necessary technical ingredients of an axiomatic approach, without, however, being axiomatic. Petsche's own analysis of the genetic component of the 1861 arithmetic treatise of the Grassmann brothers basically raises the same objections against placing this treatise inside the axiomatic tradition [Petsche 2006, 219 ff, 226 f.] as Wang's. I will return to Petsche's position below.

We could of course stop here, concluding that the 1861 treatment of arithmetic given by the Grassmann brothers is based on a mixture of genetic and axiomatic elements. None of the previously mentioned authors, however, seems prepared to do this. Webb overemphasizes axiomatics, whereas Wang and Petsche play axiomatics down.

[8] I will propose an interpretation of this problem, that makes it possible to strengthen Webb's position and weaken Wang's critique.

[9] Below I will argue that a different reading of this is possible.

[10] For details compare [Radu 2000b, Sect. I.2].

Following their arguments and comparing them with the actual work of the Grassmann brothers, however, it is not so easy to avoid the conclusion that the 1861 treatment of arithmetic is, from a methodological point of view, a hybrid, a characteristic product of an intermediate phase in the history of the foundations of arithmetic and of axiomatics, one step on the path from a through and through genetic to a through and through axiomatic approach to the foundations of arithmetic, indeed, to mathematics in general. Is it possible to reach a different conclusion? Is it possible to reach a stronger reading particularly of Hermann Grassmann's conception of the foundations of mathematics? This seems a good place to turn to Petsche's approach to the matter.

Hans-Joachim Petsche's interpretation

In his recently published Grassmann biography, Petsche explores, among other things, the place of axiomatics in Hermann Grassmann's work [Petsche 2006]. The backbone of Petsche's reconstruction of the matter is the historically well-established fact that Hermann Grassmann was influenced by the work of Schleiermacher (both directly as well as indirectly, via the work of his father Justus Grassmann and of his brother Robert). Petsche gives an interesting discussion of Schleiermacher's dialectic, which he then uses in interpreting Grassmann's work, particularly A1 and the 1861 *Lehrbuch der Arithmetik*.

Petsche's reconstruction of Schleiermacher leads him to the conclusion that one central tenet of Schleiermacher's work is the opposition between the genetic and the axiomatic approach to science.[11] He then moves on to Hermann Grassmann works, trying to prove the existence of major traces of Schleiermacher's conception. In doing this he continues a line of analysis already pursued by Albert Lewis [Lewis 1970].

Petsche's discussion of the place of axiomatics in the work of the Grassmann brothers involves at least two levels of analysis. The first level deals with the place of axiomatics in the 1861 treatise of arithmetic, when this treatise is examined in isolation from other Grassmann texts, particularly of A1. The second level concerns the significance of the 1861 treatise for the evolution of Grassmann's general conception of the foundations of mathematics.

The first narrow line of analysis is up to a point comparable to those of Webb and Wang. The great interest for the first level is dictated by the fact that, obviously, the

[11] In response to a question concerning Schleiermacher's rejection of axiomatics raised during this conference, Marie-Luise Heuser pointed out that Schleiermacher's rejection could be perhaps seen not as a general rejection of axiomatics but rather as a rejection of the account of axiomatics developed by Leibniz and pursued by Christian Wolff.

1861 treatise had a different impact on the mathematical world of his time than A1. The arithmetic treatise was studied independently of A1 and of the philosophical ideas expressed there. Petsche sees this as one of the fundamental causes of the fact that the 1861 treatise came to be regarded as an axiomatic treatment of arithmetic.

The second line of analysis leads Petsche to the conclusion that, in actual fact, the examination of the 1861 treatise in the context of the other works of Hermann Grassmann makes it possible to prove that the 1861 treatment of arithmetic was not the natural continuation of the foundational program outlined in A1, but rather the first manifestation of a radical turn. The 1844 philosophical conception led to a classification of the four mathematical disciplines (arithmetic, combinatorics, function theory, calculus of extension) according to a unifying genetic scheme, which, among other things, also led to the *Theory of Forms* as a structural-algebraic metatheory. Petsche thinks that, beginning in 1861, the general methodological conception of A1 is de facto abandoned, and replaced by a new approach in which arithmetic is developed as a part of the calculus of extension. Because of this change in theoretical design, the entire genetic scheme used in 1844 and particularly the *Theory of Forms* are said to become irrelevant for later developments [Petsche 2006, 219 ff.].

Both previous lines of analysis have implications for the assessment of the place of axiomatics in Grassmann's work. In this paper, however, I will only examine the first line of analysis in greater detail.

The first level of analysis is, as said, in a sense, comparable with that pursued by Webb and Wang. If the 1861 treatment of arithmetic is examined in complete isolation from the rest of the writings of the Grassmann brothers, then the axiomatic illusion emerges. How does it emerge and why is it just an illusion? Petsche's answer is the following one.

In their 1861 actual treatment of the arithmetic of the integers, the Grassmann brothers emphasized "a strict Euclidean approach," rigor, and symbolic proof, in a way without precedent for them.

A close look at the general introduction of the treatise indicates that, on the one hand, in 1861 the ideal of an organic unity between the symbolic form of presentation (strict mathematical synthesis) and conceptual insight (analysis) of A1 is still held high. In the introduction to the treatise we are told that the teacher should not present the theory to his students by building formula on formula (*Formel auf Formel reihen)*. Symbolic presentation and conceptual generation should go hand-in-hand.[12]

This pedagogical ideal is, however, completely separated from the exposition of the arithmetic of the integers actually given in the book. This represents a major breach

[12] "Formelentwicklung und Begriffsentwicklung (müssen) stets Hand in Hand gehen"[LA, vi].

with the path taken in A1. As said, the 1861 exposition of arithmetic is basically formal-symbolic. In this way, the ideal of organic unity between symbolic form and thinking advocated by Justus Grassmann in 1827 and by Hermann Grassmann himself in 1844 is practically abandoned in 1861:

> "While his father Justus considered dialectical progression, the identification of a guiding idea and the feeling for an organic wholeness in a mathematical concept to be fundamental especially for school purposes, far more important than any calculation or written proof, now the pedagogical emphasis is laid on the 'most rigorous method possible'. Even though the foreword emphasizes, in its further course, the teacher's obligation to bear Justus Graßmann's points in mind, this does not in any way shape the mathematical mode of presentation." [Petsche 2009, 200f.].

Here Petsche claims that in the arithmetic treatise Grassmann abandons dialectics. As a result of this, the fundamental methodological requirement stated in A1 according to which a truly scientific presentation of mathematical theory demands not just the presentation of the symbolic synthesis, but also an additional explanation capable of revealing the conceptual unity standing behind the symbolic synthesis. The necessary link between the "organic unity of a mathematical concept" and formal symbolic "written proof" appears to have been abandoned in 1861.

Does, however, Grassmann really drop the ideal of unity between conceptual content and formal proof in the 1861 arithmetic treatise? It can be argued that in the 1861 treatise the symbolic form of presentation is seen as the expression not just of constraints dictated by the quest for formal rigor and economy, but also by a new understanding of the relationship between content and form in arithmetic.

An alternative interpretation: axiomatics and self-reference

I suggest that the 1861 treatment of arithmetic can and should be interpreted as an attempt to accomplish the methodological ideal of unity of form and content, of synthesis and analysis in arithmetic. In Grassmann's view, the 1861 treatise of arithmetic was supposed to achieve a transformation of arithmetic comparable to the transformation of geometry proposed in A1. If this is correct, why has this gone unnoticed so far?

There is an important difference between arithmetic and geometry. In geometry the rift between symbolic synthesis and geometric analysis had been a constant source of difficulties, whereas no comparable difficulties emerged in dealing with natural numbers or integers. In geometry the distinction real/formal manifested itself very early. It was indeed the reason why axiomatics emerged in geometry and developed in applied science and metaphysics rather than in arithmetic. In arithmetic this distinction seems missing.

Arithmetic does not seem to have a "real" counterpart. In arithmetic the rift between synthesis and analysis is less obvious. To recognize the presence of such a rift one needs to lift the real/formal distinction to a higher level, to view it not as an ontological but rather as a through and through methodological one.

For Grassmann the "real" is not a reality given independently of thinking, but simply one side of the distinction real/formal. Drawing this distinction is, in a sense, a dialectical act. The mathematical laws of the "real" are then mathematical laws framed by the mind while observing the way in which it is operating with the distinction real/formal. The mind does not have a direct privileged access either to some world out there or to its own dealings. Knowledge generation is fundamentally a recursive process. This is clearly Grassmann's intention in 1844, even if, perhaps, the work of the historian would have been made a great deal easier, if Grassmann's explanations of the matter would have been a bit more extensive [$A1_K$, 23ff.]. Of course in arithmetic, the contribution may look less impressive because, traditionally the arithmetic of the integers had not led to difficulties of the kind registered in coordinate geometry. When confronted with the Grassmann–Peano axioms of arithmetic the beginner is in a completely different situation than in a comparable confrontation with vector algebra. The axioms of vector algebra are surprising and not so easy to interpret, whereas those of arithmetic seem trivial, indeed superfluous.

The symbolic expressions used in arithmetic seem to be immediate expressions of clear arithmetical ideas. They appear then, in a sense, as iconic expressions of thought processes. This, however, is not the pre-established harmony cherished by Leibniz. It is a dialectically constructed unity.

At least five important arguments can be brought in support of this claim. They are all based on the ideas put forward in the 1861 arithmetic treatise and, of course, on a comparison with the methodological tenets of A1:

I) Grassmann's constant emphasis on the unity of symbolic form and conceptual expression in arithmetic.
II) The two complementary accounts of the distinction between synthesis and analysis in A1.
III) The real/formal distinction as framed in A1.
IV) The construction of the sequence $\ldots e+-e+-e+-e; e+-e+-e; e+-e; e; e+e; e+e+e; \ldots$ in the beginning of the arithmetic treatise; and last but not least.
V) The emergence and wide use of recurrent definitions in the arithmetic treatise.

I will begin with a combined examination of (I), (II), and (III). According to Grassmann all that is needed in order to teach the content of the arithmetic treatise properly is to paraphrase the symbolic expressions in words [LA, VI].[13] This is a much weaker methodological requirement than that of 1844, where Grassmann insisted on the necessity of using intuitive geometric examples in the presentation of his vector algebra. This may give the impression that in 1861 Grassmann thinks that meaning follows form, that analysis is an obvious consequence of synthesis, that there is no place for dialectics in arithmetic, and finally that the real/formal distinction is irrelevant here. In order to clarify this, it is helpful to briefly compare the ideas concerning the relationship between symbolic synthesis and conceptual analysis in 1861 with those put forward in 1844. In the preface of A1 Hermann Grassmann writes:

> "Thus not only can each formula that arises in the course of a derivation be most easily expressed verbally, but in addition each step from one formula to another appears at once as just the symbolic expression of parallel act of reasoning. ... (in the calculus of extension – M.R.) the idea is no longer ... obscured but radiates through the formulas in complete clarity, the intellect grasps the progressive development of the idea with each formal mathematical development." [$A1_K$, 10]

In a footnote, Grassmann adds that his new geometric calculus leads to a perfect harmony between analysis and synthesis in mathematics. Grassmann's words are of great importance here: The difference between analysis and synthesis "vanishes" [$A1_K$, 111].

Does this not mean that, indeed, even in 1844, Grassmann already regarded symbolic expression as sufficient for a clear presentation of thinking, of course, under the condition that the required type of symbolic calculus has been found, and that it is properly expressed in words? Such a presentation is given in A2.

This ideal of symbolic clarity and correspondence between symbolic expression and thinking is distinct from the subsequent discussion of the relationship between analysis and synthesis in mathematics generally which leads Grassmann to emphasize the necessity of using concrete geometric models in the presentation of his calculus in A1. This latter requirement is not due to the lack of clarity of the symbolic side of the *Calculus of Extension*. It is just a consequence of the fact that the beginner does not have the necessary faculty of perception *(Ahnung)* of symbolic form as the creator of the theory [$A1_K$, 30 ff.].

It may therefore be argued, I think, that in A1 one can identify two distinct lines of thinking in respect to the presentation method. The first emphasizes the perfect unity between symbolic expression and thinking, between analysis and synthesis. The

[13] In A2 and in the books published in 1872 and 1891 under Robert Grassmann's name the unity of symbolic language and concept fade away. Robert Grassmann even abandons the requirement of paraphrasing symbolic expression, a significant shift in a formalist direction. But there is no trace of this in the 1861 treatise.

other, on the contrary, emphasizes not a perfect match between analysis and synthesis, but rather a necessary incommensurability, and based on that, a dialectical relationship between the two, the dialectic game being made possible by mediation through geometric applications.

Grassmann's use of "real" elements (ordinary points, line segments) etc. in A1 is the manifestation of a major methodological commitment to a self-referential account of knowledge and linked to that with a philosophical conception which places subjective agency at the basis of knowledge generation. Even if we regard geometry as an empirical science, this does not mean that we can find anything which is given in some absolute sense in geometry, something out there, waiting to be accurately matched by theoretical descriptions. Theory is no longer first-order theory, but rather second-order theory. We do not have direct access to "reality" as it may exist independently of our theories about it. All we can do is to observe the way in which the first-order real/formal distinction manifests itself, when looked upon from a second-order perspective. Knowledge is not knowledge about absolute objects and absolute ideas. It is rather knowledge about the way in which we draw certain distinctions and the observation of the consequences of that. The conception of knowledge generation and of subjective agency involved here can be best described, it seems to me, by Niklas Luhmann's approach, from which I have borrowed some of the previous terms [Luhmann 2002a, b].

The unity between symbolic calculation (synthesis) and conceptual unity (analysis) is not abandoned, but it can no longer be regarded as pre-established and once and for all given. It needs to be permanently constructed and tested through applications. Knowledge gain becomes possible due to the irritation emerging during the process of self-construction of the psychic system of the researcher. The origin of understanding, of conceptual unity, of harmony between synthesis and analysis in mathematics is, paradoxically, the lack of understanding confronting the subject in its effort to solve problems in applied science. The rather unpleasant tension between synthesis and analysis [$A1_K$, 30f.], the lack of understanding together with the constitutional striving of the subject to reach understanding, forces the subject to take on the position of a second-order observer, and thus reach understanding on a higher level of analysis.[14]

As Otte puts it the "nineteenth century ... regarded cognition as an activity following its own rules and gradually turned these rules into its object of investigation" [Otte 1989, 15]. Stated in terms of the two distinctions between analysis and synthesis present

[14] Grassmann's writings do not provide a sufficiently clear presentation of the dialectics of understanding and not-understanding comparable to that found in Luhmann's theory of systems. Even though Grassmann's explanation of the necessary rift between analysis and synthesis in mathematics goes in this direction, he gives the impression that this tension could be once and for all eliminated once the proper symbolic calculus, the proper mathematical synthesis would be in place. In this sense, Grassmann's explanations of this matter do sound like those involved in Leibniz' quest for a Universal Characteristic.

in Grassmann's work, this amounts to a sharp distinction, and at the same time, to a structural interdependence of the two. Working on the foundations of mathematics becomes not so much a logical activity of deriving the consequences of given stipulations. It becomes an experimental, indeed creative activity, of critical examination of the first principles. First principles therefore are no longer seen to describe some *given* field of objects or ideas. They are the expression of a scientific examination of the way in which axioms are stated. In a sense, the entire methodology becomes second-order and this process is well captured in Luhmann's theory. This is the new axiomatic spirit characteristic for Grassmann's work. Its best expression is A1. This approach to axiomatics may be called epistemic axiomatics or perhaps, as said, in connection with Luhmann's conception, second-order axiomatics. The shift from Euclid and Leibniz to Grassmann and indeed to Hilbert then appears as a shift from a conception of axiomatics which aspires to be able to build an absolute, static first-order axiomatics to a new dynamic approach in which axiomatics is viewed as a genuinely second-order endeavor.[15]

Even if from a general point of view the previous account of axiomatics may look attractive, and even if clear traces of such a conception are present in A1,[16] the presence of such a conception in the 1861 treatise of arithmetic is less obvious.

If my interpretation is correct, then there are two accounts of the relationship between analysis and synthesis in A1. The first account requires a perfect match between symbolic form (synthesis) and conceptual content (concept analysis expressed in words). The second is a dialectical relationship linking symbolic form (synthesis), conceptual content (conceptual expression of geometric reasoning), and geometric object. The first emphasizes unity, the second irritation. The awareness of unity is reached via irritation. This involves applications as a projection surface.

In the treatise of 1861 and in A2 and subsequent writings, the second account of the analysis/synthesis distinction seems to be, if not explicitly abandoned, at least subordinated to the former. On this reading, what happens is that beginning with 1861 the identification between analysis and synthesis is used as the sole methodological guideline for scientific presentation. If this is correct and if epistemic, mediated self-reflection involves both distinctions, then there is no axiomatics there.

---

[15] It would be interesting to discuss this in connection with the distinction between first-order and second-order logic. Here, however, I confine myself to Luhmann's terminology.

[16] As far as I can see, both Hans Niels Jahnke's [1978] discussion of A1, and Michael Otte's [1989] treatment of the subject endorse the same view. The fundamental difference between the approach provided in this text and those of Jahnke and Otte result from my reliance on Luhmann's conception as a general theoretical framework.

I think, however, that this last claim can be challenged. To achieve this, I will first have a closer look at the unity between symbolic form and conceptual expression as outlined in 1861. I will then turn to points (IV) and (V) mentioned above.

In the introduction to the arithmetic treatise, Grassmann emphasizes that even though the book presents mainly the symbolic mathematical synthesis, an appropriate use of the book in teaching requires the emphasis on conceptual unity. Grassmann claims that in the case of arithmetic this can be easily achieved and indicates that an example of the close correspondence between formal synthesis and conceptual unity is given in Sect.17 of the book. An examination of Sect.17 reveals, however, that all Grassmann has in mind is paraphrasing the formulas and the course of the proofs in plain language. Thus, the formula $a + (b - e) = a + b + -e$ is translated in "instead of adding a negative unit to a sum, it is possible to add the negative unit to the second term of the sum (my translation - M.R.)." [LA, 5].[17] He then goes on to prove this proposition and he paraphrases the proof in the same way as well [LA, 5]. It is possible to interpret this, as Petsche does, as a methodological breach with A1. At the same time, given the already-mentioned close link between symbolic expression and conceptual unity in arithmetic, it is possible to view symbolic expression as representing conceptual operations in an obvious way.

At the same time, and I now come to (IV), the arithmetic treatise contains "definition 7" which, as far as I can see, is nothing more than the construction of a concrete model of the sequence of the integers. My suggestion is that this sequence is the arithmetical counterpart of the geometric applications used in A1. Its aim is not to guarantee the existence of number, but rather to explain how first-order arithmetical experience generates the sequence in order to be able, then, in a subsequent step, to describe this first-order generation in a second-order symbolic theory, leading to the various stipulations, recurrent definitions, and theorems. As stated in A1, "discrete synthesis as equal" is used to generate the sequence. The examination of the sequence makes it possible to give an explicit symbolic expression to the steps involved in generating the sequence. If this interpretation is correct, then the role of the sequence is a merely dialectical one. Grassmann's seventh definition can then no longer be used as an argument against the axiomatic character of the arithmetic treatise unless one wishes to view axiomatics independently of the first-order, second-order distinction outlined above. Indeed, if modern axiomatics is, as I suggested above, second-order axiomatics, then the separate construction of the sequence of the integers is a necessary ingredient, even if, in the end, it can be dropped as soon as the symbolic construction has been put in place.[18]

[17] Today we would write $a + (b + (-e)) = (a + b) + (-e)$.

[18] In his axiomatization of arithmetic, Peano relies on the successor function for the construction of the sequence of the natural numbers. Why did Grassmann not use such a device himself? More importantly,

I now come to the last point of my list which concerns recurrence. This gives me the opportunity to return to Webb's ideas, which were discussed above from a different perspective. If the interpretation given so far is correct, then Grassmann's use of recurrent definitions in the arithmetic treatise gains in philosophical importance. Recurrent definitions appear as the natural mathematical expression of a general methodological conception which makes self-reference and linked with it the complementarity between the two accounts of the relationship between synthesis and analysis in mathematics the core of the process of knowledge generation. Recurrent definitions then appear as the natural expression of an axiomatic turn in arithmetic, and this, in turn, as a shift towards Hilbert. This is not so much the consequence of the fact that these definitions are mere relational stipulations, but of a much deeper reason. According to the approach proposed above, it is the expression of the fact that modern axiomatics is second-order axiomatics and recurrent definitions are the most elementary mathematical expression of this.

Instead of a conclusion

What should we then make of Grassmann's programmatic rejection of using axioms *(Axiome)* in pure mathematics stated in A1? Albert Lewis (this volume) suggests that Grassmann's rejection is aimed against building mathematics upon arbitrary assumptions introduced simply because, at a certain point in time, such assumptions appear necessary for filling the gaps in our understanding of the objects we are studying. Does this mean that Grassmann's methodological ideal is consistent with what we nowadays call "natural deduction" as Lewis suggests? Is Grassmann's rejection of axiomatics perhaps an indication that, after all, he adheres to a philosophy of mathematics inspired by Leibniz? Is Grassmann's understanding of axiomatics compatible with that of Schleiermacher? If so, is Grassmann's rejection of axiomatics an indication of a fundamental opposition between the dialectic methodology advocated by Schleiermacher and axiomatics, as suggested by Petsche, or is it rather, as suggested by Marie-Luise Heuser, the expression of a rejection of the Leibniz/Wolff understanding of axiomatics?

---

is the absence of such a function in Grassmann's approach a technical imperfection, an expression of the fact that he did not intend to propose an axiomatic treatment of arithmetic, or rather the expression of the fact that Grassmann, according to his methodology, requires the construction of a concrete object as a manifestation of first-order construction subsequent to which second order axiomatic description can proceed?

The perspective adopted in this paper makes it possible to propose an alternative interpretation. If the distinction real/formal is regarded as an absolute ontological distinction, then an axiom is the formal expression of the given reality. This point of view is characteristic for a substance logic incompatible with A1. If this is correct, and if Hermann Grassmann used the real/formal distinction in a way compatible with Luhmann's conception, as suggested above, then his 1844 rejection of axiomatics is simply a rejection of a dated conception of axiomatics and his work must be then seen as a first major step towards the emergence of a new conception. I have suggested that the new conception views axiomatics as second-order observation. The axioms do not describe some static reality but rather capture the way in which the subject reflects upon the way in which he draws the real/formal distinction.

My interpretation has a somewhat surprising consequence. As a rule, when looking for traces of modern axiomatics in the writings of the Grassmann brothers, historians turn to texts published beginning with 1862. The merely symbolic approach pursued in these texts seems to qualify them as potential candidates for axiomatic thinking. Things look different if the perspective proposed in this paper is adopted. In this case the dialectical methodology pursued in A1 appears as a fundamental ingredient of axiomatic thinking. Some historians have pointed out that in A1 and indeed even in A2, Grassmann does not pursue an axiomatic path, but in the effort to cover up the true origins of his mathematical insights (from work on applications), invents a formal symbolism, which creates the illusion of generality [Radu 2005]. The perspective adopted in this paper makes it possible to avoid such an interpretation.

Concepts and contrasts: Hermann Grassmann and Bernard Bolzano

Steve Russ

Introduction

Both Bolzano and Grassmann worked intensively for much of their lives on mathematical concepts – both technical concepts *within* mathematics such as line, number, function, continuity, extension, and broader concepts *about* mathematics such as its nature and classification, and what should be understood by a "proof". In this paper we shall compare some of the ways in which they worked with concepts because, while strikingly different, both scholars were led, through their analysis and development of concepts, often in a philosophical context, into fruitful new mathematics which was decades ahead of its time. The case is all the more interesting because of the close contextual parallels between the men which throws their differing treatment of concepts into greater relief. Furthermore, for Grassmann one of the major driving metaconcepts of his thinking about, and within, mathematics appears to have been the very concept of contrast, or of opposition (*Gegensatz*). This is a concept associated with so-called dialectic, or dialectical thinking. Examples of such oppositions, used by Grassmann, are the general and the particular, the real and the formal, equal and different, discrete and continuous. For Bolzano too there were principles – such as the Aristotelian dictum about not crossing from one "kind" of knowledge into another – which acted as a creative

H.-J. Petsche et al. (eds.), *From Past to Future: Graßmann's Work in Context*,
DOI 10.1007/978-3-0346-0405-5_11,

constraint, rather than a restriction, in his conceptual re-working of parts of mathematics. Comparing the ways in which each of Bolzano and Grassmann went about working with concepts, and developing new mathematics, may throw light on our understanding of their creative processes, of how they were able to retain and exploit both objective and subjective views of concepts, and of the mutual interactions of mathematics and philosophy in their work.

Bernard Bolzano, born in Prague in the year of appearance of Kant's *Critique of Pure Reason*, 1781, was nearly thirty years senior to Hermann Grassmann and almost exactly contemporary with his father Justus. There is no evidence, known to this author, that Bolzano or Grassmann knew of each other's mathematical work or publications, but they shared much in common. My goal is to show how a comparison of Bolzano and Grassmann with regard to their "working with concepts" in some simple cases might, if extended more widely and deeply, improve our understanding of the mathematical achievements of both scholars. So this paper is more a suggestion of a research project than an end in itself.

Some parallels of context

Both Bolzano and Grassmann had a thorough grounding in philosophy and theology while they were students, so they were familiar with abstract thinking and the making of conceptual distinctions. Nor did either of them hesitate to wade in boldly with the analysis and re-definition of concepts whether already well known or new. For example, the first three sections of the *Introduction* of [A1] are respectively headed: "Derivation of the Concept of Pure Mathematics", "Derivation of the Concept of Extension Theory" and "Exposition of the Concept of Extension Theory". On the first page of Bolzano's first publication [Bolzano 1804] we read that, "endowing all concepts of [mathematics] with the greatest possible clarity, correctness and order [...] will not only promote the *thoroughness* of education but will also make it *easier*." He goes on throughout this *Preface* to outline his principles for acceptable proofs in terms of what we have elsewhere called "conceptual correctness" [Russ 2004, 21]. Throughout their careers, and their publications, both scholars valued the development of an explicit conceptual framework and the exercise of conceptual analysis very highly. Yet such a shared attention to these "particulars of thought" (as concepts have been styled) took them in very different directions in the areas of the nature of mathematics, on how to deal with geometry, and on what makes a presentation "scientific", as the succeeding section will show.

In the opening pages of both [Bolzano 1810] and [A1] we read that mathematics should not be regarded as a "science of quantity" (as it was conventionally) and they each give the same reason: the subject of combinatorics is not covered by this definition. Instead mathematics should be regarded, they each say, as a "theory of forms". However, the verbal similarity in their formulations turns out, as we shall see, to be misleading. What was a well-grounded distinction for Grassmann, with major consequences for positioning his "extension theory", was less significant at the time for Bolzano. Indeed he returned in later life to the older definition, though on this matter the story is complicated and interesting.[1]

Broadly the same *kinds* of demands and motives, based on conceptual distinctions, appeared to be significant in guiding each man to make advances which were considerably ahead of their time and sometimes re-discovered only decades later by mathematicians quite unaware of the precedent for their results. In the case of Grassmann his developments and insights especially in logic, algebra, geometry and vector analysis, are well described in other chapters of this volume. In the case of Bolzano, his original and far-reaching contributions in logic have been described in many places, still a good starting point is [Berg 1962]. His work in the fields of analysis, theory of functions and real numbers, topology and set theory is no less remarkable and profound. For example, he carefully and deliberately constructed a simple function that is continuous everywhere and differentiable nowhere in the early 1830s – decades before the better-known example due to Weierstrass. The extensive mathematical work of Bolzano has been described briefly in [Russ 2004] and in greater detail in [Rusnock 2000] and [Sebestik 1992].

For students in the early nineteenth century, whether in Prague or in Berlin, it was natural to find questions and matters of religion, theology and philosophy to be of much more immediate and pressing importance both in everyday life, and academic life, than most of us find them today. The profound philosophical works of Kant were recent, difficult and controversial. German idealism in its many forms was being widely discussed, taught, and sometimes even prohibited. No student, of any discipline, could avoid studying and grappling to some extent with the philosophical and religious issues of the day. It was philosophy, and in particular epistemology, that critically examined *concepts* in general – their construction and emergence as well as their character and analysis. Particular interest in the concepts of mathematics – for example, those of number, function, continuity, displacement, proof and so on – led to Bolzano and Grassmann both gaining a deserved reputation for being "philosophical" mathematicians.

Theology offered a framework and grounding for ethics and values, including what it was worthwhile doing with one's life. As young men both our subjects seriously considered a vocation as a Catholic priest (Bolzano) or minister in the Protestant Church

[1] It was described in a presentation by Paola Cantù, '*Changes in Bolzano's definition of mathematics*' at a meeting in Prague in April 2010.

(Grassmann). Although Bolzano was ordained a priest he explicitly decided against the pastoral calling inseparable from work with a Church congregation. Instead he became, in 1805, Professor of *Religionswissenschaft*, at the Charles University, giving lectures and twice-weekly sermons to the students and the citizens of Prague. At least at some stages in their lives both men would have liked, and yet neither ever gained, a university chair in mathematics.

From the quotation above we have already seen how Bolzano took very seriously the importance of education. The pedagogic motive was never far away in both his theological work and in his early mathematical publications. Hermann Grassmann also, being brought up in a family with a distinguished and committed school teacher as father, was very familiar with education, and the scientific and moral issues surrounding the theory and practice of education. He himself became a school teacher, albeit one for whom mathematics was probably more important than were his pupils. Some indications of this, and the textbooks he wrote are well described in [Petsche 2009]. The sustained experience and reflection on generations of students, and their learning – both of mathematics and religion – was therefore a dominant context common to both men.

Both Bolzano and Grassmann spent a substantial part of their lives, in relatively obscure parts of Europe, developing original and powerful mathematical ideas and methods. Each knew their work was highly significant, but each failed to convince of this significance, or even interest in many cases, the leading mathematicians of their day. Curiously, it was the same young mathematician, Hermann Hankel, who first drew the public's attention to the talents of Bolzano in [Hankel 1871] *and* to the genius of Grassmann in [Hankel 1867].

So far we have gathered the common threads of background, education, motivations, talents and achievements that connected our protagonists; we now turn to how they transformed this commonality into their own quite divergent ways of working with concepts.

Some divergences of working

We shall first compare Grassmann and Bolzano, with regard to their views on the nature of mathematics and its classification, then go on to compare their work on the nature of geometry and extensive magnitude, finally we compare what they thought it is that makes for a truly "scientific" presentation of a subject (including mathematics).

The phrase "working with concepts" is vague and hard to define in general but can be illustrated here by examples from these two authors to give a reasonably clear practical understanding of what we have in mind. All quotations from Grassmann are either from the *Introduction* or *Chapter One* of Part One of [A1], in the translation by Kannenberg [$A1_K$]. The following sequence of insights leads up to Grassman's classification of mathematics into four major branches, one of them the subject of his book. They convey some idea of his progression of thought, but are naturally even more enigmatic than the full text.

The nature and classification of mathematics

> "Thought exists only in reference to an existent that confronts it and is portrayed by the thought; but in the real sciences this existent is independent, existing for itself outside of thought, whereas in the formal it is established by thought itself, when a second thought process is confronted as an existent." [$A1_K$, Sect. 1]
> "The *particular* existent that *has come to be* by thought... we call a thought form, or simply a *form* [...] pure mathematics is the *theory of forms.*" [$A1_K$, Sect. 3]
> "Each particular existent [...] comes about in one of two ways, either a simple act of *generation* [...] or a two-fold act of *placement and conjunction.* [...] the first way is the *continuous form* [...] the second way is the *discrete* [...] *form.*" [$A1_K$, Sect. 4]
> "Each particular existent becomes such through the concept of the *different* [...] and through this with the *equal* [...] that arising from the equal we call the *algebraic form*, that from the different the *combinatorial form.*" [$A1_K$, Sect. 5]
> "From the interaction of these two oppositions", says Grassmann, "arise the four species of form and the corresponding branches of the theory of forms [...]" [$A1_K$, Sect. 6].

	Discrete	Continuous
Equal (algebraic)	Number	Function theory/ intensive magnitude
Different (combinatorial)	Combination/relation theory	Extension theory/ extensive magnitude

The table above does not appear in A1, it is my summary of Grassmann's description of the "four species of form". A footnote in Sect. 6 of A1 refers to a similar classification by Hermann's father, Justus, published in the *Programme* of Stettin Gymnasium for 1827 (and translated elsewhere in this volume). In that work, undoubtedly very familiar to Hermann, there is a simpler diagram (this volume, p. 460) similar to the table above. Clearly the work of Professors (as they were titled) at secondary schools at this time went well beyond what was needed or appropriate simply for their pupils.

Compare to this what we find in [Bolzano 1810]. The following quotations are from [Russ 2004]:

> "I therefore think that mathematics could best be defined as a *science which deals with the general laws (forms) to which things must conform in their existence.*" (Sect. 8)
> "[...] let us try to derive a *logical classification* of this science into several individual disciplines [...] these laws are either so general that they are applicable to *all things completely without exception*, or not [...] the former will constitute *general mathesis*, everything else is then *particular mathesis* [...]" (Sect. 11).
> "To *general* mathesis belong *arithmetic, the theory of combinations*, and several other parts [...] chronometry, geometry etc [...] are *subordinate* to the general mathesis as a whole, as species of the genus [...] the concept of *number* will appear frequently in all these particular parts [...]" (Sect. 11).

Bolzano gives a tree-like picture [Russ 2004, 102] of his classification of branches of mathematics which involves strict dichotomies, reference to aetiology, time and space. It is a classification that is incompatible with that of Grassmann.

Returning now to Grassmann the diagram given above might suggest a neat and tidy separation of concerns. But that is not how Grassmann thought about the situation. On the one hand, the separations are not sharp. Both elements of his oppositions are present and what matters is which "prevails". Grassmann writes on his "oppositions" in shaping the four branches of mathematics:

> "The opposition between discrete and continuous is (as with all true oppositions) fluid, [...]" [$A1_K$, Sect. 4].
> "The opposition between equal and different is also fluid." [$A1_K$, Sect. 5].

On the other hand, he has no hesitation in using metaphor and analogy to help convey his vision:

> "It is thus somewhat as if the intensive magnitude is number become fluid, the extensive magnitude combination become fluid." [$A1_K$, Sect. 7].

So a provisional conclusion might reasonably be that Bolzano shows a greater detachment and objectivity in his interpretation of "theory of forms", (the "laws to which things must conform in their existence") yet it is superficial and did not really determine his subsequent classification. Grassmann's understanding is in terms of "thought" with all its subjectivity, but has a subtlety and sophistication that lends it substance and definitely determines the positioning of his "new branch of mathematics". What is striking is how dissonant and incomparable are each of their new, broader interpretations of mathematics.

What shall we do with geometry?

At an early stage Grassmann banishes geometry from pure mathematics. Perhaps this really has more to do with his delineation of "pure mathematics" than the nature of geometry.

> "[...] geometry, like mechanics, refers to a real existent; for geometry this is space. [...] the concept of space can in no way be produced by thought, but [...] is given. [...]
> Thus geometry [...] appears as an application of the theory of forms to the fundamental perceptions of the sensible world." [$A1_K$, Sect. 3]

His remarkable vision now proceeds in a highly abstract way in which geometry is no longer an integral part, but is constantly and deliberately used as a suggestive analogy.

> "[...] to obtain the extensive magnitude I relate it to the generation of the line. Here it is a generating point that assumes a continuous sequence of positions [...] we now proceed to the extension in our science if we simply replace these spatial relations by their corresponding concepts.
> First in place of the point, we here substitute the *element*, [...]
> We call the transition of the generating element from one state to another an *evolution*; and this abstract evolution of the generating element thus corresponds to the evolution in position, or motion, of a point in geometry." [$A1_K$, Chap. 1, Sect. 13]
> "[...] some geometric examples as an aid to the intuition [...]
> the plane clearly corresponds to the system of second order, and the plane is thus regarded as generated when all the points of a straight line are moved in a new direction [...] not included in it [...]
> Likewise one proceeds to the whole of infinite space as a system of third order by moving the points of the plane in a new direction... geometry cannot progress further, but the abstract science knows no limitation." [$A1_K$, Chap. 1, Sect. 16]

The treatment of geometry in [Bolzano 1804, *Preface*] begins in a much more prosaic manner with what appears to be a rather limiting principle of conceptual correctness:

> "I could never be satisfied with a completely strict proof *if it were not derived from the same concepts* which the thesis to be proved contained, but rather made use of some fortuitous, alien, *intermediate concept*. [...]
> I consider it an error in geometry that all propositions about angles and ratios of straight lines [...] are proved by means of *considerations of the plane* for which there is no cause in the *theses* to be proved. I include here the concept of *motion* [...]."

Bolzano's idea is that motion requires an "object distinguished from space" and it presupposes the geometric result needed to prove the possibility of the motion. The banishment of the concept of "plane" means angle is not a quantity (angles cannot be added without the assumption of being in the "same plane"). This leads to a curious proof (Sect. 5) of the equality of vertically opposite angles in terms of their equal "determining pieces". He develops the idea of some abstract objects determining others to replace the usual notion of congruence.

It appears that Bolzano may have thought he had "proved" in the first Part of this work the parallel postulate, though his treatment is so "non-Euclidean" that it hardly makes sense to think in these terms. In any case, he assumes the equivalent result of the existence of unequal similar triangles at an early stage. What is much more interesting, and characteristic of Bolzano, is the second Part of this work: *Thoughts concerning a prospective theory of the straight line.* Here Bolzano considers the "simplest object of geometrical consideration", a system of two points. He distinguishes and defines carefully the concepts of distance and direction associated with such a system.

At the end of this section he defines the *plane of an angle* as "[...] that object which contains all and only those points which can be determined by their relationship (their angles and distance) to the two directions R and S [of the angle]." [Bolzano 1804, II Sect. 43].

What makes a Presentation "Scientific"?

In [Bolzano 1810, II Sect. 2] Bolzano raises the question of the "true nature of scientific exposition" and declares, somewhat disarmingly, that "I must admit [...] I am not completely clear myself [about it]." But he continues, "this much seems to me certain: in the realm of truth, i.e., in the collection of all true judgements, a certain *objective connection* prevails which is independent of our accidental and *subjective recognition* of it. [...] Presenting this objective connection [...] seems to me the real *purpose* to pursue in a scientific exposition."

This is Bolzano's "ground-consequence" relation (*Abfolge*) and is expounded in great detail much later in [Bolzano 1837]. The classic example there (Sect. 162) is

1 It is warmer at X than at Y.
2 The thermometer is higher at X than at Y."

If we know either 1 or 2 then we also know the other, at least it is a basis for knowing *(Erkenntnisgrund)* the other. But objectively 1 is the ground for the consequence 2 and not conversely. In [Bolzano 1837] there are also mathematical examples. It continues to perplex commentators and there is a large, and growing, literature on the subject. See, for example, [Lapointe 2010].

Grassmann has another view on what makes a presentation scientific which is closely connected with his pedagogic motivation [$A1_K$, *Introduction* Sect. 14]. He says,

> "we characterise a method of treatment as scientific if a reader is thereby on the one hand led necessarily to recognise the individual truths, and on the other is placed in a position from which he can survey each point in the broader sweep of the development."

And in Sect. 16 we read:

> "The essence of a scientific presentation is thus an interlocked pair of approaches, one of which leads consistently from one truth to another and forms the actual content, while the other controls the treatment and thus determines the form. [...] The reader may then expect to be able to progress independently, without having to be guided along such lines of thought, regarding himself as independent of the discoverer of the truth and thus reversing the relation between himself and the author [...]"

The work [Petsche 2009] draws together convincing evidence (including Grassmann's own testimony) that attending Schleiermacher's lectures in *Dialectic* in 1831 was a deep and lasting influence on Grassmann's thinking. It is unfortunate that in English the term "dialectic" has such a wide range of disparate associations, ranging from the caricature "thesis-antithesis-synthesis" popularly attributed to Hegel, through "dialectical materialism", towards the rather gentle but powerful idea of merely a "movement of thought" – such as is characteristic of attentive conversation. According to [Frank 2005] Schleiermacher's use of the term for his lectures has more to do with the Platonic tradition of dialogue than any foundational or metaphysical sense. It seems to this author that this influence on Grassmann had as much to do with the understanding – clear in the above quotation – of what we would now call "constructivist" learning, as it had to do with the "oppositions" to which we referred earlier, important as they may be.

For example, from Grassmann's own CV of 1833 we read: "I understood that Schleiermacher showed his students how to approach every branch of science competently. He did this not by giving positive elements, but by showing how it was possible to approach every investigation correctly and continue it independently, thereby enabling the scientist to find the positive elements on his own." [Petsche 2009, 23].

This is remarkable and prescient testimony to the efficacy of a constructivist approach to learning. It was regarded by Grassmann as an integral part of a truly scientific presentation. Bolzano's ground-consequence relation has usually been interpreted as an "explanatory" approach to proofs. For example, see [Kitcher 1975] and [Mancosu 1999]. Undoubtedly a rich semantic view of knowledge and understanding is underlying both conceptions of "scientific". Bolzano was also sure that his "objective ordering" had pedagogic value. Could Grassmann's notion be relevant to our understanding of Bolzano's notion, or even vice versa? For example, there is the possibility that they could both be approaches to some "natural" ordering of the facts or theorems of a subject.

Conclusion

The way Bolzano and Grassmann worked with concepts in their mathematics was very different indeed. How should this difference be characterised? The fact of such a great difference is instructive in itself in giving an important context and perspective in assessing each of them – and indeed any other "philosophically inclined" mathematician. The scale of difference is an indication of how much choice is open to us in thinking deeply about mathematics and makes the choices we do make more significant and worth asking about. We can see both Grassmann and Bolzano in a different light having compared each one with the other. This very brief comparison suggests it would be a worthwhile project to conduct further study to compare Bolzano and Grassmann in much greater detail with regard not only to the topics touched upon here but, for example, number, intuition, axiomatics, logic and proof.

Illustration 1.3 [illegible]

45

Der chemische Unterricht auf der höheren Bürgerschule.

Von H. Graßmann, Oberlehrer an der Friedrich-Wilhelms-Schule
in Stettin.

Was den chemischen Unterricht von allen übrigen Unterrichtszweigen wesentlich unterscheidet, was ihn als ein nothwendiges Glied in das gesammte Lehrgebiet einer höheren Bürgerschule einreiht und ihm seine eigenthümliche Methode vorschreibt, das ist die enge Verbindung zwischen theoretischer Erkenntniß und praktischer Ausübung, welche in ihm, wie in keinem andern Gegenstande, möglich gemacht wird. Die theoretischen Principien lassen sich in ihm unmittelbar durch Versuche, die der Schüler selbst anstellt, bethätigen; ja so eng ist hier diese Verbindung zwischen der erkannten Wahrheit und dem selbst angestellten Versuch, daß jene ohne diesen gar nicht erkannt und dieser ohne jene nicht mit Sicherheit angestellt werden kann. So allgemein nun auch dieß eigenthümliche Wesen des chemischen Unterrichts der Idee nach anerkannt ist, so vielfach man sich auch bemüht hat, darnach die Methode des Unterrichts zu gestalten, so hat man sich doch bisher, wie ich glaube, durch die Analogie des übrigen Unterrichts davon abhalten lassen, diese Idee in ihrer ganzen Consequenz durchzuführen. Hat man auch schon längst jenes einseitige Theoretisiren, dem keine anschaulichen Versuche zur Seite gehen, auf diesem Gebiet als gänzlich unfruchtbar verworfen, so sieht man doch noch häufig, wie der chemische Unterricht analog dem physikalischen nur durch Versuche, die vom Lehrer vorgemacht werden, erläutert wird; oder wie zwar von einer gewissen Stufe an und zu gewissen Zeiten dem Schüler Gelegenheit zum Selbsterperimentiren geboten wird, ohne daß aber dieß Selbsterperimentiren das eigentliche Hauptmoment des gesammten Unterrichts wird. Ja auch da, wo das Selbstexperimentiren als der eigentliche Mittelpunct des Unterrichts erscheint, wird dessen Frucht zum Theil durch eine einseitige Heuristik, die man dabei anwendet, wieder fast gänzlich vernichtet. Es ist eine Selbsttäuschung, wenn man glaubt, den Schüler durch vorgelegte Fragen oder Aufgaben, deren Lösung er selbsterperimentirend finden soll, dahin zu führen, daß er die Reihe der chemischen Wahrheiten, wie sie in der Entwickelungsgeschichte der Wissenschaft nach und nach auftauchen, gleichsam aufs neue erfinden, die Irrthümer vergangener Jahrhunderte selbstständig überwinden soll. Der bestimmte Gang, in den man den Schüler zu diesem Zwecke einzwängt, wird, statt ihn zur Selbständigkeit zu führen, für ihn eine Fessel, die ihn unfrei macht, eine Fessel, die jedes

Illustration 11. A tract on the didactics of experimentation in chemistry class by Hermann Grassmann (1850). See also p. 499

Diversity of the influence of the Grassmann brothers

New forms of science and new sciences of form: On the non-mathematical reception of Grassmann's work

Paul Ziche

Grassmann outside mathematics

In 1907, two important texts on Kant were published, both highlighting typical aspects of a reception of Grassmannian ideas outside mathematics, mathematical logic or mathematical physics. Ernst Cassirer's article on "Kant und die moderne Mathematik" [Cassirer 1907] deals mainly with the philosophy of mathematics, but his arguments culminate in his monumental *Substanzbegriff und Funktionsbegriff* from 1910 and range far beyond the realm of traditional mathematics: he formulates a general theory of concept formation, a turn towards "order" as the most fundamental concept imaginable, a shift from substances and things towards relations and functions, and the attempt to formulate a general science of forms. In the same year, the philosopher-plus-psychologist Oswald Külpe – pupil of Wilhelm Wundt in Leipzig, professor of philosophy in Würzburg, Bonn and Munich – presents Kant's philosophy in the form of a popular survey that is motivated by the problem to bring philosophy and science (*Wissenschaft*) into a productive relationship [Külpe 1908; on Külpe see Ziche 1999; Kusch 1999]. Both authors draw important inspirations from Grassmann, and make their indebtedness to Grassmann explicit, again simultaneously, in 1910: Külpe in

H.-J. Petsche et al. (eds.), *From Past to Future: Graßmann's Work in Context*, DOI 10.1007/978-3-0346-0405-5_12,

a paper on "Erkenntnistheorie and Naturwissenschaft," where he claims Grassmann to be the discoverer of the fundamental distinction between two forms of science, namely "Realwissenschaft" and "Formalwissenschaft" [Külpe 1910, 9], and Cassirer in his *Substanzbegriff und Funktionsbegriff* where he, in virtually the same terms as Külpe, refers to Grassmann as the first discoverer of a "general *science of form*" *(allgemeinen Formwissenschaft)* [Cassirer 1910, 127].

This suggests a line of reception of Grassmannian ideas that falls outside mathematics proper, and that is related to fields such as philosophy, psychology and physiology, and that is of particular relevance at the boundaries between these fields. This type of reception starts remarkably early; it is fully present in texts by Hermann von Helmholtz[1] in the 1860s, and in Wilhelm Thiery Preyer's "Psychomathematics" from the 1870s. It also turns out to be highly consistent, rich and diversified, and thus adds an important facet to the traditional image of the delayed reception of Grassmann within mathematics.[2]

This line of reception unfolds within fields that are either newly emerging as autonomous disciplines (such as psychology or, more specifically, physiological psychology) or that have to adapt to a thoroughly changed layout of the system of scientific disciplines. The latter is the case for philosophy in particular. Philosophy had to define its task and scope anew, given its loss of control over other disciplines – most prominently, around 1900, over psychology, but the same holds for all disciplines that traditionally belonged to the philosophical faculty. This required, and stimulated, a general reflection on the forms of relations between different sciences. Ideas typical for Grassmann's mathematics, in particular its search for ultimate forms of generality or its discovery of new types of mathematical objects, are employed in this process in order to delineate and legitimate new disciplines. This may explain why there was such an intensive reception of Grassmann's mathematics particularly in those fields that were actively seeking to define their own scientific status (this context is treated extensively in [Ziche 2008]).

Grassmann in psychology and physiology

One of the earliest, and one of the most thoroughgoing – "to a large extent in the form of an almost literal transcription" [Preyer 1877, 24] – projects to adopt Grassmannian ideas in the context of empirical psychology and physiology was carried out by the

[1] On Helmholtz see, in this context [Darrigol 2003; Turner 1987; Ziche 2008, Chap. VI].

[2] See e.g., [Lewis 2004].

Jena physiologist Wilhelm Thiery Preyer[3] (1841–1897). Preyer referred to Grassmann as early as 1876, and thus almost coinciding in time with the "early Grassmannians" within mathematics named by [Lewis 2004], namely Hankel in 1867 and Schlegel in 1872. Significantly, this reference to Grassmann did not yet result in empirical research, but treated Grassmann as a source of inspiration for a general reflection on science. Grassmann, characterized as "one of the most exact and acute thinkers of our times" [Preyer 1876a, 43 ff.], is credited with a deep insight into the methodology of science: Preyer characterizes Grassmann's method as a willingness to accept facts that is based "on objectively sufficient reasons that, however, are not subjectively sufficient" [Preyer 1876a, 18]. Grassmann's mathematics is thus viewed as the paradigm example of a perfectly acceptable, absolutely pure science that can yield results, in a methodologically sound way, without having to rely on subjective conviction.

The ideals of purity and of an elementary approach continue to stand central in Preyer's engagement with Grassmann's mathematics. In 1877 he publishes his *Elemente der reinen Empfindungslehre* [Preyer 1877]. The title itself is ambiguous, as an attempt at translation clearly reveals. It could be rendered either as "Elements of a pure theory of sensations" or, equally correct, as "Elements of a theory of pure sensations". It is ambiguous in a way that shows that the typical objects of Preyer's theory and the very form of this theory are inherently linked. Preyer dedicates this work to Grassmann and had already communicated it to Grassmann in manuscript form. Grassmann himself contributes an appendix, his "Bemerkungen zur Theorie der Farbenempfindungen" that Preyer publishes, as the title states, with Grassmann's permission from a communication in a letter [H. Grassmann 1877]. Grassmann, in turn, expresses his appreciation of Preyer's work in the introduction to the second edition of the *Ausdehnungslehre*.[4] Again, he emphasizes the tendency towards increasing generality as the central achievement of Preyer: Preyer's work shows that the principles of the *Ausdehnungslehre* have applications beyond the realm of spatial structures. Grassmann and Preyer also met in person, and Engel states in his biography of Grassmann that the initiative to reprint the first version of the *Ausdehnungslehre* came, among others, from Preyer [Engel 1911, 330–333].

Although the *Elemente der reinen Empfindungslehre* goes into all sorts of empirical and technical details, it should no longer come as a surprise that the ideals of purity and of an absolutely general approach feature centrally in Preyer's theory. He promises to gain new insight into the very essence of sensation: "a new, and important, insight into the essence of sensation has been gained which one cannot but view with satisfaction, because all of a sudden a field becomes subjugated to laws that had so far obstinately

[3] On Preyer see [Preyer 1989, esp. 23–33; Ziche 2002, 2008].

[4] See also [Zaddach 1994].

resisted a treatment in terms of laws; a field, furthermore, that comprises the most certain and fundamental fact that there is, namely the fact of sensation." [Preyer 1877, 82] This amounts, as he self-consciously claims, to introducing a "new discipline" [Preyer 1877, 3] that he names "Psychomathematik", quite obviously as an alternative to Fechner's "Psychophysik". Psychomathematics is a "pure" and abstract discipline, in analogy with the transition from geometry to a "*pure theory of space*" or from algebra to a "*pure theory of time*" as envisaged by William Rowan Hamilton. Algebra as a pure doctrine of time abstracts from space and from the sensations, and in a completely analogous manner a pure theory of sensation can abstract from space and time in order to keep nothing but the perceptions in their absolute purity.

Mathematically, such a discipline faces the problem of how to combine quantities of different forms and of different dimensionality. A sensation is conceived as a two-dimensional entity, determined through a particular quality that can change on a one-dimensional scale, and through degrees of the intensity of this quality. Space and time add four additional dimensions so that a psychomathematical theory has to handle six-dimensional quantities. There are different forms of combining these various aspects of a sensation, and, again in analogy with Grassmannian models of combining mathematical entities, the different types of combination lead to different forms of results: adding two intensities in the same qualitative range leads to a new degree of intensity within the same range, adding up two qualities gives a new type of quality, and so forth. This leads to a mathematical calculus that also applies to a type of object that seems to resist quantification, namely (secondary) qualities. At the same time, this form of mathematization links qualities to well-defined processes of abstraction that can themselves be characterized mathematically.

A particularly telling example, and one that draws on yet another feature of Grassmann's mathematics, namely his interest in (generalized) spatial structures, is given in the localization of sensations. Put simply a sensation "red" is ascribed to a particular spot in my visual field, and this process lies at the basis of the perception of objects. Localizing a sensation means to add a determination in space to a sensation that has been, before that determination, characterized in purely qualitative terms. This operation can be inverted; abstraction from a sensation's being determined in space and time leaves over the pure sensation itself.

The problem of perceiving spatial structures, or of ordering impressions in space, remains one of the paradigms for the application of Grassmannian mathematics in physiology and psychology. R. H. Lotze's theory of "Lokalzeichen", local signs attached to perceptions and indicating their spatial ordering, works this idea out in the context of physiological and psychological, but also of physical and philosophical problems [Lotze 1852, 328 ff.]. It is given even more detail in the physiological psychology of Wilhelm

Wundt. Perceiving the basic spatial structure of the world requires a mapping of three-dimensional Euclidean space (if we take this type of geometry, in first approximation, as valid within the visually accessible external world) onto two curved, two-dimensional surfaces, the retinas of our two eyes. Further information comes from the nerves innervating the muscles that move our eyes and that deform our optical lens in order to bring the image into sharp focus. This information allows us to calculate the relative positions of the eyes and is indispensable for obtaining distance measurements. Such a multi-dimensional and non-Euclidean mapping can, according to Wundt, only be handled with the resources of Grassmann's mathematics [Wundt 1910, vol. 2, 727; Ziche 2008, Chap. VI].

Similar structures arise within the field of colour perception. Grassmann's contribution to the problem of a mathematical and psychological description of the different ways of combining different hues so that they cannot be distinguished perceptually is still treated prominently in current-day textbooks and research articles.[5] The purpose behind these modern versions of a Grassmannian colour theory remains to give a completely general, mathematical and axiomatized account of the perception of mixed colours.

Basic structures and operations: relations, order and abstraction

The interest in Grassmann is, within as well as outside mathematics, kindled by the urge to arrive at more and more general forms of science, and, directly related to this goal, to envisage more and more general types of objects and/or structures within these increasingly general sciences. As late as 1940, Ernst Cassirer asks what the ultimate basic phenomena or "pure forms" are, and he still takes psychology and epistemology as the most pertinent fields, referring back to Külpe [Cassirer 1995, 111–195][6]. Külpe himself strove, together with his co-workers within psychology, to introduce new types of objects into experimental psychology (such as complex, irreducible and imageless "thoughts"; see the texts in [Ziche 1999]), and with this research inspired the early Gestalt psychologists. Another node of intersection of all the fields mentioned is to be found in the texts of Rudolf Carnap, one of the founding fathers of logical empiricism: Carnap discusses the perception of space and of colour, has a strong focus on relations

[5] See for example [Krantz 1975; Pashler/Yantis 2002; Irtel 1987; Mausfeld 1998]. An interesting historical parallel is [Boring 1941].

[6] On Külpe cf. [Cassirer 1995, 141], on "pure forms" e.g., [ibid., 195].

and structures and on transforming philosophy into a scientific endeavour, and, last but not least, can be related to Cassirer's Grassmannian ideas [Richardson 1992].

The common denominator of all these enterprises can be found in the search for fundamental structures or fundamental objects. That they are no longer looking for objects in the traditional sense is of crucial importance for all the authors mentioned, and absolutely necessary for understanding their mission. Most importantly, they deny that the complex entities dealt with in psychology, physiology or philosophy might be built from simple, elementary objects. The focus on strict and pure science therefore comes, around 1900, coupled with an anti-atomistic attitude. Functions (as in Cassirer's *Substanzbegriff und Funktionsbegriff*), relations (as in all strands of modern logic) or "Ordnung/order" are the most common candidates for a non-atomistic foundation of all talk about objects.

A telling example that shows how far the relevant contexts extended is to be found in Karl Bühler's developmental psychology. Bühler, a pupil of Külpe and, in his child psychology, deeply indebted to the work of Preyer in this field (but also referring to Lotze or C. Stumpf, [Bühler 1930, preface]), explicitly aimed at a completely new form of elementarism. This he finds in relations or, alternatively, in the order that holds between the "Bestandstücke" (he obviously refrains from using the term "element"!) of experience: "The logical order of a well thought-through world-picture that adults have acquired finally rests upon the *perception of relations* between the large, small and smallest building blocks *(Bestandstücke)* of experience." [Bühler 1930, 184]. Cassirer links the concept of "order" to the transition from thinking in terms of substances to the adoption of "laws" or "functions" as the basic objects of scientific research [Cassirer 1907, 6]. Modern mathematics, in "the modern theory of manifolds, and the theory of groups" [Cassirer 1907, 8], has made it possible to give an absolutely general definition of mathematics that employs precisely those terms and theories. Following G. Itelson, Cassirer defines mathematics as the "science of ordered objects" [Cassirer 1907, 8], where "order" can be explained as "all possible qualitative relations, in so far as they are to be grasped according to a strict conceptual rule." "Order", based on the logic of relations, can accordingly also be used as the single most fundamental concept in science. Referring back to research of the type of Preyer's psychomathematics, one finds, as Cassirer notes as a rather surprising fact, that the pure and abstract concept of number as based on this notion of order extends "its dominion into the field that is normally ascribed to just the sensory intuition" [Cassirer 1907, 19]. In many contexts, Cassirer links this argument to theories in philosophy, physics and psychology (and, again and again, he mentions Külpe, e.g., [Cassirer 1913, 51–59], in the context of a theory of "order"; in [Cassirer 1927], he refers to M. Schlick and therefore to the immediate context of Carnap's work in these years).

This concept of "order" is linked to innovative areas within mathematics such as the theory of extensions or the theory of groups. But it acquires fundamental relevance in many fields besides mathematics. The theory of local signs can be understood as the solution to the problem of how to order the visual field spatially; concept formation – one of the most prominent topics in philosophy in these years – presupposes some kind of ordering of data; and entire systems of philosophy have been built on this concept, most of them, remarkably enough, with a shared interest in a revival and a thorough reshaping of a traditional philosophical discipline, namely the philosophy of nature. The best-known protagonist of a theory of order in this sense is Hans Driesch (who made "Ordnungslehre" the very title of his philosophy), but authors as diverse as Wilhelm Ostwald, Alfred North Whitehead or Bertrand Russell all gave the concept of "order" a central place in their philosophical projects (on Ostwald and the philosophy of nature in this context, see [Ziche 2009]).

Generality is increased via abstraction, and Grassmann could again serve as a model. His transition to more and more general types of combinations of quantities that at the same time preserve the essential aspects of those forms of combination that we are acquainted with from traditional mathematics, again displays close similarities with, for example, Külpe's psychological account of abstraction. Külpe devoted a series of experiments to the problem of abstraction, and – in the spirit of the anti-atomistic character of his thought – he comes to the result that abstraction cannot be understood as just leaving aside particulars. Abstraction, as revealing the essential structures underlying the particulars, is rather determined by higher-order processes; abstraction cannot be based on sensations, but is a matter of "apperception" *(Auffassung)* [Külpe 1904], that is of the (Kantian and Wundtian) faculty for integrating data into a larger whole.

New forms of science

Külpe's interest in Grassmann had always been motivated by issues in the systematization of different forms of science, and in his paper on "Erkenntnistheorie und Naturwissenschaft" he emphasizes this interest: "Thus, there emerges an active interaction between different forms of realistic determination. Physics, physiology, and psychology support one another, as have stressed, in particular, *Helmholtz* and *Mach*." [Külpe 1910, 25]. Mathematics can serve as a model for this new interaction between the disciplines, but, as Külpe also stresses, this must not result in viewing all the other disciplines as nothing but domains of an application of mathematics; the true goal of science has to be found in all types of science alike [Külpe 1910, 12].

Given this background, it becomes understandable how a reception of Grassmann's mathematics becomes relevant at the intersection between different disciplines, and why it succeeds in motivating the genesis of new types of fields or disciplines such as a new, order-based philosophy of nature.

In the years around 1900, the field of the sciences (in the broad sense of the German "Wissenschaft") undergoes far-reaching changes. In this process, heterogeneous developments converge or cross each other: institutional controversies (e.g., the "Lehrstuhlstreit" between psychologists and philosophers from 1913; the founding of several institutes for experimental psychology); the general dynamics of disciplines that gained huge momentum in the nineteenth century with the specialization of many different fields of scientific research; the more specific predicament of philosophy that had to search for a new identity after the loss of control over all the newly autonomous fields. Mathematics occupies a particular role in these processes in that mathematics continues to be, more or less unquestioned, the prototypical model (or at least one of the models) for what a science should be.

This dynamics of disciplinary developments turns out to be closely associated with changes in the objects of the disciplines. Mathematics, physiology and psychology deal with increasingly more general types of objects. Philosophy, in its turn towards methodology (on Külpe and methodology see [Wettersten 1990]) or epistemology, starts to take scientific theories or even whole classes of sciences (e.g., in separating the humanities from the natural sciences) as its objects. Philosophy also takes up the very question as to how one can arrive at a perfectly general theory of these different types of objects. Külpe's interest in the various forms of positing objects (worked out systematically in his 1912–1923) and Alexius Meinong's "Gegenstandstheorie" are typical cases in question, as is the phenomenology of Edmund Husserl. Grassmann's role in these debates is highlighted by the fact that all of these authors refer to him explicitly.[7]

In this context, a general reflection on different types of science and on inter-scientific relations becomes inseparably linked with the discovery of new objects in science. This debate continued to stick to highly traditional ideals of what characterizes a science – such as purity or generality – and mathematics remains an ideal that could safeguard these values, in particular for the new disciplines and for philosophy as the discipline that had most to lose in the movement towards ever increasing specialization. The anti-atomistic trend underlying all these developments had a very important implication: the new level of scientificity – to coin a slightly awkward translation for a rather common German term in the period under consideration, "Wissenschaftlichkeit" – should be achieved within a strictly anti-reductive framework. Again, the theory of

[7] Cf. [Meinong 1904, 152ff.]. On Husserl's reception of Grassmann see, e.g., [Hill 2000; Hartimo 2007; Ziche 2008, Chap. VI.3].

abstraction that has been adumbrated above in rather sketchy way can support this idea: if abstraction is not considered as stripping away particular characteristics, but as revealing the very essence of a particular by preserving the essential features on a higher level, it distances itself from any tendency towards reduction.

The reception of Grassmann in all these fields fits into typical developments of nineteenth century history of science and of philosophy. He thereby becomes an influential player, and a perhaps even more influential source of inspiration, in crucially important debates of this time. It may well be that the very difficulty to place his publications unequivocally within one single line of tradition, and in particular the reluctance of academic mathematics to adopt his ideas – that had been ascribed to the much lamented "philosophical" tone and form of the first *Ausdehnungslehre* – has immensely helped to bring about such a variegated, inter- and cross-disciplinary reception of his work. There was a strong inclination to read Grassmann's work not as a mathematical calculus, nor as a particular theory within mathematics, but as a truly general theory of generalization: so general, in fact, that it inherently stimulated its extension to as many fields as possible, giving them scientific status, but without reducing them to one basic type of science.

Some philosophical influences of the *Ausdehnungslehre*

Albert C. Lewis

Grassmann as philosopher

Though Grassmann has not been claimed as a philosopher by philosophers, the introduction to the first, 1844, version of the *Ausdehnungslehre* caused him to be labelled as a "philosophical mathematician" by mathematicians and mathematical philosophers such as Edmund Husserl [Husserl 2003, 101]. But this philosophical quality was read in different ways by different readers, and often not meant as flattery. Nevertheless, he provided some degree of inspiration for several philosophers in rather fundamental ways.

What is there in Grassmann's work that could be of philosophical interest? Certainly for philosophers interested in the foundation of mathematics Grassmann's stance on axioms could be a worthy object of study. Grassmann is proud of the fact that he dispenses with axioms. We might see in this the beginning of what we would today call natural deduction, but we also have to understand just what Grassmann means by axioms: he refers to Euclid and the traditional understanding of axioms as arbitrary assumptions as opposed to what Euclid termed "common notions". "Arbitrary assumptions" are antithetical to Grassmann's approach which seeks to base mathematics on

H.-J. Petsche et al. (eds.), *From Past to Future: Graßmann's Work in Context*,
DOI 10.1007/978-3-0346-0405-5_13,

the unfolding of the properties of certain fundamental concepts, such as a succession of identical entities. The general topic of the Grassmann brothers with respect to axiomatics has been investigated in [Radu 2005].

Another area of interest might be put under the heading of constructivism. Usually a distinction is made between the mathematical constructivism of Brouwer and the intuitionists, on the one hand, and the pedagogical constructivism of a tradition spanning the time from Dewey to Vygotsky, for example. The constructivist nature of Grassmann's approach has been touched upon by Hans-Joachim Petsche [Petsche 2009, 241]. Grassmann brings forms of these two strands together. He casts pedagogy as a part of the foundation of mathematics in the sense that how mathematics is presented and learned is a part of its scientific character – they are qualities making it scientific [Lewis 2004].

Grassmann's general theory of forms has probably been the most influential aspect beyond the mathematics. It is ironic that the introduction to the *Ausdehnungslehre* of 1844 was the main barrier preventing the understanding of many early readers. The general theory of forms describes a hierarchy of operations in terms of their relationship with each other. For example, the distributive property of multiplication over addition, from the right and left, is treated independently of the elements being added or multiplied.

It is noteworthy that it is the 1844 version that most influences these philosophers around 1900 (relying most likely on the 1894 edition). It is no surprise since it is precisely the introductory part of that version alone, as opposed to the later 1862 version, that addresses the nature of mathematics and that would appeal to the philosophical mind. (On the other hand, the 1862 version has tended to appeal to mathematicians of less philosophical bent.) The significance of Grassmann's approach was that he provided an attractive way to aid in addressing a principal philosophical endeavour, namely, as Whitehead put it, "the endeavour to frame a coherent, logical, necessary system of general ideas in terms of which every element of our experience can be interpreted." [Whitehead 1978, 3].

Bertrand Russell

At this conference we have learned of significant influences on Whitehead and Husserl (from C. M. Ringel and M. Hartimo, respectively). There are several additional figures around 1900 that are worth some investigation: Russell's early exposure to Grassmann, Ernst Cassirer's reliance on the *Ausdehnungslehre*, a curious American, Paul Carus, and Friedrich Kuntze.

Most of the principal Grassmann-related works of these philosophers appeared in the period 1890 to 1910. This perhaps represented the period of Grassmann's greatest height of recognition. Not coincidentally the *Gesammelte Werke* was published 1894–1911. The *Ausdehnungslehre* of 1844 appeared in the first volume of the *Gesammelte Werke* in 1894. It was probably this edition that Bertrand Russell read, at least in part, in May 1896, about the same time as reading Stout's *Analytical Psychology*, Erdmann's *Axiome der Geometrie* and Herbart's *Synechologie* [Russell 1983, 357]. It was likely Whitehead, his professor, who introduced him to Grassmann. (Whitehead's *Universal Algebra,* with its extensive use of Grassmann's ideas, appeared in 1898.) Russell's first extended attempt to deal with the issue of mathematical continuity, an unpublished paper of June 1896, declared that his chief sources were Grassmann and Hannequin, though it is not clear what was specifically Grassmannian about it. Other such early manuscripts reveal a strong interest in the geometrical application of the *Ausdehnungslehre* and it is very likely that he considered this as a possible candidate to play the role of the most fundamental description of spatial relations. In his 1897 *Essay on the Foundations of Geometry,* however, that role was played by conventional projective geometry. In an unpublished paper of 1899 entitled the *Fundamental Ideas and Axioms of Mathematics,* one of the precursors of his 1903 *Principles of Mathematics,* Russell refers to Grassmann's calculus as the opposite of projective geometry: "sum of points, not relations of distance" [Russell 1990, 270]. This remark is based on his reading of the second, 1862 version of the *Ausdehnungslehre* showing that Russell may have read more of Grassmann or, perhaps more likely, relied on Whitehead's *Universal Algebra.*

In Russell's *Principles of Mathematics* (1903) Grassmann is mentioned only once in passing and it is clear that Grassmann never held the importance for Russell that he did for Whitehead. For Russell, the *Universal Algebra* was not very relevant to the transformation that mathematics was undergoing in the aftermath of Georg Cantor's transfinite arithmetic and the beginning of set theory. Russell's *Principles* could be regarded in some sense as a philosophical counterpart to Whitehead's *Universal Algebra.* Each of these works was intended to be the first volume in a sequence which was interrupted when the two authors joined forces to produce *Principia Mathematica,* a work whose attempt to found much of mathematics – both its deductive presentation and its objects – on logic diverged from the Grassmannian spirit. As for Whitehead, there may be a stronger philosophical influence from Grassmann than has been hitherto assumed; a promising start has been made in [Dawson 2008].

Ernst Cassirer

Cassirer is the closest among those philosophers being considered here in fully expressing the mathematical spirit of the *Ausdehnungslehre* of 1844. He refers to Whitehead and Hermann Hankel (the first prominent mathematician to publicize the *Ausdehnungslehre*), but also refers directly to the *Ausdehnungslehre* of 1844 (using the 1894 edition).

The key sentence summarizing Cassirer's thesis with respect to space and geometry in *Substance and function* (1910) echoes Whitehead's dictum on the mathematical interpretation of experience: "[The system of universal geometry] shows that the pure concept on its side is prepared and fitted for all conceivable changes in the empirical character of perceptions; the universal serial form is the means by which every order of the empirical is to be understood and logically mastered." [Cassirer 1953, 111].

Cassirer appreciated that Grassmann makes explicit the fundamental shift in the nineteenth century of viewing mathematics as dealing with relations rather than magnitudes. "The logic of mathematics as Grassmann understands it is, in fact, in a strict sense logic of the source *(Logik des Ursprungs)....* [The fundamental concept of the source] is fulfilled wherever the members of a manifold are deduced from a definite serial principle and exhaustively represented by it." [Cassirer 1953, 99].

In this connection Cassirer also cites Hermann Cohen as having developed his logic of the source on the basis of principles of the infinitesimal calculus (which Cohen introduced in 1883). Cassirer was regarded as Cohen's star disciple but he eventually attained his independence from the Marburg School of which Cohen considered himself the head. While Cassirer maintained a lifelong friendship with his former teacher, commentators have speculated that he recognized that infinitesimals were not mathematically fashionable at the time and that this was one reason for his increasing distance from Cohen's approach [Skidelsky 2008, 65]. *Substance and Function* illustrates this divergence: Grassmann is relied on far more than Cohen as a mathematical resource.

The Grassmannian notions singled out by Cassirer, together with similar ideas he drew on from mathematicians and philosophers since Grassmann, lay at the heart of his substantial philosophical contribution. Cassirer was very much a student of the history of science and he is given credit for providing the first detailed reading of the scientific revolution in terms of the mathematization of nature. His earliest works were devoted to studies of Descartes and Leibniz. But it was evidently Grassmann that crystallized the notion that the development of idealistic philosophy itself could be characterized by the appreciation of the ideal forms of mathematical structure.

Paul Carus

Carus should be counted as a philosopher influenced by Grassmann even though he is not recognized as being of the same rank as the others mentioned here. Carus spent most of his career in Illinois in the US and sought to unify religion and science through his "monistic" philosophy. Nevertheless, he deserves a special place because he is one of the very few students of Grassmann in Stettin that we know something about. Unfortunately, we know little of his early life.

Carus was born in Ilsenburg/Harz, Prussia, in 1852 and obtained a doctoral degree from Tübingen in 1876 having studied mathematics, natural sciences, philology and philosophy. (A program reflecting possible influence from Grassmann.) Censorship of his religious writings got him into trouble early in life: in 1880, teaching at a military academy in Dresden, he published a small pamphlet in which the Bible is described as a work of literature rather than one containing literal truth. Given a choice of either agreeing not to write about religious matters again or leaving, he chose to leave and went to England. By 1885 he had arrived in New York where his publication of a book on Monism caught the attention of the industrial chemist Edward Carl Hegeler of Illinois. Carus moved to La Salle, Illinois, and in 1888 took over the editorship of Hegeler's journal *The Open Court*. There Carus founded the journal *Monist*, making it one of the oldest journals devoted to philosophy.[1]

The Open Court and the *Monist* reflected the lively eclectic intellect of the editor, and were important journals in the United States and beyond, known for publishing a wide range of philosophical, scientific and religious works. More of Charles S. Peirce's works, for example, were published in the *Monist* than in any other place. Ernst Mach was a frequent contributor and maintained a correspondence with Carus for over twenty-five years. In spite of a limited circulation, the two journals included among its writers and readers some of the most significant intellectual figures of the time. Besides Mach and Peirce, they included Russell, Poincaré, Alfred Binet and Max Müller.

Carus used mathematics as his main support for the viability of his principal aim: the unification of religion and science. Arguing against unrepentant atheists, Carus, as a recent biographer puts it, "favoured an argument from mathematics that actually does not depend on nineteenth-century advances in textual analysis, biology, or mass production." [Henderson 1993, 50].

Carus felt that religion was ready to overcome its superstitions and find a basis in "the scientific world-conception." To this end, Carus made a critical survey of modern philosophy and put forth a Kantian philosophy in which "there is one law only in the

[1] Main sources on Carus are [Meyer 1962; Henderson 1993]. A portrait is provided in [Petsche, Kannenberg et al. 2009, 228].

world which in its purely formal relations is the condition of all uniformities in the world." [Hay 1956, 505]. Though Carus agreed with this idea, the Kantian separation of the thing in itself, *Ding an sich*, as unknowable was not acceptable. He saw this as a barrier to having a scientific approach to topics such as God and moral law, i.e., such topics would fall outside of rational discussion as scientifically unknowable.

This topic may appear far removed from Grassmann and there is no evidence that Carus was influenced specifically in his monism by Grassmann. The *Ausdehnungslehre* was undeniably useful, however, because it incorporated the usual geometry of space but moved beyond three dimensions to arbitrarily many dimensions. This showed, for Carus, how it is possible for humans, through rational thinking, to approach ostensibly transcendent concepts; that is, in particular, the possibility of approaching religion scientifically.

Carus wrote, echoing Whitehead's dictum, "The laws of formal thought [...] are our guide which like the thread of Ariadne safely leads us through the labyrinth of the manifold sensory experiences. It is this method, and this is the only one, which frees philosophy from mysticism [...]" [Carus 1903, 52]. Carus's biographer writes: "This concept of form [...] served Carus well; it is usually credited to the influence of his professor Grassmann, but Carus never wrote of it before encountering Hegeler, who employed the idea in his philosophical thoughts." [Henderson 1993, 200].

It is hard for me to credit Henderson's implication that Grassmann was not a significant influence on Carus. Carus had published relatively little up to the time of meeting Hegeler. Also, as Henderson's biography makes clear, Carus in his torrent of writing is apt to pick up topics and sources in an ad hoc fashion, drawing upon his reservoir of experience and reading as needed. This may make it difficult to judge continuing, long-range influences on him – though Grassmann I believe was clearly a strong influence, it may not have been lasting.

In addition, it is conceivable that Hegeler's discourse on forms led Carus back to Grassmann. Henderson makes virtually no further reference to Grassmann and ignores the writings that draw explicitly on him. For Carus Grassmann produced "the best work on the philosophical foundations of mathematics from the standpoint of a mathematician." [Carus 1903, 290]. In another writing he asserts that "Grassmann has taught us to dive down to the bottom of the problems, where we can understand the origin and the whole growth of mathematics and where they are seen in connection with the other facts of reality." [Carus 1889, 1471].

Interestingly Carus in 1903 makes only one slight reference to Grassmann though the subject deals with the possibility of multi-dimensional geometry. Riemann is cited as his main mathematical source while Grassmann is mentioned only in a footnote aside in reference to naming diagrammatic figures in a consistent fashion. Such is Carus's seemingly fickle attitude towards his former teacher, at least as expressed through his published writings. Unfortunately the archival record for the Open Court establishment at Southern Illinois University in Carbondale, Illinois, USA, seems not likely to have unpublished material by Paul Carus that would shed more light on his relationship with his Stettin teacher.

Friedrich Kuntze

Kuntze (1881–1929) in a paper for the 1908 Third International Congress for Philosophy, addressed the significance of Grassmann's *Ausdehnungslehre* for Kant's transcendental aesthetic [Kuntze 1908]. Kuntze obtained his doctorate in Freiburg with Heinrich Rickert and his habilitation in 1911 in Berlin with a work on the philosophy of Salomon Maimon which was widely and well received. He was considered for a chair at Marburg in 1917 and ranked higher than Max Wundt and Heidegger, the latter deemed too young [Gerhardt, Mehring et al. 1999, 131–132]. Wundt filled that position in the end while Kuntze continued at Berlin for the remainder of his career. In his 1908 paper Kuntze uses Grassmann as a bridge to make the transition from a Kantian transcendental philosophy to science. This is part of a larger aim of Kuntze's and his conference paper is a report on work in progress.[2]

First he puts forward an argument that a passage can be made from transcendental philosophy to free ("freien") mathematics and then, secondly, a passage from free to applied mathematics which effectively accomplishes the goal. The pure *Formenlehre* of the *Ausdehnungslehre* of 1844, as the science of extensive and intensive types of connection, is identified with free mathematics. As such it governs all conceivable mental mathematical concepts. The *Ausdehnungslehre* is recognized as the abstract foundation of the theory of space of which geometry is a particular application applied to space. Thus, following Kuntze, the *Ausdehnungslehre* is the abstract science dealing with methods of our outer intuition and hence can be said to be an intermediary between transcendental philosophy and mathematics.

The second passage, from free to applied mathematics, relies on the way in which the *Ausdehnungslehre* handles abstractions of different orders. This argument, however, is not easy to follow. It is not really intended to be a full argument in any case – Kuntze admits that in this short paper he is presenting conjectures in a less than rigorous form. In conclusion he states that, even if he has not presented a complete argument for his thesis, his immediate aim was to show that "a great but outcast book, the *Ausdehnungslehre* of 1844, is something quite extraordinary for [philosophy].... As

[2] An overview of Kuntze's work is offered in [Sveistrup 1929].

Aristotle was the first to undertake the isolation of the pure forms of understanding, so Grassmann was the first to succeed in isolating the pure forms of sensuality. … If Grassmann's contribution can be put alongside Aristotle's then why cannot his name also? Thus these pages are laid with reverence and feeling on the grave of a great figure – one who lived without recognition and fame but who expected that justice from history that had been denied him in life." [Kuntze 1908, 436–437].

In Kuntze's 1908 paper we see again a recognition cognate to that of the other philosophers mentioned: the *Ausdehnungslehre* provides the abstract science dealing with methods of our outer intuition. This appears to be the common thread of interest in this selection of thinkers.

Concluding note

In the case of the turn of the twentieth century, when the philosophers described here made use of Grassmann, experimental psychology was encroaching on some of the traditional functions of philosophy. Representatives of psychology that influenced philosophy during this time included H. von Helmholtz (1821–1894), Wilhelm Wundt (1832–1920) and Gustav Fechner (1801–1887). One of the fundamental notions, however, goes back to E. H. Weber (1795–1878) and "Weber's law" which quantified the perception of change of sense stimulus, for example in the determination of the least detectable differences in sensation of heat. One contributor to this aspect of psychology, the Jena professor W. T. Preyer (1842–1897), makes an explicit connection to the *Ausdehnungslehre*. He dedicated his [Preyer 1877] to Grassmann, "founder of the calculus of extension."[3] In his "pure theory of sensations" Preyer aimed to establish a new theory in which sensations, as the objects of the theory, need not be explained by reference to other sciences. At the same time, by giving it a mathematical foundation by way of the *Ausdehnungslehre,* he intended to give it a traditional scientific status. Grassmann thus supplied, as Kuntze expressed it, a bridge between a purely mathematical application in the theory of sensations and the abstract, ideal notion of form. The large topic of Grassmann's influence in psychology has been explored in detail by Ziche in his contribution to this volume.

The philosophical topics that were suggested in an a priori fashion at the beginning of this paper are not all evident amongst the philosophers selected here. The influence of Schleiermacher on the *Ausdehnungslehre,* today generally acknowledged, was likewise not recognized. It may be that each epoch sheds its new, unique light on this rich work.

[3] On some members of Preyer's circle see the paper by Hans-Joachim Petsche in the present volume.

Grassmann's influence on Husserl

Mirja Hartimo

"Influence"

The word "influence" could be understood to refer to a causal relationship. If so, saying that "A has influenced B" is problematic for two reasons: First, such causal influence is often impossible to demonstrate, and second, seeing somebody's views as products of different kinds of influences easily undermines his or her own intellectual efforts and makes him or her look like a mindless billiard ball. Yet, everyone works in the context of preconceived ideas, which cannot be ignored. This is most emphatically so, if we are interested in the history of ideas, and in particular, Hermann and Robert Grassmann's role in it.

For these reasons I will attempt to understand an "influence" in, roughly, Collingwood's terms of problem solving: The thinker, here Edmund Husserl, is viewed as engaged in solving a problem, here providing intuitive foundations to mathematics. The influences are views that we know that the author was aware of and was able to think of as possible answers to the problem at hand. The author in question can also be thought of as being influenced by someone more or less consciously [Hyrkkänen 2009]. I will argue that when working on the end of the *Philosophy of Arithmetic* [Husserl 1970] Grassmann's *Ausdehnungslehre* influenced Husserl in this sense.

H.-J. Petsche et al. (eds.), *From Past to Future: Graßmann's Work in Context*,
DOI 10.1007/978-3-0346-0405-5_14,

The Grassmanns and Husserl

It is well known that Husserl started his career as a mathematician: he studied for six semesters with Karl Weierstrass (1815–1897) in Berlin and wrote his dissertation on calculus of variations at the University of Vienna with Leopold Königsberger (1837–1921) in 1882. However, it is not so well known that in 1876–1878, at the age of 17–18, before his studies with Weierstrass in Berlin, Husserl studied mathematics, physics, astronomy, and philosophy at the University of Leipzig [Schuhmann 1977, 4].[1] Husserl's life-long friendship with Hermann E. Grassmann, Hermann Grassmann's son, dates back to these years in Leipzig. During the winter semester 1877/78 Husserl received the *Ausdehnungslehre* (1844 edition) from Hermann E. Grassmann as a gift [Schuhmann 1977, 6].

Later on Husserl and Hermann E. Grassmann were reunited in Halle [Schuhmann 1977, 22], where Husserl moved after he had defended his *Habilitationsschrift, On the Concept of Number,* in 1887 and where Grassmann Jr. worked as a college teacher, and from 1899 onwards as *Privatdozent*, until he moved to Gießen in 1904 [Husserl 1994d, 159, 176]. In Halle Hermann E. was a regular house guest in the Husserls' household, and for many years he spent Christmas with the Husserls [Husserl 1994e, 11, 19, 149]. In addition to the 1844 edition Husserl also owned the 1862 edition of the *Ausdehnungslehre* as well as the *Lehrbuch der Arithmetik*.

At Hermann E. Grassmann's request Robert Grassmann sent two letters to Edmund Husserl in 1895 and 1896. Husserl also owned many books by Robert Grassmann,[2] but most of them are not even cut open. Only the *Denklehre* (the first part, second half of *Gebäude des Wissens*) is completely opened; of the other volumes only the prefaces are cut open.[3] Nevertheless, Hermann Ernst and Husserl seem to have been rather good friends, and their friendship continued also after the Husserls moved to Göttingen in 1901. For example in 1909 and again in 1911 Husserl visited Hermann E. Grassmann in Gießen for several days [Schuhmann 1977, 124, 157].

[1] Schuhmann reports in the Husserl-Chronik (1977) that Husserl's teachers in Leipzig were A. Mayer, Scheibner, C. Neumann, Hankel, Bruhns, G. Wiedemann, Johann Zöllner and Wilhelm Wundt. The index suggests that the mentioned Hankel is Hermann Hankel. But the Hankel in question must be the father of Hermann, Wilhelm Gottlieb Hankel (1814–1899), because Hermann Hankel (1839–1873) died three years earlier. I am grateful for this remark to Hans-Joachim Petsche.

[2] Husserl had the following books by R. Grassmann: *Formelbuch der Denklehre*, Stettin 1890, *Die Logik und die anderen logischen Wissenschaften*, Stettin 1890, *Formelbuch der Formenlehre oder Mathematik*, Stettin 1895, and the entire *Das Gebäude des Wissens* (Stettin 1890–1895) [Husserl 1994d, 159].

[3] For this information I am grateful to Carlo Ierna at the Husserl Archives, Leuven. He also informed me that the 1844 edition of the *Ausdehnungslehre* is the only work by Hermann or Robert Grassmann, which has some annotations. Most of them are in the first chapter.

The *Index Nominum* at the Husserl Archive Köln contains 23 references to the Grassmanns.[4] The references to the research manuscripts are dated to around 1887–1892. In his letters Husserl mentioned Hermann Grassmann's views to Natorp in 1897 and 1901, and to Stumpf in 1890. The rest of the references in Husserls' correspondence indicate Husserl's friendship with Hermann E. Grassmann. Thus, Husserl appears to have been occupied by Hermann Grassmann's views mainly just after having moved to Halle in 1887, after his *Habilitationsschrift*, and around the time he was writing the *Philosophy of Arithmetic* (1891). In the *Philosophy of Arithmetic* Husserl's few references to Grassmann are critical although rather appreciative. For example, while he rejects Grassmann's definition of equality from the *Lehrbuch der Arithmetik* (1861), he still refers to him as the *brilliant* Hermann Grassmann [Husserl 1970, 96; 2003, 101]. In addition Husserl also frequently refers to Hermann Hankel who is generally credited with having revealed the importance of Grassmann [Sturm, Schröder and Sohnke 1878, 2]. Thus, we may conclude that Hermann Grassmann's views were a rather central part of the intellectual context in which Husserl worked. Indeed, a conclusive exposition of Grassmann's influence on Husserl would demand a book-length treatise. For that reason, I will here focus only on a short period in Husserl's development: namely on the influence of Hermann Grassmann's *Ausdehnungslehre* on Husserl's *Philosophy of Arithmetic* (1891).

The Weierstrassian first part of the *Philosophy of Arithmetic*

Husserl's *Philosophy of Arithmetic, Psychological and Logical Investigations* [1891, hereafter *PA*] is a dialectical work. It consists of two parts: the first part focuses on "psychological" investigations of the concepts *multiplicity*, *unity*, and *number*, insofar as they are given to us authentically and not indirectly with a mediation of symbols [Husserl 1970, 6–7; 2003, 6–7]. In the second part Husserl takes up the "logical" investigations. Much of the first part derives, according to Husserl, "almost word for word," from Husserl's *Habilitationsschrift* "On the Concept of Number: Psychological Analyses," defended and printed in 1887 but never distributed to bookstores. After 200 pages of detailed "psychological" analysis, in the beginning of the second part of the book Husserl writes: "Obviously we are not on the right path. The presupposition from which we set out at the first, as from something self-evident – namely, that each arithmetical operation is an activity with and upon actual numbers – cannot correspond to the truth." [Husserl 1970, 190; 2003, 200]. The rest of the second part then pursues a

[4] I am grateful to Dirk Fonfara of the Husserl Archives, Cologne, for listing the entries for me.

different approach. In the end of the work Husserl is rather satisfied but he later claims to have been uneasy about the work already when it was published [Husserl 1994a, 490]. The published volume of the *PA* was supposed to be followed by another volume, which, however, never appeared.

The first part of the *PA* is a continuation of Weierstrass's program of arithmetizing analysis. Weierstrass is generally considered to be the father of modern analysis because of his overall usage of $\varepsilon - \delta$ definitions with which he replaced the vague natural language expressions like "approach indefinitely," or "as little as one wishes" with expressions that only used real numbers, addition and the "smaller than" relation. In his lectures in the 1870s and 1880s Weierstrass demanded stepwise demonstrations of the basic notions of analysis, beginning with the concept of number and operations on the numbers [Dugac 1973, 64–65, 73, 78].

In the first part of the *PA* Husserl complements Weierstrass's program by analyzing the content of the concept of number and the basic operations on it by using descriptive psychology he had learned from Brentano. Husserl's aim was to describe, not to define, the way in which we commonly think about the numbers and the basic arithmetical operations. Accordingly, he criticized, for example, Frege's approach as chimerical: when we say there are four apples on the table, we do not think of a complex logical definition of number [Husserl 1970, 118–120; 2003, 124–125]. Husserl is explicit about his Weierstrassian starting point still in 1890: "With respect to the starting point and the germinal core of our developments toward the construction of a general arithmetic, we are in agreement with mathematicians that are among the most important and progressive ones of our times: above all with Weierstrass, but not less with Dedekind, Georg Cantor, and many others" [Husserl 1994a, 1], Husserl wrote in an unpublished manuscript. But he soon met insurmountable problems to which he later referred to as "deleterious skepticism" [Husserl 1970, 262; 2003, 277]. The problem was that only small numbers and very easy arithmetical calculations are directly given to us, and thus analyzable in terms of the first part of the *PA*.

Husserl then realized that: "All too hastily we allowed ourselves to be guided by the common and naïve view which does not take into account the distinction between *symbolic* and *authentic* representations of number, and which does not do justice to the fundamental fact that all number representations that we possess, beyond the first few in the number series, are *symbolic*, and *can* only be symbolic. This is a fact which totally determines the character, sense, and purpose of arithmetic." [Husserl 1970, 190; 2003, 200–201]. His next approach is then based on the distinction between authentic and symbolic concepts. "A *symbolic* or *inauthentic* representation is, as the name already indicates, a representation by means of signs. If a content is not directly given to us as that which it is, but rather only indirectly *through signs which univocally characterize it*, then we have a symbolic representation of it instead of an authentic one." [Husserl

1970, 193; 2003, 205]. The Russellian sounding distinction is again something he had learned from Brentano's lectures [ibid.].

The parallel structures of symbols and concepts

Husserl then discusses symbolic representations of multiplicities and the symbolic representations of numbers. First his discussion continues along the paths of descriptive psychology: he realizes that most of the numbers are given symbolically, thus he proceeds to describe the way in which we perceive symbols and how they represent collections. I will argue below that the problem he was struggling with was the so-called principle of permanence of formal laws. In the final Chap. 13, Husserl manages to come to a resolution, but he *departs* from the method of descriptive psychology and complements it by building a theory, which *explains* why the symbolic numbers and calculations refer to conceptual numbers and conceptual thought.

Husserl's solution to the problem of extending the number domain by means of symbolic numbers relies on the idea of one-to-one correspondence between the signs (given in so-called normal form) and concepts. The concept of calculation can be understood either as a conceptual operation or else as a sense-perceptible operation which, "utilizing the system of number signs, derives sign from sign according to fixed rules, only claiming the final result as the designation of a certain concept, the one sought." [Husserl 1970, 257; 2003, 272].Whereas the former method is "highly abstract, limited, and even with the most extensive practice, laborious," the latter is "concrete, sense-perceptible, all-inclusive, and it is, already with a modest degree of practice, convenient to work with." [Husserl 1970, 257; 2003, 272]. Therefore, the rule-governed symbolic operations guarantee that the concepts corresponding to the signs exist: "The method of sensible signs is, therefore, *the* logical method of arithmetic. … [Calculation] encompasses *any symbolic derivation of numbers from numbers which is substantially based on rule-governed operations with sense-perceptible signs.*" [Husserl 1970, 257–258; 2003, 272].

While formulating his view about how to extend from the authentically given numbers to the numbers that are only symbolically given, Husserl simultaneously develops his next step, namely, generalizing from the number domain to any possible domain. Thus, he continues that "if we take into account that the mechanism of the symbolic methodology can break completely free of the conceptual substrata of its employment, yet another concept of calculation compels our attention." [Husserl 1970, 258; 2003, 273]. Husserl's second, more general, definition of calculation is that it is "*any rule-governed mode of derivation of signs from signs within any algorithmic sign-system according to the 'laws' – or better: the conventions – for combination, separation, and transformation peculiar to that system.*" [Husserl 1970, 258; 2003, 273]. This notion does not rely on the parallel structures of numbers and numerals but represents an approach Husserl claimed to elaborate later in the second volume of the work. In it the same system of symbols can serve in several conceptual systems. Therefore, solving problems by means of calculation requires interpretation: "*Conversion of the initial thoughts into signs – calculation – and conversion of the resulting signs back into thoughts.*" [Husserl 1970, 258; 2003, 273]. Whereas what Husserl calls *arithmetica numerosa* relies on the view of two parallel structures, the calculational technique of the *arithmetica universalis* is no longer identical with arithmetical cognition and is applicable in various different domains [Husserl 1970, 259; 2003, 274].

Towards the end of the chapter Husserl discusses the problem related to the fact that the same number can be meant with various different symbolizations. He refines his view of the parallel structures to saying that the one-to-one correspondence between the signs and concepts is only between what he calls *normal numbers (Normalzahlen)* and the infinitely many numbers themselves. The normal numbers can be given in countless other systematic ways: for example, the normal number 66 can be given as $18 + 48$ or as $70 - 4$. The first basic task of arithmetic is to reduce the systematic numbers into their normative forms. "Accordingly there arises, as the *first basic task of Arithmetic, to separate all conceivable symbolic modes of formation of numbers into their distinct types, and to discover for each type the methods that are reliable and as simple as possible for carrying out that reduction.*" [Husserl 1970, 262; 2003, 277].

Husserl then relates these results to the first part of the work: even though the authentic number concepts are generally not accessible to us, we can work instead with "rigorously defined, symbolic surrogate-concepts," and operations on them. "And just as the individual symbolic number stands in for a definite authentic one, so also each symbolic operation of combination stands in for a definite (although not actually executable) authentic one." [Husserl 1970, 263; 2003, 278]. Husserl is initially happy with the result: "All of the difficulties and doubts encountered in Chap. X ... we may already at this point regard as resolved." [Husserl 1970, 272; 2003, 287–288].

The problem and the influence of Grassmann

In February 1890, while Husserl was working on these problems he sent a letter to his mentor Carl Stumpf.[5] In it Husserl claims to have finished about 200 pages of proofs of *PA*. He explains that he still has 150–200 pages left and confesses that he still has no

[5] Mainly because of the late stage of proof reading for *Philosophie der Arithmetik* indicated in the letter, Dallas Willard has suggested that it is more probable that the letter is from 1891. In the Husserl-Chronik

coherent outline for the part dealing with *arithmetica universalis*.[6] However, Husserl writes: "The results which I have obtained are striking enough. The opinion by which I was still guided in the elaboration of my *Habilitationsschrift*, to the effect that the concept of cardinal number forms the foundation of general arithmetic, soon proved to be false. (The analysis of the ordinal number already made this clear to me.) By no clever devices, by no 'inauthentic representing', can one derive negative, rational, irrational, and the various sorts of complex numbers from the concept of the cardinal number. The same is true of the ordinal concepts, of the concepts of magnitude, and so on. And these concepts themselves are not logical particularizations of the cardinal concept. The fact is that 'general arithmetic' (including analysis, theory of functions, etc.) finds application to the cardinals (in 'number theory'), as well as to the ordinals, to continuous quantities, and to *n*-dimensional *Ausgedehntheiten* (time, space, color, force (*Kraftkontinua*), etc.)." [Husserl 1994a, 13].

In another manuscript from around 1889–1890 Husserl explained: "Our initial assumption, following the great mathematicians of our time, that the [cardinal] number domain is fundamental to the general arithmetic, became disputed. Some took the concept of quantity to ground arithmetic, others directed *Strecke* ... We already realized that the arithmetic that is based on the concept of number could not allow some other possible number forms, in the authentic or conceptual sense of the term, next to it. No negative, imaginary, or irrational numbers that could be created as the level of development *(Entwicklungsstufen)* or as the forms of combination of the number concepts. The number concept permits no extension to it; what is extended and allows extension is only the *arithmetical technique*." [Husserl 1983, 42–43].[7]

the letter is likewise dated to February 1891 [Schuhmann 1977, 29]. Elsewhere [Husserl 1982, 244; 1994, 157] the letter has been dated to around February 1890. The latter date has to be the correct date since in the letter Husserl tells Stumpf that he is going to teach logic in the summer semester. From *Husserl Chronik* we learn that Husserl taught logic in the summer semester 1890 [Schuhmann 1977, 27], whereas in the summer semester 1891 Husserl lectured on ethics [ibid., 30]. Moreover, Husserl congratulates Stumpf for having finished the proofs for *Tonpsychologie*, which appeared in 1890.

[6] Husserl apologizes that he did not answer Stumpf's letter before because "I was by this date to have finished a section of my book *Philosophy of Arithmetic*, and that required every minute" [Husserl 1994a, 12]. He also says that he has now finished some 200 pages of proofs. The first part of the *Philosophy of Arithmetic* published in the Husserliana series is 193 pages long. The first chapter of the second part ends on page 221. In the first chapter of the second part Husserl discusses symbolic representing of numbers on the basis of inauthentic representing of groups of objects.

[7] Die Supposition, von der wir ausgingen, großen Mathematikern unserer Zeit folgend, daß das Anzahlengebiet das für die allgemeine Arithmetik fundamentale sei, wurde bestritten. Die einen legten den Begriff der Quantität der Arithmetik zugrunde, die anderen den der gerichteten Strecke. ... Wir erkannten schon, daß eine Arithmetik, welche die Zahlbegriffe zum Fundament hat, nicht etwa neben diesen noch andere Zahlformen, dies Wort im eigentlichen und begrifflichen Sinn genommen, zuläßt. Keine negativen, imaginären, gebrochenen Zahlen lassen sich nachweisen, die als Entwicklungsstufen oder Kombinationsformen

These quotes suggest that Husserl realized that the extended number domain is always given only indirectly by means of symbols. Thus, instead of extending the number domain, one should rather talk about extending the arithmetical technique. The realization suggested to him to detach the arithmetical technique from the conceptual domains and thus to allow the possibility of applying the arithmetical technique in any kinds of domains such as that of vectors. For Husserl this is justified by "Prinzip der Permanenz":

"*Principle of Permanence:* When, by virtue of the peculiarity of the concepts founding the algorithm, certain algorithmic operations cannot be executed in full generality without arriving at contradictory (*widerstreitende*) constructions of concepts, then one extends the algorithm so that it becomes detached from the conceptual foundation and is thus thought of as [purely] conventional. This takes place so that one adds on a trial basis every such construction to the algorithmic domain and adds a convention that the old laws remain valid also for the symbolized objects (signs), then the old laws should be executable in full generality. In any case, one has to prove the consistency of the extended algorithm. [Husserl 1982, 33].[8]

The principle was at the time much discussed by Hermann Hankel, but initially formulated by Peacock. Hankel had presented it in his *Theorie der complexen Zahlensysteme* (1867) that had been inspired by Grassmann's system of complex numbers [Tappenden 1995, 327; Monna 1973, 71].[9] Husserl had been aware of Hankel's principle at least since his Habilitation disputation: he had prepared to defend the thesis "Hankel's *Principle of the permanence of formal laws* in arithmetic is neither a *metaphysical* nor a *pedagogical* (*hodegetisches*) principle" in front of Cantor in 1887.

der Anzahlbegriffe entstehen könnten. Der Anzahlbegriff läβt keinerlei Erweiterungen zu; was erweitert wird und Erweiterung zuläβt, ist nur die *arithmetische Technik*. [Husserl 1983, 42–43].

[8] "*Prinzip der Permanenz*: Wenn vermöge der Besonderheit der einen Algorithmus begründenden Begriffe gewisse der algorithmischen Operationen nicht in voller Allgemeinheit ausführbar sind, ohne daβ man auf widerstreitende Begriffsbildungen kommt, so erweitert man den Algorithmus, nachdem man ihn von der begrifflichen Grundlage losgelöst und als einen konventionellen gedacht hat, dadurch, daβ man jede solche Bildung versuchsweise dem algorithmischen Gebiete adjungiert und die Konvention hinzufügt, daβ auch für die durch sie symbolisierten Gegenstände (Zeichen) die alten Gesetze gültig bleiben, also die alten Gesetze unbeschränkt ausführbar sein sollen. Man muβ dann in jedem Fall die Konsistenz des erweiterten Algorithmus nachweisen." [Husserl 1982, 33].

[9] Hankel formulated his "*Prinzip der Permanenz der formalen Gesetze*" as follows:
"If two forms expressed in the general symbols of universal arithmetic are equal to each other, then they will also remain equal when the symbols cease to represent simple magnitudes, and the operations also consequently have another meaning of any kind." Translation is from Bottazzini 1986, 212. In German: "Wenn zwei in allgemeinen Zeichen der 'Arithmetica universalis' ausgedrückte Formen einander gleich sind, so sollen sie einander auch gleich bleiben, wenn die Zeichen aufhören, einfache Grössen zu bezeichnen, und daher auch die Operationen einen irgend welchen anderen Inhalt bekommen." [Monna 1973, 69–70].

Husserl's principle requires that the extension is shown to be consistent. To him, this was the case if every equation between the signs can be, by using the definitions of the signs, reduced to an identity [Husserl 1982, 35]. Thus, Husserl's view presented in Chap. 13 of the *PA* is an elaborated form of Hankel's principle of permanence to justify the move to general arithmetic. It is curious that Husserl comes close to the modern notion of computation. That it was influenced by Grassmann is suggested not only by the fact that this was the time Husserl was studying Grassmann's views, but also by the way in which Husserl continues his letter to Stumpf: "Arithmeticians who – now with hesitation, and now decisively – explain numbers as signs, allow themselves to be guided merely by the study of algebraic formalisms. These mathematicians (Grassmann above all) have brought to evidence the possibility of deriving the whole algorithm of arithmetic and analysis by means of mere sign definitions ($1 + 1 = 2$, $2 + 1 = 3$, etc.; $a^*a = a^2$, $(\sqrt{a})^2 = a$, etc., all understood in the sense of mere equivalences of signs upon paper). This occasioned their identification of number and sign." [1994b, 158–159; Husserl 1994a, 14].

Husserl then proceeds to say that this is quite alright: there should be nothing to wonder at in the fact that "a system of signs and operations with signs can replace a system of concepts and operations with judgments, where the two systems run rigorously parallel" [ibid.]: "Upon this fact syllogistic and the logical calculus rest, as does the much more subtle system of ordinary arithmetic. To calculate is not to think (infer), but rather is to derive sign from sign systematically, in conformity with set rules. The sign which is obtained at the end is interpreted, and thus the desired thought results. So we have a method of deriving a judgment from given judgments, not by means of actual inferring, but rather through a rule-governed procedure in which, from the arithmetical symbols for the data, the symbol for the result is obtained in a mechanical fashion." [Husserl 1994a, 14].

Thus, Husserl's solution of reducing the symbolic representations by means of equations to their normal signs (*Normalzeichen*) is influenced by Grassmann. Admittedly, later in the letter Husserl also claims that his result formally agrees with Helmholtz, whom he had severely criticized before. Grassmann also influences Husserl in his attempts to generalize arithmetic to vectors. This means that he wanted to allow extending the arithmetic also beyond the quantitative domains (he had remarked already in 1889 that the *Ausdehnungslehre* extends outside the domain of quantity [Husserl 1983, 256]). Later in his *Logical Investigations* (1900–1901), Husserl explains, for example, how the possibilities of generalizing formal arithmetic beyond the field of quantity made him see "that quantity did not at all belong to the most universal essence of the mathematical" [Husserl 2001, 1] and thus helped him to adopt a purely abstract structural view of mathematics and logic.

Husserl, however, never came to a final resolution about the general arithmetic. He meant to discuss it in the second volume of the *PA*: "The higher level symbolic methods, quite different in nature, which constitute the essence of the general arithmetic of cardinal numbers, are reserved for the second volume, where that arithmetic will appear as one member of a whole class of arithmetics, unified in virtue of the homogeneous character of identically the same algorithm." [Husserl 2003, 300]. The second volume never appeared, but in the years to come Husserl engaged extensively in a study of different kinds of theories of geometry, formal arithmetic, and theory of manifolds. Husserl later explicitly admitted that the *Ausdehnungslehre*, together with Gauss's introduction to general complex numbers, influenced him: "During the years 1886/93 I engaged in the theories of geometry, formal arithmetic and theory of manifolds, at times with exclusive dedication. The result of these investigations was the preface of my *PA* (1891) (cf. the remark on Gauss's report on the second edition on the biquadratic residues, W. W. Part II), and also many important musings with his demonstrations. I also made, influenced by Grassmann's *Ausdehnungslehre* and Gauss's *Introduction to General Complex Numbers* (l. c.), plane as certain continuous double series, space as certain three dimensional series etc. In the general complex numbers (such as representation of $re^{\varphi i}$) I tried to find sufficient arithmetical expressions for the order relations on plane and likewise in corresponding complex numbers of higher order for the smooth manifolds of higher order. To gain complete clarity and rigor I went back to the concept of number and recognized the necessity to distinguish between cardinals, continuous magnitudes, discrete and continuous ordinals, etc. I analyzed the concept of series, particularly limited, unlimited, open and cyclical series, looking for the necessary and sufficient criteria with which one could decide whether an infinitely large manifold is linearly or cyclically ordered, etc. In all that I also consider distance and direction the basic elements of a topological manifold. One almost wants to attach more originality to direction, since it is already included in the asymmetry of certain relations (right – left, higher – lower, in general if a ρ b, b ρ' a, such that: when a ρ b not b ρ a) that are the basic building blocks of all series and orderings." [Husserl 1994c, 80–81].[10]

[10] "In der Zeit von 1886/93 habe ich mich um die Theorie der Geometrie, der formalen Arithmetik u. Mannigfaltigkeitslehre sehr viel, periodenweise mit ausschließlicher Hingabe, bemüht. Davon giebt die Vorrede meiner Philosophie der Arithmetik 1891 entfernte Kunde (cf. Den Hinweis auf Gauss' Anzeige zur 2. Abhandlung über biquadratische Reste, W.W. Bd. III), und zwar auch Kunde von manchen wichtigen Berührungen mit Ihren Ueberzeugungen. Auch ich fasste, beeinflußt durch Grassmann's Ausdehnungslehre und Gauss' Einführung der gemeinen complexen Zahlen (l. c.), die Ebene als eine gewisse stetige Doppelreiche, den Raum als eine gewisse stetige 3fache Reiche u.sw. In den gemeinen complexen Zahlen (bezw. Auch in der Darstellung $re^{\varphi\iota}$) suchte ich die adäquaten arithmetischen Ausdrücke für die Ordnungsverhältnisse der Ebene nachzuwiesen und ebenso in entsprechende complexen Zahlen höherer Ordnung die arithmetischen Ausdrücke für die ebenen Mannigfaltigkeiten höherer Ordnung. Um zu voller analytischer Klarheit u. Strenge durchzudringen, gieng ich auf den Begriff der Zahl zurück, erkannte die Notwendigkeit

Accordingly, one can find an interesting and long text by Husserl dated to around 1892/93 in which he discusses an approach to geometry on the basis of the notion of *Strecke* [Husserl 1983, 348–389].

Conclusion

Husserl's approach in the first part of the *PA* (1891) is a continuation of the Weierstrassian program of arithmetizing analysis by means of Brentano's descriptive psychology. However, it soon became clear to him that it was not possible to cover all of mathematics with this approach. Instead he came to portray arithmetic as a double structure of computable signs and conceptual thought, and the general arithmetic as the calculational technique applicable also in domains such as geometry, mechanics, and physics. His research was influenced by Grassmann's *Ausdehnungslehre*, which thus was instrumental in Husserl's transition from the Weierstrassian genetic starting point where everything was erected on the concept of number to the view of general arithmetic where the same form can be found in various domains, and eventually to his adoption of the modern structural view of mathematics.

der Scheidung zwischen Anzahlen, stetigen Größenzahlen, discreten und stetigen Ordnungszahlen u.dgl. Ich zergliederte die Begriffe Reihe, näher: begrenzte u. unbegrenzte, offene u. cyclische Reihe, ich suchte nach den notwendigen u. hinreichenden Kriterien, nach welchen von einer unendlichen Mannigfaltigkeit von Elementen entschieden werden kann, daß sie eine orthoid oder cyclisch geordnete Mannigfaltigkeit sei u.s.w. Bei all dem galten auch mir Abstand u. Richtung als die Grundelemente topoider Mannigfaltigkeiten." [Husserl 1994c, 80–81].

Ernst Abbe's reception of Grassmann in the light of Grassmann's reception of Schleiermacher

Hans-Joachim Petsche

The reception of Grassmann in Göttingen and Jena

The university towns of Göttingen and Jena were the starting points for the German reception of Grassmann's *Extension Theory*, which set in during the last 30 years of the nineteenth century.

In 1869, Grassmann's son Justus began his university studies in Göttingen and offered a copy of his father's *Extension Theory* of 1862 [A2] to the local professors Moritz Abraham Stern and Alfred Clebsch. This marked the beginning of the reception of Grassmann's work by professional mathematicians.

But not very many specialists are aware of the fact that, as early as 1861, Ernst Abbe enthusiastically studied the *Extension Theory* of 1844 [A1]. Based in Jena, Abbe pointed Grassmann out to Gottlob Frege [Dathe 1993, 2005] and, most likely, also to William Thierry Preyer, a physiologist. Preyer then introduced many psychologists to Grassmann, among them Wilhelm Wundt [Ziche 2002, 2008]. One might also surmise that Hermann Hankel learned about Grassmann from Abbe before he even read

H.-J. Petsche et al. (eds.), *From Past to Future: Graßmann's Work in Context*,
DOI 10.1007/978-3-0346-0405-5_15,

William Rowan Hamilton [Petsche 2009, 81sqq].[1] Back in Göttingen, the young Felix Klein first learned about Grassmann from Hankel.[2]

During the winter semester of 1876/1877, Abbe began to refer directly to Grassmann's new calculus in his lectures on physics [Reich 1996, 200]. Also, in the 1870s, Abbe, Frege, and Preyer met up every Sunday with like-minded friends at the house of the Snell family to discuss philosophy, the natural sciences, and politics [Auerbach 1918, 162ff.]. "In Jena, especially for the people who were in contact with Abbe," Uwe Dathe remarks, "Grassmann's insights were discussed vividly." [Dathe 1993, 42].

While the line of reception which began in Göttingen around 1870 focused on the mathematical side of Grassmann's work and the *Extension Theory* of 1862, the Jena group felt attracted to the mathematical and philosophical aspects of "the edition of 1844, which contained more philosophical thinking" [ibid., 43].

Mathematics, philosophy and experimentation: Abbe's scientific interests

What was it that made Abbe the perfect reader and propagator of Grassmann's *Extension Theory*?

Ernst Abbe was born on 23 January 1840 in the town of Eisenach. His father was a worker in the textile industry.[3] Thanks to his father's initiative, the gifted young man was given the chance to study at the Universities of Jena (1857/1859) and Göttingen (1859/1861). In Jena, Ernst Abbe studied mathematics and physics with Hermann Schaeffer and Karl Snell, philosophy with Kuno Fischer and Ernst Friedrich Apelt, while also attending lectures on history and botany. In Göttingen, his program included physics lectures by Wilhelm Eduard Weber, mathematics with Bernhard Riemann, Johann Benedict Listing, Moritz Abraham Stern, and Ernst Christian Julius Schering. Abbe's interest in philosophy had remained, and he attended courses by Rudolf Hermann Lotze. A friendship had developed between Hermann Hankel and Abbe, and both shared not only a great gift for mathematics, but also a serious interest in philosophy.

---

1 Early letters of Abbe's indicate that this may be the case [Abbe 1986].

2 See Mirja Hartimo's contribution to the present volume, which discusses the importance of the friendship between Hermann Grassmann junior and Edmund Husserl for readings of Grassmann in philosophy (Ernst Cassirer, Friedrich Kuntze, and Paul Natorp). The contribution by Albert C. Lewis to the proceedings discusses Friedrich Kuntze.

3 The following biography of Abbe is mostly based on [Abbe 1986] and [Auerbach 1918].

In 1861, with Weber as his supervisor, Abbe completed a dissertation in physics and worked as a lecturer for the Frankfurt (Main) Physics Society. Since he did not obtain a permanent position, Abbe moved to Jena in autumn of 1862 and set out on his habilitation. After completing it successfully in August 1863, he started teaching at the university. After years without a secure position, he was named professor in 1870 and received a full professorship in 1878. He would remain connected to the University of Jena until his death in 1905. Abbe's connection to the university's mechanical engineer, Carl Zeiss, would prove to be decisive for the further course of his life. Using Abbe's designs, Carl Zeiss built the instruments the physicist needed for his lectures. Zeiss encouraged Abbe to go beyond his usual teaching duties and become a scientific collaborator in the Carl Zeiss mechanical engineering company. The chemist Otto Schott, a specialist for glass, joined Zeiss and Abbe. This was the scientific input that would create the basis for a business venture in the optical industry: the small beginnings of the world-famous company Carl Zeiss Jena.

Ernst Abbe's profound knowledge of mathematics – the notes he had taken of Riemann's lectures circulated among friends and were used by Hankel and even Riemann himself [Abbe 1986, xxiii] – and his extraordinary competence in experimental physics – proven by his dissertation under Weber's supervision and his work for the Frankfurt Physics Society – were unique. Combined with an explicit interest in the philosophy of science,[4] these aspects of Abbe's personality merged into an inimitable synthesis.

Abbe's first encounter with Grassmann's *Extension Theory* of 1844

Considering Abbe's wide range of interests, it comes as no surprise that he was quite attracted to the approach of Grassmann's A1, which combined mathematics and philosophy. But how did he first come across Grassmann, and why was Abbe so intrigued by him?

Abbe's letters to his life-long friend Harald Schütz contain important clues. They include remarks concerning a man by the name of Alexander Crailsheim. Crailsheim, so it seems, acted as a catalyst in Abbe's reception of Grassmann.

In a letter dated 14 December 1861, Abbe informed his friend that he had met one Dr. Crailsheim in Frankfurt, a medical doctor who in his spare time was apparently a fervent and very competent mathematician. "What makes this man so interesting to me," Abbe continued, "is the way he approaches mathematics. He shares our goals (as I am proud to say), and he always aims to leave isolated theorems behind and

[4] Not to forget his acute sense of social responsibility.

create a complete structure in his mind instead. He is especially interested in methodological problems, and he is not afraid to ask the most fundamental epistemological questions [...]. Among other things, he has worked on a methodological structure for arithmetic, something I am very keen to learn more about. To make a long story short, I am very happy to have encountered somebody who knows a thing or two about the subject and with whom to have a civilized discussion on mathematics. [...] Clearly he is also familiar with philosophy and its developments, as his approach to mathemat. would suggest." [Abbe 1986, 154].

This quotation illustrates Crailsheim's and Abbe's mindsets very well: both were interested in a type of methodological thinking which aimed to shed light on fundamental questions and on the relationships between the individual mathematical disciplines. This point of view favored "civilized discussions on mathematics" and an interest in philosophy.

Only one week later Abbe wrote another letter to Harald Schütz[5], praising Grassmann's *Extension Theory*, which he had "discovered thanks to Crailsheim." And Abbe went on to say: "My curiosity was aroused by another publication by the same author, which I received recently and which remained quite unclear to me, even though, judging by the introduction, its content seemed quite relevant. This text referred to the other book I mentioned above, which – according to its author – deals with the same issue in a different (heuristic) way. I immediately ordered the book. Directly after that, Crailsheim gave me the book and recommended it to me, saying that it was an excellent piece of work which, unfortunately, as so often, had gone unnoticed and unrecognized. A few days ago, I had a very enjoyable look at the book, at least at the first chapters, and I share Crailsheim's view that this is an extremely important work. This is the basic point the author is trying to elaborate (if I have grasped this correctly): he aims to *remove* a certain segment from the total structure of geometry, namely insights which are a lot less specific than the theorems of geometry because these insights are not in any way linked to spatial entities. They would only *apply* to spatial entities as *special* cases, thereby creating geometry. I think this is a brilliant approach [...], which is why I am looking forward to studying the book. But the introductory chapters have already made me very happy. Here the auth. studies the relationships between the individual mathemat. disciplines and draws some conclusions which, as I must say after giving it some serious thought, seem to be right on the mark." [Abbe 1986, 159–160].

This allows us to draw the following conclusions:

1 Crailsheim encouraged Abbe to study A1.
2 The new publication Abbe had received "remained quite unclear" to him.

[5] Letter from Ernst Abbe to Harald Schütz, 21 December 1861. In: [Abbe 1986].

3 It was in this book that Abbe discovered a reference to A1, which "deals with the same issue in a different (heuristic) way."
4 The philosophical aspects of Grassmann's introduction to A1 fascinated Abbe and he believed that Grassmann's project of developing generalized extension theory which was meant to go beyond geometry was a "brilliant idea."[6]

Before making a detailed analysis of the question of "heuristic form," we will have to make some remarks on Alexander Crailsheim.

Alexander Crailsheim: Grassmann's contemporary and Abbe's inspiration

Alexander Crailsheim was born on 18 May 1806 – three years earlier than Grassmann – in Frankfurt (Main). His father was Abraham Crailsheim (1759–1840).[7] Apparently, his father was a very talented mathematician and a total autodidact.[8] He was an employee of the "Rothschild bank, a mathematician for difficult problems" [Varrentrapp 1880, 252], and worked as a private teacher of mathematics. The geographer Karl Ritter was among his pupils.[9]

Alexander Crailsheim attended a private school before, in 1820, he began his secondary-school education at the Frankfurt "Gymnasium." His graduation certificate speaks of his gift for languages and an exceptional talent for mathematics. In 1826, a year before Grassmann, Alexander Crailsheim enrolled at the University of Berlin.

He attended lectures on logic and metaphysics, physics, geography and chemistry, while also showing an intense interest in medicine. He attended Hegel's lectures on the history of philosophy. In 1830, he finished a dissertation in medicine and became a medical doctor in Frankfurt in 1831. Shortly thereafter, Alexander Crailsheim traveled to Warsaw and got involved in the Polish struggle for independence.

---

[6] Two remarks must be made here: On the one hand, it is very improbable that the book Abbe received (maybe even from Grassmann?) by the end of 1861 was Grassmann's *Textbook of Arithmetic* ("Lehrbuch der Arithmetik," LA), even though the letters seem to say so. Instead, this must have been the *Extension Theory* of 1862 (A2), which was already on the market in late 1861. Only in A2 can we find references to A1. But, on the other hand, Grassmann does not use the term "heuristic" in A1 or A2. He does use the term repeatedly in the introduction to LA. Therefore we must suspect that Crailsheim, who was attempting to find a foundation for arithmetic as well, had also discovered LA and had interpreted Grassmann's remarks on heuristics as an indication as to what A1 was about and that Crailsheim had communicated his interpretation to Abbe.

[7] The following biographical information on Crailsheim is taken mainly from [Varrentrapp 1880].

[8] It seems that he learned French just from studying Laplace.

[9] We still possess a review of a small mathematical treatise written for merchants [Anon. 1798].

Ridden by typhus, he returned to Frankfurt in November 1831 after the failure of the Polish uprising.

Apart from his work as a doctor, Crailsheim published a number of critical reviews in the "Halle'sche Jahrbücher" (beginning in 1838) and in the "Jahrbücher für wissenschaftliche Kritik" (beginning in 1843). His academic work was that of a left-wing Hegelian. In this context, Georg Varrentrapp remarked: "His academic work allowed him to display the uniqueness of his mind and education: mathematical precision, combined with a philosophical worldview and knowledge of the natural sciences. ... From the early days of his university studies until his death, his mind had always been on philosophy and mathematics, and he spent as much time as he could thinking about related problems [...]" [Varrentrapp 1880, 254].

Young Abbe literally came to the same conclusion: "Crailsheim is a great man," he told Harald Schütz in January 1862.[10] "[...] he knows much about philosophy and is very aware of current philosophical discussions, despite the fact that his entire free time is taken up by mathematics. He is a pupil of Hegel (whom he encountered in Berlin) and also – despite some reservations – an admirer – not an unconditional follower, but he certainly believes that there is historical value to Hegel's work. So here we have a *second* case in which he has proven himself to be a man who believes in serene and serious thinking and who is not caught up in lopsided convictions or scholarly systems – rather, he knows his *Kant* very well and appreciates this philosophy very much – while he also accepts Hegel [...].

It is interesting to learn how Crailsheim developed an interest in mathematics: As a medical doctor, he was aiming to solve problems of physiology, for which he was trying to find as rigorous an approach as Du Bois, Helmholtz, etc. This ambition led him to mathemat. physics, and as a consequence he discovered mathematics. [...] Now he spends as much time as his profession as a medical doctor will permit on mathematics, showing tireless dedication to these kinds of problems." [Abbe 1986, 164–165].

Summing up, we may conclude that it was Grassmann's contemporary Alexander Crailsheim, like Grassmann well-educated in mathematics and philosophy, but also on the sidelines of institutionalized scientific research, who introduced Ernst Abbe to Grassmann's *Extension Theory*. And while Crailsheim spent all of his free time searching for a deeper understanding of mathematics, Abbe had also acquired excellent mathematical, physical, and philosophical knowledge from Riemann, Weber, Apelt, Fischer, and Lotze.

Was it a coincidence that a pupil of Hegel had introduced Abbe to Grassmann's approach?

[10] Letter from Ernst Abbe to Harald Schütz, dated 1/4 January 1862. In: [Abbe 1986].

It seems obvious that *a mathematical mindset aware of Hegelian philosophy* constituted a perfect point of departure when it came to understanding Grassmann's concept of mathematics, which was of a heuristic, architectonical nature and linked to Friedrich Schleiermacher's concept of dialectics.

This thesis entails a double assumption: Namely,

a) That it was possible to understand a Schleiermacherian concept of mathematics by relying on Hegel, and
b) That Grassmann's concept of mathematics was strongly influenced by Schleiermacher.

Hegel, Schleiermacher and Robert Grassmann's opinion

Robert Grassmann, Hermann's brother, made some statements which confirm our first assumption. Robert Grassmann was prone to German nationalism, even in a scientific context. Therefore, he chose the German word "Erspähungslehre" to express what Schleiermacher called a heuristic in his *Dialectic.* Hermann also recurred to the term heuristic in his introduction to LA. Concerning the history of "Erspähungslehre," however, Robert said the following in his *Theory of Thinking* ("Denklehre"):

> "*In modern history* the concept of Erspähungslehre has received little attention; it was only in the 19th century that the concept was revived by no less than two noteworthy philosophers, namely Hegel and Schleiermacher [...]" [R. Grassmann 1890b, 511].
>
> "Hegel's formal approach is well defined; he begins with an opposition and attempts to grasp it as a unity, and he discovers unity [...] when one side of the opposition transforms into the other; he sublates the opposition. This is why he never discovers stable truths and concepts. ... Therefore, the preciseness and clarity of science has been lost" [ibid., 511].

In Schleiermacher, Robert Grassmann found the same point of departure as in Hegel. But Robert believed that Schleiermacher's approach to dialectics was more promising:

> "Schleiermacher also wanted to grasp the unity of the opposition, and his search led him to an opposition that runs crosswise.[11] If two things are supposed to form

[11] This should remind us of the schematic representation of the mathematical disciplines, which Hermann Grassmann, continuing his father's approach, developed in the introduction to the *Extension Theory* of 1844.

> a unity, then [...] each of the two must contain both sides of the opposition, not in a parallel, but in a crosswise constellation" [ibid., 512].

Robert Grassmann told his readers that both philosophers chose dialectical oppositions as their points of departure. But while in Hegel each of the opposing sides transforms into the other, Schleiermacher develops them by connecting them crosswise. In Robert's eyes, there could be no doubt that Hegel's approach to dialectics was completely flawed. Even though he granted Hegel and Schleiermacher the honor of having revived and pursued "Erspähungslehre," Robert also made one thing very clear: "Schleiermacher's great achievement is that he discovered and understood the basic idea of Erspähungslehre or theory of scientific speculation, the highest branch of logic, and introduced it to science [...]" [R. Grassmann 1890a, 82–83].

Following in Schleiermacher's footsteps and fighting Hegel, Robert spent almost half a century working on his *Edifice of Knowledge* (*Gebäude des Wissens*).[12]

Robert Grassmann's words prove that the reception of Hegel could propel the reader towards a Schleiermacherian concept of mathematics. But this should not distract us from the fact that Hermann's opinion on Schleiermacher was not just more nuanced than his brother's, but also that the two brothers interpreted very important points quite differently. While Robert limited Schleiermacher's achievements to the crosswise opposition, Hermann also took into account the dialectic of heuristics and architectonics, the dissociation of mathematics and sense perception, the synthetic and constructive understanding of mathematics, the relationship between empirics and theory, the representational and processual dimension of knowledge, and the essential complementarity of dialectics and mathematics in all knowledge (which means that – as opposed to Robert – he did not make dialectics a part of the "sciences of logic").[13]

Schleiermacher's influence on the work of Hermann Grassmann

But what about the assumption that ideas from Schleiermacher's *Dialectic* permeate Grassmann's *Extension Theory*?

This thesis has been strongly supported by Albert Lewis [Lewis 1977] and, in [Petsche 1979], with no connection to Lewis, by the author of the present article. But it seems to have lost much of its substance since Gert Schubring [Schubring 1996c] called Schleiermacher's influence on Grassmann's A1 into question, emphasizing, instead, the

---

[12] For more information on Schleiermacher's influence on the *Edifice of Knowledge*, see Ivor Grattan-Guinness: Discovering Robert Grassmann (1815–1901), in the present book.

[13] See, for more information [Petsche 2009, 221–249].

great philosophical influence of Robert Grassmann and, through this connection, of Jacob Friedrich Fries. Given the fact that [Ziche 2008, 344] also supported the view that Schleiermacher's impact on A1 is "hard to reconstruct in detail" and that Schubring categorically opposed this possibility in [Schubring 2008, 2009] as well, this assumption will require closer examination.

Firstly, we will have to note that Fries is never mentioned in the work of Hermann Grassmann or in the countless volumes of Robert Grassmann's scientific edifice. We will also have to note that Schleiermacher was the only modern philosopher Robert Grassmann truly accepted. Robert Grassmann's *Edifice of Knowledge* gives us countless affirmative references to Schleiermacher.

Therefore, it is safe to say that if Robert was an influence on his brother's philosophical views, this would have established a connection to Schleiermacher, but certainly not to Fries. Apart from these considerations, many first-hand sources point to the great influence Schleiermacher's views on the philosophy of science exerted on Hermann Grassmann:

1 In a curriculum vitae dated 17 December 1831, which Hermann Grassmann wrote for his first examination as a teacher, he stated that his university studies of theology began with a feeling of fascination for August Wilhelm Neander. "Soon however," he went on to say, "I turned more to Schleiermacher, and I was so stirred by his lectures that I would not miss a single one of them, though I had concluded my study of theology" [Petsche et al. 2009, 126]. This proves that Schleiermacher's importance did not remain confined to theology.

2 In the curriculum vitae Hermann composed for an examination in theology, dated 23 March 1833, he gave more details on the inspiration he had received from Schleiermacher and how it had gone beyond the frame of theology. As he had already stated in his first curriculum vitae, which emphasized that Schleiermacher played a dominant role even after Grassmann had completed his studies of theology, he explained: "Just as little will I claim here that I had made Schleiermacher's views completely my own (since indeed much of them I did not understand); in the mean time he still exerted such a powerful influence on my development, I have so infinitely much to thank him for spiritually [...]. Yet only in the last year did Schleiermacher attract me completely; and although by that time I was more concerned with philology, still only then did I realize what one can learn from Schleiermacher for every science, since he did not so much provide positive answers, as he made one skilled in attacking every investigation from the correct side and continuing independently, and thus to stand in a position to find the positive answer oneself" [Petsche et al. 2009, 145–146]. In this quotation, Grassmann made it very

clear that Schleiermacher had important things to say on the methodic and methodological aspects of scientific work.

3 In a biographical outline of his brother's life, which has been published only recently, Robert Grassmann remarked: "At Michaelmas 1827, [...] he entered the University of Berlin [...]. In theology it was especially Schleiermacher who attracted the young student. His ingenious dialectic method exerted the greatest influence on the young Grassmann" [ibid., 203–204]. For the first time, we learn about Schleiermacher's "ingenious dialectic method" and Hermann's fascination for it.

4 And Robert went on to say: "Among the Greek authors he was chiefly attracted by Plato, who stood so near Schleiermacher's direction, and [in 1832, during his university studies,] he read several of his Dialogues" [ibid., 205].

"In the year 1840, together with his brother Robert, he studied the recently published *Dialektik* of Schleiermacher" [ibid., 211].
"In the year 1845 Grassmann studied Schleiermacher's *Aesthetik* and connected to it his own elaborations" [ibid., 213–214].
In 1847, he studied "Schleiermacher's *Theory of the State*" [ibid., 216].

Hermann Grassmann studied Schleiermacher more closely and over a longer period of time than any other mathematician, physicist, philologist, or philosopher.

5 We have another description of Grassmann's life, written by his son Justus, recently discovered as well.[14] Justus Grassmann remarked that his father "[apparently] was a follower of Schleiermacher, since his unpublished papers contain notes and drafts referring to his lectures and books."

This remark indicates for the first time that Grassmann's unpublished papers not only dealt with mathematics and physics (a part which unfortunately has disappeared), but also that there were "notes and drafts" concerning Schleiermacher's work on what would later be considered the humanities and liberal arts.

6 In his biography of Grassmann, Victor Schlegel (1878) remarked: "The *Dialectic* [1839], by his admired master Schleiermacher, had just been published, and it attracted him strongly, briefly sending him in a different direction, which he began to work on with his brother Robert. They spent the

[14] This biographical outline was discovered in the Röhl family documents, which refer to the Grassmann family and are now deposited in the department of autographs of the Ksiaznica Pomorska in Sczcezin. It was first published on the occasion of the 2009 Grassmann conference.

following year (1841) working on a philosophical theory of language, the results of which he published in an *Outline of German Grammar* and the *Guidelines for Teaching German*." [Schlegel 1878, 10].

Erika Hültenschmidt [1996] confirmed this assessment and presented an in-depth analysis of how Schleiermacher's *Dialectic* influenced the two works mentioned above.

7 For Robert, the experience of having studied the *Dialectic* with his brother turned out to be the decisive impulse that, beginning in the winter of 1844/1845, pushed him towards a philosophical theory of thinking, which was supposed to provide the basis for all sciences. According to Robert, Schleiermacher was "the most important critic [...] we have seen in modern times." [R. Grassmann 1890a, 82].

In the present book, Ivor Grattan-Guinness even demonstrates for the first time that the conceptual structure of the more than 20 volumes of Robert Grassmann's *Edifice of Knowledge* is in fact based explicitly on Schleiermacher's dialectic. Nothing suggests that Fries might have been an influence.

In other words, we have no reason to assume that the Grassmann brothers' interest in Schleiermacher's *Dialectic* diminished in any way after 1841.

8 "According to his 'Dialectic' Berlin 1839," Robert commented on Schleiermacher, "only two academic fields can show us the idea of knowledge. Both of these fields deal with the idea of knowledge, that is to say, the mutual relationships between thinking and existence. Dialectics, which deals with the oppositions within unity, does so in the conceptual frame of the general, whereas mathematics, which only deals with equal and unequal magnitudes, does so in the conceptual frame of the particular. According to him, all true thinking is scientific thinking depending on to what extent dialectics and mathematics are a part of it (Sects. 344–346). Mathematics is closer to the empirical, dialectics closer to the speculative form. The empirical process always precedes the speculative process, contextualizing it. Schl. is completely on the mark in these theorems; but, as he remarked himself, he lacked knowledge of mathematics" [ibid., 83]. Even a superficial glance at the introduction of A1 will show that Hermann Grassmann shared this view.

9 The biography of Grassmann, written by the author of the present article [Petsche 2009, 244–246], offers information proving that the introduction to A1 is linked directly to the general concept of Schleiermacher's *Dialectic*.

10 Robert Grassmann's objection that Schleiermacher lacked knowledge of mathematics[15] can no longer be upheld today without further comment. Recent scholarship has shown that Schleiermacher studied mathematics intensely at an early point in his life. As early as 1898, in an article entitled "Schleiermacher als Mathematiker," Hans Borkowski published one exemplar of what was probably a series of letters (1791) to Count Friedrich Ferdinand Alexander zu Dohna, in which Schleiermacher stated his views concerning the relationship between arithmetic and geometry. Unfortunately, these letters have been lost.

Johannes Michael Dittmer [Dittmer 2001] presented a comprehensive analysis of Schleiermacher's knowledge of mathematics and the natural sciences, and of how his philosophy of science systematically relied on this knowledge.

Finally, Inken Mädler [Mädler 1997, 199–295] offers slightly less than a hundred pages on Schleiermacher's concept of mathematics, including previously unpublished parts of Schleiermacher's notes on geometry [ibid., 369–374]. Continuing the line of thought on which [Lewis 1977] had set out, Mädler proves that Grassmann's A1 was in complete accord with Schleiermacher's concept of dialectics [Mädler 1997, 259–271].

11 It is quite probable that Schleiermacher had already influenced Grassmann's father and his uncle. The biography and the collection of autographs and unknown documents offer more on this [Petsche 2009, 246–247; Petsche et al. 2009, 4–8].

This list of formal and literal references to Schleiermacher speaks for itself, and it most likely proves once and for all that Schleiermacher influenced Grassmann decisively.

Heuristics and architectonics in the work of Schleiermacher and the Grassmanns

Finally, we will have to make some remarks on "heuristics," something Grassmann himself felt lacking in his introduction to LA and which he promised to make up for later [Grassmann 1861, vi–vii]. Abbe, who had (initially) remained somewhat mystified by the rigorous, mathematical way of presenting the innovative ideas of A2, would

[15] Gert Schubring [Schubring 2008] renewed this objection.

certainly have shared Carl Scheibert's attitude, which in 1861 Scheibert expressed in a letter to Ludwig Wiese[16]: "But I am not at all inclined to learn more about the way mathematics has been developing recently [and of which the] textbook written by my brother-in-law, Professor Grassmann in Stettin, is a typical example. I will try to put it mildly: This way of thinking seems to turn a mind into dried fruit, [...] while I consider systematic learning, which progresses along the lines of dialectically developing concepts, to be the pedagogical approach to mathematics." [Schulze 1906, 91].[17]

Concerning such a heuristic, Schleiermacher remarked in his *Dialectic*: "Every given pattern of thinking entails the need for a different and innovative pattern of thinking, which would allow us to define whatever has remained in the dark. The first paradigm is the extensive, the second the intensive aspect of combination, and we will have to operate within the oscillating movement between the two. The first paradigm's way of proceeding, which leads us from familiar to new knowledge, we call the heuristic. The second paradigm's, which connects isolated and disparate elements of knowledge, we call the architectonic." [Schleiermacher 1839, 179–180].

Whoever studies A1 will quickly realize that Hermann Grassmann was aiming to establish exactly this kind of oscillation. But after his publication had been ignored, he probably followed his brother's strict advice and – in LA as well as in A2 – limited himself to developing the architectonics of his theory.

From this point of view, Hermann Grassmann's contribution to axiomatics, which deviated from Schleiermacher's approach, was caused by the ignorance of the mathematical audience and by his surrender to his brother's mindset.

This interpretation is supported by the fact that Robert Grassmann, who approved of Schleiermacher in all other points, vehemently rejected Schleiermacher's disapproval of axiomatics. Schleiermacher emphasized that he had "nothing to do with the procedures brought forward by those who believe that, by developing a system of propositions which supposedly contains the essence of knowledge and therefore develops it, they are engaging in the theory of knowledge or metaphysics." Robert vehemently criticized Schleiermacher for this: Schleiermacher's error, according to Robert, arose from the fact that he had known nothing about "the laws of the pure theory of thinking and of the theory of recognition" [R. Grassmann 1890e, 164–165]. These laws, of course, represented the groundwork of his *Edifice of Knowledge...*

Concerning their temperaments, fields of activity and worldviews, the Grassmann brothers interacted in a complementary way. Accordingly, they valued heuristics and architectonics in different ways.

---

[16] Letter from Carl Scheibert to Ludwig Wiese, dated 1 May 1861. In: [Schulze 1906].

[17] Previously, Scheibert had attended Grassmann's private lectures on A1.

Abbe was enthusiastic about the amalgamation of (philosophic) heuristics and (mathematical) architectonics, which he had discovered in A1. According to a letter to Harald Schütz, dated June 1862, he spent every spare minute he had studying A1.[18]

> "I must say that this is very enjoyable work. This man has certainly produced a brilliant piece of work, and I find it hard to believe that it has remained completely unknown. My satisfaction grows larger with every page I read. The more I manage to penetrate and understand this utterly unique and in fact quite un-mathematical way of looking at things, the more I realize how adequate and justified it really is. [...] But this is not merely about shedding light on the philosophical background of mathematics. Rather, the author convincingly demonstrates that there really are things to be gained. He applies the principles of his "new analysis" to a large number of actual problems – be they geometrical, mechanical or purely arithmetical – and thereby uncovers completely new *methods* everywhere, which produce new theorems in an incredibly efficient and transparent way. And even if we should already be familiar with these theorems, our procedures of generating them presuppose a long and complicated list of concepts and formulas." [Abbe 1986, 214/215]

Appendix

Like Ernst Abbe before him, Max Herzberger (1899–1982) belonged to a group of scientists who worked on optical calculations for the company Carl Zeiss Jena. After fleeing from Nazi Germany, Albert Einstein helped him find work in the United States with Eastman Kodak. In the 1960s, he became a university professor in Zurich and New Orleans. He died on 9 April 1982. As the organizers of the Grassmann Bicentennial Conference learned from Ursula Bellugi, Herzberger's daughter, he left an unfinished manuscript which bears the remarkable title: "The ideas of Hamilton and Grassmann."

Acknowledgment

I would like to thank Ivor Grattan-Guinness for his generous help and advice.

---

[18] Letter from Ernst Abbe to Harald Schütz, 21 June 1862. In: [Abbe 1986].

On the early appraisals in Russia of H. and R. Grassmann's achievements[1]

Boris Biryukov and Zinaida Kuzicheva

Starting in the first quarter of the nineteenth century, one can observe a revival of mathematical activity in Russia, which followed a relative decline caused by Leonhard Euler's death. This revival was accompanied by growth in the number and quality of scientific investigations, as well as by the appearance of good educational materials from the hands of native mathematicians. Translations into Russian of works of prominent foreign scientists were published. In addition a considerable role in the distribution of scientific knowledge was played by reviews of Russian and foreign publications and distinguished scientists' biographies which appeared in periodic publications, for example in the *Journal of the Ministry of Mass Enlightenment* which appeared regularly from 1832 until October 1917. In addition, new scientific journals were founded. Thus Viktor Viktorovitch Bobynin founded a journal *Physical–Mathematical Sciences, Past and Present.* It appeared in two parallel editions: *Scientific papers section* (four issues a year) and *Scientific news, criticism and bibliographies* (monthly). The first issue of the journal came out in 1885 and in its "Scientific papers section" a short biographical essay on H. Grassmann translated by V. V. Bobynin was published [Sturm, Schröder and

[1] The work is supported by grant RFBR 09–06–00299.

H.-J. Petsche et al. (eds.), *From Past to Future: Graßmann's Work in Context*,
DOI 10.1007/978-3-0346-0405-5_16, © Springer Basel AG 2011

Sohncke 1879]. Within a year a translation of this essay came out as a standalone edition [Bobynin 1886a]. But to begin with, we need to say a few words about the founder of this journal.

Viktor Viktorovich Bobynin (1849–1919) was descended from a noble family of modest means. He graduated from a gymnasium in Tula and then studied at Moscow University in the department of mathematics. In 1882, having defended a thesis on the topic "Mathematics in Ancient Egypt," he was employed by Moscow University as an associate professor. Here he delivered lectures on the history of mathematics and began the study of the history of mathematics and astronomy in Russia in the seventeenth century, an area which had not attracted investigators before him. He is an author of more than 550 scientific works in all. Nevertheless for 35 years he remained a mere associate professor; he was granted the title of professor only in 1917. It was his desire to introduce the Russian reader to Grassmann's works and his propaganda of the new scientific trend – mathematical logic – is most interesting for us.

Let us now return to the "essay." H. Grassmann was a mathematician with broad philosophical and philological interests. He began his mathematical works with philosophical introductions that are exceptionally hard to comprehend, whether one considers the style or the contents, and as a consequence did not receive much acclaim from contemporaries.

The article about H. Grassmann translated by Bobynin is noteworthy for introducing new facts to the Russian scientific community as well as for pointing out his original philosophic–mathematical principles. In particular, provided for this purpose were some biographical facts along with a list of Grassmann's works, not only mathematical, accompanied by short authors' commentaries. Four sections are highlighted in the list: works on physics, mathematics, philology, and other topics. A reader can perceive some outstanding results fathered by H. Grassmann in the areas of mathematics, physics, and the study of Sanskrit.

Let us briefly focus on H. Grassmann's mathematical achievements, mainly using for reference – just as did the authors of the essay translated by Bobynin – his works *Geometrical Analysis*, *Arithmetic,* and *Trigonometry*. The latter two books are both reckoned by the authors as falling into the pedagogic-didactic field. The authors point out the extraordinary wealth of their contents but do not show enough regard for the deep methodological concepts contained in *Arithmetic.* As to the virtues of the latter work the authors suggest "realization of the tendency in every case to reduce the development of algebraic theorems to the simplest, [...] to take as a basis only such definitions as are free from any arbitrary elements, completely unambiguous and determinate. To this end, Grassmann first defines the addition of a unit (as a second summand) to a number, in that he thinks of the 'fundamental sequence' derived from a unit *e*as given by the prescription that specifies for each number n the name of the next higher as

$m = n + 1$, so that $me = ne + e$; thereupon he defines the addition of any number to a given ae by reducing the addition of $(b + 1)e$ to that of be according to the stipulation $ae + (b + 1)e = (a + b)e + e$ [...] and proceeding from these two rules, using only the transition from n to $n + 1$ as a basis, Grassmann proves the associativity and commutativity of addition and of all properties of the types of reckoning of the first degree. He does the same by introducing the negative unit and for a fundamental sequence prolonged backwards."[2] [Sturm, Schröder and Sohncke 1879, 31–32].

The above entails a recursive introduction of the operations of addition and multiplication, which are defined on all integers. The authors are wrong to distinguish a recursive definition of addition and multiplication of positive and of negative integers: H. Grassmann primarily deals with integers (not only natural numbers).

Nowadays we know that it is impossible to extend this approach – in the manner in which it is used by Grassmann – to real numbers. The three German authors and V. V. Bobynin *perceived* this, but of course could not describe it. All the more should we note the validity of their comments. The description of the theory of irrational numbers by H. Grassmann "is obviously surpassed with regard to the rigor or rather the completeness of the foundations by the works of Müller, Heine, and Georg Cantor."[3] But from the authors' point of view this disadvantage is compensated by the variety of the material included in *Arithmetic*. We have yet to return to the appraisal of the methodological value of *Arithmetic*, but now let us say a few words on the evaluation of other works of Grassmann by the authors of the essay.

As the authors of the essay point out, the *Linear Extension Theory* [A1] can definitely be considered Grassmann's principal mathematical achievement. An operation of addition of points is agreed to be its starting point; it is defined by Grassmann via the addition of line segments. Any point can be produced from four other, mutually independent points. These four points, "from which the rest are derived are called units by

[2] "... es verwirklicht die Tendenz, die Entwickelung der algebraischen Sätze überall auf die *einfachsten* – bei den ganzen Zahlen z.B. auf die nur auf die Einheiten bezüglichen – Voraussetzungen zurückzuführen, zu Grunde nur solche Definitionen zu legen, die frei von jedem willkürlichen Element, vollkommen unzweideutig und bestimmt sind. *Grassmann* definirt zu dem Ende zuerst das Hinzufügen einer Einheit (als zweiten Gliedes) zu einer Zahl, indem er die aus einer Einheit e abzuleitende, Grundreihe' gegeben denkt durch eine Vorschrift, welche zu jeder Zahl n den Namen der nächst höheren $m = n + 1$ finden lehrt, so dass $me = ne + e$ [...]. Er definirt hierauf das Hinzufügen irgend einer Zahl zu einer gegebenen ae, indem er die Addition von $(b + 1)e$ zurückführt auf die von be zufolge der Festlegung, dass $ae + (b + 1)e = (a + b)e + e$ bedeuten solle, und schreitet von diesen beiden einfachen Grundlagen lediglich unter Anwendung des Schlusses von n auf $n + 1$, streng recurrirend, weiter zum Beweis der Associativität der Addition [...] sowie überhaupt des Commutationsgesetzes und sämmtlicher Eigenschaften der Rechnungsarten erster Stufe – auf für die durch Einführung einer negativen Einheit nach rückwärts fortgesetzte Grundreihe."

[3] "die hinsichtlich Strenge oder besser Vollständigkeit der Begründung allerdings durch die Arbeiten von J.H. Müller, Heine und Georg Cantor überholt ist." [Sturm, Schröder and Sohncke 1879, 32]

H. Grassmann, and those derived from them via this addition are called *extensive magnitudes* of the first order. ... In this way one obtains *the principal domain* of 4th order, and points, lines and spaces form the subordinate domains of orders 1, 2, 3 respectively"[4], [Sturm, Schröder and Sohncke 1879, 7–8]. Both the German authors and Bobynin indicate that in the 1862 work H. Grassmann is considering n identity elements: "From n units $e_1, \ldots, e_n$, themselves standing in no numerical relation, is formed by multiplication and summation the extensive magnitude:

$$a = a_1e_1 + a_2e_2 + \ldots + a_ne_n.$$

The formation of a *calculus with such hypercomplex [...] numbers* is the fundamental idea on which Grassmann's research is based."[5]

Note also that in "Geometrical Analysis" Grassmann defines an inner product of linear magnitudes (namely vectors) which is nowadays called a scalar product and also generalizes it to all cases: of two flat domains, of a flat domain and a line segment, of point-like magnitudes [HGW11, 350].

He then demonstrates applications of his methods to mechanical problems. These problems do not pose any difficulty in themselves; the author's goal is different here. Thus he uses a familiar method to deduce a formula for the velocity of a point, but his form of recording it is noteworthy. Let p and g be points and let g be fixed; let $a_1, a_2, \ldots$ be constant, mutually orthogonal line segments, $T_1 T_2, \ldots$ any functions of time, and $p = g + a_1T_1 + a_2T_2 + \ldots$. Then

$$\frac{\mathrm{dp}}{\mathrm{dt}} = a_1\frac{\mathrm{dT_1}}{\mathrm{dt}} + a_2\frac{\mathrm{dT_2}}{\mathrm{dt}} + \ldots, \text{[HGW11, 360]}.$$

Given all this it is absolutely clear that just as in these formulas in other formulas and in discussions about them the *number* of line segments is not significant. Although not speaking of this directly he points out that all these deductions can be carried out *purely algebraically* without using any *geometrical intuitions.*

[4] "werden die 4 Punkte, aus denen die übrigen abgeleitet werden, Einheiten genannt und die aus ihnen durch diese Addition abgeleiteten *extensive Grössen* 1. Stufe. [...] Man erhält auf diese Weise ein *Hauptgebiet* 4. Stufe, und die Punkte, Geraden, Ebenen bilden untergeordnete Gebiete 1., 2., 3. Stufe."

[5] "Also aus n selbst in keiner Zahlbeziehung stehenden Einheiten $e_1, \ldots .e_n$, entsteht durch Vervielfachung und Summirung die extensive Grösse: $a = a_1e_1 + a_2e_2 + \ldots + a_ne_n$

Die Aufstellung des *Calcüls mit solchen hypercomplexen* [...] *Zahlen* ist der Grundgedanke, von welchem *Grassmann's* Untersuchungen ausgehen" [Sturm, Schröder and Sohncke 1879, 9]

Over the years some papers dedicated to H. Grassmann would appear in our country in encyclopedias, and works would also be written with an analysis of his mathematical works, for example, the papers of Veniamin Fedorovitch Kagan (1896–1953). Now let us turn our attention to logical works.

In the nineteenth century, mathematical logic was a new scientific discipline. One of the first Russian authors in this area was astronomer and mathematician Platon Sergejewitsch Poretzky (1846–1907) who worked in the University of Kazan. He was the first person in Russia to deliver a course of lectures on mathematical logic.

In April 1880 the first meeting of the Section of Physics and Mathematics of the Naturalist Society in the Imperial University of Kazan took place. The decision to found such a section had been made in February of the same year. During the second meeting of this section, Professor Alexander Vasilievitch Vasiliev drew the attention of those gathered to the work of George Boole, *An investigation of the laws of thought: on which are founded the mathematical theories of logic and probabilities* [Boole 1854]. During the next meeting on the 17th of May, Poretzky gave a speech entitled "Explanation of the foundations of mathematical logic in a form as clear and understandable as possible" (published in 1881) [Poretzky 1881]. According to Poretzky, the formal side of mathematical logic is given more thoroughly than in Boole's treatise in the 1877 work of E. Schröder *Der Operationskreis des Logikkalkuls* [Schröder 1877]. The lecturer bases his synopsis on the above-mentioned works of George Boole and Ernst Schröder, drawing attention to the explanation of the basic principles of the science. He introduces multiplication and addition on the set of classes; the supplement of a is denoted a_1, the universal and empty classes – 1 and 0, respectively. Concerning the inverse operations – subtraction and division – Poretzky writes: "Our conclusion will be the following: the actions of subtraction and division sometimes can be applied to our classes of cases, sometimes cannot, that is they can be applied, but it would be better if they are not applied at all. The truth is that the operations of addition and multiplication of classes are themselves to some extent self-inverse; thus subtraction can be replaced by a multiplication of a sort and division – by addition." [Poretzky 1881, 6]. So to "subtract" class b from the sum $a + b$ one only needs to multiply $a + b$ by a, because $(a + b)a = a$. To restore the multiplier a given the product ab we do not need division of class ab by b, but summation of ab with a, because $ab + a = a$. In the same manner, using the properties of the supplement and the laws of absorption, Poretzky proves the laws of De Morgan. Speaking of the equality relation, it is understood as an identity, that is "both its parts are really forms of one and the same class" [ibid., 8]. The equalities can be added together, multiplied, negated and still one will get true equalities given true initial ones, but one can neither subtract nor divide them, because these procedures are superfluous in logic.

Regrettably, as is widely known, the work of H. Grassmann found little response from his contemporaries. His brother Robert's work also remained unnoticed, although this is not absolutely correct for R. Grassmann's works in logic and the theory of values, which were developed by the two brothers together. Thus, Schröder gave high praise to R. Grassmann's "Logic" immediately upon its appearance in 1872. For us, however, it is important that Robert's logical work was quickly noticed in Russia. P. S. Poretzky refers to it among other works on mathematical logic in an 1884 research article "On methods of solving logical equalities and on the inverse method of mathematical logic," [Poretzky 1884], in which he lists the following articles with a short resume for each:

1 *G. Boole.* An investigation of the laws of thought… (1854)
2 *A. Macfarlane.* Principles of the algebra of logic. (1879)
3 *R. Grassmann.* Die Begriffslehre oder Logik (1872)
4 *E. Schröder.* Der Operationskreis des Logikkalkuls (1877)
5 *W. S. Jevons.* The principles of science. (Poretzky uses a Russian translation published in 1881.)

We should also mention the fact that Poretzky had a discussion with Schröder concerning the question: What exactly does it mean "to solve a logical equality."

Regretfully they came to no agreement.

Speaking of R. Grassmann's work, Poretzky observes that it does not even have any mention of logical equalities and their solution. This, according to Poretzky, is a significant flaw. The case is that at that time many scientists, including Poretzky, who were investigating the problems of algebraic logic were especially interested in solving logical equalities. Truth be told however in R. Grassmann's *Logic* published in 1890 [R. Grassmann 1890b] there is a paragraph entitled "Die Logischen Gleichungen und Formeln," but his treatment of this subject does not follow the tradition founded by Boole.

In 1885 the editors of the journal *Physical–Mathematical Sciences, Past and Present* published the first part of "Attempts towards a mathematical account of logic" composed by V. V. Bobynin; a standalone reprint appeared in the following year [Bobynin 1886b]. In this part works of G. Boole and R. Grassmann are laid out. The second part of "Attempts" was dedicated to Schröder's work mentioned above. It appeared in 1894.

Let us focus on how Bobynin evaluates in general the mathematical–logical concepts of R. Grassmann. Firstly, there is an overall account of R. Grassmann's *Logic*: "It shows a treatment, original and absolutely independent from the works of Boole reviewed above, of one and the same subject. Apparently Grassmann was completely unfamiliar with the work of Boole. Coinciding with him in the main goal of his effort – a mathematical expression of logical operations, – he veers away in the most important of the secondary goals. While Boole is mainly concerned with eliminating linguistic imperfections, Grassmann almost exclusively worries about the improvement of logic as a science." As for the question on the place of mathematical logic in the scientific system, Bobynin cites R. Grassmann's words on the relation of logic and mathematics: "The theory of concepts or logic forms is the second branch of the theory of forms or mathematics" [Bobynin 1886b, 30–31]. Thus, according to R. Grassmann, mathematics *precedes* logic.

R. Grassmann perceived that, excluding his brother's "Arithmetic" and also Schröder's *Lehrbuch der Arithmetik und Algebra* [Schröder 1873], "*all the other* treatments in their main parts connect with so-called proofs some doubtful conclusions, which prove nothing. Nearly all expositions of the arithmetic *(Zahlenlehre)*, notwithstanding those that have appeared lately, base their proofs on logical conclusions, whereas even today there exists not a single scientific treatment of logic, in spite of the fact that mathematics needs no applications of logical conclusions and can be postulated without any logic at all, using only propositions on unequivocal values, their equality and inequality"[6] [R. Grassmann 1895, III]. This passage is cited by Vladimir Petrovitch Sheremetyevskiy in 1898 in his "Outline of the history of mathematics." Here we need to give an explanation. In this year the first volume of a two-volume edition of Hendrik A. Lorentz's *Elements of Higher Mathematics* was published, translated, or rather revised, by Sheremetyevskiy. The revision was such that the initial volume of Lorenz's work increased twofold. The second part of the first volume called *Outlines of the History of Mathematics* fully belongs to Sheremetyevskiy. (In 1940 these "Outlines" were published as a standalone, edited by Adolph-Andrei Pavlovich Yushkevich and supplemented with his remarks [Sheremetyevskiy 1940]). The last (80th) paragraph of the "Outlines" is dedicated to a short review of the works on the foundations of mathematics wherein the above reflections of R. Grassmann on the relation between logic and mathematics are given.

Later, Russian scientists often turned to the work of the brothers Grassmann. Thus, in 1913 Alexander Vasilievitch Vasiliev organized the publication of a series of digests under the title "New ideas in mathematics"; in all ten issues appeared [Vasiliev 1913]. In the first issue of this series one can find a translation of the "Introduction" from the *Ausdehnungslehre* of H. Grassmann. We will give an evaluation of H. Grassmann's methods and creative contribution, which are given in Veniamin Fedorovitch Kagan's paper "Theoretical developments in mathematics" [Kagan 1913]. But first, a few words about the author of this paper.

---

[6] "so bieten sämmtliche andere Darstellungen der Zahlenlehre in ihren grundlegenden Abschnitten bei ihren sogenannten Beweisen die bedenklichsten Zirkelschlüsse und Trugschlüsse, welche nichts beweisen und nur geeignet sind, die Leser an unwissenschaftliches Denken zu gewöhnen und sie zu verwirren."

V. F. Kagan from 1898 had been working in the University of Novorosiysk as an associate professor. In 1908 he defended as a magisterial thesis a work "The Foundations of Geometry." He was one of the founders of "Higher Women's Courses" in Odessa; he also organized the famous publishing house "Mathesis" where he took part in the editing of a series of books. In the Soviet period V. F. Kagan moved to Moscow and from 1922 (1923 according to other sources) until the end of his life was a professor of the Metropolitan University, had become a doctor of physical–mathematical sciences and had received the title of Honorable Scientist of the RSFSR.

The long paper by V. F. Kagan contains an analysis of H. Grassmann's works, his contribution to the research of the foundations of arithmetic as well as geometry. Furthermore, he highlights a *genetic* approach of Grassmann to constructing mathematical theories, which we will discuss in more detail later. Note that Kagan's interest in the development of numbers was not arbitrary – it was connected with his investigations of the foundations of geometry. In his paper Kagan turns twice to the development of arithmetic. First, he points out in H. Grassmann's construction a part that deals with natural numbers and only then does he move on to integers; thus he goes in the direction opposite to the original. For us it is most interesting that Kagan accurately uncovers the specification of the construction of Grassmann's "Arithmetic." That is, an interpretation of numbers as words in a fixed alphabet; a generation of numbers via choosing the next number in an infinite number series; a bijection between a number and the number which follows afterwards (a property which arises from the process of building a numerical system); the pairwise difference of all number-words; a recursive procedure for defining basic operations and an inductive (meaning "perfectly inductive") procedure for proving theorems; and a systematic usage of nominal definitions. V. F. Kagan notices that Grassmann not only discovered that in the arithmetic of the natural sequence all proofs can be carried out via perfect induction, but also showed that all the basic definitions can be introduced recursively.

It was a notable feature of Kagan's narration of Grassmann's concepts that it demonstrated the logical role of all the main definitions and also expressions introduced via a "that which" operator, nowadays called defined descriptions. A basis upon which these descriptions lay Kagan calls "the principle of free notation" and explains it thus: "If we are introducing a new symbol or term which earlier had no meaning, than we can agree to understand under this symbol or term any previously defined object" [Kagan 1913, column 413]. In H. Grassmann's *Textbook of Arithmetic*, [LA], this principle is applied, for example, when the operation of subtraction is introduced.

If we look at Kagan's work as a reconstruction of Grassmann's number theory then many of Kagan's inaccuracies strike the eye. Thus, it is not true that H. Grassmann's natural number theory is a special construction. On the contrary it is included in the theory of the "fundamental sequence" and integer arithmetic. Analysis, characteristic for natural numbers, appears in H. Grassmann's construction only *after* the construction of the fundamental sequence, the integer sequence and the basic operations defined for this sequence. But Kagan presents the case as if in the work of the German mathematician the natural number theory precedes the integer theory. Truth be told however, in a footnote in the end of Sect. 20 it is noticed that "Grassmann really operates with a two-sided natural sequence unrestrictedly extended in the positive, as well as in the negative direction," yet these words leave a reader wondering where one should assign zero.

In Sect. 23 "Relative (positive and negative) numbers" V. F. Kagan again refers to Grassmann's arithmetic – this time to his integer theory, reconstructing his recursive definitions of operations. But Kagan's interpretation deviates from the original. In a given equality system that defines addition, one may distinguish equalities referring on the one hand to natural numbers and on the other to negative integers. This does not coincide with Grassmann's idea – to introduce an operation of addition for all integers; moreover a significant fact that the German mathematician proceeds from a more general structure – the fundamental sequence – remains unnoticed.

Kagan's conclusion: "Grassmann's achievement consists in the fact that he built a pure scientific arithmetic of the natural sequence and by doing that laid a foundation not only for scientific arithmetic, but *for all calculus*," [Kagan 1913, column 419], matches the real contribution neither of Hermann nor of Robert to the foundations of arithmetic. It was more than modest: It surely could not serve as a foundation for calculus, because one needs much more powerful arithmetic for that – second-degree arithmetic, that is logically speaking a theory including quantors which refer not only to objects but also to predicates. The Grassmanns were not armed with such logical theory. But still V. F. Kagan's reconstruction was a significant event for Russian historical–methodological development.

Later Boris Vladimirovitch Biryukov and Lyubov Gavrilovna Biryukova undertook a substantial philosophical–methodological analysis of the works of the brothers Grassmann. One may point in connection with this to a comprehensive work, the first part of which was published in 1982, the second – in 1984.

Hermann Grassmann's work and the Peano School

Paolo Freguglia

Introduction

I would like to present a historical reconstruction and analysis of the theoretical development which, in the context of Peano's works and of the school of Peano, led from H. Grassmann's legacy to the realization of vector calculus and the theory of homographies.

It should be remembered that Hermann Grassmann (1809–1877) published his *Ausdehnungslehre* in 1844, a work full of philosophical reflections, written in language that had little to do with the mathematical mentality. In Germany this work made no impression on the mathematical world; but in Italy Giusto Bellavitis read it and began an exchange of letters with Grassmann. It was also appreciated by Luigi Cremona.

When, in 1862, Grassmann published a second edition in which he devoted considerable space to geometrical interpretations and applications, it was no more successful than the first. Grassmann's fundamental idea is proposed in the following definition:

"5. DEFINITION. I define as an *extensive magnitude* any expression that is derived from a system of units (none of which need be the absolute unit) by numbers, and I call the numbers that belong to the units the *derivation numbers* of that magnitude; for

H.-J. Petsche et al. (eds.), *From Past to Future: Graßmann's Work in Context*,
DOI 10.1007/978-3-0346-0405-5_17,

example the polynomial

$$\alpha_1 e_1 + \alpha_2 e_2 + \dots$$

or

$$\sum \alpha e \quad \text{or} \quad \sum \alpha_i e_i$$

where α_1, $\alpha_2, \dots$ are real numbers and $e_1, e_2, \dots$ form a system of units, is an extensive magnitude, specifically the one derived from the units $e_1, e_2, \dots$ by the numbers $\alpha_1, \alpha_2, \dots$ belonging to them.

In general I will reserve the expression *magnitude* for these two species of them only. If the extensive magnitude can be derived from the *original units,* then I call that magnitude an extensive magnitude of *first order.*

Remark 1. From elementary mathematics we assume the laws of calculation for numbers, and also for the so-called "named numbers" {benannten Zahlen}, that is, for the extensive magnitudes derived from a single unit; but only for the case that that unit is an original unit." [$A2_K$, 4].

Grassmann introduces algebraic operations (addition, product) among extensive magnitudes. Actually, Grassmann proposes an abstract and general theory about the magnitudes. These are seen as the bases of the more general, and not necessarily geometrical, mathematical thought. Peano was to give to Grassmann's ideas a Euclidean interpretation which broadens the very bases of Euclidean geometry. Indeed, first he introduces oriented segment and other concepts which do not belong to Euclidean geometry. Moreover, the interpretation by Peano of the product among extensive magnitudes ("geometrical formations") has projective meanings.

When from the academic year 1885–1886 to that of 1888–1889 Giuseppe Peano (1858–1932) held the post of lecturer in "Geometrical applications of infinitesimal calculus "at the University of Turin, he was well aware of the problems regarding geometric calculus, so that, when in 1887 he published his lectures in a book entitled *Geometrical applications of infinitesimal calculus,* he had in mind Bellavitis, Möbius, Hamilton and Grassmann. In particular, in this first treatise on the subject, he gave importance to Bellavitis' manner of expression, in part because of the influence of his colleague and master Genocchi[1], who was linked to Bellavitis by friendship and respect. But it was in 1888 that Peano published the basic work on these topics: *Geometric Calculus according to H. Grassmann's Ausdehnungslehre preceded by the operations of deductive logic,* a work

[1] The Genocchi-Bellavitis letters are analyzed in G. Canepa, P. Freguglia "Alcuni aspetti della corrispondenza Giusto Bellavitis – Angelo Genocchi", *Angelo genocchi e i suoi interlocutori scientifici* (a cura di A.Conte e L.Giacardi), Deputazione subalpina di storia patria, Torino, 1991.

crucial also for the history of logic. Here he shows that he is decidedly convinced by Grassmann's approach, making reference to the 1844 edition of the *Ausdehnungslehre.*

The student of Peano who devoted himself above all to the studies of geometric calculus was Cesare Burali Forti (1861–1931); but Filiberto Castellano (1860–1919), Tommaso Boggio (1877–1963) and Mario Pieri (1860–1904) also took an interest in the subject.

Peano's geometric calculus

According to Peano and to Burali-Forti, the co-ordinates method constitutes a numerical intermediation for the study of geometrical objects and their properties, while geometric calculus proposes absoluteness and conciseness, and the approach through it is immediate and direct for the study of geometrical problems. However, this calculus does not exclude the use of co-ordinates.

In the volume of 1888 Peano presents Grassmann's ideas in an original way: as we have already said, he gives a Euclidean interpretation to the fundamental Grassmannian notions, by limiting his considerations on the nature of the system of unities, not beyond the three dimensions. At first, he introduces the notion of *geometrical formation* so:

$$\sum_{i=1}^{n} m_i \tau_i^q$$

where m_i are real numbers and τ_i^q are q – hedrons, with $1 \leq q \leq 4$, which are Peano's interpretations of Grassmann's unities. That is, we have for the system of unities the following possibilities:

1-hedron	point
2-hedron	segment
3-hedron	triangle
4-hedron	tetrahedron

With Peano, we call the geometrical formations which have as the system of unities points (and only points) of the first kind (or degree), of the second kind if the unities are segments, of the third kind if the unities are triangles and finally of the fourth kind if the unities are tetrahedrons. We will in general denote a geometrical formation by F_q, where q expresses the kind of formation considered. Even if the possibility of geometrical formations with $q \geq 5$ is not contemplated by Peano, it should be an easy and natural generalization. Veronese, who was a contemporary of Peano, should have done it. But Peano considers only a traditional vision of geometry. Between two geometrical formations we can establish the operation of algebraic addition, which complies with the rules of the algebra of polynomials. But conceptually the more important operation is the *alternated product* [Peano 1888, 110–111], which is introduced by Peano (and by [Burali-Forti 1926, 5–6]) thus:

If we have two geometrical formations in 3D, F_r and F_s, the alternated product is a product which complies with the rules of the algebra of polynomials, but without changing the order of the letters which denote points. If $r + s \leq 4$ the product is called *progressive* [Peano 1888, 30] and expresses the geometrical operation of projection. If $r + s > 4$ the product is called *regressive* [ibid., 107] and represents the geometrical operation of section. In the plane case, 2D, in the definition we must replace respectively $r + s \leq 3$ and $r + s > 3$. The alternated product is not commutative.

For instance, a segment is represented by the product AB, if A and B are two points which determine the segment. But a segment can be represented by the formation of the first kind B − A. We also have that: AB = −BA and AA = 0. Some particular progressive products, equalized to 0, have interesting geometrical interpretations:

ABCD = 0: points A, B, C, D are coplanar
ABC = 0: points A, B, C are in a straight line
AB = 0: points A and B coincide

If α is a plane and A a point, the expression:

Aα = 0 means that A lies on the plane α

If a, b, c are straight lines:

abc = 0 means that three straight lines a, b, c have a point in common.

etc.

By means of our geometric calculus, Peano and Burali-Forti are able to show theorems of projective geometry. Even Bellavitis [Bellavitis 1854, 13] proposed the applications of his equipollence calculus to elementary geometry and to projective geometry. Peano does not utilize figures. He considers the figures which we find in the traditional treatises of synthetic (elementary and) projective geometry as heuristic representations. Indeed, a figure influences the actual proof of a theorem and the solution of a problem. Instead, Peano and Burali-Forti consider only chains of identities of our calculus and subsequently they interpret the last expression geometrically. Hence, according to our mathematicians, a theorem is an interpretation, a model of an identity which concludes a sequence of geometrical calculus. Now, for instance, think of Peano's exposition of Menelaus–Ptolemy theorem [Peano 1888, 47]. To simplify, we will explain this theorem by utilizing a figure. The reader can see that the role of this figure to prove the theorem is superfluous. But first it is necessary to introduce the following lemma:

Lemma 1. *If* C *is a point which belongs to the straight line* AB, *then we can write:*

$$\mathrm{C} = x\mathrm{A} + y\mathrm{B} \tag{1}$$

where

$$x = \frac{\mathrm{CB}}{\mathrm{AB}} \quad \text{and} \quad y = \frac{\mathrm{AC}}{\mathrm{AB}}$$

The Menelaus–Ptolemy theorem says:

Theorem 2.1. *If* $\mathrm{AB} \cdot \mathrm{B'C'} = \mathrm{C'}$; $\mathrm{BC} \cdot \mathrm{B'C'} = \mathrm{A'}$; $\mathrm{AC} \cdot \mathrm{B'C'} = \mathrm{B'}$ *then*

$$\mathrm{AC'} \cdot \mathrm{BA'} \cdot \mathrm{CB'} = \mathrm{BC'} \cdot \mathrm{CA'} \cdot \mathrm{AB'}$$

that is:

$$\frac{\mathrm{AC'} \cdot \mathrm{BA'} \cdot \mathrm{CB'}}{\mathrm{BC'} \cdot \mathrm{CA'} \cdot \mathrm{AB'}} = -1 \tag{2}$$

Proof. In virtue of the previous Lemma, because $\mathrm{C'} \in \mathrm{AB}$, $\mathrm{B'} \in \mathrm{AC}$ and $\mathrm{A'} \in \mathrm{BC}$ we have:

$$\mathrm{A'} = \left(\frac{\mathrm{A'C}}{\mathrm{BC}}\right)\mathrm{B} + \left(\frac{\mathrm{BA'}}{\mathrm{BC}}\right)\mathrm{C}$$

$$\mathrm{B'} = \left(\frac{\mathrm{B'A}}{\mathrm{CA}}\right)\mathrm{C} + \left(\frac{\mathrm{CB'}}{\mathrm{CA}}\right)\mathrm{A}$$

$$\mathrm{C'} = \left(\frac{\mathrm{C'B}}{\mathrm{AB}}\right)\mathrm{A} + \left(\frac{\mathrm{AC'}}{\mathrm{AB}}\right)\mathrm{B}$$

Now if we apply the progressive product A′B′C′ we have:

$$\mathrm{A'B'C'} = \left[\left(\frac{\mathrm{A'C}}{\mathrm{BC}}\right)\left(\frac{\mathrm{B'A}}{\mathrm{CA}}\right)\left(\frac{\mathrm{C'B}}{\mathrm{AB}}\right) + \left(\frac{\mathrm{BA'}}{\mathrm{BC}}\right)\left(\frac{\mathrm{CB'}}{\mathrm{CA}}\right)\left(\frac{\mathrm{AC'}}{\mathrm{AB}}\right)\right]\mathrm{ABC} \tag{3}$$

Hence Peano says: "if ABC is different from zero, a necessary and sufficient condition so that the three points A′, B′, C′ are in a straight line (that is when $\mathrm{A'B'C'} = 0$), is to

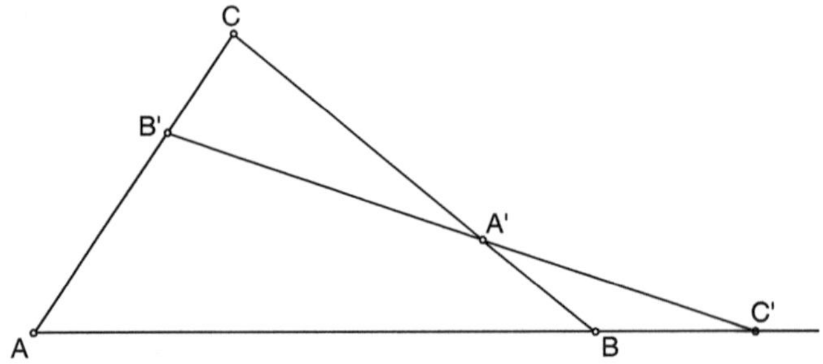

Fig. 1. Menelaus-Ptolomy's theorem graphic representation

equalize to zero the coefficient [in the square brackets of (3), that is the (2)] of ABC$''$. This proposition is equivalent to saying: if the points A$'$, B$'$, C$'$, which lie respectively on the sides BC, CA, AB of the triangle ABC, are in a straight line, then (3) is true and vice versa. In this proposition the symbol ≪.≫ denotes both the progressive product (geometrical operation of section) and the arithmetical product.
Peano also analyses Desargues' classic theorem [ibid., 92] (plane case of homological triangles) and Pascal's theorem. At this point we will examine Desargues' theorem.

Theorem 2.2. *The points* T, U, V, *where the corresponding sides* BC, B$'$C$'$; CA, C$'$A$'$; AB, A$'$B$'$ *respectively intersect, are collinear only if the straight lines* AA$'$, BB$'$, CC$'$ *intersect at the point* S.

Proof. Independently of the drawing (Fig. 2), we obtain the following steps.
We can start from the identity:

$$\mathrm{AB.CD} = (\mathrm{ABD}).\mathrm{C} - (\mathrm{ABC}).\mathrm{D} \tag{4}$$

(4) says that the product AB.CD (the result of this product is a fourth kind formation on the plane) between the segments AB and CD (respectively formations of second species) is regressive and hence it expresses an intersection on the plane, i.e., a point, which is represented by the right-hand expression of (4). We shall denote the points with capital letters and the segments with small letters. Thus (4), putting p in the place of CD, will be written as follows:

$$\mathrm{AB.p} = \mathrm{Ap.B} - \mathrm{Bp.A} \tag{5}$$

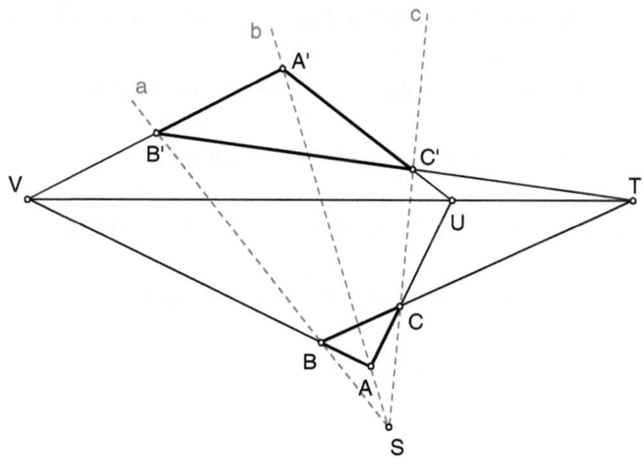

Fig. 2. Desargues' homological triangles theorem graphic representation

Now we will take into consideration the following expression:

$$(BC.a)(CA.b)(AB.c) \tag{6}$$

Applying (5)–(6) we obtain consecutively:

$$\begin{aligned}
&(Ba.C - Ca.B)(Cb.A - Ab.C)(Ac.B - Bc.A)\\
&= (Ba.Cb.CA + Ca.Ab.BC)(Ac.B - Bc.A)\\
&= Ba.Cb.Ac.CAB - Ca.Ab.Bc.ABC\\
&= Ba.Cb.Ac.ABC - Ca.Ab.Bc.ABC\\
&= (Ba.Cb.Ac - Ca.Ab.Bc)ABC
\end{aligned}$$

Hence we have the identity:

$$(BC.a)(CA.b)(AB.c) = (Ba.Cb.Ac - Ca.Ab.Bc)ABC \tag{7}$$

Dually, i.e., putting capital letters in the place of the small letters and vice versa, we obtain the true expression:

$$(bc.A)(ca.B)(ab.C) = (bA.cB.aC - cA.aB.bC)abc \tag{8}$$

Multiplying (7) by abc and (8) by ABC and adding member to member, we obtain:

$$\mathrm{abc(BC.a)(CA.b)(AB.c) + ABC(bc.A)(ca.B)(ab.C) = 0} \tag{9}$$

If in (9) we put $\mathrm{a = B'C', b = C'A', c = A'B'}$ we will have:

$$\mathrm{A'B'C'(BC.B'C')(CA.C'A')(AB.A'B' + ABC(A'A.B'B.C'C)) = 0} \tag{10}$$

Assuming that (the triangles) ABC and A′B′C′ are different from zero, (10) leads us to:

$$\mathrm{(BC.B'C')(CA.C'A')(AB.A'B') = 0}$$

and

$$\mathrm{(AA'.BB'.CC') = 0}$$

In conclusion and interpreting this, we have (see Fig. 2):

$\mathrm{(BC.B'C')(CA.C'A')(AB.A'B') = TUV = 0}$, i.e. "*the points* T, U, V, *where the corresponding sides* BC, B′C′; CA, C′A′; AB, A′B′ *respectively intersect, are collinear*", if $\mathrm{(AA'.BB'.CC') = abc = 0}$, i.e. "*the straight lines* $\mathrm{AA', BB', CC'}$ *intersect at the point* S".

While Bellavitis' calculus of equipollence is more directly connected with elementary (plane) synthetic geometry, Peano and Bellavitis use their calculus in a more compact way. For instance, Pascal's hexagon theorem [ibid., 95] is presented only through the expression (see Fig. 3):

$$\mathrm{(AB.DE)(BC.EX)(CD.AX) = 0} \tag{11}$$

and in the interpretation of (11) the proof of the theorem also remains. In fact (11) is a second degree monomial for X (number of times X appears). Because X is an unknown variable and A, B, C, D, E, X are points, then (11) is a second degree equation, i.e. a conic equation which passes through the previous points. Besides, (4) is identically equal to zero if X = A, or B, or C, or D, or E. Therefore, when X = F the six different points A, B, C, D, E, F belonging to the conics, determine a hexagon inscribed in these same conics. We can observe that: (AB.DE) = P, (BC.EF) = R and (CD.AF) = Q. In virtue of (11) we have: PQR = 0, i.e. the points P, Q, R are aligned. Hence:

Theorem 2.3. *If a hexagon* ABCDEF *is inscribed in a conic then the three intersections* P, Q, R *of opposite sides belong to the same straight line, i.e.* P, Q, R *are aligned, and vice versa.*

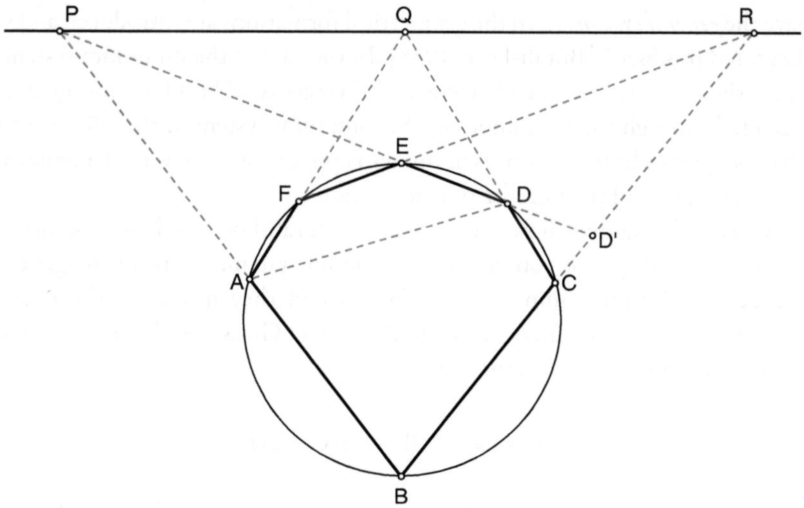

Fig. 3. Pascal's hexagon theorem graphic representation

If we write (4) in lower case, we obtain, by means of duality, Brianchon's theorem.

Hence Peano and his student propose a very synthetic geometric analysis.

Toward the minimum system

In Chap. IX (first pages) of his 1888 treatise, Peano presents the definition of the *linear system*, that is, according to our language, he establishes the notion of *vector space* (on **R**); and he also introduces the notion of *linear transformations*. Still in the context of the Grassmann–Peano trend, Burali Forti and Roberto Marcolongo developed their studies about vector calculus. Marcolongo was not, in the strict sense, a student of Peano, but with Burali-Forti he had a systematic scientific collaboration. We must consider the following important works of these mathematicians:

- *Elementi di calcolo vettoriale con numerose applicazioni alla geometria, alla meccanica e alla fisica-matematica* (1st edition 1909, second enriched ed. 1921)
- *Analyse vectorielle générale, I. Transformations linéaires* (ed. 1912), *II. Applications à la Mécanique et à la Physique* (ed. 1913)

In the 1909 book they present vector calculus as a structure of vector space with the operations of scalar product and vector product. This is the *minimum system*, while "we have the *general system* when the geometrical formations are introduced and we use the alternated product." [Burali-Forti 1938]. In the case of the minimum system, they introduce the notions of gradient, rotor and divergence. The Hamilton quaternions are presented through the techniques of the minimum system in the edition of 1920. The sixth chapter of both editions is devoted to applications to electrodynamics and to Maxwell equations and the Lorentz transformations.

Now we will examine how, according to [Burali-Forti and Marcolongo 1909, 27–45], the general system contains the minimum system, i.e. from the general system we deduce the minimum system. To this end they introduce the notions of bi-vector and of tri-vector (inherited from Peano and Grassmann). The first is seen as an alternated product of two vectors, so:

$$\hat{u} = \vec{v}\vec{w} = (\mathrm{B} - \mathrm{A})(\mathrm{C} - \mathrm{A})$$

A bi-vector is a geometrical object determined by three elements: a modulus, a plane position and an orientation. mod $\hat{u}$ is the area of the parallelogram determined by $\vec{v}$ and $\vec{w}$. These two vectors also determine a plane position and the direction depends on the property $\vec{u}\vec{w} = -\vec{w}\vec{u}$. We define *index of a bi-vector* $\hat{u}$, and we denote it by $|\hat{u}$, a vector such that:

1. $\mathrm{mod}\,|\hat{u} = \mathrm{mod}\,\hat{u}$
2. If $\hat{u} \neq 0$ the direction of $|\hat{u}$ is normal to the plane position of $\hat{u}$.
3. The orientation of $|\hat{u}$ is such that the number $\mathrm{O}\hat{u}(|\hat{u})$ is either positive or zero.

If $\vec{v}$ is a vector, we define *index of a vector* $\vec{v}$, and we will denote it thus: $|\vec{v}$, the bi-vector $\hat{u}$ which has $\vec{v}$ as index.

A tri-vector is, in its turn, a alternated product of three vectors. If we consider three vectors:

$\vec{u}$, $\vec{v}$, $\vec{w}$, we can show that the product $\mathrm{O}\vec{u}\vec{v}\vec{w}$ is a real number (which represents the affine volume of the oriented parallelepiped constructed through the vectors $\vec{u}$, $\vec{v}$, $\vec{w}$). In particular we put $\Omega = \mathbf{Oijk}$, where **i**, **j**, **k** determine the usual unitary orthogonal right system. Ω is called unitary tri-vector.

Hence we can establish the following definitions:

Definition 3.1. *If* $\vec{u}$ and $\vec{v}$ *are two vectors, we will call scalar product (or inner product) the operation:* $\vec{u} \times \vec{v} = \frac{\vec{u}(|\vec{v})}{\Omega}$

We can show that: $\vec{u} \times \vec{v} = \mathrm{mod}\vec{u} \cdot \mathrm{mod}\vec{v} \cdot \cos(\vec{u}, \vec{v})$

Definition 3.2. *If $\vec{u}$ and] $\vec{v}$ are two vectors, we will call vector product (or outer product) the operation:*

$$\vec{u} \wedge \vec{v} = \big| \, (\vec{u}\vec{v})$$

So, from the general system of geometrical calculus we have obtained the minimum system.

The theory of homographies is well presented and analyzed in the treatise *Analyse vectorielle générale*. Burali-Forti and Marcolongo define a (vector) homography as a linear operator which transforms vectors into vectors.

Certainly, also from a foundational point of view, the Grassmann–Peano statement shows its importance. But nevertheless, as we have seen, the goal of geometrical calculus, or vector calculus or homographies goes beyond a "philosophical" justification of the bases of geometry. In fact the applications to mathematical physics and to the geometry of transformations is very important and, notwithstanding the matrix techniques, even today proposable. Of course, suitable adaptations could be made. Analogously, *mutatis mutandis*, as the reappraisal of the quaternion techniques has been made. In the Preface (to *Analyse vectorielle générale*) [Burali-Forti and Marcolongo 1912, IX, Footnote 4] again explain that their calculus is not, in itself, seen as tachygraphical in comparison with the concepts of calculus which are introduced by means of Cartesian co-ordinates. In fact, geometrical synthetic calculus is, according to our mathematicians, "intrinsic, or absolute, or autonomous". In our opinion what is important is its absoluteness, because this calculus presents concepts and procedures which disregard the particular Cartesian co-ordinate system. We may say that we have an "intrinsic absoluteness".

Conclusion

Peano's geometrical calculus theory is a general theory which is of intrinsic mathematical interest and which is also applied to mechanics and to physics. Peano's contributions, which come from an elaboration of Grassmann's ideas, consist of a Euclidean interpretation of relative concepts. Moreover, in this context, Peano demonstrates important fundamental theorems of projective geometry. For this reason, Peano's geometrical calculus has an implicit foundational interest. In our opinion, the protophysical role of Euclidean geometry in Peano's works is essential and decisive. It is true – of course – that he distinguishes position geometry from Euclidean geometry, but this is not negative; on the contrary, from a theoretical point of view, it is appropriate.

Did Gibbs influence Peano's "*Calcolo geometrico secondo l'Ausdehnungslehre di H. Grassmann . . .*"?

Filippo Demonte-Barbera

Introduction

Last year I suggested to Aragno, an Italian publisher, that a new edition of "*Calcolo geometrico secondo l'Ausdehnungslehre di H. Grassmann preceduto dalle operazioni della logica deduttiva*" [Peano 1888] by Giuseppe Peano (1858–1932) be published. He accepted my proposal [Peano 2010]. This book had never been previously reprinted until 2000, when an English translation by Lloyd C. Kannenberg appeared [Peano 2000]. In 2002, Clara Silvia Roero published a CD of the *Opera Omnia* of Peano [Roero 2002] containing a copy of the original edition.

The English translation and Kannenberg's previous translations of Grassmann's two editions of the *Ausdehnungslehre* [$A1_K$; $A2_K$] signify a reawakening of interest in this subject, together with the developments and applications of Geometric Algebra (by David Hestenes and Sobczyk 1984) and of Invariant Theory (by Gian-Carlo Rota and his school [Barnabei, Brini and Rota 1985]).

While I was researching the existing literature on *Calcolo geometrico etc.*, I came across a review published in the online *Zentralblatt MATH* by Roger Cooke [Cooke

H.-J. Petsche et al. (eds.), *From Past to Future: Graßmann's Work in Context*,
DOI 10.1007/978-3-0346-0405-5_18, © Springer Basel AG 2011

1993], concerning a biography of William Hamilton [Polak 1993] written by Lev Solomonovich Polak (1908–2002). At the end of the review Cooke wrote:

> "*Chapter 4, on quaternions, contains an interesting section entitled 'How Gibbs tricked Peano'. (It appears that Peano's 'exposition' of Grassmann's multilinear algebra is really based on a reading of manuscripts of Gibbs, which were sent to Peano's teacher G. Basso. In the preface to this manuscript Gibbs claims that his ideas are closer to those of Grassmann than to the concepts of quaternions).*" [Cooke 1993]

At first I thought I may have a scoop, then I got a copy of Polak's book and started my research: this paper is the result. Mention of the main works by Peano on the subject of Geometric Calculus must be made before we start to analyze what Polak said. These are as follows:

- A textbook of Peano's lectures containing some preliminary concepts about Geometric Calculus and in which he mainly refers to Bellavitis, but also to Hamilton, Möbius, and Grassmann. In this book we also find other important contributions [Peano 1887].
- The most important work, which is the subject of this article [Peano 1888]. This is what Polak makes reference to. Hereafter it will be referred to as *Calcolo geometrico*.
- A shorter, more elementary pamphlet [Peano 1891a] that was translated into German [Peano 1891b].
- A short essay for academic mathematicians giving the main concepts and entities of Geometric Calculus, with reference to other areas of mathematics [Peano 1896]. It was translated into Polish [Peano 1897a] and German [Peano 1897b].
- Vector Theory to be used for Elementary Geometry, with the help of Mathematical Logic [Peano 1898].

What Polak said, and related comments

Lev Solomonovich Polak was an outstanding figure in physics and physical chemistry, founder of the Russian school of plasma chemistry and a historian of physics. He was forced to spend almost 20 years in labor camps and "internal" exile. However, he was always scientifically active. His interest in Hamilton started early with his PhD thesis (1933): "*W. R. Hamilton and the Principle of Stationary Action.*" His works and those of his school were widely recognized both in his country and abroad [Asinovskii et al. 2002].

The contents of Polak's book referring to Peano and Gibbs do not seem to have received further mention after Cooke's review. I found only a comment by Hans-Joachim Petsche in his biography of Grassmann [Petsche 2006, 241], who in a note sums it up as "*wenig überzeugend*" (hardly convincing). The lack of references on Polak's opinion regarding Peano did not make my research any easier.

In Polak's book, written in Russian, the table of contents is also in English. Chapter 4, "*Saga on Quaternions*," has a Sect. 6, "*How Gibbs 'deceived' Peano*" [Polak 1993, 233–236]. The following is a summary of the contents of this section with my comments.

Polak's starting point

Polak begins: "The interest of Italian mathematicians in vector calculus without doubt stemmed from G. Peano," and he goes on to say that in 1888 Peano published "*Principles of Geometric Calculus according to the Ausdehnungslehre of H. Grassmann preceded by the Exposition of the Operations of Deductive Logic*," which was translated into German under the title "*Geometric Calculus*" in 1891.

We should note that the title of this book [Peano 1888] contained only the words underlined in the previous paragraph. The work translated into German in 1891 was *Gli elementi di calcolo geometrico* [Peano 1891a], rendered as "*Die Grundzüge des geometrischen Calculs*" (i.e., "*The Principles of Geometric Calculus*").

Polak on Grassmann and Peano

Then Polak goes on to say that Junius Massau (1852–1909), a contemporary of Grassmann and Peano, wrote: "*Inasmuch as Grassmann is obscure and incomprehensible, Peano is clear and understandable*," while Hubert Kennedy, our contemporary, wrote: "*Peano made no claim for ideas contained in it, but there can be no doubt that the extreme clarity of his presentation, in contrast with the notorious difficulty of reading Grassmann's work, helped to spread Grassmann's ideas and make them more popular.*" [Kennedy 1980, 22].
And he goes on to say:

> "If we compare Peano's book [. . .] with the *Ausdehnungslehre*, we can see what little there is in common between these works [. . .]" and
> "what could be 'translations' of Grassmann's work, Peano attributes to Möbius [. . .] only the X symbol is taken from the *Ausdehnungslehre* for the scalar product of vectors, which is called *inner product*."

The fact that Peano is clear, and Grassmann difficult, does not prove that Peano was not in accord with Grassmann's theory. The words "*no claim for ideas contained in it*" confirm the coherence of his thinking with Grassmann. *Calcolo geometrico* is a work about a calculus applied to geometry, based on the *formations* established by Grassmann, which broaden the *Barycentric Calculus* of Möbius.

August Ferdinand Möbius (1790–1868) developed his *Barycentric Calculus* [Möbius 1827] before Grassmann's *Ausdehnungslehre*, and Peano recognized this fact. The reference to Möbius [Peano 1896, 953] mentioned by Polak only concerns the sign of a tetrahedron orientation [Möbius 1885, 41].

In *Calcolo geometrico*, the X symbol for the inner product, introduced by Grassmann [H. Grassmann 1847], is never used for the scalar product that Peano defines and uses without designating it, except once, where he says that "*one also calls it the* inner product *of the two vectors*" [Peano 1888, 88]. And he uses $\boldsymbol{a} \perp \boldsymbol{b}$ (in 2D) and $\boldsymbol{a}|\boldsymbol{b}$ (in 3D), with two different symbols (in 2D and 3D) deriving from Grassmann's *Ergänzung* and the related *inner product* $\boldsymbol{a}|\boldsymbol{b}$. Prior to this, Peano used [Peano 1887] X for the scalar product, but called it "*prodotto di due segmenti*" (since in this book vectors were called '*segmenti*'). He used X again later [Peano 1891a] and called it "*prodotto* (interno o geometrico)."

Polak's unconvincing consideration

Hence, Polak states: "At this point some valid questions arise: Why did Peano think that he was exposing the 'geometric calculus according to Grassmann'? [...] Based on which works did he come to know about vector methods? The first work ('Elements of Geometric Calculus' equivalent to the 'Ausdehnungslehre' of Grassmann) becomes less and less credible as we continue reading Peano's later works, according to the following considerations:

1 The obvious sequence should have been as follows: Peano read Grassmann's works, was struck by his ideas, started to think about them, developed his version of the theory, was sidetracked from the original idea [...]. In fact, it was the contrary: in every new work of Peano new details of Grassmann's theory always appeared together with definitions and concepts belonging to the 'Ausdehnungslehre', and finally in 1896 Peano complained about how difficult it was to read Grassmann's works."

This is a weak point. When Polak says that "new details of Grassmann's theory [...] appeared," this may be true for *Calcolo geometrico* with reference to the previous work

[Peano 1887], but not for the following works, which are generally less complete (as far as the *Ausdehnungslehre* is concerned) than *Calcolo geometrico*. I will come back to Peano's complaint of 1896.

2 "In 1891, when the first volume of H. Grassmann's work appeared, Peano wrote a review. Of course you would expect that from this review you should know Peano's opinion of Grassmann's works [...]. Unfortunately, the review shows only a superficial knowledge of Grassmann – all Peano did was translate the first paragraphs of Engel's foreword [...]"

The first volume of Grassmann's works [HGW11] appeared in 1894 and was reviewed by Peano [Peano 1894] in 1894 (not in 1891). In this write-up he put Grassmann into his historical context, saying that "*Grassmann should be given credit for having developed a very general and organic system of operations and including the previous ones as a special case*" (i.e., those of Bellavitis, Chelini, Möbius, De Saint Venant). Then Peano explained the importance of Grassmann's work, even though it was not a success (one reason being the philosophical language), and compared it to the success of the work of Hamilton (who was already well known and used clear mathematical language). And he added that Grassmann had "*more general and simpler basic concepts with a more compact notation*" than Hamilton. He continued by stressing the importance of Grassmann's methods for the applications, notwithstanding resistance from the older generations.

No translation of Engel's foreword was offered by Peano, but the contents on the front page of the book (in German) were reproduced under the title of the review.

3 "The symbols and some terms of Peano's 'Geometric Calculus' (i, j, k, ∇, 'vector') give evidence to the fact that Peano knew the works of the quaternion school. But many other considerations also convincingly show that he had not read Hamilton's works [...]"

In *Calcolo geometrico* ∇ does not appear and, mostly, the capital letters I, J, K are used as an example of generic 3D linearly independent vectors. ∇ only appears later [Peano 1891a, 37–40].

Peano owned a copy of Hamilton's "*Elemente der Quaternionen*" in German [Hamilton 1882], which is presently in the Giovanni Ricci Università di Milano Mathematical Library *with handwritten notes* by Peano. The title page and p. 146 (with Peano's handwritten notes) are reproduced in Fig. 1.

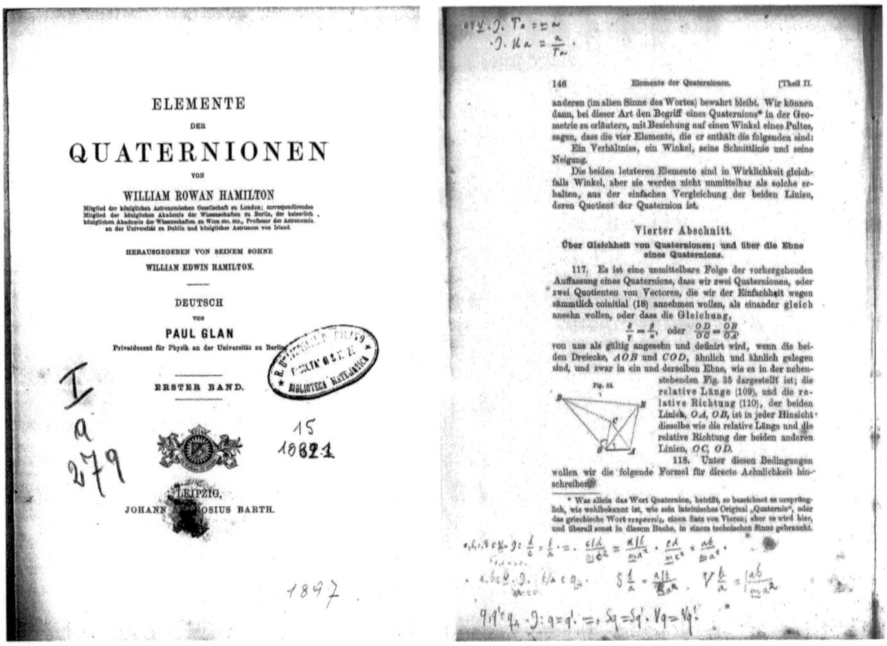
ELEMENTE
DER
QUATERNIONEN
VON
WILLIAM ROWAN HAMILTON
HERAUSGEGEBEN VON SEINEM SOHNE
WILLIAM EDWIN HAMILTON.
DEUTSCH
VON
PAUL GLAN
ERSTER BAND.
LEIPZIG,
JOHANN AMBROSIUS BARTH.

146 Elemente der Quaternionen. [Theil II.

anderen (im alten Sinne des Wortes) bewahrt bleibt. Wir können dann, bei dieser Art den Begriff eines Quaternions* in der Geometrie zu erläutern, mit Beziehung auf einen Winkel eines Pultes, sagen, dass die vier Elemente, die er enthält die folgenden sind:

Ein Verhältniss, ein Winkel, seine Schnittlinie und seine Neigung.

Die beiden letzteren Elemente sind in Wirklichkeit gleichfalls Winkel, aber sie werden nicht unmittelbar als solche erhalten, aus der einfachen Vergleichung der beiden Linien, deren Quotient der Quaternion ist.

Vierter Abschnitt.

Über Gleichheit von Quaternionen; und über die Ebne eines Quaternions.

117. Es ist eine unmittelbare Folge der vorhergehenden Auffassung eines Quaternions, dass wir zwei Quaternionen, oder zwei Quotienten von Vectoren, die wir der Einfachheit wegen sämmtlich coinitial (18) annehmen wollen, als einander gleich ansehn wollen, oder dass die Gleichung,

$$\frac{\delta}{\gamma} = \frac{\beta}{\alpha}, \quad \text{oder} \quad \frac{OD}{OC} = \frac{OB}{OA}$$

von uns als gültig angesehn und definirt wird, wenn die beiden Dreiecke, AOB und COD, ähnlich und ähnlich gelegen sind, und zwar in ein und derselben Ebne, wie es in der nebenstehenden Fig. 35 dargestellt ist; die relative Länge (109), und die relative Richtung (110), der beiden Linien, OA, OB, ist in jeder Hinsicht dieselbe wie die relative Länge und die relative Richtung der beiden anderen Linien, OC, OD.

118. Unter diesen Bedingungen wollen wir die folgende Formel für directe Aehnlichkeit hinschreiben:

* Was allein das Wort Quaternion, betrifft, so bezeichnet es ursprünglich, wie wohlbekannt ist, wie sein lateinisches Original „Quaternio", oder das griechische Wort τετρακτύς, einen Satz von Vieren; aber es wird hier, und überall sonst in diesem Buche, in einem technischen Sinne gebraucht.

Fig. 1. Peano's copy of Hamilton's "*Elemente der Quaternionen*," presently in the Giovanni Ricci Università di Milano Mathematical Library. *Left*: Title Page. *Right*: p. 146 with handwritten notes by Peano

4 "In a later debate (1909–1913) Peano's disciples – Marcolongo and Burali-Forti – admitted that the Italian vector calculus school was 'indebted to Gibbs for many things'."

Already in 1907, Burali-Forti and Marcolongo (only the first can be called Peano's disciple) appreciated Gibbs, when in one of their papers "*For the Unification of Vector Notations*" [Burali-Forti and Marcolongo 1907c, 325], they wrote: "*[. . .] the essence of Gibbs' notation is perfect and even if, as far as form is concerned, somebody called it hermaphrodite,*[1] *we have to admit that it has the power to create the whole Hamiltonian system by itself, as we have proven, and also the one by Grassmann, as we will see.*"

[1] This is a reference to Peter Guthrie Tait (1831–1901), who in the 'Preface' to his "An Elementary Treatise on Quaternions" (3rd ed. Cambridge, 1890) wrote "Even Professor Willard Gibbs must be ranked as one of the retarders of Quaternion progress, in virtue of his pamphlet on Vector Analysis; a sort of hermaphrodite monster, compounded of the notations of Hamilton and of Grassmann" [Tait 1890, VI].

But the problem of the influence of Gibbs on Burali-Forti and Marcolongo is different to that of the possible influence of Gibbs on Peano. I will come to this later.

At this point Polak comes to a sudden conclusion:

> "The previous considerations suggest the idea that Peano had knowledge of the vector calculus in '*Elements of Vector Analysis*' by Gibbs."

This assertion looks like a *non sequitur*, since in the section *How Gibbs "deceived"* Peano, before the sentence by Burali-Forti and Marcolongo, Polak never mentioned either Gibbs or the "*Elements of Vector Analysis*" and no consideration (in Polak or elsewhere) suggests that Peano knew of the vector calculus contained in it (even though this was possible). Furthermore, *Calcolo geometrico* and the following works are formulated in a completely different way to Gibbs' work, for example Peano writes about and makes use of *bivectors* and *trivectors*, terms that he introduced, which do not appear in Gibbs.

Then Polak continues:

> "Gibbs sent reprints of his works to Turin and amongst the addressees G. Basso (1842–1895), one of Peano's professors, appeared. Otherwise Peano could have got copies of the unpublished works (in a more complicated way) via Lüroth, a German mathematician who was his friend and with whom Gibbs corresponded."

In the biography of Gibbs by Lynde Phelps Wheeler, we have an APPENDIX IV – *Gibbs' Mailing List of Reprints* [Wheeler 1951, 235–248] with all the addresses to whom Gibbs sent reprints. This list "*comprises [...] 507 names*" with "*11 from Italy*" and is the result of a work by the author, who merged "*four lists of names*" by Gibbs, that "*are all in his own handwriting.*" An excerpt from this list regarding the Italians and Jakob Lüroth (1844–1910) is given in Table 1, showing the reprints sent to each person. **Basso, Lüroth** and the relevant items **8.** (pp. 1–36 of "*Elements of Vector Analysis*"), **9.** (pp. 37–83), and **V.** (*the whole text*) are in **bold.**[2]

From Table 1 we see that no copies of the reprints (8., 9., V.) of Gibbs' "*Elements of Vector Analysis*" (printed but NOT PUBLISHED)[3] were sent to Giuseppe Basso in Turin.

[2] The list reproduces the names as they are spelled in Wheeler (from Gibbs' manuscripts). One name is spelled in an evidently wrong way: *Faei, de Bruno* instead of *Faà di Bruno (..., Francesco).*

[3] As can be seen from Fig. 2, Gibbs' pamphlets on vector analysis were "printed but NOT PUBLISHED." But he used them for his students and circulated them worldwide. In his biography, Wheeler shows that this hesitation on Gibbs' part was due to some doubts he had about the notations.

Table 1. Gibbs' reprints sent to Italians and to Lüroth

		1	2	3	4	5	6	7	**8**	N	**9**	**V**	A	C	Q	O	R	T	**M**	K
SaintRobert, Paul de	Turin			X	X	X		X					X							
Cremona, Luigi	Rome						X		**X**	X	**X**		X	X	X	X	X	X	X	X
Brioschi, Francesco	Milan						X			X		**X**	X	X	X	X	X	X	X	X
Betti, Enrico	Pisa								**X**	X	**X**		X	X	X	X	X	X		
Battaglini, G.	Rome												X					X		
Basso, G.	**Turin**									X				X	X					
Faei, de Bruno													X							
Denza F.	Moncalieri															X				
Schiaparelli, J. V.	Milan															X				
Celoria, J.	Milan															X				
Tacchini, P.	Rome															X				
Lüroth, J. Prof.	**Freiburg, Baden**								**X**		**X**		X	X	X	X	X	X		X

1. Graphical Methods in Thermodynamics of Fluids
2. A Method of Geometrical Representation
3. On the Equilibrium of Heterogeneous Substances I
4. On the Equilibrium of Heterogeneous Substances II
5. On the Equilibrium of Heterogeneous Substances Resumé
6. On the General Equations of Dynamics
7. On the Vapor-Densities
8. Vector Analysis, pp. 1–36 [1881]
N. Optics, I, II, III
9. Vector Analysis, pp. 36–83 [1884]
V. Vector Analysis, whole
A. Multiple Algebra, Buffalo Address
C. Optics, V
Q. Optics, VI
O. Orbits
R. Rudolf Clausius
T. Quaternions and Ausdehnungslehre
M. Quaternions and Algebra of Vectors
K. Quaternions and Vector Analysis

The only recipients of 8. and 9. were L. Cremona in Rome and E. Betti in Pisa, while F. Brioschi (who founded the Politecnico di Milano) received V.

J. Lüroth received both 8. and 9., but in Peano's remaining correspondence (far from being complete), the only letter mentioned from Peano to Lüroth (1897) is related to the delay in publishing *Rivista di Matematica* [Luciano and Roero 2008], with no reference to Gibbs' works.

Following a specific research effort in the academic libraries of Turin and Milan (and a web search of all Italian public libraries), only the copy of V. sent to Francesco

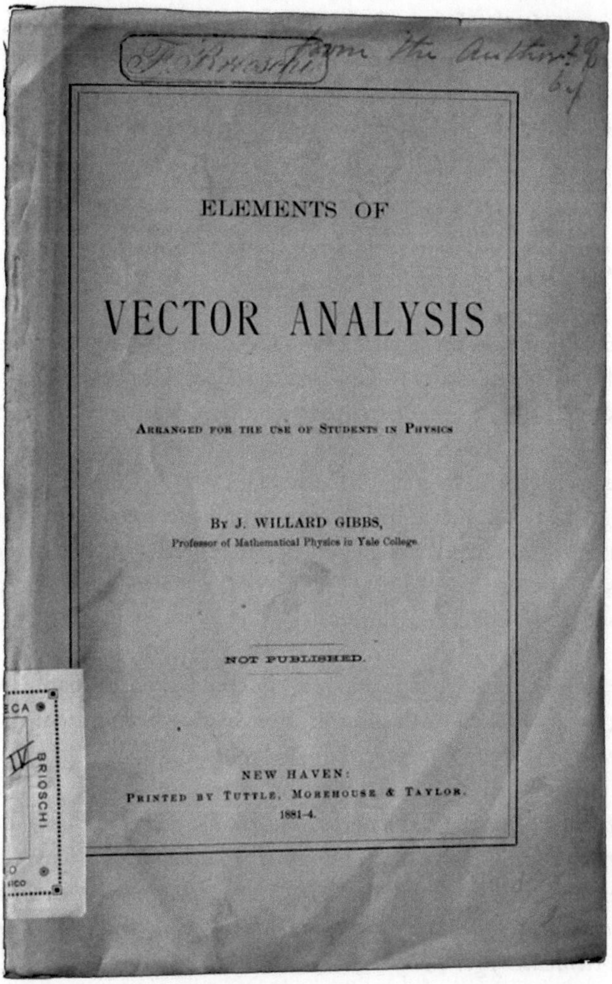

Fig. 2. The front cover of the pamphlet Gibbs sent to Brioschi (in Milan)

Brioschi (1834–1897) seems to exist today in the Biblioteca Centrale di Ingegneria of the Politecnico di Milano. The front cover of this copy is shown in Fig. 2.[4]

Hence, regarding the last two of Polak's sentences, I can say that he has not proven in any way that Peano was acquainted with the vector calculus in "*Elements of Vector Analysis*" by Gibbs. Nor has he proven that Peano received a copy of the reprints of

---

[4] Probably the handwritten words *from the author by* are written by Gibbs (or one of his collaborators), since the shape of the lower case handwritten *r* was (and is not) common in Italy.

"*Elements of Vector Analysis*" from Basso (who never did receive them), or via Lüroth (who received them, but there is no proof that he sent them to Peano).

Furthermore, *Calcolo geometrico* and all other works by Peano on this subject are not related to Gibbs' vector works, yet are fully coherent with Grassmann, despite their originality.

Was Peano 'deceived' by Gibbs?

In the following, Polak goes on to say: "Peano was misled by Gibbs' '*Foreword*' in which he wrote that his work was closer to Grassmann's ideas than to the Quaternion Theory [. . .] Peano decided that '*Elements of Vector Analysis*' were actually a compendium of Grassmann."

"Peano researched the history of mathematics [. . .] Leibniz [. . .] Möbius [. . .] Grassmann [. . .], he never mentioned the source of his vector methods – maybe because he did not want to admit his error when he understood that he had made a mistake?"

These assertions are an elaboration of the previous ones, but do not supply additional proof. Then Polak continues:

> "[. . .] we should think that Peano was indebted to Grassmann's work [. . .] the following words written by Peano confirm what has been stated:
> '*If knowledge of this work (Ausdehnungslehre) was delayed [. . .] the reason [. . .] is due [. . .] to the nebulous and metaphysical form [. . .] that puts readers off from the start. I too, in the study of this work, discovered the power of the new method only by analyzing applications, particularly the ones published in the* 'Kurze Übersicht über das Wesen der Ausdehnungslehre' (1845). *Starting from these applications I was able to reconstruct (!) the theory and to provide the definitions of the entities introduced using only elementary geometry*'.
> These words are the only direct reference (up to 1890) to the work of Grassmann and prove that Peano '*reconstructed*' Grassmann's study and presented his reconstruction as the true theory of Grassmann."

These are the words [Peano 1896, 953] to which Polak hinted previously, when in 2.3.1, he wrote: "[. . .] in 1896 Peano complained about how difficult it was to read Grassmann's works."

But, shortly after the publication of *Calcolo geometrico*, we read in a review [Loria, 1888]: "*If the new book by Mr Peano has been stimulated by the ideas of Grassmann, it is so original that in no way may it be defined as a new presentation of the Ausdehnungslehre. [. . .] it will be useful in spreading the methods that are hidden in the works of Grassmann.*"

Hence, it was clear from the beginning that Peano had *reconstructed* the work of Grassmann in an *original* way.

Finally, at the end of the section "*How Gibbs 'deceived' Peano*," Polak comes to the following conclusion:

> "It is common opinion that Grassmann's '*Ausdehnungslehre*' contributed to influence the formation of contemporary vector calculus through Peano. However, also this second source of vector methods goes back to Hamilton's quaternions."

I will comment on this conclusion after the following section concerning Burali-Forti and Marcolongo.

Burali-Forti and Marcolongo and the Italian Vector School

Cesare Burali-Forti (1861–1931), assistant and collaborator of Peano, then professor at the Military Academy of Turin, and Roberto Marcolongo (1862–1943), from Rome and professor at the University of Messina, began their collaboration in 1907/1908 by writing five notes entitled "*Per l'unificazione delle notazioni vettoriali.*" In these notes [Burali-Forti and Marcolongo 1907a–c, 1908a, b] they proposed an international standard for vector notations (to be presented at the International Congress of Mathematicians in Rome in 1908) in which, among other things, they introduced

$\mathbf{a} \times \mathbf{b}$ for the *scalar product*; $\mathbf{a} \wedge \mathbf{b}$ *for the vector product.*

Their proposals for unification failed internationally, but were successful in Italy (and in France).

In previous works they had followed the Peano/Grassmann approaches and their notations regarding vectors, but at the time the Gibbs–Heaviside approach was deemed useful in Italy for applications, mainly in the field of electromagnetism (also due to the fact that Galileo Ferraris (1847–1897) had used the Heaviside version). And as we have previously seen, they appreciated Gibbs' approach.

In 1909, they published "*Elementi di calcolo vettoriale*," stemming from Peano's *Geometric Calculus* but coherent with Gibbs, apart from the symbols.

And in "*Omografie vettoriali*" (1909), on linear transformations of vectors, they introduced a new treatment, different from the *dyadics* of Gibbs and completely avoiding the use of ∇.

In the following years they (and others, such as T. Boggio and P. Burgatti) published further works on *calcolo vettoriale* and its applications.

But the polemic, self-assured and nationalistic attitude of Burali-Forti in defense of his notations made him enemies in all fields (Quaternionists, Gibbs–Heaviside school, Grassmannians), not to mention the unfortunate "*Espaces courbes. Critique de la relativité*" [Burali-Forti and Boggio 1924], in which, together with contributions from Boggio regarding *calcolo vettoriale* for *n*D and curved spaces, an ill-founded attack on relativity took place. Burali-Forti never understood it and probably felt that his work was competing with the success of the Tensor Analysis of Ricci and Levi-Civita, used by Einstein. Furthermore, the book contained an error that has only recently been discovered [Parra and Sallent 2002].

Conclusion

The Italian school of *Calcolo Vettoriale* founded by Burali-Forti and Marcolongo, derived from Peano/Grassmann, is an elementary system similar to that of Gibbs (but with different notations). So, Polak's first conclusion:

> "It is common opinion that Grassmann's '*Ausdehnungslehre*' contributed to influence the formation of contemporary vector calculus through Peano."

is correct.

Furthermore, given the fact that Gibbs developed his system starting from Hamilton's quaternions and that in some way Burali-Forti and Marcolongo followed it up, Polak's second conclusion:

> "However, also this second source of vector methods goes back to Hamilton's quaternions."

is conceivable.

But the title of the section "*How Gibbs 'deceived' Peano*," together with all related considerations, is definitively wrong, since the assumptions made to justify it are groundless.

Acknowledgments

Thanks are due to Hans-Joachim Petsche for encouraging me to write this paper, to Marco Boella for helping me to get a copy of Polak's book from the *Det Kongelige Bibliotek* (Copenhagen) and to Sally J. Pitt for reviewing the English text.

Rudolf Mehmke, an outstanding propagator of Grassmann's vector calculus

Karin Reich

Textbooks on the history of mathematics normally do not mention the mathematician Rudolf Mehmke (1857–1944), and you cannot find his name in smaller dictionaries of mathematicians. Mehmke's contributions to mathematics would be worthy of greater recognition however, because they were very original and of high quality. Mehmke did not work in the field of pure mathematics, but rather in the field of applied mathematics, and the history of applied mathematics has not been investigated as intensively as the history of pure mathematics.

One of Mehmke's main disciplines was vector calculus: he was among the main propagators of Grassmann's "Extension Theory". If you consider the reception of Grassmann's extension theory, then Mehmke is one of the outstanding pioneers.

Biography

Rudolf Mehmke was born in Lauterberg in the "Harz" mountains (between Brunswick and Göttingen) on August 28, 1857. In 1873 he began his studies at the "Polytechnical College" in Stuttgart,[1] but he continued to study mathematics at the universities of

[1] The so-called "Königlich Württembergische Polytechnikum" in Stuttgart was founded in 1840.

H.-J. Petsche et al. (eds.), *From Past to Future: Graßmann's Work in Context*,
DOI 10.1007/978-3-0346-0405-5_19,

Berlin and Tübingen. The reason for the change was that, at this time, only universities had the right of granting degrees. In Berlin Mehmke had heard, among others, lectures of Karl Weierstraß (1815–1897)[2] and in Tübingen Paul Du Bois-Reymond (1831–1889) became his main tutor. Du Bois-Reymond had succeeded Hermann Hankel (1839–1873) in 1874, who is well known as being among the very first who recognized the importance of Grassmann's extension theory. In 1880 Mehmke submitted his thesis with the title "Application of Grassmann's Extension Theory to the Geometry of Circles within the Plane" [Mehmke 1880]. Therein he mentioned: "In this work I will try to apply a part of the methods and principles, contained within Grassmann's extension theory in a free way to geometry of circles within the plane. The impetus for this investigation was given by Nr. 394f from the 'Extension Theory' of 1862. The multiplication of circles introduced here corresponds to the inner multiplication of Grassmann" [Mehmke 1880, V].[3] In a further paper Mehmke intended to apply the exterior multiplication to the geometry of circles and to improve the theory of pencils and nets of circles, but unfortunately this continuation was not written.

In the same year Mehmke graduated he became "repetent" (assistant) at the "Polytechnical College" in Stuttgart where he stayed until 1884. Then he became professor of mathematics at the "institute of technology" in Darmstadt[4] where he spent the next ten years. When he left Darmstadt in 1894, Hermann Wiener (1857–1939) succeeded him.

In 1894 Mehmke came back to Stuttgart once again, this time as professor of descriptive and projective geometry, where he stayed until 1922. The former "Polytechnical College" had been since 1890 called the "Technische Hochschule" (institute of technology). There were two professors of mathematics, Mehmke for geometry and Carl Reuschle (1847–1909) for analysis and analytical geometry; Reuschle was succeeded by Georg Faber (1877–1966), who left Stuttgart in 1912, then followed Wilhelm Kutta (1867–1944) [Böttcher and Maurer 2008, 96–100, 157–161, 166–170]. When Mehmke retired in 1922 a third chair of mathematics was founded, Friedrich

[2] In the university archive of Stuttgart there exist several manuscripts of lectures of Weierstrass, written by Mehmke [Reich 1993, 266f], but unfortunately they were written in a special kind of shorthand.

[3] "In der vorliegenden Arbeit ist der Versuch gemacht, einen Theil der in Grassmann's Ausdehnungslehre enthaltenen Methoden und Principien in freier Weise auf die Geometrie der Kreise in der Ebene anzuwenden. Den Anstoss dazu gaben die Nr. 394 ff der Ausdehnungslehre von 1862. Die hier eingeführte Multiplication von Kreisen entspricht der, inneren' Multiplication Grassmann's. In einer bereits begonnenen Fortsetzung dieser Arbeit denkt der Verfasser auch die Anwendung der ,äusseren' Multiplication auf die Kreisgeometrie zu zeigen und mit Hilfe derselben namentlich die Theorie der Kreisbüschel und Netze weiter auszuführen".

[4] The so-called "Polytechnische Schule" in Darmstadt was founded in 1868; since 1877 it was a "Technische Hochschule".

Pfeiffer (1883–1961) taught higher mathematics and differential equations, later differential geometry [Böttcher and Maurer 2008, 183–187]. Mehmke was succeeded in 1924 by Gustav Doetsch (1892–1977) who became the new professor for descriptive geometry [Böttcher and Maurer 2008, 188–197]. As professor emeritus Mehmke still gave regular and special courses and held seminars; it was only during the Third Reich that he stopped teaching. Mehmke died on 16 November 1944 in Stuttgart. At this time, at the end of the Second World War, there was virtually no one able to take notice of his death [Böttcher and Maurer 2008, 131–143].

It has to be mentioned that during his entire life as a scientist Mehmke was professor at "polytechnical colleges" or "institutes of technology", he was never a university teacher. Furthermore, there was a great difference between mathematicians at universities and mathematicians at "institutes of technology": at universities, the foundations of mathematics, for example in geometry (Hilbert) and in analysis (Weierstrass), played a dominant role, but at "institutes for technology" applied mathematics was much more important.

Lectures

Mehmke gave lectures in nearly all fields of mathematics, mostly in applied mathematics, including mechanics, descriptive, synthetic, analytical and non-Euclidean geometry, algebraical curves and surfaces, higher algebra, kinematics, graphical calculation, determinants, perspective, graphical calculation within mechanics and complex numbers, practical methods for the solution of numerical equations and graphical and numerical integration of partial differential equations. It is most interesting that he taught Grassmann's extension theory, i.e., vector calculus, regularly in Stuttgart, in Darmstadt and again when back in Stuttgart after 1904/1905.

Here his three first lectures on Grassmann shall be mentioned *expressis verbis* [Reich 1993, 273]:

> 1881/1882
>
> "Application of Grassmann's extension theory. Course I, addition, multiplication for segments of straight lines, points, straight lines, planes, circles and spheres. With numerous applications to mechanics, lower and higher geometry. Course II, application to higher analysis, higher geometry and mechanics."
>
> 1882/1883
>
> "Application of Grassmann's extension theory to partial differential equations and higher dynamics"; "Grassmann's extension theory. Calculation with geometrical magnitudes: addition and multiplication of segments of straight lines, points, straight lines and so on. With numerous applications to geometry and mechanics."

1883/1884
"Application of Grassmann's extension theory to hydrodynamics."[5]

The lecture from the year 1881/1882 was most probably the very first lecture on Grassmann's extension theory in the history of mathematics.

During the ten years in Darmstadt Mehmke did not change the terminology, he always spoke of "Grassmann's *Ausdehnungslehre*" and not of vector calculus.

It was only after 1888 that Grassmann's geometrical magnitudes were generally called vectors, bivectors, trivectors etc. The technical term "vector" was introduced by William Hamilton (1805–1865); since 1833, when he had first discovered quaternions, he designated the three complex components of quaternions as "vectors". William Kingdon Clifford (1845–1879) was the first who spoke of "vectors" in the case of Grassmann's extension theory, and there were several others who used the term "vector" afterwards in referring to Grassmann. A real breakthrough for "vectors" was achieved by Giuseppe Peano (1858–1932), an admirer and adherent of Grassmann's extension theory. In his book *Calcolo geometrico secondo l'Ausdehnungslehre di H. Grassmann* [Peano 1888], which was quite widespread, he spoke generally of vectors and also introduced the terminology bivectors, trivectors, and so on, in the case of higher dimensions [Reich 1996b, p.201f].

Back in Stuttgart Mehmke gave no lectures on Grassmann for ten years, but beginning in 1904/1905 he came back to this field, and from then on talked regularly about "vector and point calculation" (Vektoren- und Punktrechnung). During the years 1904/1905 to 1922 Mehmke held 20 lectures about "vector and point calculation".

Scientific publications and instruments

Mehmke's list of publications covers about 130 items [Böttcher and Maurer 2008, 135–143; Reich 1993, 279–285], not including his contributions to compilations, his reviews and his exercises which he published in the *Jahresbericht der Deutschen*

---

[5] 1881/82 "Grassmann'sche Ausdehnungslehre. Kurs I: Addition und Multiplication von Strecken, Punkten, Geraden, Ebenen, Kreisen und Kugeln. Mit zahlreichen Anwendungen auf Mechanik, niedere und höhere Geometrie". "Kurs II: Anwendungen auf höhere Analysis, höhere Geometrie und Mechanik".

1882/83 "Anwendung der Grassmann'schen Ausdehnungslehre auf partielle Differentialgleichungen und höhere Dynamik", "Grassmann'sche Ausdehnungslehre. Rechnung mit geometrischen Grössen: Addition und Multiplication von Strecken, Punkten, Geraden u.s.w. Mit zahlreichen Anwendungen auf Geometrie und Mechanik".

1883/84 "Anwendung der Grassmann'schen Ausdehnungslehre auf Hydrodynamik".

Mathematiker-Vereinigung and in the *Archiv für Mathematik und Physik*. He was co-editor of the *Zeitschrift für Mathematik und Physik* during the years 1897–1899 with Georg Cantor (1845–1918), and also during the years 1901–1917 with Carl Runge (1856–1927). Mehmke published in a variety of languages: English, Russian, Polish and Volapük. The latter was a constructed language and there was a small community that was fascinated by it. Mehmke was the only one who published mathematical articles in Volapük.[6]

Mehmke also invented special instruments and apparatus, some of which were exhibited at the great mathematics exhibition that took place in 1893 in Munich, where 18 different objects of his were exhibited [Reich 1993, 269f].

Vector commission

In the meantime, in 1903 a vector commission was founded. This occurred in Kassel, where in September the congress of the Society of German Scientists and Physicians (Gesellschaft deutscher Naturforscher und Ärzte) took place. At this meeting the engineer Ludwig Prandtl (1875–1953) gave a lecture with the title: "About a unified terminology of vector calculus within technical and physical classes" which was published in the *Jahresbericht der Deutschen Mathematiker-Vereinigung* [Prandtl 1904]. In this paper Prandtl recommended the kind of vector calculus that was introduced by Josiah Willard Gibbs (1839–1903), i.e.,

$\mathbf{a} \cdot \mathbf{b}$: inner product
$\mathbf{a} \times \mathbf{b}$: outer product
$\nabla \cdot$: divergence
∇ x: curl

The terms inner and outer product, however, were those used by Grassmann, while Prandtl's definitions were those of Gibbs.

At this time Prandtl was very successful as far as his career was concerned. In 1901 he became professor for mechanics at the Institute of Technology in Hannover, and in 1904 professor of technical physics at the University of Göttingen. During the years 1907–1947 he was professor for applied mechanics at Göttingen. He is well known as the founder of hydro- and aerodynamics.

---

[6] It has to be mentioned that also Giuseppe Peano was attracted by constructed languages, from Volapük to Interlingua. In 1908 Peano became President of the "Akademie Internationale de Lingua Universal" and remained so for the rest of his life [Kennedy 1974, 443].

The organizer of this vector commission was the famous mathematician Felix Klein (1849–1925), who, since 1886, was professor of mathematics at the University of Göttingen. The aim of the commission was to unify the different directions of vector calculus which had been developed and established in the past. The result, which Klein had expected, was one kind of terminology, one kind of definition and one kind of symbolism. This commission had three members: the engineer Ludwig Prandtl, the physicist Arnold Sommerfeld (1868–1951) and the mathematician Rudolf Mehmke. Sommerfeld, who earlier had been an assistant of Felix Klein, became in 1897 professor of mathematics at the "Mining College" (Bergakademie) in Clausthal. In 1900 he became professor of mechanics at the Institute of Technology in Aachen and in 1906 professor of theoretical physics at the University of Munich, where he stayed until his retirement in 1938. Sommerfeld was one of the most influential physicists. In 1895 he was editor of Volume V, "Physics", of the *Encyklopädie der mathematischen Wissenschaften*. In 1904 Sommerfeld published a paper with the title "Notation and terminology of electromagnetic magnitudes". There exists a draft of this paper, dating from 1901. It was the basis for all the authors who contributed to Volume V, "Physics", of the *Encyklopädie*. Sommerfeld made use of the "vectorial" product and the "scalar" product,[7] terms which had been introduced by Gibbs [Reich 1996a, 320–326].

The vector commission, however, was not successful: none of the three scientists was looking for a compromise, each hoped to be able to convince the others of his own method [Reich 1996a, 326–337].

Mehmke's main publications on vector calculus

Mehmke's paper, "Comparison between vector analysis according to the American direction and the German–Italian direction" [Mehmke 1904], was a reply to Prandtl's paper, published somewhat earlier in the same journal [Prandtl 1904]. Prandtl, in contrast to Mehmke, did not support Grassmann's extension theory, but instead Josiah Willard Gibbs's kind of vector calculus, the so-called American or physical direction.

According to Mehmke the terminology of these two main directions was summarized in the following way:

[7] Most of the adherents of Grassmann's extension theory used the terms "inner" and "outer" product instead.

German–Italian school	Gibbs
$\vert\mathbf{a}$ (complement of the vector $\mathbf{a}$)	Does not exist
$\mathbf{ab}$ (bivector)	Does not exist
$\mathbf{a}\vert\mathbf{b}$ (inner product)	$\mathbf{a}\cdot\mathbf{b}$
$\vert\mathbf{ab}$ (vectorial product = complement of the bivector)	$\mathbf{a}\times\mathbf{b}$
$\mathbf{abc}$ (trivector)	$\mathbf{a}\cdot\mathbf{b}\times\mathbf{c}$
$\mathbf{ab}\vert\mathbf{c}$	$(\mathbf{a}\times\mathbf{b})\times\mathbf{c}$
$\mathbf{a}\vert\mathbf{bc}$	Does not exist
$\mathbf{ab}\vert\mathbf{cd}$	$(\mathbf{a}\times\mathbf{b})\cdot(\mathbf{c}\times\mathbf{d})$
$\mathbf{ab}^2$	$(\mathbf{a}\times\mathbf{b})^2$
$\mathbf{ab}\cdot\mathbf{cd}$	$(\mathbf{a}\times\mathbf{b})\times(\mathbf{c}\times\mathbf{d})$

After a ten-year pause, Mehmke resumed lecturing on vector calculus during the winter semester 1904/1905. From then on he continuously held lectures on vector calculus. He also immediately planned to write a textbook. In a letter dated December 7, 1904, to the Scottish mathematician Alexander Macfarlane (1851–1913) Mehmke mentioned: "During this semester I am giving a lecture on vector and point calculus (3 h of lectures and 2 h of exercises during the week) and I intend to publish it in the next year."[8] It was not the next year, but in 1913 when Mehmke published his textbook *Lectures on point and vector calculus* (Mehmke 1913) which he had mainly written for the younger generation who was still able to accept the new ideas of Grassmann: "As the oral lectures were primarily thought to be of interest to young mathematicians, the printed version was also written primarily for the young students, in whom the old theories are not yet firmly rooted by means of long habits and do not prevent improvement" [Mehmke 1913, III][9]. As sources Mehmke mentioned: "As far as the foundation of the point calculus is concerned, in the first half I essentially adopted Peano, but in the second half, as well as in the case of vector calculus, I adopt more the original ways of Grassmann" [Mehmke 1913, IV][10].

[8] Archive of the University of Stuttgart, Nachlass Mehmke, shelfmark SN6.II.49.

[9] "So wie die mündlichen Vorträge an junge Mathematiker gerichtet waren, wendet sich auch die gedruckte Ausarbeitung in erster Linie an die studierende Jungend, bei der das Alte noch nicht infolge langer Gewohnheit so festgewurzelt ist, daß es dem Fortschritt hindernd im Wege stünde".

[10] "Was die Begründung der Punktrechnung betrifft, so habe ich in der ersten Hälfte mich im wesentlichen an Peano angeschlossen, während ich in der zweiten Hälfte wie bei der Vektorenrechnung mehr den ursprünglichen Wegen Graßmanns folgen werde."

This textbook was only the first of two volumes. The second volume did not appear because of the First World War and its aftermath. Among the papers which Rudolf Mehmke left and which are now preserved in the archive of the University of Stuttgart, there are many manuscripts and papers on vector calculus, among them a typescript of 160 pages, which is probably the manuscript of the intended second volume.[11]

Relativity theory

In 1905, Albert Einstein (1877–1955) published his special theory of relativity. At that time Einstein was unfamiliar with vector calculus. It was Hermann Minkowski (1864–1909), who, in 1908, was the first to present the special theory of relativity in four dimensions and by means of a special kind of vector and matrix calculus. But the American kind of vector calculus, which was developed in three dimensions, could not be adopted in four dimensions without larger modifications. There were several propositions for these modifications: from Max Abraham (1875–1922) in 1909, from Gilbert Lewis (1875–1946) in 1910, from Arnold Sommerfeld also in 1910, and from Max von Laue (1879–1960) in 1911 [Reich 1994, 168–184]. In addition, Ludwig Silberstein (1872–1948) tried to present an improved version by means of quaternions in 1912 and 1913.

The most important paper was that of Sommerfeld, who introduced two kinds of vectors with four components, linear vectors and three-dimensional spatial vectors, and vectors with six components, the so-called vectors of the second kind, together with all the different multiplications which were now possible [Reich 1994, p.178–181]. Sommerfeld distinguished between scalar, vectorial and tensorial products. Tensorial products were the result when two vectors of the second kind, i.e., with six components, were multiplied together. The result was a so-called tensor with 16 components, which could be reduced to ten components by symmetry.

Mehmke's correspondence

Surely the surviving correspondence of Rudolf Mehmke is by no means complete. There are many places where letters are kept: Berlin, Chemnitz, Frankfurt, Gießen, Göttingen, Harvard and especially in Stuttgart. There are 137 known letters, not

[11] Archive of the University of Stuttgart, Nachlass Mehmke, shelfmark SN6, 3.31.

including the letters from and to Carl von Bach (1847–1931) which are preserved in Chemnitz. The latter belong to the voluminous posthumous works of Bach, which have yet to be put in good order.

The known letters to and from Mehmke were written between 1883 and 1934. There are many years where no letters have survived, and in most cases less than ten letters for one year are known. But there is one year which is exceptional, the year 1904. From this year there exist 66 letters from and to Mehmke, all dealing with vector calculus; this was the year when Mehmke published his "Comparison" [Mehmke 1904]. Unfortunately most of the letters written by Mehmke himself only exist in shorthand and need a specialist to decipher.

As far as vector calculus is concerned, Mehmke's main correspondents were:

Ludwig Prandtl, 30 letters
Arnold Sommerfeld, 10 letters
Sebastian Finsterwalder, 11 letters
Richard von Mises, 6 letters
Richard Gans, 4 letters
August Gutzmer, 4 letters
Jakob Lüroth, 4 letters
Carl Runge, 4 letters
Edwin Bidwell Wilson, 4 letters
Alexander MacFarlane 3 letters
Emil Müller, 3 letters
Rudolf Heinrich Weber, 3 letters
Max Tolle, 2 letters
Max Abraham, 1 letter
August Föppl, 1 letter
Giuseppe Peano, 1 letter
Paul Stäckel, 1 letter.

Mehmke attended the third International Congress of Mathematicians, which took place in August 1904 in Heidelberg. There he not only met, among others, Sommerfeld and Prandtl, but also the Scotsman Alexander MacFarlane, who had published several papers on vector calculus, and the American mathematician Edwin Bidwell Wilson (1879–1964). Wilson had edited in 1901 *A textbook for the use of students of mathematics and physics founded upon lectures of J. Willard Gibbs*, which was re-edited in 1907 and in 1913. Wilson's textbook was responsible for vector calculus in the form of Gibbs becoming widespread, because Gibbs's *Elements of vector analysis,* which had appeared in a very small issue in 1881–1884 in New Haven, was a rare book. Wilson had lent a copy of Willard Gibbs's original to Mehmke who promised to send the book back as soon as possible.[12] So it was only at the end of 1904 that Mehmke became familiar with Gibbs's original publication.

Nearly all of the above-mentioned letters are located in the Archive of the University of Stuttgart. It is hardly by accident that such a great part of the surviving letters concern vector calculus, it is very probable that Mehmke kept exactly these letters.

Some excerpts of the letters follow:

> Emil Müller (1861–1927), who had written a thesis on Grassmann's methods in 1898, had become professor of mathematics at the institute of technology in Vienna in 1902. He was one of the most important supporters of Grassmann's extension theory. On September 3, 1903 he wrote to Mehmke: "In my opinion Grassmann's method of notation should not only be introduced into vector calculus but also into the theory of determinants and into analytical geometry; I am sure that it will be successful as soon as people are convinced that it is not a mere formalism."[13]

> Sebastian Finsterwalder (1862–1951), who in 1891 became professor of mathematics at the institute of technology in Munich, belonged among the few mathematicians whom Mehmke could convince of the superiority of vector calculus according to Grassmann. Finsterwalder was right when he suggested: "For me it is most important what will happen in the future. I doubt that it will be possible to forestall the American direction by means of logical argumentation. Unfortunately, this direction has been too successful within physics. On the one hand, we are dealing with abstract, legal claims.
> On the other hand, the fight for supremacy. All I can imagine is that the future will bring two separate approaches, in peaceful coexistence. [...] The terminology which the most important papers will rely on will certainly prevail."[14]

---

[12] Archive of the University of Stuttgart, Nachlass Mehmke, letters from Wilson to Mehmke, August 18, 1904 and February 17, 1906, shelfmark SN6.II.19 and SN6.II.5; letters from Mehmke to Wilson, August 22, 1904 and January 28, 1906, shelfmark SN6.II.17 and SN6.II.2.

[13] Archive of the University of Stuttgart, Nachlass Mehmke, shelfmark SN6.II.53: "Meiner Meinung nach sollte die Graßmann'sche Bezeichnungsweise nicht nur in der Vektorenrechnung sondern auch in der Determinantentheorie und analytischen Geometrie angewandt werden, ja ich hege die Überlegung, daß sie sich Bahn brechen wird, wenn man nur erst einsieht, daß es sich um mehr als einen bloßen Formalismus handelt."

[14] Archive of the University of Stuttgart, Nachlass Mehmke, shelfmark SN6.II.11, letter of March 21, 1904 to Mehmke: "Von größter Wichtigkeit scheint mir die Frage, was soll in Zukunft geschehen? Ich zweifle, dass es gelingen wird, die amerikanische Schule mit Gründen der mathematischen Logik zu verdrängen. Dazu kann sie mit allzugrossem Recht auf ihre physikalischen Erfolge hinweisen. Es steht hier gleichsam auf der einen Seite das papierne Recht der Juristen, auf der anderen das Recht der Eroberung. Ich kann mir nichts Anderes vorstellen als dass vorderhand eine schiedliche Trennung und ein friedliches Nebeneinander

Most of the physicists, however, were adherents of Gibbs's methods; so August Föppl (1854–1924) wrote to Mehmke on March 22, 1904: "I don't doubt the importance of the reasons which you mention in favour of your notation, especially not in those cases where it may be quite useful to distinguish between a bivector and its complement. In the case of mechanics, however, which alone is interesting for me, I think the distinction is not necessary and cannot be desirable since it only makes the presentation more difficult."[15]

Summary

The correspondence of Rudolf Mehmke documents an important part of the history of vector calculus. Because vector calculus had several roots, there existed eventually a jungle of variants. Nearly every mathematician, every physicist and every engineer who used vector calculus, had his own special system, which, of course, he thought to be the very best. Mehmke was not exceptional in this, but he went even further, he wanted to convince his colleagues that his method was superior. He was like a missionary in favour of Grassmann. In his time, Mehmke fought a losing battle and received a flood of negative responses. Also his textbook published in 1913 was not successful. With Mehmke the attempt failed to introduce a vector calculus following Grassmann's extension theory. This happened in spite of good arguments, especially when the problem came up of developing a four-dimensional vector calculus for the special theory of relativity. Mehmke's point of view was not relevant for a physicist in the category of Arnold Sommerfeld or for an engineer like Ludwig Prandtl. This meant that physicists as well as engineers retained the so-called American direction of vector calculus. The correspondence of Mehmke is a nearly perfect example that not the very best method is the winner; the method that wins has the most powerful supporters, and these were the physicists and the engineers. Philip Frank (1884–1966) correctly commented:

Platz ergreift. [...] Den schliesslichen Sieg wird zweifellos jene Bezeichnung davontragen, in welcher die bedeutendsten Arbeiten publiciert werden."

[15] Archive of the University of Stuttgart, Nachlass Mehmke, shelfmark SN6.II.18: "Ich verkenne übrigens keineswegs die Bedeutung der Gründe, die Sie für Ihre Bezeichnungen anführen, namentlich auch, daß es für manche Anwendungen recht nützlich sein wird, zwischen dem Bivector und seiner Ergänzung zu unterscheiden. Für die Mechanik, die mich dabei allein interessiert, halte ich allerdings die Unterscheidung nicht für notwendig und – im Interesse der Einfachheit der Darstellung – daher auch nicht für wünschenswert."

"We will use the four-dimensional vector calculus introduced by Arnold Sommerfeld to prove the statements. The calculation would be much shorter and less complicated if we used Grassmann's extension theory, because then the mathematical elegance of Minkowski and the clearness of Sommerfeld would be combined. I do not, however, want to stress the reader once again with an unknown symbolism because I want to be as understandable as possible; therefore I will continue to use Sommerfeld's terminology and concepts."[16] [Frank 1911, 599f].

[16] "Wir bedienen uns zum Beweise der von Sommerfeld eingeführten vierdimensionalen Vektorenrechnung. Zwar ließe sich die Rechnung durch Heranziehung und konsequente Verwendung der Grassmannschen Ausdehnungslehre, die es gestatten würde, Sommerfelds Anschaulichkeit mit Minkowskis mathematischer Eleganz zu verbinden, noch wesentlich kürzer und durchsichtiger gestalten; doch will ich dem physikalischen Leser, dem in der letzten Zeit in dieser Hinsicht ohnehin schon sehr viel zugemutet wurde, mit einer fremdartigen Symbolik verschonen und mich im Interesse der allgemeinen Verständlichkeit möglichst an die Sommerfeldsche Terminologie und Begriffsbildung halten."

Robert and Hermann Grassmann's influence on the history of formal logic

Volker Peckhaus

Introduction

It is one of the tasks of this conference, as it was of the Sesquicentennial Conference in Lieschow 15 years ago [Schubring 1996a], to make sure that Hermann Günther Grassmann's work gains the recognition it deserves. What is astonishing is that his contributions to the development of modern mathematics were widely ignored among his contemporaries. Asking for the reasons, two answers are very popular:

1 Grassmann's style of writing, in particular his and his brother's attempts to restrict mathematical terminology to German words. They differed significantly from contemporary standards of mathematical texts. Such terminological choices were heavily criticized already by Felix Klein. He, although favourably discussing Grassmann's contributions to mathematics in his lectures about the development of nineteenth century mathematics, presented him as an outsider and the Grassmannians as a sect. They glorified their

H.-J. Petsche et al. (eds.), *From Past to Future: Graßmann's Work in Context*,
DOI 10.1007/978-3-0346-0405-5_20,

master as a martyr, using a terminology and methods of calculation differing from common usage in order to set themselves apart from common mathematics and to retract Grassmann from a comparative competition with contemporaries [Klein 1926, 182].

2 Grassmann's philosophical approach to his subjects was rather odd in the eyes of most mathematicians of the time. Klein wrote that access to both versions of the *Ausdehnungslehre* was extremely difficult. They were almost unreadable. Klein gives the reason: "The first *Ausdehnungslehre* deduces from the most general philosophical concepts without any formulas" [ibid., 175].

Being a philosopher myself, I would be the last to complain about philosophical approaches. On the contrary, nineteenth century mathematics was a subject of revolutionary change towards structural mathematics that affected many philosophical aspects: questions concerning the internal structure of mathematics and the relations between its subdisciplines, the role of mathematics within a programme of explaining the world scientifically, which concerns in particular the relation between pure mathematics and applications, and even the role of mathematics as a methodological model for certainty and exactness. These changes required, as David Hilbert (1918) later called it, a deepening of foundations [Hilbert 1996, 1109], and this required philosophical efforts, leading thereby to new disciplines on the reflective side of science: philosophy of mathematics on the one side and modern mathematical logic on the other.

I will argue that Hermann Günther Grassmann and his brother Robert Grassmann played a significant role in this reorganization of mathematics, not only by anticipating these developments tacitly, but it is rather possible to determine some sort of causal influence of Hermann Günther Grassmann's General Theory of Forms and of elaborations presented in his arithmetic and in his brother's *Formenlehre oder Mathematik* [R. Grassmann 1872a–f] on the emergence of the German variation of the algebra of logic. With this, there are also influences on modern abstract algebra, universal algebra, semantics, model theory, and other fields [Mehrtens 1979, 12, 24–28; Peckhaus 1997, Chap. 6].

With respect to the charge of having approached mathematics too philosophically, the following should be considered: Grassmann himself distinguished between the philosophical and the mathematical method of presentation. The philosophical method essentially proceeds from the general to the particular, whereas the mathematical method proceeds from the simplest concepts to complex ones. The philosophical method works by distinguishing oppositions aiming at a division of the whole into parts [A1, 28].

General theory of forms

In accordance with this kind of philosophical method, Grassmann opened his *Ausdehnungslehre* with a "survey of a general theory of forms", understood as a series of truths being equally related to all branches of mathematics. The general theory of forms was astonishing and audacious for his contemporaries, as Gert Schubring stressed. However, it is a characteristic contribution for the emergence of pure mathematics to regard abstract operations without taking any concrete meaning into account [Schubring 2009, 182].

The general theory of forms only contains the general concepts of equality and inequality, of connection and separation [A1, 33]. The notion of connection is set without definition. It is an operation applied to two elements. Brackets indicate the order of these connecting operations in forming complexes. They indicate, for example, commutativity or associativity. Every synthetic connection of two elements *a* and *b* is accompanied by two analytic or "separating" operations which lead back to *a* or *b* respectively if applied to the result of connecting *a* and *b*. Grassmann, then, introduces a second connecting operation (with its inverses) which he regards as a connection of higher level. Both connecting operations are distributive. The resulting algebraic structure is dependent on the features given to these operations. It is similar to the calculus of operations proposed by Duncan F. Gregory in 1840 [Gregory 1840] which was further developed by George Boole in applying it to logic [Boole 1847]. It also precedes modern abstract algebra, in particular group theory. Grassmann was the first who, as Birjukova and Birjukov pointed out, formulated an abstract algebraic theory of groupoids, semigroups, quasigroups, groups, commutative groups and rings [Birjukova and Birjukov 1997].

The general theory of forms was not only applied to the *Ausdehnungslehre*, but in Grassmann's *Lehrbuch der Arithmetik* from 1861 [LA] to arithmetic as well. The transparent dependence between the features of the connecting operations and the resulting structure was further elaborated by the two brothers to an architecture of mathematics, published by Robert Grassmann in six short pamphlets in his *Formenlehre oder Mathematik* [R. Grassmann 1872a–e].

In the theory of quantities (*Grösenlehre*) [R. Grassmann 1872a, 8, 26] Grassmann introduces the letters *a*, *b*, *c*, ... as syntactical symbols for arbitrary quantities. The letter *e* stands for special quantities: elements or, in the germanophile terminology of Grassmann "Stifte" ("pins"), quantities which do not emerge from other quantities by applying connecting operations. Beside braces to indicate the order of connections he introduces the equality sign $=$, the inequality sign $\neq$ (Grassmann himself used a stylized z) and the general sign $\circ$ for designating connecting operations. As special connecting operations he treats "joining" ("fügen") or adding (symbol $+$), and "weaving" ("weben") or multiplying (juxtaposition or symbols $\cdot$, $\times$). These connections can either occur as inner connections (if $e \circ e = e$), or as outer connections (if $e \circ e \neq e$).

The different results of connecting pins with themselves give the criteria for distinguishing between the special parts of the theory of quantities. The "theory of concepts or logic" ("Begriffslehre oder Logik") is the first part, "the most simple and, at the same time, the most inward part", as Grassmann calls it [ibid., 13]. Inner joining $e + e = e$ and inner weaving $ee = e$ are valid. In the "theory of binding or theory of combinations" ("Bindelehre oder Combinationslehre") as the second part of the theory of forms, inner joining $e + e = e$ and outer weaving $ee \neq e$ are valid; in the "theory of number or arithmetic" ("Zahlenlehre oder Arithmetik") it is outer joining $e + e \neq e$ and inner weaving $ee = e$, respectively $1 \times 1 = 1$ and $1 \times e = e$. In the "theory of the exterior or theory of extensions" finally, the "most complicated and most exterior" part of the theory of forms, outer joining $e + e \neq e$ and outer weaving $ee \neq e$ are valid [ibid., 12f.].

Logical interpretation

In our context we are most interested in the logical part. Its significance can already be deduced from the distinctions mentioned above. In laws of inner weaving we find Boole's "Law of Duality" also called "Boole's Law", but we have also the inner joining $e + e = e$, a formula which is not valid in Boole's system of logic because of his exclusive interpretation of the logical "or". It was, however, introduced already with William Stanley Jevons's modifications of the Boolean calculus published in his *Pure Logic* of 1864 [Jevons 1864].

In the beginning of the "theory of concepts or logic" the syntactical elements have to be given a "way of reading" (*Lesart*) [R. Grassmann 1872c, 7–8], i.e. they have to be interpreted logically. Now "quantity" means everything that can become an object of reasoning in a definite way. "Pins" are now understood as primordial objects of thinking which have not been derived from other objects. Two objects are equal if they can be substituted for each other *salva veritate*; they are unequal if such substitution is impossible. Addition is read "and", meaning, however, adjunction, i.e. the logical "or". "Multiplication" is read "times", meaning conjunction, the logical "and". Grassmann introduces the signs $<$ and $>$ to express sub- or superordination of concepts [R. Grassmann 1872c, 12]. The sign $\leq$ means that a concept is equal or subordinated to another [ibid., 13]; it is equivalent to Ernst Schröder's (later) subsumption or inclusion. In the theory of judgements, Grassmann expresses this relation in a shorter way by using the angle sign $\angle$ [ibid., 24]. The sign "T" stands for "all" or the totality, the sum of all pins [ibid., 15]. The following laws are valid: $a + T = T$ and $a \cdot T = a$. Zero is interpreted "as the lowest concept which is subordinated to all other concepts". The laws $a + 0 = a$ and $a \cdot 0 = 0$ are valid. Finally, Grassmann introduces the "not" or negation as complement by the laws $a + \bar{a} = T$ and $a\bar{a} = 0$ [ibid., 16].

The similarities between Grassmann's calculus and that of Boole, at least in Jevons' modified version, are indeed striking. They have motivated the following benevolent judgement by John Venn, published in the "Historic Notes" for the second edition of his *Symbolic Logic* of 1894: "Robert Grassmann's scheme was published in 1872, under the title of *Begriffslehre oder Logik*. Like a number of other writers on this subject he seems to have worked out his results in complete ignorance of all that had already been done before him. The work is, however, systematic; and he seems to have been one of the first, after Boole, – with the exception of Peirce, – to realize what was wanted for a complete scheme of Symbolic Logic" [Venn 1894, 489].

Influences

Hermann Günther Grassmann's theory of arithmetic was taken up by Ernst Schröder, then professor for mathematics at the Pro- und Realgymnasium in Baden-Baden, in his own *Lehrbuch der Arithmetik und Algebra* published in 1873 (for the Grassmanns' influence on Schröder cf. also [Peckhaus 1996, 2004]). According to the subtitle of this work, Schröder treated "the seven algebraical operations", i.e., the three "direct" operations of adding, multiplying, and raising to a higher power, and their inverses subtracting, dividing, extracting the roots and forming the logarithms (in the beginning only applied to natural numbers). Schröder defines (pure) mathematics as the "theory of numbers" [Schröder 1873, 2], and by this definition he deviates from the traditional view of mathematics as the theory of quantities. Although Schröder calls the objects algebraically connected "numbers", he leaves open what kind of objects they are. Hence, the structure erected needs an interpretation. The foundation of this theory is close to that of Grassmann.

For the introduction of natural numbers (in the formal part without 0) Schröder refers to the ability to count things which are distinct in our imagination [ibid., 3]. Each of these things to be counted is called a unity [ibid., 4]. Schröder introduces the connecting operations which can be applied to numbers in two different ways, in an "independent" manner, i.e., regarding the combination of units (e.g., $4 = 1 + 1 + 1 + 1$), and, explicitly according to Grassmann's model, in a "recurrent" manner, i.e., by reference to definitions already given (e.g., $4 = 3+1 = [2+1]+1 = [1+1+1]+1$). As Schröder tells us, the recurrent procedure cannot be recommended for beginners of only average ability, although it "originated in the need for greater rigor in order to simplify the presuppositions taken as starting point as far as possible, and to restrain the means of inferences used" [ibid., 51].

The approaches to arithmetic by Grassmann and Schröder share characteristic features: both start with implicit definitions of connecting operations with arbitrary objects. The numbers of arithmetic are defined by applying these operations to units, presupposing the ability of man to count. Grassmann and Schröder were sceptical towards an axiomatic order of theories (for Grassmann's scepticism cf. [Petsche 2006, 221, 226, 266]). Although the structures produced could be reconstructed according to the model of Hilbertian modern axiomatics, it has to be stressed that Grassmann and Schröder used the traditional notion of axiom as a truth which cannot be proved but which also does not need any proof because of its evidence. Grassmann and Schröder did not found their structures on evident truths, but on definitions that can be changed at will, producing however, different structures. Schröder later wrote, following C.S. Peirce, that principles (or axioms) are only definitions in disguise [Schröder 1895, 68]. He accepted only one axiom for mathematics which made him an object of mockery [Frege 1884, VIII]: the axiom of the inherence of signs. It gives us the certainty "that in all our arguments and inferences the signs inhere in our memory, and even more on the paper." [Schröder 1873, 16–17]. Axioms like this follow neither the Euclidean, nor the Hilbertian style. They provide the necessary conditions for doing mathematics.

Schröder's theory of arithmetic was imbedded in a universal algebraic programme. He wanted to create a theory of structures which he called "absolute algebra". He referred to it already in a section of his *Lehrbuch der Arithmetik und Algebra* [Schröder 1873]. It was further developed in his school programme dissertation *Über die formalen Elemente der absoluten Algebra*, published in 1874 – to mention only the first two relevant writings. The formal, and in its last stage of development, "absolute" algebra is founded on a generalization of the notion of number. Schröder proceeds from the existence of an "unlimited manifold (*Mannigfaltigkeit)* of objects (of any kind)" [Schröder 1874, 3] called "domains of numbers" (*Zahlengebiete*). Examples for such "objects constituting a manifold", called "general numbers", are "proper names, concepts, judgements, algorithms, numbers [of pure mathematics], symbols for dimensions or operations, points, systems of points, or any geometrical object, quantities of substances, etc" [ibid.].

For such domains of numbers, Schröder postulates the existence of connecting operations "with the help of which two objects can be connected to form a third one which is to be counted to the same manifold" [ibid.]. In particular, he investigates the non-commutative "symbolical multiplication", $c = a.b = ab$, with its related inverse reverse operations "measurement" (*Messung*), $b.(a : b) = a$, and "division" (*Teilung*), $\frac{a}{b}.b = a$. The application of these operations to general numbers leads to the most simple equations ("fundamental relations", *Fundamentalrelationen*) which are now the object of a four-step programme of "formal", and finally, "absolute" algebra [Schröder 1873, 293–294]:

1 Formal algebra compiles all assumptions that can serve for defining connectives for numbers of a domain.
2 Formal algebra compiles for every premiss or combination of premisses the complete set of inferences ("separation" of resulting formulas from the complete stock of combinatorially possible formulas).
3 Formal algebra investigates which particular domain of numbers can be constructed by the operations defined.
4 Formal algebra has finally to decide "what geometrical, physical, or generally reasonable meaning these numbers and operations can have, which real substratum they can be given" [ibid., 294].

In the framework of this absolute algebra, Schröder's algebra of logic appears to be an algebraic "Operationskreis", in his terminology a domain of numbers in which more than one connecting operation is working: logical addition and logical multiplication with their inverses [Schröder 1877], later additionally relative addition and multiplication.

It should be clear from this that both Grassmann's and Schröder's approach to mathematics can be interpreted as universal-algebraical programmes in which the symbols have to be interpreted according to their fields of application. When Schröder wrote the logical portions of his *Lehrbuch* he was not aware of Grassmann's *Formenlehre*. He got hold of it, however, when he was writing a later section. It provoked a voluminous footnote, running over three pages:

> "The author of the work under consideration uses in the part devoted to logic the $+$ sign for the collective combination, and regards it virtually as an *addition* – one could say as a 'logical' addition – which has beside the properties of the usual (numerical) addition the basic property $a + a = a$ additionally.
>
> [...]
>
> Especially interesting and new for me was [...] the role the author assigns to *multiplication* in the domain of logic. Whereas the *sum* of two concepts is interpreted as the whole of the individuals belonging to the one or the other of these concepts, the *product* of these concepts is such concepts as comprise the marks of both. Thus, the real extensional addition is opposed to an intentional addition or *addition* of marks as multiplication. This procedure can indeed not be surprising if one takes into account that the basic features of addition and of multiplication are essentially the same, that both operations have an already fixed relation

to one another only in the usual arithmetic, and that one has therefore in new fields from the beginning the choice between the two conceptions.—" [Schröder 1873, 145–147, footnote].

Robert Grassmann's considerations are thus the source for Schröder's first acquaintance with the essentials of the algebra of logic. As it becomes clear from a remark in *Über die formalen Elemente der absoluten Algebra* (1874), he learned about Boole's calculus only after the publication of the *Lehrbuch*. From Grassmann's book he borrows the symbolization of logical connecting operations which he had not taken into account before. He states the analogy of these operations to the respective operations in arithmetic, particularly the arithmetic of the numbers 0 and 1. At this stage, he still criticizes the denotation of logical operations with mathematical signs in those contexts that comprise mathematical and logical considerations. Later, he would not hesitate to use the signs + and ·, regarded, however, as signs for algebraical operations interpreted logically. Of decisive importance for Schröder's change of his working field seems to be the discovery of the duality of logical addition and logical multiplication which he elaborated to perfection in his *Operationskreis des Logikkalkuls* of 1877. At the same time this discovery opened the scope of his programme of abstract algebra, since then logic could be regarded as a representative of a simple structure of abstract algebra.

Acknowledgement

I would like to thank Charlotta Lax for her efforts to improve the English of this paper.

Hermann Grassmann's contribution to Whitehead's foundations of logic and mathematics

Jacques Riche

Introduction

The title could sound presumptuous to a reader familiar with Gert Schubring's volume [Schubring 1996a] devoted to Hermann Grassmann. There, we learn from Ivor [Grattan-Guinness 1996] and Volker [Peckhaus 1996] that, if Herman Grassmann contributed to logic at all, his contribution was not as substantial as that of his brother, Robert.

Moreover, although Herman did influence Whitehead's views on mathematics, this cannot be obvious if one restricts Whitehead's work on the foundations of mathematics to his production during the years of collaboration with Bertrand Russell. This period was only one step in the overall development of Whitehead. There is a single Whitehead, and there is a unity in his thought that covers his entire work. Moreover, this unity is deeply rooted in the nineteenth century's cultural, intellectual and scientific context.

The foundations the title refers to thus extend far beyond the foundational issues in logic and mathematics that appeared toward the end of that century. They include questions raised by complex and hypercomplex numbers, the new algebras and the new

H.-J. Petsche et al. (eds.), *From Past to Future: Graßmann's Work in Context*,
DOI 10.1007/978-3-0346-0405-5_21,

geometries. Moreover, various perspectives and interests related to education, ethics and philosophy of nature that were shared by Grassmann and Whitehead are not foreign to these questions. This is what we intend to suggest in the next few pages.

A. N. Whitehead's *Treatise on Universal Algebra*

Alfred North Whitehead (1861–1947), the logician and author of *Principia Mathematica* with Bertrand Russell, was also a well-known philosopher of nature and, in the last phase of his life at Harvard, simply a famous philosopher. Before that, he had been a mathematician at Cambridge where, from the end of the 1880s, Grassmann's *Ausdehnungslehre* had been part of his teaching.

In 1898, he published a *Treatise on Universal Algebra with applications* [Whitehead 1898]. Volume one of this work was also the first book essentially devoted to Grassmann to be published in Britain. The preface explains Whitehead's goal of investigating all systems of symbolical reasoning related to ordinary algebra, including Grassmann's algebra, Hamilton's quaternions, Boole's symbolic logic and Peirce's associative algebras. His approach relied on a "*generalized conception of space* [...] *in the belief that the properties and operations involved*" can lead to "*a uniform method of interpretation of the various algebras*" [Whitehead 1898, v]. These would therefore appear as a symbolical system and analysis engine of reasoning with respect to the abstract notion of space.

The style and presentation of Herman Grassmann's *lineale Ausdehnungslehre* in 1844 [A1], the theory of linear form, i.e., "*the collection of elements into which a generating element is transformed by a continuous evolution*" [$A1_K$, 47], the first part of his *Wissenschaft der extensiven Grösse*, had been a major obstacle to the reception of the work. Introduced in its full title as a new mathematical discipline, A1 started with general and foundational considerations, more philosophical than mathematical, that were certainly not on the minds of practitioners of mainstream mathematics. Indeed, as Victor Schlegel remarks in the preface of his *System der Raumlehre* [1872, vi], at the time of publication, imaginary and impossible quantities were still considered with suspicion, non-Euclidean geometries were not unanimously accepted, and mathematicians were not ready to show interest in the n-dimensional *lineale Ausdehnungslehre* and its unusual operations on unusual objects.

This failure convinced the author to rewrite A1 entirely, stripping it of its misunderstood philosophical content and form of presentation while, at the same time, completing the originally projected second part. This work, *Die Ausdehnungslehre, Vollständig und in strenger Form bearbeitet*, [A2], came out in 1862.

Following this publication, Grassmann's theory attracted some attention, but certainly not to the point that he himself had wished, i.e., not that just a few procedures, but that the domain itself treated in the book be taken into consideration.

These difficulties being known, one could wonder why Whitehead also began, in the first book of his *Treatise*, with a similar long and verbose introduction to the foundations of its subject. He had an ambitious program in mind, namely to show the conceptual unity of the new algebras. Indeed, in the new algebras which no longer consider quantities and numbers, he sees "*bodies of propositions, inter-related by deductive reasoning, and based upon conventional definitions which are generalizations of fundamental conceptions*" [Whitehead 1898, viii]. This is the beginning of a science that applies to almost everything. He thus sees the ideal of mathematics in the construction of a calculus that supports the reasoning in all compartments of thought or experience, such that "*all serious thought which is not philosophy, or inductive reasoning, or imaginative literature, shall be mathematics developed by means of a calculus*" [ibid.]. This thus requires a definition of a calculus and of its object.

Following a presentation of the algebra of symbolic logic, the rest of the volume is devoted to Grassmann's algebra as applied to space in descriptive geometries, linear and metric, Euclidean and non-Euclidean.

Whitehead borrowed the title of his work from Sylvester, and several reviewers complained that the content was not universal algebra. Nevertheless, it has been argued elsewhere that the title is quite appropriate to the content and to the intention of the book [Riche 2007]. Whitehead's view of extension theory will not be considered here. It is a variation of A1 and A2. More may be learned from other authors contributing to these proceedings, including David Hestenes, Alvin Swimmer and John Browne, as well as others who have written specifically on Whitehead's geometric algebra, for example, Henry George Forder [1941] or Stephen Blake [2005].

A picture of A. N. Whitehead by D. Emmett

We may rely on the authority of the British philosopher Dorothy Emmett (1904–2000), a former student of Whitehead, to summarize in a few words the leading ideas in her teacher's cosmology, his so-called "Process Philosophy".

Throughout Whitehead's work, Emmett sees two sides: "*an interest in formal schemes of logical relations, built on the model of a scheme of mathematical postulates, and an interest in the concrete many-sidedness of experience.*" [Emmet 1948, 266]. Indeed, Whitehead "*seeks the forms in the facts*" of a world of events seen as a process from which the formal patterns can be abstracted. And Whitehead held not only that events could be related by abstractions "*but also that there can be a general scheme of formal relationships defining the logical structure of all that exists.*" [Emmet 1948, 274].

Emmett thus quotes that startling remark of Whitehead in his last paper: "*We must end with my first love – Symbolic Logic. When in a distant future the subject has expanded, so as to examine patterns depending on connections other than those of space, number and quantity – when this expansion has occurred, I suggest that Symbolic Logic, that is to say, the symbolic examination of patterns with the use of real variables, will become the foundation of aesthetics. From that stage, it will proceed to conquer ethics and theology."* [Whitehead, quoted in Emmet 1948, 274].

One may add a few specific themes or concepts to complete this extremely short characterization of Whitehead: permanence amid change; abstract scheme and concrete experience; uniform and universal relatedness in the general scheme, that is, its logical structure and the process of the world supplying it with empirical value. One can then consider the ambiguity in the term "value" as a way to confer an ethical dimension to the assignation of empirically observable or real content to the logical variables of the scheme.

Structure and method: From Leibniz to the Grassmanns and A. N. Whitehead

These characteristics can be used as threads through the work of Whitehead in search of an abstract and general structure explaining all that exists. They are not foreign to H. Grassmann himself and, actually, to the whole Grassmann family. They were inherited from Leibniz.

Leibniz's thesis

In 1666, G. Leibniz submitted a nine page "*Disputatio Arithmetica de Complexionibus*" [Leibniz 1666a] to the Philosophy Faculty of the Academy of Leipzig. This short piece is extracted from the thesis he defended shortly after, his "*Dissertatio de Arte Combinatoria*" [Leibniz 1666c]. At the time, the idea of a *Mathesis Universalis*, a universal science, was popular and, among mathematicians and scientists, the famous German Jesuit Athanasius Kircher, author of an "*Ars Magna Sciendi sive Combinatoria*", was a source of inspiration. It is there that Leibniz found his "*ars inveniendi*", his combinatory theory that he sees as universally applicable and which allowed him to go further than the current mathematical method based on analysis and synthesis, i.e., deconstruction and reconstruction, and to reach his art of invention and judgment.

Divisibility

Briefly Leibniz starts from the metaphysical category of Being and its affections. These are either quality, quantity, or relation; they are respectively absolute and relative modes of Being, i.e., Being with respect to itself, to its parts and to other Beings.

Relation is either a union, i.e., it relates the parts in a unit, thus forming a whole, or it is a *convenientia*, i.e., some gathering of elements. From the union, unity as well as number can be abstracted. Quantity is thus the number of parts; and in all things, quantity and number coincide. Nevertheless, the relation or the ratio of quantity to number remains extrinsic as long as the number of parts is not known. This is, according to Leibniz, where the *Analytica Speciosa* originates. It was mainly developed by Descartes and later collected by van Schoten and Bartholinus in their their edition of Descartes' *Principia matheseos universalis* [Descartes 1661]. Analysis is thus the doctrine of ratios and proportions, or of unknown quantity, while arithmetic is that of known quantity or numbers.

Numbers having that universal property of being in any way incorporated in some union of Beings, they belong to metaphysics – as long as metaphysics is accepted as the doctrine that concerns all kinds of Beings. Indeed, *Mathesis* is no longer a single discipline but several disciplines whose particular objects are treated independently of their particular quantity, and in a unified way. For example, arithmetic and analysis concern Quantity in Beings; geometry geometry concerns Quantity in bodies, their spatial extension.

Combinatorics

Leibniz's combinatorics is based on the notion of "*complexions*", i.e., the decomposition of a whole into its smaller parts and their combinations. The parts vary with respect to each other and with respect to the whole. These variations are of two sorts: the combinations ("*complexions*") and the positions ("*situs*") of the parts. Although they both belong to metaphysics, the quantity of variation belongs to number and arithmetic. The variation of position belongs to the figures. In the "*situs*", the positions of the parts with respect to the whole can be absolute; this introduces order. Or they can be relative to the other parts, and this introduces neighbourhood.

These general ideas are then systematically defined precisely. Obviously, the complexions amount to the usual n to k combinations, $C(n,k)$, of n elements or, as Leibniz writes them, the *com2/3/4/nations*, i.e., the rows of Pascal's triangle or some permutations of elements.

Soon after the defence, in a letter to Christian Daum, Leibniz admitted that it is to please the audience rather than to follow his own ideas that he mentioned Kircher and

Lulle and entitled his dissertation *de Arte Combinatoria*. Because, "*it is not so much the variation of conjunctions of things but the combination of these same things by the method of permutation of positions that matters. Moreover, in the operation of combination, we can decide which are the most useful and the best transpositions* [...] *Here is the key to all sciences whose precepts contain the Universal Logic of discovery.*" [Leibniz 1666b, 5].

As Leibniz wrote in his thesis, his *analysis situs*, his calculus of invention, different from algebra in its operations and use of symbols, explains the *situs* immediately and accurately. This is the geometric calculus that H. Grassmann will develop.

With this new calculus, Leibniz discovered the operational character of thought and the premises of the notion of group as it will be developed later by Joseph-Louis Lagrange and Evariste Galois in their treatment of algebraic equations on the basis of the invariance of permutations of their roots. One may remark here that it is in groups, their form and invariance, that Henry Poincaré saw the essence of mathematical reasoning.

What did Hermann learn from his father Justus?

The publication of some of Leibniz' works during the seventeenth century renewed the interest in his ideas and in his combinatorics. This is obvious with Christian Wolff, Gottfried Ploucquet and Johann Heinrich Lambert, but it is more so with Carl Friedrich Hindenburg in his *Beschreibung einer ganz neuen Art,...,* [Hindenburg 1776], with Heinrich August Töpfer in *Combinatorische Analytik und Theorie der Dimensionszeichen in Parallele gestellt* [Töpfer 1793], or in Christian August Semler's *Versuch über die combinatorische Methode, ein Beitrag zur angewandten Logik und allgemeinen Methodik* [Semler 1811].

From his Crystallonomy

Of course, this "neue Art" influenced Justus Grassmann, the father of Hermann and Robert. In his *Zur physischen Krystallonomie und geometrischen Combinationslehre* [J. Grassmann 1829], or in his *Combinatorische Entwicklung der Krystallgestalten* [J. Grassmann 1833], it allowed him to devise simpler methods and rediscover independently earlier ones. For example, while the respective positions of the faces and cleavage planes of a crystal were usually determined by sets of specific equations corresponding to the planes of a specific crystal, J. Grassmann considered normals to the surfaces, planes and cleavages of the crystal and their projection on a circumscribing sphere. Doing so, the analytical treatment was replaced by spherical trigonometry, an approach which does not make any reference to the specific crystal it is intended to represent. This method will be advertised much later by William H. Miller in his *On the Crystallographic Method of Grassmann, and on its employment in the investigation of the general geometric properties of Crystals* [Miller 1868].

Obviously, combinatorics and geometrical methods introduced some improvement in crystallography, in the classification of the various forms of crystals as well as in their structural theory. In A1, as an application of his own method, Hermann will show that given three edges (or boundary planes) of a crystal not in one plane, every other edge can be represented by a rational linear combination of the given ones.

From his philosophy

Justus Grassmann also developed the philosophical interests of his sons. Jakob Friedrich Fries, author of *Die mathematische Naturphilosophie nach philosophischer Methode bearbeitet: Ein Versuch,* [Fries 1822], who discusses the issue of the foundations of mathematics, pure and applied, may have been influential, but not as much as Friedrich Schleiermacher.

We know from A. C. Lewis [Lewis 1996] that Schleiermacher's *Dialektik* [Schleiermacher 1839] had impressed the Grassmann brothers. Here, we also want to emphasize Schleiermacher's method and his ethics, because *Dialektik* is the continuation of his earlier *Grundlinien*, a critique of ethics at the time [Schleiermacher 1803], and because his philosophy, in many respects, sounds rather "Whiteheadian". A single quotation from the last pages of the *Grundlinien* may suffice to give the tone and to clarify what we mean here: "*The way we represent the nature of any science to ourselves depends on the nature of our own moral consciousness, so that, in general, the true idea of a system of human knowledge without which no science can be complete nor true, depends also on* [...] *the full consciousness of the highest laws and of the true nature of humanity.*" [Schleiermacher 1803, 352].

The goal of Schleiermacher in the *Grundlinien* was to furnish ethics with a scientific foundation that would give it a degree of certainty equal to that of arithmetic or geometry. Having missed his goal, he looked for a method that could provide the expected basis. Among the methods he contemplates, the heuristic method is the method that will be echoed in the introduction of A1, in the derivation and the exposition of the notions of pure mathematics and of extension theory. Actually, this method amounts to the construction of knowledge, mathematical or otherwise, in what we consider in this section a structural way, from the combination of elementary parts.

In the process of building up knowledge, the opposition of reason and nature determines the existence of two main sciences: the sciences of the mind and the sciences of nature. They both have a speculative and an empirical side. In the sciences of reason, these are ethics and history. Ethics is constituted by the good, virtue and duty. Virtue is the union of nature and reason; seen objectively, this union of nature and reason is the good; and duty is action directed towards the understanding of nature.

Any knowledge is rooted in the *Mannigfaltigkeit* (the diversity) of appearances and in the unity of the world. *Dialektik* that supervises and organizes the process of knowledge also insures its unity. Of course, didactics, a major interest of Grassmann as well as of Whitehead, is included in that process. Knowledge is now part of our moral duties; being virtuous will make us good and secure the unity of the world.

Parenthesis on prizes

Proof of the renewed interest for Leibniz, the Jablonowskische Gesellschaft der Wissenschaften of Leipzig, whose vocation was to stimulate the cooperation of European researchers with scientific questions and prizes, in 1844, on the occasion of the celebration of the 200th anniversary of Leibniz's birthday two years later, asked: "to evaluate and to develop the geometric calculus invented by Leibniz or to construct one equivalent to it (my translation - J.R.)." [Cited in HGW11, 461].

That celebration was also the occasion to inaugurate the Royal Society of Sciences of Saxony. At the end of the ceremony, it was announced that the price of 48 ducats for solving the problem was attributed to Hermann Grassmann, school teacher in Stettin. The anonymous reviewer of his *Preisschrift*, the *Geometrische Analyse* [Grassmann 1847], for the *Archiv der Mathematik und Physik*, (1847 IX, 501–2), recommended the essay to his readers, noting that it is related to A1, and that it could not be understood without an understanding of the latter.

A coincidence? Perhaps. If H. Grassmann was the lucky (and only) candidate for the Jablonowski price, A. N. Whitehead will be the unlucky candidate in the second Lobachevsky Prize in 1900. According to the prize regulations, a draw would decide in case several works submitted were found to be of equal value. Even though the report of R. Ball [Ball 1901] on Whitehead's *Treatise* is far from being impressive, it happened that the commission of the Société Physico-Mathématique de Kasan found of equal scientific merit two of the three submissions, those of W. Killing and of A. N. Whitehead. Space permitting, it would be interesting to speculate on this equivalence [Hawkins 2000, 180]. A preference for one or the other would indeed be difficult to justify, given the highest value of the two proposals. In any case, the draw decided that the Lobachevsky prize was thus luckily accorded to W. Killing for his *Einführung in die Grundlagen der Geometrie* [Killing 1893–98] and his other memoirs relative to non-Euclidean geometry and to the theory of transformation groups.

The new geometries

B. Riemann's *Habilitationschrift* of 1854, *On the hypotheses which lie at the bases of geometry*, was published posthumously in 1867 [Riemann 1854] and translated into English by W. Clifford in 1873 [Riemann 1873]. Any reader of Riemann and Grassmann will easily notice the close relationship of ideas of the former with ideas of A1. To some extent, Riemann adds to Grassmann an empirical dimension and a measure theory. According to F. Klein [Klein 1895, 167], Riemann's main goal had always been "*to find a general mathematical formulation for the laws underlying all natural phenomena*".

In his Presidential Address to the British Association in 1869, the mathematician J. J. Sylvester [Sylvester 1869, 650–661] reported on some ideas of W. Clifford advertising a fourth dimension of space. This prompted a controversy in the pages of *Nature* [Sylvester 1869, 714–731] on the precise meaning of space as a form of intuition and a priori concept in the philosophy of Kant, a philosophy which was used as a shield by the defenders of the traditional Euclidean tri-dimensional space of our everyday experiences.

This sort of debate on geometry was part of the intellectual environment of the young Whitehead. Moreover, his domains of research and, later, of teaching, directly confronted him with the question of space and time: applied mathematics and algebra, the algebras of Grassmann or those of Hamilton that had permitted and boosted the recent developments of physics with the calculus of vectors. In following Grassmann and incorporating his foundations in the first book of his *Treatise*, Whitehead was inheriting a Kantian legacy passed down to Grassmann by Schleiermacher and Schopenhauer.

In order to briefly characterize this philosophical heritage that can be traced in Whitehead's work, a quotation from E. Cassirer may suffice. In [Cassirer 1923], a book that also investigates the epistemological foundations of relativity theory, relying explicitly on Grassmann, he wrote, for example, that "*the real elements of mathematical calculus are not magnitudes but relations*" [Cassirer 1923, 99], and also, that the system of Grassmann – which he calls universal geometry – "*shows that the pure concept on its side is prepared and fitted for all conceivable changes in the empirical character of perceptions; the universal serial form is the means by which every order of the empirical is to be understood and logically mastered*" [ibid., 111]. These are ideas that will reappear explicitly in Whitehead's work.

Courses in Cambridge

It is in the courses of Homersham Cox in Cambridge that the student Whitehead first heard of Grassmann's work. For example, [Cox 1882] compares the non-Euclidean geometries to ordinary geometry on the basis of the parallel axiom. Relying on works of Cayley, Sylvester and Klein, Cox arrives at systems of forces whose properties in ordinary geometry can be deduced from three simple rules: addition of forces is commutative, associative and transitive (when applied along a straight line). Only experience can teach us that forces combine in this way. But even if it was not the case, it would be possible to introduce, as H. Grassmann did in A1, purely geometrical quantities supposed to follow these laws.

Elsewhere, Cox begins a paper [Cox 1883a] with general remarks on the new algebras, those of Hamilton, Grassmann and Boole. These are systems of symbols whose formal rules of combination guarantee the validity of the results independently of any geometrical or arithmetical interpretation of the symbols. Here, in an original contribution, and following a geometrical method analogous to that of A1, much simpler but longer than the analytical method, Cox applies theses principles to the treatment of circles, their addition and multiplication.

As last example, we may consider Cox's application of the theory of quaternions and of the theory of extension to space [Cox 1883b]. Following Hamilton and Grassmann, the author establishes a pure algebraical calculus whose laws coincide with those of geometry. That is, algebra being seen as the calculus of one-dimensional space, it is included in a two-or-more dimensional geometrical calculus. Next, incorporating Cayley's and Klein's ideas on distance into Grassmann's calculus, he shows that distance can be introduced in three different ways, thus showing that there are three different kinds of uniform geometry: ordinary geometry, spherical geometry, and non-Euclidean geometry.

These are some of the themes that Whitehead will later consider in his own work.

Whitehead's early geometrical works

For example, in a paper on geodesic geometry [Whitehead 1897–98], Whitehead shows that the surfaces in any non-Euclidean space can be reduced to the geodesic geometry of straight lines in a Euclidean space.

First, he makes an important clarification about the notion of "radius of space curvature" that has misled philosophers as well as many mathematicians "*into the belief that some peculiar property of flatness is to be ascribed to Euclidean space*" [Whitehead 1897–98, 276]. Indeed, Beltrami's theorem refutes that belief because planes or other sorts of space can be represented as surfaces in that space and Euclidean geometry itself can be interpreted as geodesic geometry in a non-Euclidean space.

Using Grassmann's method, and starting from Beltrami's theorem, Whitehead investigates the properties of surfaces in the three types of geometries which are such that their geodesic geometry is that of straight lines in a Euclidean space. Extending the theorem to non-Euclidean space, he shows that surfaces of constant curvature are the only surfaces for which geodesics can be transformed into straight lines when the surface is represented point for point on a plane.

Although the mind can contemplate the various sorts of space, our geometrical interpretation of phenomena cannot make use of the idea of a fourth dimension because we have no intuition of it. This finding will lead Whitehead to develop his own approach to physics in his later work. Classical physics distinguishes space and time because it has no way of representing four-dimensional events. Einstein's theory of relativity relies on two principles, the equivalence of inertial frames and the constancy of the speed of light, from which Lorentz transformations relating space to time are deduced. But the physical content of Einstein's theory can be deduced without these two principles as Whitehead showed in deducing Lorentz's transformations from the uniformity and symmetry of space-time and from the symmetry and transitivity of the transformations.

The laws of nature are deduced from experience. They must not only correspond to experience in observation, but they have to explain nature as it is perceived. This will be achieved with the help of Whitehead's method of extensive abstraction, a direct adaptation of Grassmann's theory of extension [Whitehead 1929; Ringel 2008].

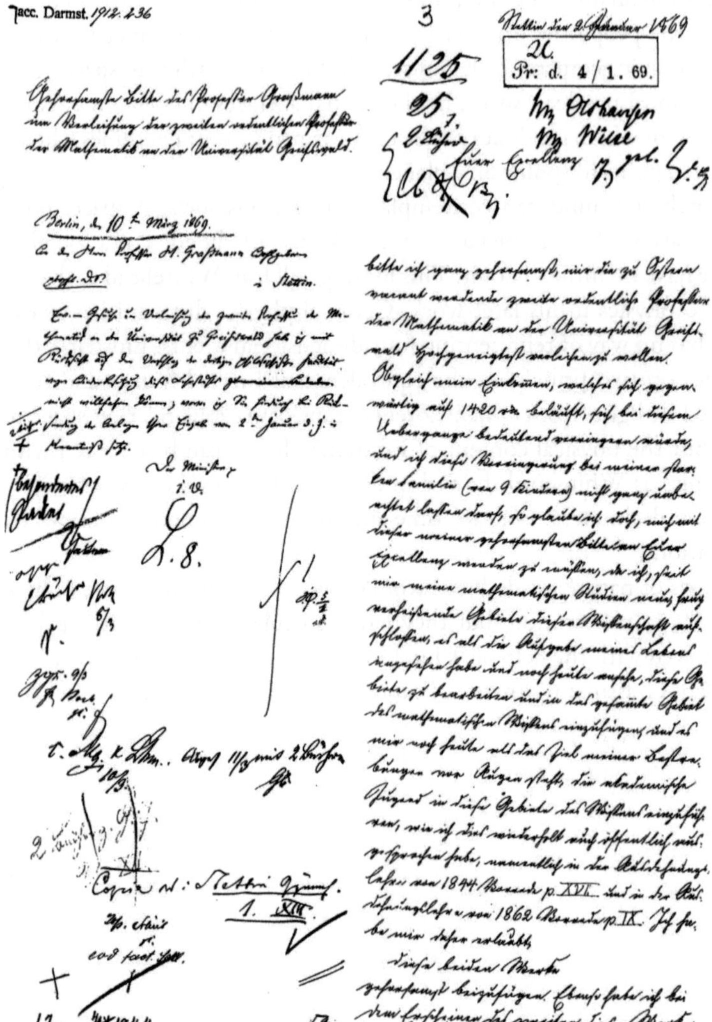

Illustration 12. Hermann Grassmann's petition from 2 January 1869 to minister von Mühler, in which he applied for the second professorship of mathematics at the University of Greifswald (with handwritten comments by the ministry and a draft of the negative reply). The transcription and translation are given on p. 499

[Staatsbibliothek zu Berlin – Preußischer Kulturbesitz, Handschriftenabteilung, Sammlung Darmstaedter]

Present and future of Hermann Grassmann's ideas in mathematics

Grassmann's legacy

David Hestenes

In a previous conference honouring Hermann Grassmann's profound intellectual contributions [Schubring 1996a], I cast him as a central figure in the historical development of a *universal geometric calculus* for mathematics and physics [Hestenes 1996]. Sixteen years later I am here to report that impressive new applications in this tradition are rapidly developing in computer science and robotics as well as physics and mathematics. Especially noteworthy is the emergence of *Conformal Geometric Algebra* as an ideal tool for computational geometry, as it fulfils at last one of Grassmann's grandest goals and confirms the prescience of his mathematical insight. Geometric Calculus has finally reached sufficient maturity to serve as a comprehensive geometric language for the whole community of scientists, mathematicians and engineers. Moreover, its simplicity recommends it as a tool for reforming high-school mathematics and physics, as Grassmann had envisioned.

My purpose here is threefold: to extend my previous account of Grassmann's pivotal role in the evolution of Geometric Algebra (GA) to place him in a broad historical context; to survey landmarks in the recent development of Geometric Calculus that demonstrate its current vigour and broad applicability; to explain precisely what extensions of Grassmann's system were needed to meet his ambitious goals.

H.-J. Petsche et al. (eds.), *From Past to Future: Graßmann's Work in Context*,
DOI 10.1007/978-3-0346-0405-5_22,

Evolution of Geometric Algebra and Calculus

For present purposes, *Geometric Algebra* and *Geometric Calculus* can be regarded as synonymous, with algebra regarded as a tool for calculation. (I use capitals to emphasize specific meanings for the terms "geometric algebra and calculus" employed here.) Technically speaking, *Geometric Calculus* is the broader term, referring to an extension of *Geometric Algebra* to differentiation and integration, including differential geometry and differential forms [Hestenes and Sobczyk 1984]. However, our main interest here is the underlying algebraic structure.

A family tree of major steps in the evolution of GA is laid out in Fig. 1. We have space here for only a few comments about it. The main line proceeds directly from Euclid through Grassmann and Clifford to the beginning of Geometric Calculus [Hestenes 1966]. Perhaps Descartes should be included in this sequence, because analytic geometry was surely a crucial input to Grassmann's thinking, though Grassmann drew directly on Euclid to create a coordinate-free algebra of geometric concepts.

I have included Boole in an unconnected bubble in Fig. 1 as a reminder that extracting logical structure from natural language has much in common with Grassmann's program to put synthetic geometry in algebraic form.

A branch of Grassmann followers including Peano, Whitehead and Forder has been pruned from the Family Tree, because they did not significantly advance Grassmann's agenda or contribute to the emergence of GA.

Ironically, W. K. Clifford [Clifford 1878], the mathematician exhibiting the deepest understanding of Grassmann's system and advancing it in a major way, is seldom mentioned as a follower of Grassmann in historical accounts, though Clifford himself could not have been more explicit or emphatic in his claim to be following Grassmann in developing what he called "*geometric algebra*"! Mathematicians too have overlooked Clifford's link to Grassmann, and, to this day, treat Clifford algebra as a completely separate algebraic system.

Élie Cartan [Cartan 1922] incorporated Grassmann's outer product into his calculus of *differential forms*. Though it placed Grassmann's name into mainstream mathematics, it so diluted his ideas that Engel called it "Cartanized Grassmann". In another irony, Cartan [Cartan 1937] also employed a matrix form of Clifford algebra in his "theory of spinors", but he failed to recognize its relation to Grassmann algebra and differential forms.

In physics, Clifford algebra was rediscovered in the matrix algebras of Pauli and Dirac where it plays an essential role in quantum mechanics. Finally, in 1959 the many threads in Fig. 1 converged to a rebirth of Grassmann's vision of a universal geometric algebra with powerful applications to physics. The midwife of the rebirth was a set of lecture notes by Marcel Riesz [1958]. Let me describe briefly what happened.

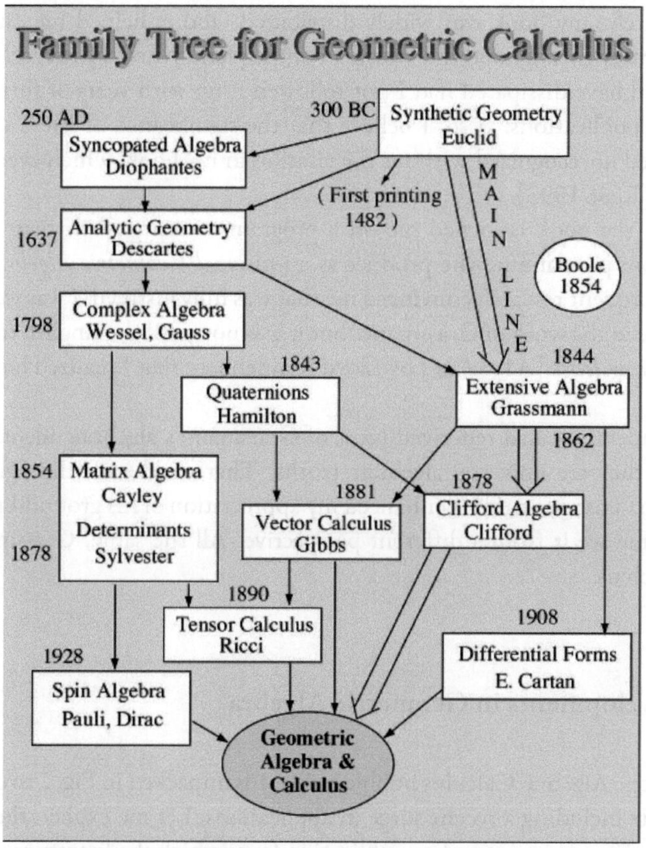

Fig. 1. The main mathematical progenitors of Geometric Calculus

I was a graduate student in 1959 intensively studying the alternative mathematical systems used in physics, including the Feynman trace calculus in quantum electrodynamics and tensor calculus in general relativity. I was lucky to get a thorough introduction to differential forms, including its intuitive origins, in a course on differential geometry by Barret O'Neil, as there was no good book on the subject in English at the time. I was consciously concerned with questions relating the structure of these mathematical systems to the structure of the physical world. One day Riesz's notes appeared on the new-book shelf of the UCLA library. The impact on me was immediate and striking! By the time I was half way through the first chapter I was convinced that Clifford algebra was the key to unifying mathematical physics. During the next few years I worked out the framework for a fully geometric unification. The result was published in my book *Space Time Algebra* [Hestenes 1966].

Fortunately, my book was widely distributed, and it helped me establish many fruitful contacts throughout the world in subsequent years. However, I believe its impact would have dissipated had I not followed it up with years of further research, lectures and publications. Also, I believe that the significance of Riesz's notes would have remained unrecognized without the citation in my book, which eventually led to publication [Riesz 1993].

Though the book launched me on a program to unify mathematical physics, I refrained from proclaiming the product as a *universal Geometric Algebra and Calculus* until subsequent research convinced me that was fully justified. I was well aware that its roots were in the work of Grassmann, but it was not until the English translations of the *Ausdehnungslehre* [$A1_K$, $A2_K$] by Lloyd Kannenberg that I realized how deep those roots were.

Independently, I had rederived most of Grassmann's algebraic identities (as have others), for they are universal algebraic truths. That was better than getting results directly from Grassmann, for it enhanced my appreciation of his ground-breaking work and helped me see it from a different perspective. All the same, Grassmann still has much to teach us.

Recent developments in Geometric Algebra

The Geometric Algebra/Calculus bubble in Fig. 1 is unpacked in Fig. 2 to outline major developments including a recent surge in applications. Let me explain the significance of each box with reference to a key publication from which the literature can be traced. The reader is invited to correlate my explanations with the figure as I proceed.

To fulfil the promise of Geometric Calculus as a comprehensive mathematical language for all of physics, my book *Space Time Algebra* launched me along *three main lines of research and development* that were clearly demarcated and consolidated within the next two decades.

The first line was a straightforward reformulation of classical physics in terms of GA. It produced the first comprehensive coordinate-free treatment of Newtonian mechanics, including rotational dynamics [Hestenes 1985]. Both Grassmann and Clifford had a similar goal, but comparison with their work shows what a difference a century of science can make. A similar reformulation of classical electrodynamics was equally straightforward and enlightening [Baylis 1999].

The second line of research emerged from reformulating the Dirac equation in terms of GA. This revealed a hidden geometric structure in quantum mechanics, including a hitherto unrecognized geometric interpretation for the unit imaginary relating

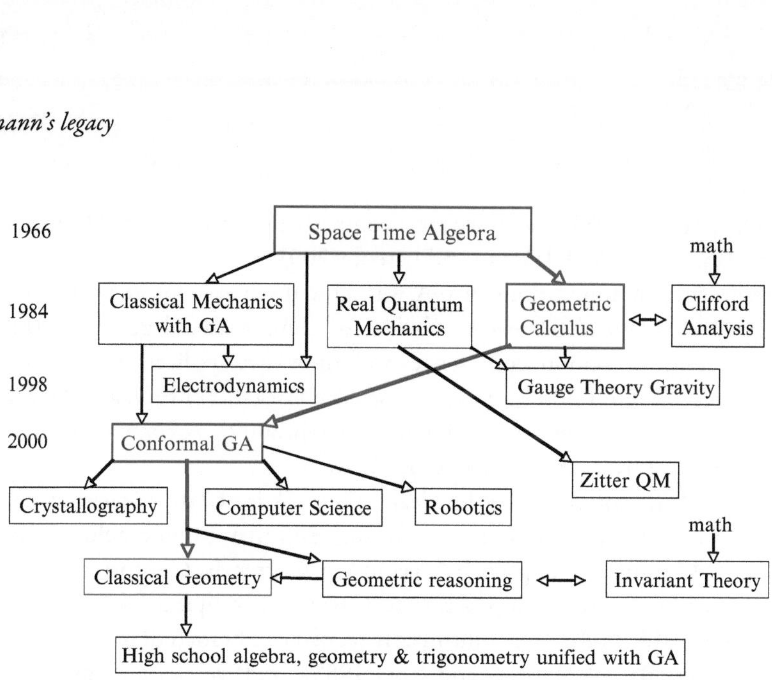

Fig. 2. Development of Geometric Algebra and Calculus

it unequivocally to electron spin [Hestenes 1967]. I call the reformulation of quantum mechanics in these terms *Real Quantum Mechanics.* Though it has not yet been recognized in the physics mainstream, research on its implications is still underway [Hestenes 2009].

The third line of research and development was to produce a self-contained system of mathematical tools sufficient for addressing any problem in physics without resorting to alternative mathematical formalisms. The result was a book that defines the domain of *Geometric Calculus* [Hestenes and Sobczyk 1984]. The most innovative features of this book are, perhaps, its concepts of *vector manifold, vector derivative and geometric integration theory* (which generalizes Cartan's differential forms).

Only a handful of people worked with GA until Roy Chisholm organized in 1984 an international conference on "Clifford Algebras and their Applications in Mathematical Physics" [Chisholm and Common 1986] that brought together a wide range of mathematicians and physicists with overlapping interests. In particular, it revealed a strong connection between *Geometric Calculus* and an emerging new branch of mathematics called *Clifford Analysis.* The conference (with its published proceedings) was so successful that it has been repeated every four years and is still going strong.

In 1988 a group of theoretical physicists at Cambridge University picked up the threads of Real Quantum Mechanics and Geometric Calculus. The most important result was a new approach to Einstein's General Relativity called *Gauge Theory Gravity* [Lasenby, Doran and Gull 1998]. One by-product was extension of the GA approach to Lie Groups [Hestenes and Sobczyk 1984] and to spin representations for all the classical groups [Doran et al. 1993]. The single most comprehensive treatment of GA for physics is now available in the book [Doran and Lasenby 2003].

A resurgence in applications of GA was ignited when I pointed out the unique advantages of *Conformal Geometric Algebra* for computational geometry [Hestenes 2001]. It led directly to more conferences and innovative applications in crystallography [Hestenes and Holt 2007], computer science [Dorst, Fontijne and Mann 2007] and robotics [Bayro-Corrochano and Scheuermann 2009]. A compact review of the current state of the art is given in [Hestenes 2010].

Conformal GA appears to be the ideal framework for *Classical Geometry* as envisaged by [Klein 1939]. We describe some of its surprising new insights below to complete and vindicate Grassmann's approach to geometry. Currently, Conformal GA is at the centre of an ambitious research program to master the complexities in advanced *geometric reasoning* [Li 2008]. This is intimately related to a branch of mathematics called *invariant theory* [Barnabei, Brini and Rota 1985], and we can anticipate a fruitful interaction if not a merging of these mathematical domains in the future.

Finally, it is worth mentioning that Grassmann's objective to reform the elementary mathematics curriculum is more feasible now than ever before [Hestenes 2003a]. Steps in this direction are underway, but this is not the place to discuss the complexities of education reform.

Products in Geometric Algebra

Geometric Algebra today is very close to Grassmann's original algebraic system. To make the connection, we establish the correspondence of current notation, nomenclature and definitions with Grassmann's in his second Ausdehnungslehre [A2]. It suffices to consider multiplication. Grassmann's various products were so well conceived that they are still in use today, and, as we shall see, they can all be reduced to the fundamental geometric product in GA.

We begin with an *n-dimensional real vector space* $\mathcal{V}^n = \{a, b, c, \ldots\}$, as defined by Grassmann in perfect accord with the modern concept. As I believe Grassmann would have it, we use the term "vector" only in this strict algebraic sense. Geometric interpretation will be addressed as a separate matter.

Grassmann's antisymmetric ***outer product*** of two vectors (sometimes called the *join*) is denoted by

$$a \wedge b = -b \wedge a \leftrightarrow [ab] = -[ba],$$

with the modern notation on the left corresponding to Grassmann's notation on the right. The *outer product* of k vectors produces *a k-blade* or *k-vector* A:

$$a_1 \wedge a_2 \wedge \ldots \wedge a_k \equiv A \quad \leftrightarrow \quad [a_1 a_2 \ldots a_k] \equiv [A]\,.$$

The integer k is called the *grade* of A.

The outer product of n vectors generates a *pseudoscalar*:

$$a_1 \wedge a_2 \wedge \ldots \wedge a_n = \lambda I \leftrightarrow [a_1 a_2 \ldots a_n] = \lambda,$$

where λ is a scalar and I is the unit pseudoscalar, which has the properties $I \neq 0$, $a \wedge I = 0$ for every vector a. Here we see a subtle design flaw in Grassmann's system, for he identifies pseudoscalars with scalars. This is not a logical mistake, but it complicates the rest of his system (as we see below) and may have kept him from discovering the ultimate simplification with the geometric product.

Grassmann's *regressive product*, sometimes called the ***meet*** [Brini and Teolis 1996; Zaddach 1996; Hestenes and Ziegler 1991], is defined for arbitrary blades A and B in Grassmann's ingenious way by

$$A \vee B = (A^* \wedge B^*)^* \quad \leftrightarrow \quad [A|B] = |\,[|A|B]\,,$$

where the ***dual*** A^* corresponding to Grassmann's *supplement* $|A$ is defined by

$$A^* \equiv \tilde{A} \cdot I = \tilde{A} I \leftrightarrow |A,$$

with the ***reverse*** of A defined by

$$\tilde{A} = (a_1 \wedge a_2 \wedge \ldots \wedge a_k)^{\sim} = a_k \wedge \ldots \wedge a_2 \wedge a_1 = (-1)^{k(k-1)/2} A.$$

The dual is often defined without the reverse, which contributes only a sign. The reverse is included here to agree with Grassmann's definition. The definition of dual employs the *inner product* (denoted by a centre dot), or better, the *geometric product*, both of which are defined below. Grassmann's *supplement* is based on a concept of orthogonality that amounts to presuming a Euclidean inner product.

For a pair of vectors, Grassmann's *scalar-valued inner product* is expressed by

$$a \cdot b = b \cdot a \quad \leftrightarrow \quad [a|b] = [b|a]\,.$$

Similarly, Grassmann's regressive product defines a scalar-valued inner product for any pair of blades with the same grade.

Now let us start all over again to define *Geometric Algebra.* Like Grassmann, we introduce an associative (and, of course, distributive) product on the vector space $\mathcal{V}^n = \{a, b, c, \ldots\}$, but now we define it by the simple rule that the square of every vector is a scalar. Explicitly, we write

$$a^2 = aa = \varepsilon |a|^2,$$

where scalar $|a| \geq 0$ is called the *magnitude* of a, and its *signature* ε is *positive* ($\varepsilon = 1$), *negative* ($\varepsilon = -1$) or *null* (if $|a| = 0$). For any two vectors, we now *define* the ***inner product*** by

$$a \cdot b \equiv \frac{1}{2}(ab + ba) = b \cdot a.$$

It is easy to prove that this symmetric product is scalar-valued and is the usual Euclidean inner product if both vectors have positive signature. We also assume that the inner product is *nondegenerate,* which means that every vector has a non-vanishing inner product with some other vector.

Now we define the ***outer product*** by

$$a \wedge b \equiv \frac{1}{2}(ab - ba) = -b \wedge a.$$

Adding the last two equations, we see inner and outer products as symmetric and antisymmetric parts of a single ***geometric product***

$$ab = a \cdot b + a \wedge b.$$

The definition of the outer product is easily generalized to the antisymmetrized geometric product of any number of vectors to give us k-blades, precisely equivalent to those defined above. The inner product can also be generalized to give us

$$aA = a \cdot A + a \wedge A.$$

This decomposes the geometric product of a vector with a k- blade into a $(k-1)$-blade $a \cdot A$ and a $(k+1)$-blade $a \wedge A$. In other words, the inner product is a grade-lowering operation complementary, or better, dual to the grade-raising outer product. That symmetry is perfectly expressed by the (easily proved) identity

$$a \cdot (AI) = (a \wedge A)I.$$

Thus, duality interchanges the roles of inner and outer products.

Now we can formulate the meet as $A \vee B = A \cdot B^*$, the inner product of one blade with the dual of another. This has been used to formulate projective geometry in terms of GA [Hestenes and Ziegler 1991]. That approach is remarkably similar to Grassmann's treatment of projective geometry, but a better approach is described below.

From the vector space $\mathcal{V}^n$ the geometric product generates a ***Geometric Algebra*** $\mathcal{G}^n = \mathcal{G}(\mathcal{V}^n)$. Defining the outer product introduces a *grading* in the algebra that decomposes it into a sum of linear subspaces $\mathcal{G}^n_k$ of homogeneous grade:

$$\mathcal{G}^n = \mathcal{G}(\mathcal{V}^n) = \sum_{k=0}^{n} \mathcal{G}^n_k,$$

including the scalars $\mathcal{G}^n_0 = \mathcal{R}$ as a 1-dimensional subspace. The geometric product also induces a signature (r, s) on the vector space that expresses decomposability into an r-dimensional subspace of vectors with positive signature and an s-dimensional subspace of vectors with negative signature, so that $n = r+s$. This is incorporated in the notation for the algebra by writing $\mathcal{G}^{r,s} = \mathcal{G}(\mathcal{V}^{r,s})$.

Further details about GA are available in publications already mentioned. My purpose here has been to show how perfectly GA incorporates Grassmann's ideas for an *Algebra of Extension*. Indeed, I submit that GA is a next step, perhaps a final step in the evolution of a *universal geometric calculus* first envisaged by Leibniz [Crowe 1967; Grassmann 1995c]. Let me summarize the value added by this last step to GA.

In the first place, we have seen that GA reduces the algebraic structure to a single geometric product that mixes grades. Grassmann himself recognized that such a product is needed to incorporate quaternions into his algebraic system, but he did not push his analysis far enough to recognize its fundamental role [Hestenes 1996]. Indeed, the geometric product is essential not only to incorporate quaternions [Hestenes 1999], but the whole theory of spinors and spin representations in group theory [Doran et al. 1993]. Below, we see a simple application of the geometric product to congruence.

But what about the strong claim to *universality*!? How can that be justified? Matrix algebra already plays the role of a universal "arithmetic for higher mathematics". Indeed, from its beginning in the middle of the nineteenth century, matrix algebra was a major competitor to Grassmann's system, and it soon swamped his voice. Over the next century matrix algebra was cultivated by legions of mathematicians and physicists to become the dominant mathematical tool in use today. From his Ausdehnungslehre of 1862 it is clear that Grassmann understood the issue deeply (he had already declared his aim to suppress the use of coordinate systems in geometry), but instead of critiquing matrix theory, he set out to show how to handle linear transformations without coordinates. The significance of Grassmann's approach to linear algebra went largely unrecognized, but GA has reinvigorated it with new tools and fully assimilated matrix algebra and all its capabilities [Hestenes 1991; Hestenes and Sobczyk 1984]. Without delving into details, it is worth noting that every element of a real matrix can be expressed as the inner product of a pair of vectors; thus

$$a_{ij} = a_i \cdot b_j,$$

so the inner product plays an essential role in matrix representation. Contrary to common opinion, though, the inner product in GA does not limit its applicability to metric spaces. Rather, it serves the general role of *contraction* (in the sense of that term in tensor algebra). For example, every linear form, that is, every linear mapping φ of vectors into scalars, can be expressed as a contraction; thus

$$\varphi : a \quad \rightarrow \quad \varphi(a) = a \cdot b.$$

Of course, this does not preclude one from the very useful practice of defining a metric tensor by

$$g(a, b) = a \cdot b.$$

One last point about matrices: We could "generalize" to complex matrices by introducing complex numbers as scalars. However, that is not advisable, because GA has better (geometrically significant) ways to deal with "complex structure".

As to other algebraic systems, it is well known that every associative algebra has a matrix representation. With matrix algebra incorporated into GA, it follows that every associative algebra can be represented in GA. Even non-associative products can be represented in GA. This has been demonstrated explicitly for the octonian product [Lounesto 1997]. If there is an algebraic system that cannot be neatly represented in GA as it stands, then it is likely that GA can be generalized to include it.

Perhaps the ultimate justification for a universality claim is the fact that GA has by far the broadest range of applications to physics and engineering of any single mathematical system [Hestenes 2003b], as amply documented by the review in this paper. Furthermore, that review supports an important observation about GA as a language. The discussion of multiplication in the present section is essentially about defining the *grammar of GA*. But there is far more to a language than its grammar! You need to know how to express important ideas in the language. Thus, creation of *GA as a language* required development of a huge superstructure of definitions, constructions, proofs and calculations to cover classical physics, quantum mechanics, general relativity and engineering applications. Some of it came by insightful translation from other mathematical systems. Some of it involved genuine new insights from exploiting the unique features of GA.

No one has ever been more attuned to the relation of grammar to language than Grassmann. He refined and extracted the geometric structure inherent in synthetic geometry and expressed it in algebraic form. Then he separated the algebra from geometric interpretation to create a general theory of algebraic structures with unlimited dimensions. Besides freeing algebraic structures from the limitations of geometric intuition, Grassmann realized that the same structures can be given many different geometric interpretations. In the following we examine one of the most striking examples of that fact. It involves all the crucial features of GA that have been added to Grassmann's original system: *the geometric product, the pseudoscalar, signature and null vectors.*

Conformal Geometric Algebra

The *Conformal Geometric Algebra* $\mathcal{C}(\mathcal{E}^n)$ for the n-dimensional Euclidean space $\mathcal{E}^n$ is defined by $\mathcal{C}(\mathcal{E}^n) \equiv \mathcal{G}(\mathcal{V}^{n+1,1})$. The points of $\mathcal{E}^n$ are represented by null vectors in $\mathcal{V}^{n+1,1}$. Remarkably, the remaining vectors in $\mathcal{V}^{n+1,1}$ represent the hyperplanes and hyperspheres of $\mathcal{E}^n$, and their geometric products generate the entire group of conformal transformations on $\mathcal{E}^n$, including the *Euclidean Group* of rigid displacements and rotations. An example is given in the next section.

To demonstrate the power and convenience of Conformal GA, we consider the simplest case of the *Euclidean plane* $\mathcal{E}^2$ with its algebra

$$\mathcal{C}(\mathcal{E}^2) \equiv \mathcal{G}(\mathcal{V}^{3,1}).$$

Immediately we encounter an astounding fact: The vector space $\mathcal{V}^{3,1}$ is precisely the standard *Minkowski model* for spacetime, and $\mathcal{G}(\mathcal{V}^{3,1})$ is (except for a trivial difference in sign) precisely the *Spacetime Algebra* that has been so extensively applied to characterize spacetime geometry and physics [Doran and Lasenby 2003; Hestenes 1966].

Thus, the *Conformal Algebra of the Euclidean plane is isomorphic to the Geometric Algebra of Spacetime*! They differ only in geometric interpretation – and a wider difference in interpretation one can hardly imagine! Before describing the geometric interpretation of $\mathcal{C}(\mathcal{E}^2)$ in the next section, let us examine the skeleton of the algebra.

Let $\{e_0, e_1, e_2, e_3\}$ be an orthonormal basis in $\mathcal{V}^{3,1}$. The inner product specifies the signature by:

$$e_0{}^2 = -1, \quad e_1{}^2 = e_2{}^2 = e_3{}^2 = 1,$$

and orthogonality by: $e_\mu \cdot e_\nu = 0$ for $\mu \neq \nu$. The cone of null vectors is depicted in Fig. 3.

The outer product generates a basis of six *bivectors*:

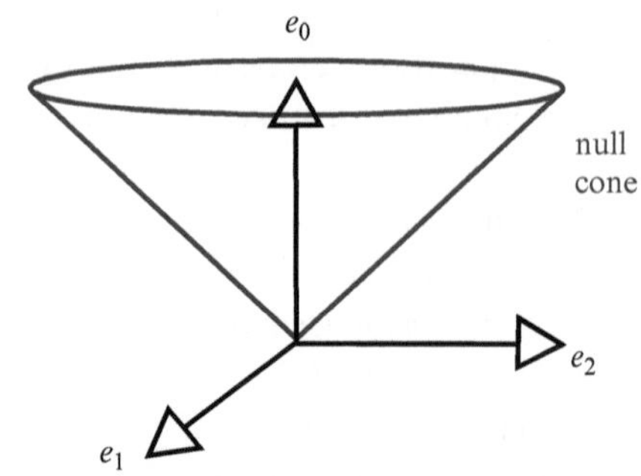

Fig. 3.

$$e_1 e_0, \quad e_2 e_0, \quad e_3 e_0, \quad e_1 e_2, \quad e_2 e_3, \quad e_3 e_1,$$

four *pseudovectors*: $e_0 I$, $e_1 I$, $e_2 I$, $e_3 I$, and a unit pseudoscalar: $I = e_0 e_1 e_2 e_3$.

The ladder of subspaces in $\mathcal{G}(\mathcal{V}^{3,1})$ is depicted in Fig. 4, with an indication of the dimension of each k-vector subspace.

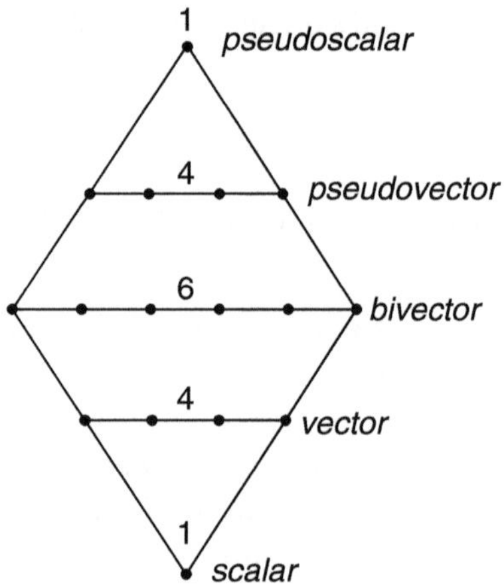

Fig. 4.

The algebra of ruler and compass

As envisioned by Leibniz and conceived by Grassmann, the ultimate goal of geometric calculus is perfect correspondence between algebraic structures, synthetic descriptions in natural language and construction of geometric figures, as summarized by:

Algebraic forms ⇔ Synthetic descriptions ⇔ Geometric figures

Grassmann came close to meeting this goal, but he was roundly criticized by two of his strongest supporters among mathematicians, Friedrich Engel and Felix Klein, for an inadequate treatment of projective infinity and ideal (imaginary) figures [Klein 1939]. My purpose here is to show how Conformal GA surpasses all expectations in completing Grassmann's program and answering his critics. It is sufficient to consider representations of geometric forms in the algebra of the Euclidean plane. Some elementary calculations will be omitted to concentrate on interpretation of results.

Every *point* p is a null vector, $p^2 = 0$ with *weight* $-p \cdot \infty = 1$, where the symbol ∞ designates the *point at infinity*, with $\infty^2 = 0$. The inner product determines the distance d_{21}between two points:

$$d_{21}{}^2 = (p_2 - p_1)^2 = -2p_2 \cdot p_1 \geq 0,$$

which vanishes if the points coincide, thus justifying the representation of points by null vectors. Grassmann himself made an intuitive distinction between points and the difference between points like $p_2 - p_1$ [H. Grassmann 1995c]. Now the distinction is encoded as an algebraic difference between two kinds of vectors, null and non-null. We shall see that this distinction has some surprising new implications. Before continuing, though, it may be worth confirming the familiar properties of a Euclidean triangle determined by points p_1, p_2, p_3. The sides of the triangle can be represented by vectors $p_{ij} = p_i - p_j$, which implies the triangle equation $p_{21} + p_{32} + p_{13} = 0$, whence the familiar *law of cosines*

$$p_{21}{}^2 + p_{32}{}^2 + 2p_{21} \cdot p_{32} = p_{13}{}^2.$$

This can be regarded as an implicit definition for the cosine of the included angle, where $p_{12} \cdot p_{32} = d_{12}d_{32} \cos \theta_{13}$. Less familiar is the fact that the trigonometric *law of sines* is expressed by

$$p_{21} \wedge p_{32} = p_{32} \wedge p_{13} = p_{13} \wedge p_{21},$$

with *directed area* A for the triangle given by

$$\begin{aligned} 2A \equiv p_{12} \wedge p_{32} &= (p_1 - p_2) \wedge (p_3 - p_2) \\ &= p_1 \wedge p_3 - p_1 \wedge p_2 + p_3 \wedge p_2 = (p_1 \wedge p_2 \wedge p_3) \cdot \infty, \end{aligned}$$

and magnitude specified by

$$4\,|A|^2 = (p_{12} \wedge p_{32}) \cdot (p_{32} \wedge p_{12}) = p_{32}^2 p_{12}^2 - (p_{12} \cdot p_{32})^2 = d_{32}^2 d_{12}^2 \sin^2 \theta_{13}.$$

Now things get more interesting.

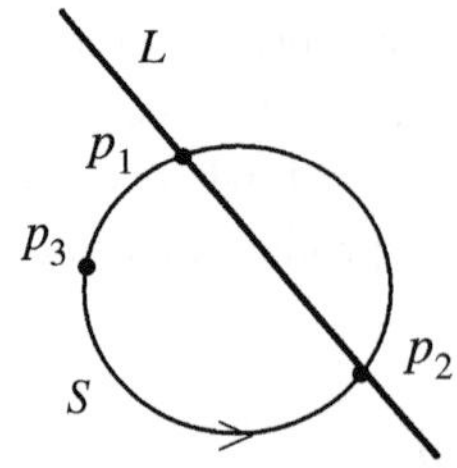

Fig. 5.

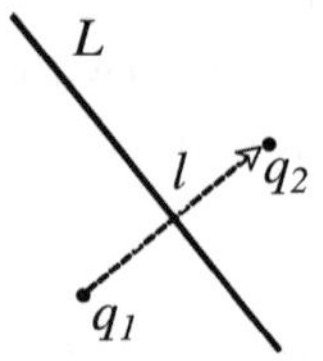

Fig. 6.

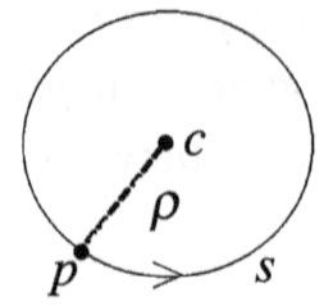

Fig. 7.

A *circle* S is generated by the product of three points: $S = p_1 \wedge p_2 \wedge p_3$. This algebraic form can be interpreted as an instruction for drawing a circle through three given points, as depicted in Fig. 5. Each *line* L is a circle through ∞, as expressed by $L = p_1 \wedge p_2 \wedge \infty$ and also depicted in Fig. 5. Circles and lines are *oriented*, as expressed by a change in sign induced by interchanging the order of points; thus, $-S = p_3 \wedge p_2 \wedge p_1$ and $-L = p_2 \wedge p_1 \wedge \infty$. An orientation for the circle, but not the line, is indicated in Fig. 5. In many applications, such as projective geometry, the orientation is not of interest, so the sign can be ignored.

It is often more convenient to represent geometric objects by their duals. The *dual forms* for both lines and circles are non-null vectors. In particular, the dual of line L is the *normal* for the line $l = L^* = LI$. The normal for a line is distinguished from other vectors by being orthogonal to the point at infinity; that is,

$$L \wedge \infty = 0 \quad \Leftrightarrow \quad l \cdot \infty = 0.$$

This property implies that the normal can be expressed as the difference between two points: $l = q_2 - q_1$, as depicted in Fig. 6. Conversely, the vector difference between any two points is the "perpendicular bisector" of the line through those points. The line has a magnitude given by

$$L^2 = (p_2 - p_1)^2 = l^2 = \left(q_2 - q_1\right)^2 > 0,$$

where the points are those depicted in Figs. 5 and 6. Consequently, L can be interpreted as a line segment, though it does not specify a location of the segment along the line.

The dual of circle S is a vector

$$S^* = s = c - \frac{1}{2}\rho^2\infty,$$

which, like the normal for a line, is the difference between points; but, in this case, one point is the centre of the circle c, while the other is the weighted point at infinity. Indeed, it is readily verified that $c^2 = 0$ and $s^2 = \rho^2$. To verify that scalar ρ is the radius of the circle, note that the condition for a point p to lie on the circle is

$$p \wedge S = 0 \quad \Leftrightarrow \quad p \cdot s = 0.$$

The usual equation for a circle then follows easily:

$$(p - c)^2 = -2p \cdot c = \rho^2.$$

Thus, we have discovered the surprising fact that the specification of a circle by its centre and radius (in Fig. 7) is dual to its specification by three points (in Fig. 5).

One advantage of Conformal GA is that projective geometry is fully integrated with metrical geometry. For example, the *incidence relations* among points and lines are expressed directly by the meet product. Thus, the incidence (intersection) of a point p with line L is expressed by

$$p \vee L = p \cdot L^* = p \cdot l = 0,$$

as depicted in Fig. 8. Parallelism of lines depicted in Fig. 9 is expressed by

$$L_1 \vee L_2 = 0 \quad \Leftrightarrow \quad l_1 \wedge l_2 = 0.$$

The intersection of lines at a point depicted in Fig. 10 is expressed by

$$p \vee (L_1 \vee L_2) = p \cdot (l_1 \wedge l_2) = 0.$$

The projective incidence relations for intersecting objects generalize automatically to metrical relations for objects in arbitrary positions. This could be called *metric incidence*. Thus, as depicted in Fig. 11, the distance δ between a point and a line is given by

$$p \cdot \hat{l} = \pm\delta, \quad \hat{l}^2 = 1,$$

where the sign depends on the orientation of the line, so it distinguishes on which side of the line the point lies.

Fig. 8.

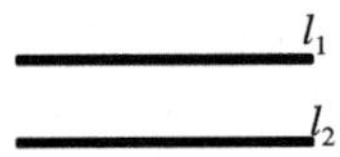

Fig. 9.

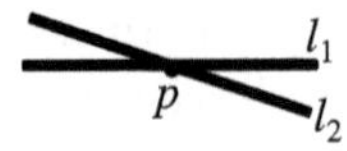

Fig. 10.

As depicted in Fig. 12, the distance between a point and a circle is given by

$$s \cdot p = c \cdot p + \frac{1}{2}\rho^2 = \frac{1}{2}\left[\rho^2 - d^2\right],$$

where d is the distance from the point to the centre of the circle. Clearly, the sign of $s \cdot p$ specifies whether the point is inside or outside the circle. It is therefore a topological property expressing relative orientation. More generally, if signs are retained but magnitudes are ignored, we have a generalization of projective geometry to include orientation that is very useful in geometric computation [Stolfi 1991].

Since a point can be regarded as a circle with zero radius, we have the obvious generalization to a distance between two circles:

$$s \cdot s' = \left(c - \frac{1}{2}\rho^2\infty\right) \cdot \left(c' - \frac{1}{2}\rho'^2\infty\right) = \frac{1}{2}(\rho^2 + \rho'^2 - d^2).$$

Now we are prepared to give a response to the critique of Engel and Klein that Grassmann's algebra cannot handle *ideal geometric forms*. The sum of two points gives us something that we have not seen before:

$$\bar{s} \equiv \frac{1}{2}(p_1 + p_2) = c + \frac{1}{2}\rho^2\infty.$$

This can be regarded as an *imaginary circle*, because $\bar{s}^2 = -\rho^2$. Imaginary circles arose first as complex solutions of quadratic equations, and then demanded a geometric interpretation (Klein 1939). We have here a new possibility.

It is an old idea that *lines can be regarded as points at infinity*. Accordingly, we can represent lines by their normals, and define a *circle at infinity* as the join of two normals with the circle s defined before:

$$\overline{S} = l_1 \wedge l_2 \wedge s = (l_1 \wedge l_2)s.$$

Its dual form is the vector $\bar{s} = \overline{S}^*$. Since $\overline{S}^2 = (l_1 \wedge l_2)^2 s^2$, we can normalize with $(l_1 \wedge l_2)^2 = -1$, so $\bar{s}$ has the same *centre* c and *radius* ρ as the *real circle* s.

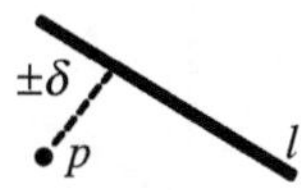

Fig. 11.

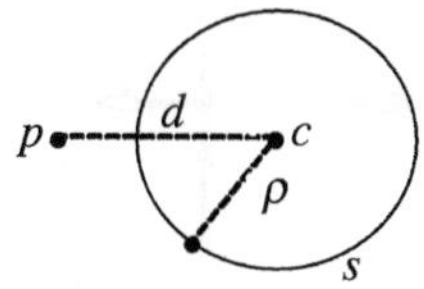

Fig. 12.

Though the circle centre is a *real point* (a null vector), there can be no real point p on the imaginary circle, because the equation $\bar{s}\cdot p = c\cdot p - \rho^2/2 = 0$ has no null vector solutions. However, the equation $\bar{s}\cdot l = c\cdot l = 0$ does have imaginary points as its solution. Indeed, the solution set consists of all lines with weight ρ passing through the circle centre. Thus, we see that *an imaginary circle is just a representation of a real circle by the family of lines through its centre.*

As far as I know, this is a completely new perspective on imaginary circles, and it illustrates how Conformal GA can provide spectacular rejoinders to objections by Grassmann's critics.

All the geometric constructions so far have involved only inner and outer products. It remains to demonstrate advantages of using the geometric product directly. Let n be the unit normal for a given line. Then it is easy to prove [Hestenes 2003b] that the *reflection* of every point p (hence every geometric object) across that line is specified by the transformation

$$p \quad \mapsto \quad p' = -npn,$$

as shown in Fig. 13. Following this with a second reflection across a line with unit normal m, the net result is a *rigid displacement*:

$$p \quad \mapsto \quad p'' = Dp\tilde{D},$$

where $D = mn$ with reverse $\tilde{D} = nm$. If the two lines are parallel, the displacement is a *translation* through twice the distance δ between the lines (Fig. 13). If the lines intersect at a point, the displacement is a *rotation* about that point through twice the angle between the lines (Fig. 14). Thus, the well-known synthetic description of translations and rotations in terms of reflections through lines (or planes in $\mathcal{E}^3$) has been reduced to the simple geometric product of vectors. One consequence is considerable simplification in the treatment of crystallographic symmetries [Hestenes and Holt 2007].

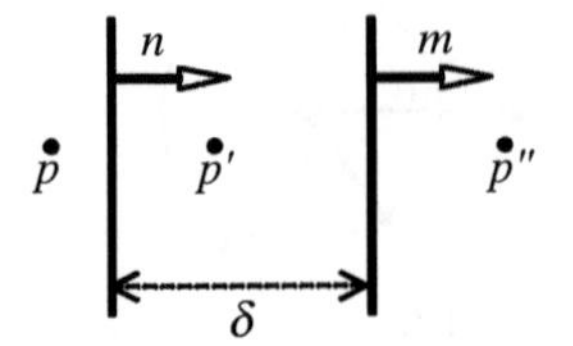

Fig. 13.

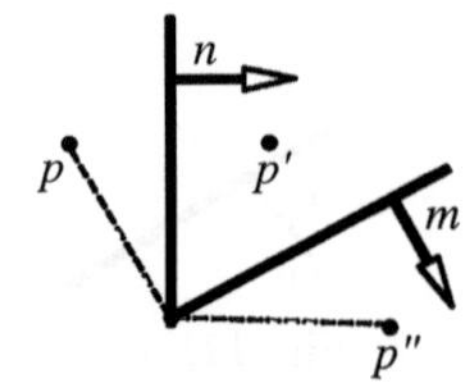

Fig. 14.

The present formulation for rigid displacements, hence of *congruence*, applies without change in form to symmetries in $\mathcal{E}^n$. In $\mathcal{E}^3$ it provides the foundation for powerful engineering applications. For example, in rigid body dynamics it unifies translational and rotational equations (Newton's and Euler's Laws) into a single equation of motion [Hestenes 2010].

I believe Grassmann would be greatly pleased!!

On Grassmann's regressive product

Henry Crapo

A new mathematical discipline

Grassmann's plan was to develop a purely formal algebra to model natural operations on geometric objects: *flat* or *linear* pieces of space of all possible dimensions. His approach was to be *synthetic*, so that the symbols in his algebra would denote geometric objects themselves, not just numbers (typically, coordinates) that could be derived from those objects by measurement. His was not to be an algebra of numerical quantities, but *an algebra of pieces of space*.

In the *analytic* approach, so typical in the teaching of Euclidean geometry, we are encouraged to assign "unknown" variables to the coordinates of variable points, to express our hypotheses as equations in those coordinates, and to derive equations that will express our desired conclusions.

The main advantage of a *synthetic* approach is that the logic of geometric thought and the logic of algebraic manipulations may conceivably remain parallel, and may continue to cast light upon one another. Grassmann expressed this clearly in his introduction to the *Ausdehnungslehre*[1]:

---

[1] This quotation is from the Foreword to the first edition, which we denote A1, on page 10 of the English translation [A1$_K$]. We owe special thanks to Lloyd Kannenberg for his fine English translation of Grassmann's mathematical works.

H.-J. Petsche et al. (eds.), *From Past to Future: Graßmann's Work in Context*,
DOI 10.1007/978-3-0346-0405-5_23, © Springer Basel AG 2011

"Thus not only can each formula that arises in the course of a derivation be most easily expressed verbally, but in addition each step from one formula to another appears at once as just the symbolic expression of a parallel act of abstract reasoning. The methods formerly used require the introduction of arbitrary coordinates that have nothing to do with the problem and completely obscure the basic idea, leaving the calculation as a mechanical generation of formulas, inaccessible and thus deadening to the intellect. Here, however, where the idea is no longer strangely obscured but radiates through the formulas in complete clarity, the intellect grasps the progressive development of the idea with each formal development of the mathematics." [$A1_K$, Foreword]

We may also suggest a contemporary rationale for the synthetic approach. After completion of a synthetic calculation:

- There is no need to *climb back up* from scalars to statements of geometric incidence.
- There are no real numbers, necessarily subject to round-off errors.
- There is no division by zero.
- There are no drawings, fraught with their own approximations of incidence.
- There are precise warnings as to particular positions of degeneracy.

So the synthetic approach is tailor-made for machine computation.

An algebra of pieces of space

Grassmann emphasizes that he is building an *abstract* theory that can then be *applied to real physical space*. He starts not with geometric axioms, but simply with the notion of a skew-symmetric product of letters, $ab + ba = 0$. He finds in practice, with his geometric model, that this product is distributive over addition and modifiable by scalar multiplication. By skew symmetry (and $2 \neq 0$), $aa = 0$.

A k-extensor is a product of k points. Two k-extensors are *equal* if they represent pieces of the same projective subspace, of the same orientation and $(k-1)$-dimensional volume. Figure 1 shows how any piece of plane is equal, as an extensor, to a triangular area with one edge along a given line anywhere in its plane.

Since any two projectively distinct 3-extensors (planes) in 3-space are representable as triangles in distinct planes, but sharing a common edge, the addition of planes in 3-space is reduced to the addition of points on a line, as in Fig. 2.

Grassmann studies the addition of oriented line segments, being careful to distinguish between the *intensive magnitude* required to create a displacement and the

Fig. 1. The 3-extensors *abc* and *a′bc′* are equal

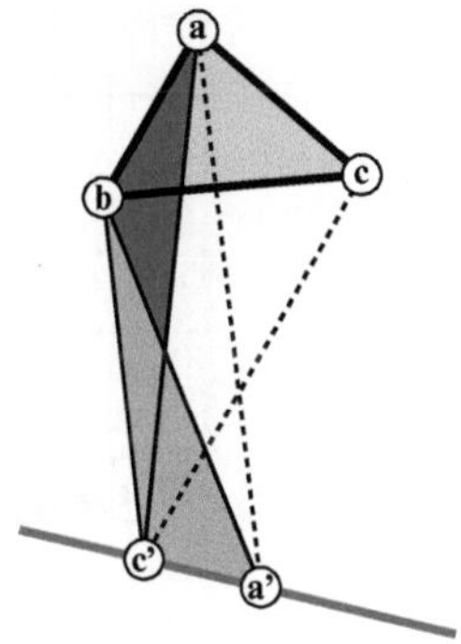

Fig. 2. A sum of cospatial 3-extensors, $pqr + pqs = 2pqt$

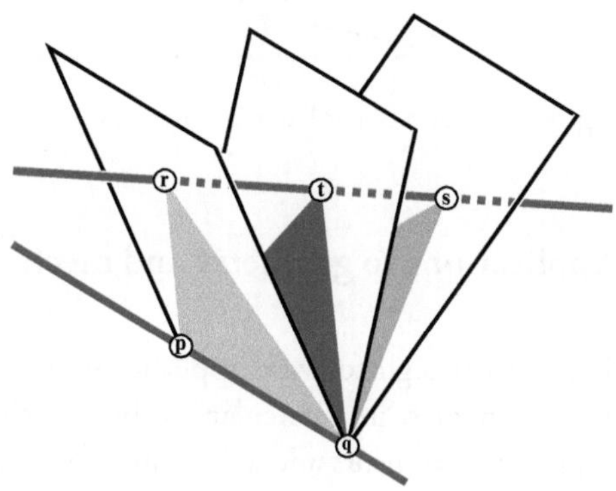

extensive magnitude thereby created, the set of points traversed under such a movement. He then defines a *product* of displacements.

Grassmann proposes to *apply* this abstract algebra to geometry via an *interpretation*:

- The individual letters may be understood as *points*.
- The center of gravity of several points can be interpreted as their sum.
- The displacement between two points as their product.
- The surface area lying between three points as their product.
- The region (a pyramid) between four points as *their* product.

He is delighted to note that his approach is independent of the dimension of overall space in question: "[T]he origins of this science are as immediate as those of arithmetic; and in regard to content, ... the limitation to three dimensions is absent. Only thus do the laws come to light in their full clarity and generality, and are their essential interrelationships revealed." [$\mathrm{A1_K}$, Foreword]

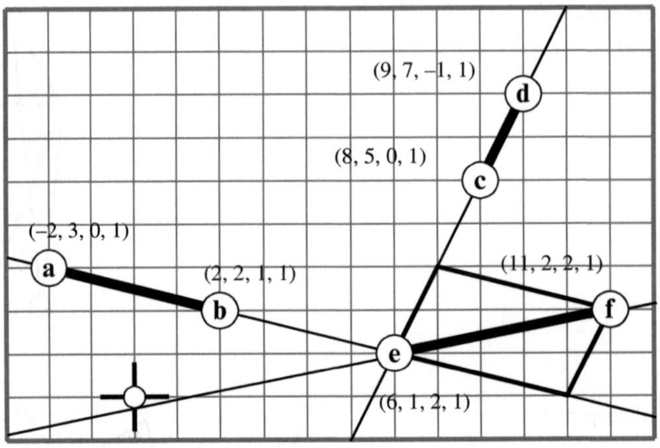

Fig. 3. Addition of coplanar 2-extensors

Applications to geometry and mechanics

The most straightforward application of Grassmann's new discipline was to geometry and mechanics, in particular to the additive group of displacements in space, where coplanar 2-extensors add as coplanar forces do in physical systems.

For one concrete exercise, let us step forward 150 years, and carry out the tensor calculations necessary in order to add two coplanar line segments *ab* and *cd*, to find the point *e* of intersection of the lines they span, and the coordinates of the common plane *Q* on which they lie. This problem is interesting, in that the exterior product is itself of no avail, the exterior product of two coplanar lines being zero.

Forces represented by the line-bound vectors *ab* and *cd*, as in Fig. 3 are added by sliding the directed line segments along their lines of action until the ends *a* and *c* coincide. (Note that this figure is not in the *xy*-plane of the drawing: points *a* and *c* are in that plane, but *d* is below, *b*, *e*, *f* above.)

So you know I'm not bluffing, here are the detailed calculations:

	1	**2**	**3**	**4**
a	−2	3	0	1
b	2	2	1	1
c	8	5	0	1
d	9	7	−1	1

	12	**13**	**14**	**23**	**24**	**34**
ab	−10	−2	−4	3	1	−1
cd	11	−8	−1	−5	−2	1
ab + cd	1	−10	−5	−2	−1	0

	1	**2**	**3**	**4**
e	6	1	2	1
f	11	2	2	1

	12	**13**	**14**	**23**	**24**	**34**
ef	1	−10	−5	−2	−1	0

Note that $ab + cd = ef$, an equality of 2-extensors.

Two such coplanar lines in a space of higher dimension, determine *the plane Q they span* and *their point e of intersection*, lying in that plane. The exterior product does not help, since the lines are not independent, so their exterior product $(ab)(cd)$ is zero. But all the required information is present in the homogeneous coordinates of the lines *ab* and *cd*. In particular, we can show that since *abcd* is zero,

$$a \otimes bcd - b \otimes acd - c \otimes abd + d \otimes abc = 0$$

so,

$$a \otimes bcd - b \otimes acd = c \otimes abd - d \otimes abc$$

are two valid expressions for a point collinear with both a, b and c, d. Since *bcd* and *acd* represent the same plane, they differ only by a scalar multiple. Shifting these scalars onto the points *a* and *b*, we find a point somewhere on the line *ab* tensored with a 3-extensor representing the plane Q. Similarly shifting scalars in the expression $c \otimes abd - d \otimes abc$, we find a point on the line *cd* tensored with a 3-extensor representing the same plane Q. So each of these two expressions is equal to the tensor product of the point *e* of intersection and the plane Q containing the two lines.

To see more clearly how to recover the homogeneous coordinates of the point of intersection and common plane of two coplanar lines, we now redo this calculation in a contemporary algebraic language: in terms of letter/place algebra.

For any four points, not necessarily coplanar, a, b, c, d (negative letters) and places i, j, k, l (negative places), the following four displayed sums are equal:

$$\left(\left(\begin{matrix} a \\ bcd \end{matrix} - \begin{matrix} b \\ acd \end{matrix} \right) \middle| \begin{matrix} i \\ jkl \end{matrix} \right)$$

$$(+a_i b_j c_k d_l - a_i b_j c_l d_k + a_i b_k c_l d_j - a_i b_k c_j d_l + a_i b_l c_j d_k - a_i b_l c_k d_j)$$
$$-(+a_j b_i c_k d_l - a_j b_i c_l d_k + a_k b_i c_l d_j - a_k b_i c_j d_l + a_l b_i c_j d_k - a_l b_i c_k d_j)$$

$$(+a_i b_j c_k d_l - a_i b_j c_l d_k + a_j b_i c_l d_k - a_j b_i c_k d_l)$$
$$-(a_i b_k c_j d_l - a_i b_k c_l d_j + a_k b_i c_l d_j - a_k b_i c_j d_l)$$
$$+(a_i b_l c_j d_k - a_i b_l c_k d_j + a_l b_i c_k d_j - a_l b_i c_j d_k)$$

$$\left(\begin{matrix} ab \\ cd \end{matrix}\middle| \left(\begin{matrix} ij \\ kl \end{matrix} - \begin{matrix} ik \\ jl \end{matrix} + \begin{matrix} il \\ jk \end{matrix}\right)\right)$$

In the specific example at hand, the calculation of the tensor product, supposed to represent $e \otimes Q$, where Q is the plane spanned by ab and cd, yields the matrix

	123	**124**	**134**	**234**
1	−102	−54	30	6
2	−17	−9	5	1
3	−34	−18	10	2
4	−17	−9	5	1

which does not look very promising ... until you realize that the tensor is *reducible*, the matrix of coordinates has *rank* 1, all its 2×2 minors are zero, and it is expressible as the tensor product of a 1-extensor representing the point e of intersection, with a 3-extensor $Q = -\frac{1}{2}abc$ representing the plane of the 4 points a, b, c, d.

	123	**124**	**134**	**234**	**e**
1	−102	−54	30	6	**6**
2	−17	−9	5	1	**1**
3	−34	−18	10	2	**2**
4	−17	−9	5	1	**1**
Q	**−17**	**−9**	**5**	**1**	

This calculation is the expression, in homogeneous coordinates, of an example of Grassmann's regressive product. Bear in mind that this factorization is unique up to a scalar factor that can be shifted from one factor to the other. In fact, that is the whole point! – and led to the invention of the *tensor product*.

Fortunately, many speakers at the Grassmann Bicentennial Conference dealt with related issues, and there were splendid opportunities to exchange views. To mention a few of the more closely related talks:

- John Browne: *Grassmann, geometry and mechanics.*
- David Hestenes: *Grassmann's legacy.*
- Hongbo Li: *Projective geometric theorem proving with Grassmann–Cayley algebra.*
- Paola Cantù: *Grassmann's epistemology: multiplication and constructivism.*
- Not forgetting, from the 1994 Grassmann Colloquium, the paper of Andrea Brini and Antonio Teolis on Grassmann–Cayley algebras [Brini and Teolis 1996].

The regressive product

It troubled Grassmann that the outer product of pieces of space is zero whenever the subspaces they span have a common point.

"Now since this definition characterizing the product as "outer" is not inherent in the concept of a product as such, it must be possible to retain the general concept of a product and yet give up that definition or replace it with another. In order to discover this new definition we must investigate the different degrees of dependence, since according to this new definition the product of two dependent magnitudes can also have a non-trivial value." [$A1_K$, 125]

So the regressive product is *not "just the linear dual of the exterior product"*. It *modifies and extends* the concept of exterior product to yield a non-trivial value for the product of dependent magnitudes. To think it is "simply the dual exterior product" is a common misreading and misconception, especially today. This misreading may be largely due to the fact that Grassmann retracted the essential features of this proposed theory of the regressive product before the publication of the 1862 *Ausdehnungslehre* [A2; $A2_K$].

Subordinate form

Given any two extensors A (rank a) and B (rank b), as in Fig. 4, together generating a subspace E of rank e, thus having a common subspace C of rank c, with $a + b = c + e$, represent B as an exterior product $B = CD$ where D is a relative complement of C in B [$A1_K$, 126].

Grassmann decides that "the concept of the regressive product is still dependent on the values of the orders, since ... two products cannot yet be considered equal so long as their factors have unequal orders. This dependence of the concept on the values of the orders introduces a limitation that spoils the simplicity of the concept and is repugnant to the analytic treatment." [$A1_K$, 132]

He refines the definition of product, to state that "the value of a regressive product consists in the common and [i.e., together with the] nearest covering system of the two factors, if the order of the factors is given, apart from [a scalar] quantity to be distributed multiplicatively on the two systems." [$A1_K$, 131]

Thus, he sets in [$A1_K$, 132], for the pair of extensors indicated in Fig. 4,

$$A \circ CD = AD \circ C$$

for the regressive product $A \circ B$, where C represents the common system of A and B, and $B = CD$. Here we have changed the tensor product symbol $\otimes$ to $\circ$ because the above expression is not an equality in the tensor algebra of the exterior algebra, but rather expresses the fact that the regressive products of $A \otimes CD$ and $AD \otimes C$ are equal. (Note that $A \otimes CD = AD \otimes C$ is not an equality in the tensor algebra itself because $A \otimes CD$ and $AD \otimes C$ have different grades.)

Note carefully that Grassmann states that the regressive product of two extensors is *equal* to its associated flag representation, $A \circ CD = AD \circ C$. He also figures out the *sign law* for [skew] commutativity of the regressive product in terms of the *supplementary numbers* $a - c$ and $b - c$

$$A \circ B = (-1)^{(a-c)(b-c)} B \circ A$$

It is fascinating that he managed to get this right!

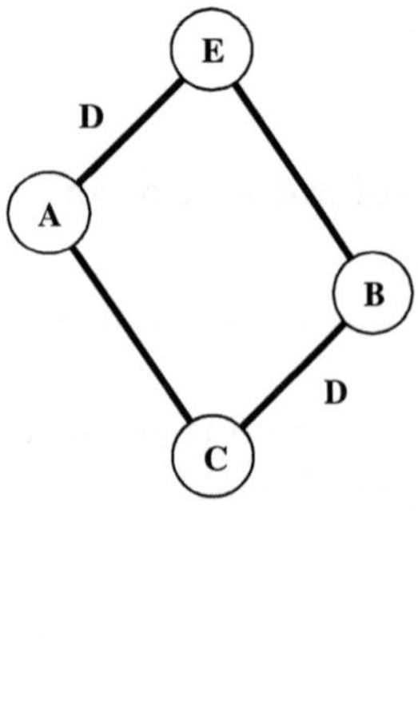

Fig. 4. Expression of a regressive product in subordinate form

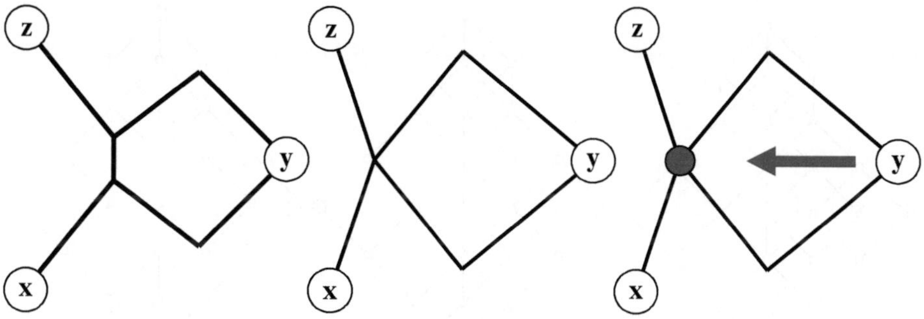

Fig. 5. Nonmodularity, modularity, and projection in a lattice interval

Modular lattices

In any lattice, for any pair of elements in the order $x < z$, and for any element y,

$$x \vee (y \wedge z) \leq (x \vee y) \wedge z,$$

as in Fig. 5. A lattice is *modular* if and only if, under these same conditions, equality holds:

$$x \vee (y \wedge z) = (x \vee y) \wedge z.$$

Thus, in a modular lattice, there is a natural method of *projection* of an arbitrary lattice element into an arbitrary lattice interval (on the right, in Fig. 5). The lattice of subspaces of any vector space is modular, so any calculus of linear pieces of space will use *the logic of modular lattices.*

Nonassociativity of the geometric product

Dedekind constructed the free modular lattice on three generators, as shown in Fig. 6. We show the result of projecting b into the lattice intervals $0 < c < 1$, to obtain the chain $b \wedge c < b \vee c$, then projecting a into the intervals $0 < b \wedge c < b \vee c < 1$ to obtain the 3-chain $a \wedge b \wedge c < q < a \vee b \vee c$, where q is the lattice element

$$(a \vee (b \wedge c)) \wedge (b \vee c).$$

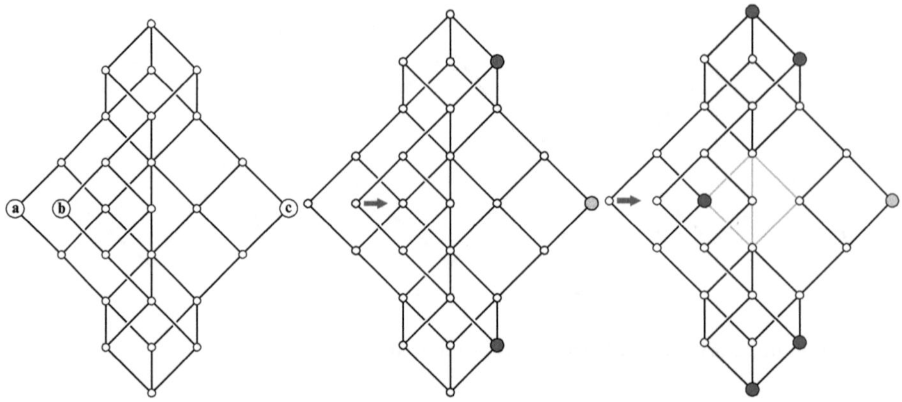

Fig. 6. The free modular lattice on generators a, b, c, with geometric products $b \circ c$ and $a \circ (b \circ c)$

Note that this process leaves a trace of the order in which the factors were combined! Try this with the simple case of three collinear points a, b, c. Then

$$a \circ (b \circ c) = a \circ (1 \circ bc) = 1 \circ a \circ bc,$$

but

$$(a \circ b) \circ c = (1 \circ ab) \circ c = 1 \circ c \circ ab.$$

Parentheses are not necessary in the final flags, because flag products are associative. All products in which a enters last will be the same, up to sign:

$$a \circ (b \circ c) = (b \circ c) \circ a = -(c \circ b) \circ a.$$

Multiplication of flags

The regressive product of two flag products *is*, however, *well defined*. Grassmann showed this to be true, *an incredible feat, given the tools he had at hand!*

He says: "Instead of multiplying by a product of mutually incident factors one can progressively multiply by the individual factors, and indeed in any order." [$A1_K$, 136]

Garrett Birkhoff, in the first edition of his *Lattice Theory* [Birkhoff 1940], proved that the free modular lattice generated by two finite chains is a finite *distributive* lattice. *This changes the whole game.* This gives rise to the fact that Grassmann's geometric

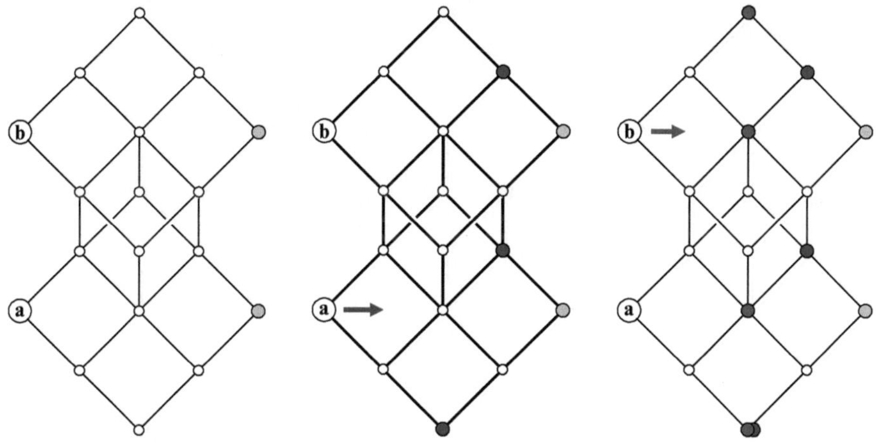

Fig. 7. Projecting a then b into the 2-chain $c < d$

product is *associative* as long as it is applied only to tensor products of extensors that fall into two chains.

In Fig. 7 we show the process of projecting the 2-chain $a < b$ into the 2-chain $c < d$, forming the geometric product $b \circ (a \circ (c < d))$. This results in the same 4-chain

$$a \wedge c < (a \vee c) \wedge (b \wedge d) < (b \wedge d) \vee (a \vee c) < b \vee d,$$

as when we project d then c into the 2-chain $a < b$, as in Fig. 8, forming the geometric product $c \circ (d \circ (a < b))$. The geometric product of these two 2-chains is well defined as running up the *backbone* of the distributive lattice.

For longer chains the calculation is carried out in an analogous distributive lattice (see Fig. 9), easily characterized as the lattice of order ideals in a direct product of two chains, less the top and bottom elements 1 and 0.

Where did this leave Grassmann?

By the end of his chapter on the regressive product, Grassmann seems rather disheartened. He admits clearly that "the multiplicative law of combination *[associativity]*... is not generally valid for the mixed product of three factors." [$A1_K$, 139]

He includes a footnote to say that "cases can be found in which our law still finds its application" via the results available for the product of an extensor by a flag of extensors,

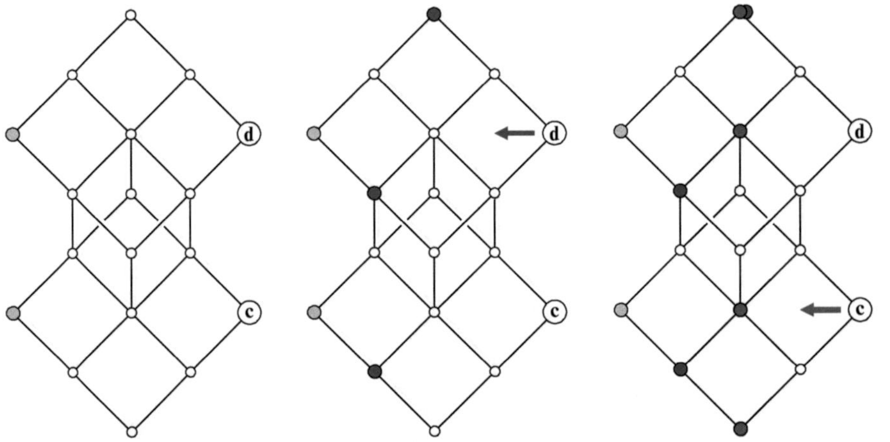

Fig. 8. Projecting d then c into the 2-chain $a < b$

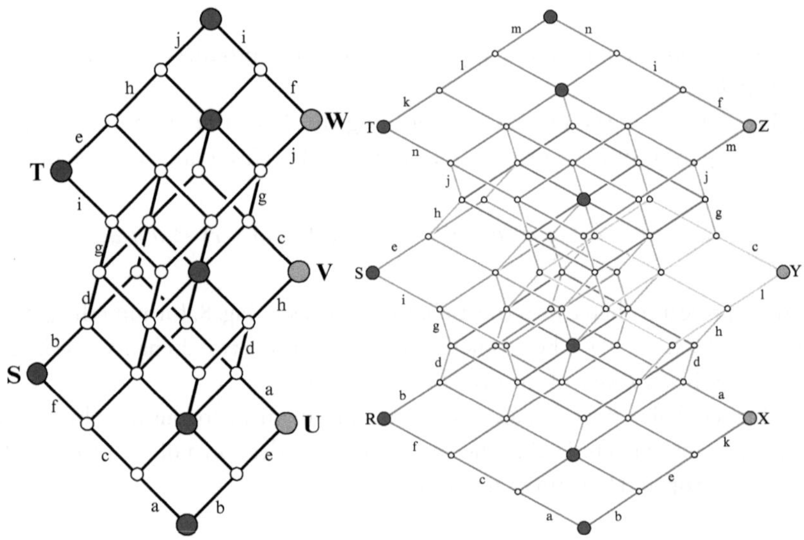

Fig. 9. The products of an n-chain with a 3-chain, for $n = 2, 3$

but concludes, with a certain degree of disillusion, that "these cases are so isolated, and the constraints under which they occur so contrived, that no scientific advantage results from their enumeration."

He is quite satisfied with the extreme cases: for joins of independent extensors and for meets of co-independent extensors, having investigated the duality between these two cases, and having proven they are associative. He concludes with the note: "the theoretical presentation of this part of extension theory now appears as completed, excepting consideration of the types of multiplication for which the law of combination [associativity] is no longer valid." [$A1_K$, 143]

Where does this leave us?

Clarification of these questions involving the regressive product of two flags was the subject of joint work with Andrea Brini, Francesco Regonati, and Bill Schmitt, last year in Bologna.

There exists a major divide between

- The lattice theory of products whose definition depends on the ranks of joins and intersections, products that do not respect the additive structure of the exterior algebra, but do provide flag representations of geometric quantities.
- The algebraic theory of products of *fixed intersection degree*, which act as bilinear operators on the exterior algebra, but do not yield flag representations.

This sort of separation between theoretical contexts is one small example of the sort of problem that frequently arises when one tries to construct algebraic representations of geometric structures and operations, or geometric representations of physical phenomena. Compare the situation where Grassmann is building a universal algebra from a physical, or at least geometric, structure, with that when Hestenes and Li are adjusting the signatures of a quadratic form in order to get a hold on certain geometric properties.

This is a perfect problematic for category theorists. We should really have a categorical definition of an *algebraic representation* or a *geometric interpretation*, by which we can hope to judge the proper domain of application and the reliability of a theoretical framework purporting to link algebra with geometry, or with physics. (Consider the great variety of proposed links between particle physics and geometries of all sizes and shapes.) In principle, this should do for *écritures* and applications of geometry what André Joyal's theory of *species of structures* has done for algebraic combinatorics.

Technically speaking, there is a perfectly acceptable calculation of products with fixed intersection degrees in terms of the signed Boolean algebra. I gave a short talk on this at the 1990 ICM in Kyoto. Bernard Mourrain liked the idea, and went on to develop the idea of a complete geometric product, additively combining the results obtained for different degrees of possible intersection [Mourrain 1991]. This geometric product is now part of Hongbo Li's arsenal [Li 2008] for geometry theorem proving.

The fact that the p-relations [Hodge and Pedoe 1954] holding among dependent sets of vectors, say $\{a, b, c, d\}$, as coplanar projective points, (the *brackets* being determinants [scalars]),

$$[abc]d - [abd]c + [acd]b - [bcd]a = 0,$$

is the starting point for the definition of the Grassmann–Cayley algebra (or Peano space) [Barnabei, Brini and Rota 1985] of *joins and meets*. That such expressions hold equally in arbitrary dimension if rewritten as tensor products,

$$abc \otimes d - abd \otimes c + acd \otimes b - bcd \otimes a = 0,$$

this latter expression being a homogeneous slice of the coproduct of the dependent set *abcd*, leads to the definition of the *Whitney algebra of a matroid* [Crapo and Schmitt 2000], as a *lax Hopf algebra*, quite generally applicable in combinatorial geometry. The Whitney algebra of a matroid has a well-defined *geometric product*, an analogue of Grassmann's regressive product.

Andrea Brini and Francesco Regonati have recently established Gian Carlo Rota's insistent affirmation that the Whitney algebra is to be written super-algebraically as an algebra in negative letters and positive places, the latter to code tensor product position.

When this work is combined with the extraordinary synthesis [Bravi and Brini 2001; Brini, Regonati and Teolis 2001] of Clifford algebra and Grassmann–Cayley algebra, all made super-symmetric, already achieved by Andrea Brini and Francesco Regonati, and Paolo Bravi, I feel there may finally be the makings of a banquet that can truly be termed *geometric algebra*.

Giving Hermann Grassmann the final word

"Even if this work as well should lie idle yet another seventeen years or more without influencing the living development of science, a time will come when it will be drawn forth from the dust of oblivion and the ideas laid down here will bear fruit." (From *Die Ausdehnungslehre von 1862* [$A2_K$, xvii].)

That time may yet come, even for the regressive product! *N'est-ce pas?*

Let me take this opportunity to express my thanks to Professor Petsche and his colleagues who have organized this wonderful opportunity to share our discoveries in the marvelous work of Hermann Grassmann. *Happy birthday!*

Projective geometric theorem proving with Grassmann–Cayley algebra

Hongbo Li

Introduction

Grassmann–Cayley algebra was invented by Grassmann and Cayley in the nineteenth century [$A1_K$]. It is an algebra equipped with two products: the exterior product (outer product), and the dual of the exterior product called the meet product. Geometrically, this algebra provides an invariant language for the synthetic projective geometry on the incidence relations among points, lines and other "flat" objects. The algebra of invariants associated with this algebra is the so-called bracket algebra, or the algebra of determinants [White 1975].

In the era of computer machinery, the two classical algebras were revitalized by the need of powerful algebraic manipulations for geometric computing [Rota and Stein 1973; Barnabei, Brini and Rota 1985; Bokowski and Sturmfels 1989; Hestenes and Ziegler 1991; White 2005], and have found important applications in emerging fields like combinatorics, robotics, computer algebra, etc.

In the field of automated theorem proving, inspired by Wu's coordinate approach to automated geometric reasoning [Wu 1984, English translation in 1994],

This work was completed with the support of NSFC 10871195 and 60821002/F02.

H.-J. Petsche et al. (eds.), *From Past to Future: Graßmann's Work in Context*,
DOI 10.1007/978-3-0346-0405-5_24,

Grassmann–Cayley algebra and bracket algebra were used for coordinate-free geometric reasoning [Crapo and Richter-Gebert 1994; Richter-Gebert 1995; Li and Wu 2003; Li 2008]. These methods improve the elegance of algebraic proofs by reducing the size of expressions produced in the proofs, with the cost of reduced range of applicability.

The state of art of the invariant approach to projective incidence geometry can be found in [Li and Wu 2003] and [Li 2008]. Powerful algebraic manipulations such as Cayley expansion, contraction, rational transformation and Cayley factorization are proposed in these works, which reduce the size of middle expressions in algebraic proofs to such a degree that no theorem in projective incidence geometry tested so far has not been given a *binomial proof*, i.e., a proof throughout which the number of terms in any expression produced by algebraic manipulations is at most two. Such proofs can be dubbed *optimal algebraic proofs*, as no proof of smaller size in terms of algebraic expressions can be found. Without the new techniques, only a minority of theorems in projective incidence geometry can be given binomial proofs.

With the advent of metric structure and metric geometry, algebras with more complicated structures are needed. In Euclidean geometry, *conformal Grassmann–Cayley algebra* [Li 2008] extends the representation ability of the meet product and the outer product to not only include circles and spheres of various dimensions, but allow symmetry generators to participate in the two products. In other words, a flat or round geometric object can be represented by the span of not only lower dimensional objects incident to it, but also symmetries of the object such as reflection invariance, inversion invariance, and antipodal inversion invariance.

In this paper we barely have enough space to introduce classical Grassmann–Cayley algebra and it applications in obtaining binomial proofs for theorems in projective incidence geometry. Since there is no room left for introducing new developments of this algebra in the setting of metric geometry, some references given in this paper can be resorted to for further reading.

Classical Grassmann–Cayley algebra

Let $\mathbb{K}$ be a number field of characteristic $\neq 2$. An nD projective space over $\mathbb{K}$, denoted by $\mathbb{PK}^n$, is composed of all 1-D linear subspaces in $\mathbb{K}^{n+1}$, each being called a *projective point*. *Projective transformations* are linear transformations in $\mathbb{K}^{n+1}$ modulo $\mathbb{K}-\{0\}$.

A projective space has a natural grading structure. *Projective lines* are 2-D linear subspaces in $\mathbb{K}^{n+1}$, *projective planes* are 3-D linear subspaces in $\mathbb{K}^{n+1}$, and *projective r-planes* are $(r+1)$-D linear subspaces in $\mathbb{K}^{n+1}$. The *dimension* of a projective r-plane refers to r. The number r ranges from 0 to n. A projective point is a 0-plane.

Projective incidence geometry is the part of projective geometry on incidence relations among points, lines, planes, etc. A geometric object is *incident* to another geometric object if one is part of the other.

A projective point as a 1-D linear subspace can be represented by a non-zero vector of the subspace, and the representation is unique up to scale. An algebraic representation is said to be *homogeneous* if it is unique up to scale. In projective geometry, all algebraic representations must be homogeneous.

The introduction of Grassmann algebra is stimulated by the need to represent high-dimensional "flat" objects by the generating of low dimensional ones. To start with, a projective line is determined by any two points incident to it, say points $\mathbf{a}, \mathbf{b}$ each represented by a vector. The intention is to construct a product, denoted by "$\wedge$", of $\mathbf{a}$ and $\mathbf{b}$, so that $\mathbf{a} \wedge \mathbf{b}$ represents the line. The homogeneity of the representation requires the product be linear with respect to both $\mathbf{a}$ and $\mathbf{b}$. If $\mathbf{a}$ and $\mathbf{b}$ represent the same point, they cannot determine a line, and the element zero is proposed to represent the failed result:

$$\mathbf{a} \wedge \mathbf{a} = 0. \tag{1}$$

Furthermore, for three points $\mathbf{a}, \mathbf{b}, \mathbf{c}$, the plane incident to them can be constructed as the plane incident to either line $\mathbf{a} \wedge \mathbf{b}$ and point $\mathbf{c}$, or point $\mathbf{a}$ and line $\mathbf{b} \wedge \mathbf{c}$. Algebraically, this corresponds to the associative extension of the "$\wedge$" product:

$$(\mathbf{a} \wedge \mathbf{b}) \wedge \mathbf{c} = \mathbf{a} \wedge (\mathbf{b} \wedge \mathbf{c}). \tag{2}$$

That $\mathbf{a} \wedge \mathbf{b}$ represents the line incident to $\mathbf{a}$ and $\mathbf{b}$ means if there is any point $\mathbf{c}$ in the projective space, then $\mathbf{c}$ is on the line if and only if $\mathbf{c} \wedge (\mathbf{a} \wedge \mathbf{b}) = 0$. In other words, $\mathbf{c}$ is on a line if and only if it cannot span a plane with the line.

Definition 1. Let $\mathcal{V}^n$ be an nD vector space over $\mathbb{K}$. The outer product "$\wedge$" is defined for any finite tuple of elements in $\mathcal{V}^n$ by the following three and only three requirements: multilinearity, nilpotency (1), and associativity (2). The Grassmann space $\Lambda(\mathcal{V}^n)$ is the vector space spanned by all kinds of outer product results from $\mathcal{V}^n$, and the Grassmann algebra denoted by the same symbol $\Lambda(\mathcal{V}^n)$ is the Grassmann space equipped with the outer product.

In Grassmann algebra, the incidence relation between a point and an r-plane can be expressed by their outer product being zero. However, the outer product between a line and an r-plane being zero, where $r > 0$, no longer corresponds to the incidence between the line and the r-plane. We need the dual of the outer product, the "meet product", to represent the latter incidence relation. In other words, we need to extend Grassmann algebra to an algebra equipped with both the outer product and the meet product.

First we take a look at the concept "duality", or, more accurately, *Hodge duality*. Let $\mathbf{e}_1, \ldots, \mathbf{e}_n$ be a basis of $\mathcal{V}^n$. Let

$$(i_1 < \ldots < i_k, j_1 < \ldots < j_{n-k})$$

be a *bipartition* of the sequence $1, 2, \ldots, n$ into two non-overlapping subsequences. The *dual operator* "$\sim$" is the linear extension of the following operation upon the basis elements $\{\mathbf{e}_{i_1} \wedge \ldots \wedge \mathbf{e}_{i_k} \mid i_1 < \ldots < i_k\}$ of $\Lambda(\mathcal{V}^n)$:

$$(\mathbf{e}_{i_1} \wedge \ldots \wedge \mathbf{e}_{i_k})^{\sim} := \operatorname{sign}(i_1 \ldots i_k, j_1 \ldots j_{n-k}) \mathbf{e}_{j_1} \wedge \ldots \wedge \mathbf{e}_{j_{n-k}}, \tag{3}$$

where "sign" denotes the sign of permutation with respect to the original sequence of subscripts $1, 2, \ldots, n$. This operator is basis-dependent. It changes a basis element of $\Lambda(\mathcal{V}^n)$ to another basis element so that the two sequences of basis vectors in the two elements form a permutation of the original sequence of basis vectors with positive sign.

Example 1. In $\mathcal{V}^3$ a plane incident to three points $\mathbf{a} = (a_x, a_y, a_z, 1), \mathbf{b} = (b_x, b_y, b_z, 1)$, $\mathbf{c} = (c_x, c_y, c_z, 1)$ is represented by a "3-vector":

$$\alpha(-\mathbf{e}_2 \wedge \mathbf{e}_3 \wedge \mathbf{e}_4) + \beta(\mathbf{e}_1 \wedge \mathbf{e}_3 \wedge \mathbf{e}_4) + \gamma(-\mathbf{e}_1 \wedge \mathbf{e}_2 \wedge \mathbf{e}_4) + \delta(\mathbf{e}_1 \wedge \mathbf{e}_2 \wedge \mathbf{e}_3), \tag{4}$$

where $(\alpha, \beta, \gamma, \delta)$ are 3×3 subdeterminants of $\begin{pmatrix} a_x & b_x & c_x \\ a_y & b_y & c_y \\ a_z & b_z & c_z \\ 1 & 1 & 1 \end{pmatrix}$. The dual of the plane is $\alpha \mathbf{e}_1 + \beta \mathbf{e}_2 + \gamma \mathbf{e}_3 + \delta \mathbf{e}_4$.

We introduce the dual of the outer product by considering a simple example. In $\mathcal{V}^3$ a line passing through points $(x_1, y_1, z_1, 1)$ and $(x_2, y_2, z_2, 1)$ is represented by their outer product:

$$\begin{aligned}&(x_1\mathbf{e}_1+y_1\mathbf{e}_2+z_1\mathbf{e}_3+\mathbf{e}_4)\wedge(x_2\mathbf{e}_1+y_2\mathbf{e}_2+z_2\mathbf{e}_3+\mathbf{e}_4)\\&=(x_1y_2-x_2y_1)\mathbf{e}_1\wedge\mathbf{e}_2+(z_1x_2-z_2x_1)\mathbf{e}_3\wedge\mathbf{e}_1+(y_1z_2-y_2z_1)\mathbf{e}_2\wedge\mathbf{e}_3\\&\quad+(x_1-x_2)\mathbf{e}_1\wedge\mathbf{e}_4+(y_1-y_2)\mathbf{e}_2\wedge\mathbf{e}_4+(z_1-z_2)\mathbf{e}_3\wedge\mathbf{e}_4\\&=\left(\begin{vmatrix}x_1&x_2\\y_1&y_2\end{vmatrix},\begin{vmatrix}z_1&z_2\\x_1&x_2\end{vmatrix},\begin{vmatrix}y_1&y_2\\z_1&z_2\end{vmatrix},\begin{vmatrix}x_1&x_2\\1&1\end{vmatrix},\begin{vmatrix}y_1&y_2\\1&1\end{vmatrix},\begin{vmatrix}z_1&z_2\\1&1\end{vmatrix}\right).\end{aligned}$$

The resulting 6-tuple is called the *Plücker coordinates* of the line.

A line is also the intersection of two planes. For line

$$\begin{cases}a_1x+b_1y+c_1z+d_1=0,\\a_2x+b_2y+c_2z+d_2=0,\end{cases}\tag{5}$$

by extracting the subdeterminants of the coefficient matrix of the two linear equations, we get the *dual Plücker coordinates* of the line as follows:

$$\left(\begin{vmatrix}a_1&a_2\\b_1&b_2\end{vmatrix},\begin{vmatrix}c_1&c_2\\a_1&a_2\end{vmatrix},\begin{vmatrix}b_1&b_2\\c_1&c_2\end{vmatrix},\begin{vmatrix}a_1&a_2\\d_1&d_2\end{vmatrix},\begin{vmatrix}d_1&d_2\\b_1&b_2\end{vmatrix},\begin{vmatrix}c_1&c_2\\d_1&d_2\end{vmatrix}\right).\tag{6}$$

Dual Plücker coordinates are related to Plücker coordinates by the dual operator. Let there be two planes $\mathbf{abc}$ and plane $\mathbf{a}'\mathbf{b}'\mathbf{c}'$ each being determined by three incident points, the intersection of the two planes is then

$$\begin{cases}\begin{vmatrix}a_y&b_y&c_y\\a_z&b_z&c_z\\1&1&1\end{vmatrix}x-\begin{vmatrix}a_x&b_x&c_x\\a_z&b_z&c_z\\1&1&1\end{vmatrix}y+\begin{vmatrix}a_x&b_x&c_x\\a_y&b_y&c_y\\1&1&1\end{vmatrix}z-\begin{vmatrix}a_x&b_x&c_x\\a_y&b_y&c_y\\a_z&b_z&c_z\end{vmatrix}=0,\\[2ex]\begin{vmatrix}a'_y&b'_y&c'_y\\a'_z&b'_z&c'_z\\1&1&1\end{vmatrix}x-\begin{vmatrix}a'_x&b'_x&c'_x\\a'_z&b'_z&c'_z\\1&1&1\end{vmatrix}y+\begin{vmatrix}a'_x&b'_x&c'_x\\a'_y&b'_y&c'_y\\1&1&1\end{vmatrix}z-\begin{vmatrix}a'_x&b'_x&c'_x\\a'_y&b'_y&c'_y\\a'_z&b'_z&c'_z\end{vmatrix}=0.\end{cases}\tag{7}$$

Replacing the coefficients in (5) with those in (7), then substituting them into (6), we can get the dual Plücker coordinates of the line. After applying the dual operator, we get the Plücker coordinates of the line as follows, where there are two equal forms for the result:

$$
\begin{aligned}
&((\mathbf{a}\wedge\mathbf{b}\wedge\mathbf{c})^{\sim}\wedge(\mathbf{a}'\wedge\mathbf{b}'\wedge\mathbf{c}')^{\sim})^{\sim}\\
&=\begin{vmatrix} a_x & b_x & c_x & c'_x\\ a_y & b_y & c_y & c'_y\\ a_z & b_z & c_z & c'_z\\ 1&1&1&1\end{vmatrix}\mathbf{a}'\wedge\mathbf{b}'-\begin{vmatrix} a_x & b_x & c_x & b'_x\\ a_y & b_y & c_y & b'_y\\ a_z & b_z & c_z & b'_z\\ 1&1&1&1\end{vmatrix}\mathbf{a}'\wedge\mathbf{c}'+\begin{vmatrix} a_x & b_x & c_x & a'_x\\ a_y & b_y & c_y & a'_y\\ a_z & b_z & c_z & a'_z\\ 1&1&1&1\end{vmatrix}\mathbf{b}'\wedge\mathbf{c}'\\
&=\begin{vmatrix} c_x & a'_x & b'_x & c'_x\\ c_y & a'_y & b'_y & c'_y\\ c_z & a'_z & b'_z & c'_z\\ 1&1&1&1\end{vmatrix}\mathbf{a}\wedge\mathbf{b}-\begin{vmatrix} b_x & a'_x & b'_x & c'_x\\ b_y & a'_y & b'_y & c'_y\\ b_z & a'_z & b'_z & c'_z\\ 1&1&1&1\end{vmatrix}\mathbf{a}\wedge\mathbf{c}+\begin{vmatrix} a_x & a'_x & b'_x & c'_x\\ a_y & a'_y & b'_y & c'_y\\ a_z & a'_z & b'_z & c'_z\\ 1&1&1&1\end{vmatrix}\mathbf{b}\wedge\mathbf{c}.
\end{aligned}
\tag{8}
$$

For vectors in $\mathcal{V}^4$, we set $[\mathbf{abcc}'] := \begin{vmatrix} a_x & b_x & c_x & c'_x\\ a_y & b_y & c_y & c'_y\\ a_z & b_z & c_z & c'_z\\ 1&1&1&1\end{vmatrix}$. In general, for vectors in $\mathcal{V}^n$, we set the *bracket notation*

$$[\mathbf{a}_1 \dots \mathbf{a}_n] := \det(a_{ij})_{i,j=1..n}. \tag{9}$$

The Plücker coordinates (8) of line $\mathbf{abc} \cap \mathbf{a}'\mathbf{b}'\mathbf{c}'$ are called the *meet product* of the coordinates $\mathbf{a}\wedge\mathbf{b}\wedge\mathbf{c}$ and $\mathbf{a}'\wedge\mathbf{b}'\wedge\mathbf{c}'$ of the two planes. In the notation of brackets, the meet product can be written as

$$
\begin{aligned}
&((\mathbf{a}\wedge\mathbf{b}\wedge\mathbf{c})^{\sim}\wedge(\mathbf{a}'\wedge\mathbf{b}'\wedge\mathbf{c}')^{\sim})^{\sim}\\
&\quad=[\mathbf{abcc}']\mathbf{a}'\wedge\mathbf{b}'-[\mathbf{abcb}']\mathbf{a}'\wedge\mathbf{c}'+[\mathbf{abca}']\mathbf{b}'\wedge\mathbf{c}'\\
&\quad=[\mathbf{ca}'\mathbf{b}'\mathbf{c}']\mathbf{a}\wedge\mathbf{b}-[\mathbf{ba}'\mathbf{b}'\mathbf{c}']\mathbf{a}\wedge\mathbf{c}+[\mathbf{aa}'\mathbf{b}'\mathbf{c}']\mathbf{b}\wedge\mathbf{c}.
\end{aligned}
\tag{10}
$$

Extending this simple example to the general case, we abstract the following definition.

Definition 2. In $\Lambda(\mathcal{V}^n)$ the *meet product* of two elements $\mathbf{A}$ and $\mathbf{B}$ is

$$\mathbf{A}\vee\mathbf{B} := (\mathbf{A}^{\sim}\wedge\mathbf{B}^{\sim})^{\sim}. \tag{11}$$

An equivalent definition is based on the bracket notation: If $\mathbf{A}_r = \mathbf{a}_1 \wedge \ldots \wedge \mathbf{a}_r$ and $\mathbf{B}_s = \mathbf{b}_1 \wedge \ldots \wedge \mathbf{b}_s$, then

$$\begin{aligned} \mathbf{A}_r \vee \mathbf{B}_s &= \sum_{(n-s,r+s-n)\vdash \mathbf{A}_r} [\mathbf{A}_{r(1)}\mathbf{B}_s]\mathbf{A}_{r(2)} \\ &= \sum_{(r+s-n,n-r)\vdash \mathbf{B}_s} [\mathbf{A}_1\mathbf{B}_{s(2)}]\mathbf{B}_{s(1)}, \end{aligned} \tag{12}$$

where $(u, v) \vdash \mathbf{A}_{u+v}$ denotes partitioning a sequence of $u + v$ elements into two subsequences of length u, v respectively, with $\mathbf{A}_{u+v(1)}$ being elements of the first subsequence multiplied using the product comprising $\mathbf{A}_{u+v}$, and $\mathbf{A}_{u+v(2)}$ being the second subsequence of elements multiplied in the same way. It is assumed that $\mathbf{A}_{u+v(1)}$ carries the sign of permutation of the bipartition.

It can be proved that for different choices of bases of $\mathcal{V}^n$, the corresponding meet products differ only by scale. In the sense of homogeneous representation, we can say that the meet product is basis independent.

By (11), the meet product can be extended by associativity to any finite number of elements in $\Lambda(\mathcal{V}^n)$. As the dual to the outer product, the meet product is also multilinear and nilpotent. Algebraically, the outer product of linear subspaces represents their span, while the meet product represents their intersection.

Definition 3. The Grassmann–Cayley algebra over a vector space $\mathcal{V}^n$ is the Grassmann space $\Lambda(\mathcal{V}^n)$ equipped with two products: the outer product and the meet product.

Example 2. In $\mathbb{PK}^3$ a line as the intersection of two planes $\mathbf{C}, \mathbf{D}$ can be represented by $\mathbf{C} \vee \mathbf{D}$. Line $\mathbf{A}$ is on plane $\mathbf{C}$ if and only if $\mathbf{A} \vee \mathbf{C} = 0$.

Example 3. In $\mathbb{PR}^2$ three lines $\mathbf{12}, \mathbf{1'2'}, \mathbf{1''2''}$ concur if and only if the point of intersection $\mathbf{a} = (\mathbf{1} \wedge \mathbf{2}) \vee (\mathbf{1'} \wedge \mathbf{2'})$ is on line $\mathbf{1''} \wedge \mathbf{2''}$:

$$[((\mathbf{1} \wedge \mathbf{2}) \vee (\mathbf{1'} \wedge \mathbf{2'}))\mathbf{1''}\mathbf{2''}] = (\mathbf{1} \wedge \mathbf{2}) \vee (\mathbf{1'} \wedge \mathbf{2'}) \vee (\mathbf{1''} \wedge \mathbf{2''}) = 0, \tag{13}$$

where the first equality follows (12). If the three lines do not concur, then in the affine plane supporting the three lines,

$$(\mathbf{1} \wedge \mathbf{2}) \vee (\mathbf{1'} \wedge \mathbf{2'}) \vee (\mathbf{1''} \wedge \mathbf{2''}) = \frac{\text{signed area of triangle } \mathbf{a1''2''}}{\text{signed area of quadrilateral } \mathbf{11'22'}}. \tag{14}$$

Theorem proving in projective incidence geometry with Grassmann–Cayley algebra

In $\Lambda(\mathcal{V}^n)$ the *bracket* of a sequence of n vectors is defined to be the determinant of their coefficients (9). Given a finite set of symbolic vectors $\mathbf{a}_1, \ldots, \mathbf{a}_m$ of $\mathcal{V}^n$, where $m > n$, all possible brackets formed by their subsequences as indeterminates generate a polynomial ring. However, there are non-trivial algebraic relations among these indeterminates, called *syzygy relations*. It is a classical result that all relations among the brackets are generated algebraically by the following *Grassmann–Plücker relations*.

Let $i_1, \ldots, i_{n+1}$ and $j_1, \ldots, j_{n-1}$ be two subsequences of $1, \ldots, m$ that may overlap each other. Then

$$\mathbf{a}_{i_1} \wedge \ldots \wedge \mathbf{a}_{i_{n+1}} = 0 \tag{15}$$

because any $n+1$ vectors in $\mathcal{V}^n$ are linearly dependent. By (12),

$$\begin{aligned} 0 &= (\mathbf{a}_{i_1} \wedge \ldots \wedge \mathbf{a}_{i_{n+1}}) \vee (\mathbf{a}_{j_1} \wedge \ldots \wedge \mathbf{a}_{j_{n-1}}) \\ &= \sum_{k=1}^{n+1} (-1)^{k+1} [\mathbf{a}_{i_k} \mathbf{a}_{j_1} \ldots \mathbf{a}_{j_{n-1}}] \mathbf{a}_{i_1} \wedge \ldots \wedge \check{\mathbf{a}}_{i_k} \wedge \ldots \wedge \mathbf{a}_{i_{n+1}}. \end{aligned} \tag{16}$$

The resulting quadratic bracket polynomial in (16) is called a *Grassmann–Plücker syzygy*, and this polynomial being equal to zero is called a *Grassmann–Plücker relation*.

Definition 4. The *nD bracket algebra* generated by m symbolic vectors $\mathbf{a}_1, \ldots, \mathbf{a}_m$ is the quotient of the polynomial ring whose indeterminates are antisymmetric subsequences of the vectors, modulo the ideal generated by the *Grassmann–Plücker syzygies* of the m vectors.

Grassmann–Cayley algebra and bracket algebra are closely related to each other by the so-called *Cayley expansions* and *Cayley factorizations*.

Cayley expansion: Rewrite an expression of outer products and meet products as an expression involving only brackets and outer products. In particular, the definition (12) is itself a Cayley expansion formula.

The purpose of Cayley expansion is to eliminate the meet product. It is a procedure of simplification in this regard. Usually one term is rewritten as more than one term in the procedure of Cayley expansion. Furthermore, usually there are several different Cayley expansion results from the same expression.

For example, in $\Lambda(\mathcal{V}^3)$,

$$\begin{aligned}(\mathbf{a}_1 \wedge \mathbf{a}_2) \vee (\mathbf{b}_1 \wedge \mathbf{b}_2) \vee (\mathbf{c}_1 \wedge \mathbf{c}_2) \\ &= [\mathbf{a}_1\mathbf{b}_1\mathbf{b}_2][\mathbf{a}_2\mathbf{c}_1\mathbf{c}_2] - [\mathbf{a}_2\mathbf{b}_1\mathbf{b}_2][\mathbf{a}_1\mathbf{c}_1\mathbf{c}_2] \\ &= [\mathbf{a}_1\mathbf{a}_2\mathbf{b}_2][\mathbf{b}_1\mathbf{c}_1\mathbf{c}_2] - [\mathbf{a}_1\mathbf{a}_2\mathbf{b}_1][\mathbf{b}_2\mathbf{c}_1\mathbf{c}_2] \\ &= [\mathbf{a}_1\mathbf{a}_2\mathbf{c}_1][\mathbf{b}_1\mathbf{b}_2\mathbf{c}_2] - [\mathbf{a}_1\mathbf{a}_2\mathbf{c}_2][\mathbf{b}_1\mathbf{b}_2\mathbf{c}_1].\end{aligned} \tag{17}$$

The first expansion in (17) is by distributing $\mathbf{a}_1, \mathbf{a}_2$ among $\mathbf{b}_1 \wedge \mathbf{b}_2$ and $\mathbf{c}_1 \wedge \mathbf{c}_2$, the second by distributing $\mathbf{b}_1, \mathbf{b}_2$ among $\mathbf{a}_1 \wedge \mathbf{a}_2$ and $\mathbf{c}_1 \wedge \mathbf{c}_2$, and the third by distributing $\mathbf{c}_1, \mathbf{c}_2$ among $\mathbf{a}_1 \wedge \mathbf{a}_2$ and $\mathbf{b}_1 \wedge \mathbf{b}_2$.

Cayley factorization: Rewrite a polynomial of brackets as an expression involving no addition and subtraction, but only two products: the outer product and the meet product. This is a procedure resulting in a (rational) monomial form. The purpose is on the one hand to unify different bracket polynomials into the same expression in Grassmann–Cayley algebra, on the other hand to obtain geometric interpretation for the polynomial of brackets, as already shown in (14). The result of Cayley factorization is usually not unique. However, the transformation group among different factorization results is often easy to describe.

Below we show a simple example of theorem proving in projective incidence geometry with the interaction of Grassmann–Cayley algebra and bracket algebra.

Example 4. [2-D Desargues' Theorem] In the projective plane there are two triangles $\mathbf{123}$ and $\mathbf{1'2'3'}$. If the three lines $\mathbf{11'}, \mathbf{22'}, \mathbf{33'}$ concur, then the three pairs of lines $(\mathbf{12}, \mathbf{1'2'}), (\mathbf{13}, \mathbf{1'3'}), (\mathbf{23}, \mathbf{2'3'})$ intersect at three points that are collinear.

As shown in Fig. 1, let $\mathbf{a} = \mathbf{12} \cap \mathbf{1'2'}$, $\mathbf{b} = \mathbf{13} \cap \mathbf{1'3'}$ and $\mathbf{c} = \mathbf{23} \cap \mathbf{2'3'}$; let $\mathbf{d}$ be the point where lines $\mathbf{11'}, \mathbf{22'}, \mathbf{33'}$ concur. The configuration can be constructed as follows:

Free points: $\mathbf{1}, \mathbf{2}, \mathbf{3}, \mathbf{d}$.
Collinear points: point $\mathbf{1'}$ on line $\mathbf{d1}$, point $\mathbf{2'}$ on line $\mathbf{d2}$, point $\mathbf{3'}$ on line $\mathbf{d3}$.
Intersections: $\mathbf{a} = \mathbf{12} \cap \mathbf{1'2'}, \mathbf{b} = \mathbf{13} \cap \mathbf{1'3'}, \mathbf{c} = \mathbf{23} \cap \mathbf{2'3'}$.
Conclusion: $\mathbf{a}, \mathbf{b}, \mathbf{c}$ are collinear.

In Grassmann–Cayley algebra, we have the following algebraic representation of the geometric theorem:

Hypothesis $(\mathbf{1} \wedge \mathbf{1'}) \vee (\mathbf{2} \wedge \mathbf{2'}) \vee (\mathbf{3} \wedge \mathbf{3'}) = 0$.
Conclusion $[\{(\mathbf{1} \wedge \mathbf{2}) \vee (\mathbf{1'} \wedge \mathbf{2'})\}\{(\mathbf{1} \wedge \mathbf{3}) \vee (\mathbf{1'} \wedge \mathbf{3'})\}\{(\mathbf{2} \wedge \mathbf{3}) \vee (\mathbf{2'} \wedge \mathbf{3'})\}] = 0$.

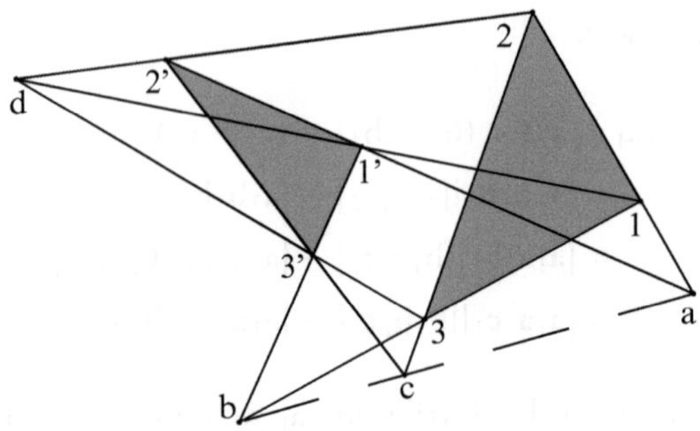

Fig. 1. Desargues' theorem

Proof of Desargues' Theorem. Simplify the conclusion expression by Cayley expansions and Cayley factorizations.

$$
\begin{aligned}
& [\{\underline{(\mathbf{1}\wedge\mathbf{2})}\vee\underline{(\mathbf{1}'\wedge\mathbf{2}')}\}\{(\mathbf{1}\wedge\mathbf{3})\vee(\mathbf{1}'\wedge\mathbf{3}')\}\{(\mathbf{2}\wedge\mathbf{3})\vee(\mathbf{2}'\wedge\mathbf{3}')\}] \\
\overset{binomial}{=} & \{(\underline{\mathbf{1}}\wedge\mathbf{2})\vee(\underline{\mathbf{1}}\wedge\mathbf{3})\vee(\mathbf{1}'\wedge\mathbf{3}')\}\{(\mathbf{1}'\wedge\underline{\mathbf{2}'})\vee(\mathbf{2}\wedge\mathbf{3})\vee(\underline{\mathbf{2}'}\wedge\mathbf{3}')\} \\
& -\{(\mathbf{1}\wedge\underline{\mathbf{2}})\vee(\underline{\mathbf{2}}\wedge\mathbf{3})\vee(\mathbf{2}'\wedge\mathbf{3}')\}\{(\underline{\mathbf{1}'}\wedge\mathbf{2}')\vee(\mathbf{1}\wedge\mathbf{3})\vee(\underline{\mathbf{1}'}\wedge\mathbf{3}')\} \\
\overset{monomial}{=} & [\mathbf{123}][\mathbf{1}'\mathbf{2}'\mathbf{3}'](-[\mathbf{11}'\mathbf{3}'][\mathbf{232}']+[\mathbf{131}'][\mathbf{22}'\mathbf{3}']) \\
\overset{factor}{=} & -[\mathbf{123}][\mathbf{1}'\mathbf{2}'\mathbf{3}'](\mathbf{1}\wedge\mathbf{1}')\vee(\mathbf{2}\wedge\mathbf{2}')\vee(\mathbf{3}\wedge\mathbf{3}').
\end{aligned}
$$

Explanation of the proof. The first step is a Cayley expansion leading to a binomial result. Let $\mathbf{A}_i, \mathbf{B}_j$ be bivectors (outer products of vector pairs), then $[(\mathbf{A}_1\vee\mathbf{B}_1)(\mathbf{A}_2\vee\mathbf{B}_2)(\mathbf{A}_3\vee\mathbf{B}_3)]$ has three different expansions by distributing $\mathbf{A}_i, \mathbf{B}_i$ among the other two meet products for $i = 1, 2, 3$:

$$
\begin{aligned}
& [(\mathbf{A}_1\vee\mathbf{B}_1)(\mathbf{A}_2\vee\mathbf{B}_2)(\mathbf{A}_3\vee\mathbf{B}_3)] \\
& \quad = (\mathbf{A}_1\vee\mathbf{A}_2\vee\mathbf{B}_2)(\mathbf{B}_1\vee\mathbf{A}_3\vee\mathbf{B}_3)-(\mathbf{B}_1\vee\mathbf{A}_2\vee\mathbf{B}_2)(\mathbf{A}_1\vee\mathbf{A}_3\vee\mathbf{B}_3) \\
& \quad = (\mathbf{A}_1\vee\mathbf{B}_1\vee\mathbf{B}_2)(\mathbf{A}_2\vee\mathbf{A}_3\vee\mathbf{B}_3)-(\mathbf{A}_1\vee\mathbf{B}_1\vee\mathbf{A}_2)(\mathbf{B}_2\vee\mathbf{A}_3\vee\mathbf{B}_3) \\
& \quad = (\mathbf{A}_1\vee\mathbf{B}_1\vee\mathbf{A}_3)(\mathbf{A}_2\vee\mathbf{B}_2\vee\mathbf{B}_3)-(\mathbf{A}_1\vee\mathbf{B}_1\vee\mathbf{B}_3)(\mathbf{A}_2\vee\mathbf{B}_2\vee\mathbf{A}_3).
\end{aligned}
\tag{18}
$$

Choosing any of them leads to much the same succeeding proving procedure. In the above proof, the first expansion is used.

The second step is a set of Cayley expansions each leading to a monomial result. Each *monomial expansion* is of the following form:

$$(\underline{\mathbf{1}} \wedge \mathbf{2}) \vee (\underline{\mathbf{1}} \wedge \mathbf{3}) = [\mathbf{113}]\mathbf{2} - [\mathbf{213}]\mathbf{1} = [\mathbf{123}]\mathbf{1}. \tag{19}$$

The last step is a Cayley factorization inverse to Cayley expansion (17). The computed result is stronger than the original theorem, because it not only provides the information that the hypothesis implies the conclusion, but tells how the conclusion expression depends on the hypothesis expression *quantitatively*.

Furthermore, the computed result leads to the following term-rewriting rule that provides an elegant way of applying Desargues' Theorem in geometric computing:

$$\begin{aligned} &[\{(\mathbf{1} \wedge \mathbf{2}) \vee (\mathbf{1}' \wedge \mathbf{2}')\}\{(\mathbf{1} \wedge \mathbf{3}) \vee (\mathbf{1}' \wedge \mathbf{3}')\}\{(\mathbf{2} \wedge \mathbf{3}) \vee (\mathbf{2}' \wedge \mathbf{3}')\}] \\ \rightarrow\ &-[\mathbf{123}][\mathbf{1}'\mathbf{2}'\mathbf{3}'](\mathbf{1} \wedge \mathbf{1}') \vee (\mathbf{2} \wedge \mathbf{2}') \vee (\mathbf{3} \wedge \mathbf{3}'). \end{aligned} \tag{20}$$

Conclusion

In this paper we introduce Grassmann–Cayley algebra with applications in theorem proving in projective incidence geometry. Proofs produced by this algebra have the following prominent features:

- Robust binomial, i.e., the maximal number of terms throughout any proof remains two no matter how Cayley expansions vary, as long as the expansion candidates produce results of the same number of terms.
- Algebraic manipulations are easy.
- For a geometric theorem, the hypothesis and the conclusion expressions are interrelated quantitatively, which in some sense generalizes the original geometric theorem.
- Geometric theorems can be expressed as single algebraic identities that are easy to implement and apply.

Grassmann, geometry and mechanics

John Browne

Introduction

Many of the capabilities of Grassmann's algebra are still not well-enough known in the scientific community. This paper discusses just two of them: the capability to distinguish algebraically between points and vectors in space, and the capability to compute intersections of geometric entities with a simple product operation. The first may be little known due to the mainstreaming of Gibbs' three-dimensional vector algebra; the second due to Grassmann's choice of notation for his product operations. We explore two applications.

In the first application we look at how distinguishing between points and vectors enables us to model positioned linear (and multilinear) entities, for example, lines in geometry and forces and momenta in physics. As a consequence of this more faithful representation of force and momentum we discover that Newton's Second Law relating the motion of a system of masses to a system of forces is independent of dimension, requires no specific metric for its formulation, and unites what Gibbs' algebra splits: the linear and angular components of such a system. A Grassmannian approach shows that the commonly accepted dichotomy between forces and moments is purely an artifact of the point-less algebra traditionally used to describe it.

In the second application we look at how combining the capability to model *positioned* linear and multi-linear entities in geometry with the capability to compute intersections of these entities subsumes projective geometry in a way which no longer

H.-J. Petsche et al. (eds.), *From Past to Future: Graßmann's Work in Context*,
DOI 10.1007/978-3-0346-0405-5_25,

requires any entities-at-infinity. These entities-at-infinity are simply the "free" elements of the algebra: vectors, bivectors, . . . , multivectors. This Grassmannian approach to projective geometry, being algebraically founded, enables proofs to be demonstrated straightforwardly using a computer algebra system.

Although the notation in this paper follows (where possible) modern convention, the capabilities discussed are to be found throughout Grassmann's original *Ausdehnungslehre* [$A2_K$]. Applications of these capabilities may also be found in early interpreters of his work, for example Alfred North Whitehead [Whitehead 1898] and Edward Wyllis Hyde [Hyde 1890], and more recent ones, for example Alvin Swimmer [Swimmer 1996].

Grassmann, Hamilton, and Gibbs

In his *Ausdehnungslehre* [$A2_K$], Hermann Grassmann discovered and created an *algebra*. And being an algebra, its entities were naturally devoid of ***interpretation***. Nevertheless, his own discovery process and the penchants of his contemporaries sought a geometric interpretation [Crowe 1967; Petsche 2009].

The most natural geometric interpretation for the fundamental entities of the algebra was as *points*. Since the algebra was a linear algebra, it would only be meaningful under this interpretation if all types of linear combinations of points were geometrically meaningful, that is, scalar multiples, sums, and differences. A scalar multiple of a point may simply be interpreted as a (geometric) point with a "weight" attached, much like an idealized mass point in physics. A sum of points (or weighted points) would then result in a weighted point at their center of mass, as described by Möbius in his *Der barycentrische Calcul* [Crowe 1967, 48]. But under this interpretation of sum, a difference of two points was not a finite point. Nevertheless, if a difference of two points was added to another point, the result was a new point!

It is to Sir William Rowan Hamilton that we must attribute the English term for the difference of two points. In his *Lectures on Quaternions* [Hamilton 1853, 15] he states:

> "... I regard the symbol B-A as denoting "the step from B to A:" namely, that step by making which, from the given point A, we should reach or arrive at the sought point B; and so determine, generate, mark or construct that point. This step (which we shall always suppose to be a straight line) may also, in my opinion, be properly called a vector; or more fully, it may be called "the vector of the point B, from the point A:" because it may be considered as having for its office, function, work, task or business, to transport or carry (in Latin *vehere*) a moveable point, from the given or initial position A, to the sought or final position B."

Thus, a geometric interpretation of the fundamental entities of a linear algebra as points necessarily involves a second type of entity – *vectors*. Grassmann was of course well aware of this. Points and vectors are concomitant parts of the one geometric interpretation. Points model *position*. Vectors model *direction*.

Gibbs' three-dimensional vector algebra lost the point completely because it could not deal with the four-dimensional linear space required. Grassmann's algebra subsumes Gibbs' vector algebra, and it allows for points. Why not adopt the system that allows for both?

Interpreted spaces

Consider a four-dimensional linear space whose entities are interpreted as points, and suppose a basis of the space is given by four independent points.

Another basis could be formed by fixing one of the points (P_1, say) and subtracting it from the remaining three,

$$\{P_1, P_2 - P_1, P_3 - P_1, P_4 - P_1\}.$$

As Hamilton observed, these basis point differences can be interpreted as vectors. We denote them conventionally by e_i. The remaining fixed point in the basis we shall call the *origin*, and denote it $\mathbb{O}$. To describe an n-dimensional vector space which also contains points thus requires an $(n+1)$-dimensional linear space,

$$\{\mathbb{O}, e_1, e_2, e_3, \ldots, e_n\}.$$

To construct a Grassmann algebra from a linear space, simply generate a new linear space by taking the *exterior product* of all its basis elements in all combinations, and keeping only those that are essentially different.

A Grassmann algebra constructed from a linear space endowed with the interpretation above will be called a *bound vector n-space* and its elements will be called *entities* to emphasize this interpretation. Figure 1 shows a palette of the 2^4 basis entities of a bound vector 3-space generated by the *GrassmannAlgebra* package [Browne 2009].

A basis for a bound vector 3–space

$\underset{0}{\Lambda}$	$\underset{1}{\Lambda}$	$\underset{2}{\Lambda}$	$\underset{3}{\Lambda}$	$\underset{4}{\Lambda}$
1	$\mathbb{O}$	$\mathbb{O} \wedge e_1$	$\mathbb{O} \wedge e_1 \wedge e_2$	$\mathbb{O} \wedge e_1 \wedge e_2 \wedge e_3$
	e_1	$\mathbb{O} \wedge e_2$	$\mathbb{O} \wedge e_1 \wedge e_3$	
	e_2	$\mathbb{O} \wedge e_3$	$\mathbb{O} \wedge e_2 \wedge e_3$	
	e_3	$e_1 \wedge e_2$	$e_1 \wedge e_2 \wedge e_3$	
		$e_1 \wedge e_3$		
		$e_2 \wedge e_3$		

Fig. 1. Basis elements of a bound space

Points and weighted points

A point P can now be redefined as the sum of the origin $\mathbb{O}$ and a vector x, say. The vector x is called the *position vector* of the point P,

$$P = \mathbb{O} + x.$$

The *sum* of a point and a vector is another point,

$$(\mathbb{O} + x) + y = \mathbb{O} + (x + y).$$

The *difference* between two points is a vector because the origin cancels,

$$(\mathbb{O} + x) - (\mathbb{O} + y) = x - y.$$

A multiple mP of a point P by a scalar m is called a *weighted point*. Weighted points can model point masses. A sum of point masses gives a point mass whose mass is equal to the total mass, and whose position is equal to the center of mass,

$$\sum_i m_i P_i = \sum_i m_i (\mathbb{O} + x_i) = \mathbb{O} \sum_i m_i + \sum_i m_i x_i,$$

$$MP_G = \left(\sum_i m_i\right)\left(\mathbb{O} + \frac{\sum_i m_i x_i}{\sum_i m_i}\right).$$

Vectors are interpreted to have direction but no position. Points are interpreted to have position but no direction. Multiplying a vector by a scalar does not change its direction. Multiplying a point by a scalar does not change its position.

Bound vectors and bivectors

From the definition above, a bound vector space will contain exterior products of points, of vectors, and of points and vectors. These products divide into two families: the *free* entities composed only of vectors, and the *bound* entities composed of points, or points and vectors. The free entities are naturally interpreted as having a (multidimensional) direction but no location. The bound entities are naturally interpreted as located through their points in the direction of their vectors. A given bound entity may be expressed as a product only of points, or as a product of points and vectors – as long as it contains at least one point.

The exterior product of two vectors x and y is called a *bivector*, or 2-*blade* (Fig. 2). The exterior product of m vectors is called an m-vector, or m-blade. Like vectors, m-vectors are viewed as located nowhere.

The exterior product of a point and a vector, or of two points, is called a *bound vector* (Fig. 3),

$$P \wedge x = P \wedge (Q - P) = P \wedge Q.$$

We say that a point or a vector is *in* an entity iff its exterior product with the entity is zero. Entities may be used to define geometric objects. For example, all the points and vectors in a bound vector define a line. Bound vectors can then be viewed as located *anywhere* along their line (Fig. 4),

$$(Q - P) \wedge x = 0 \qquad \Longrightarrow \qquad Q \wedge x = P \wedge x.$$

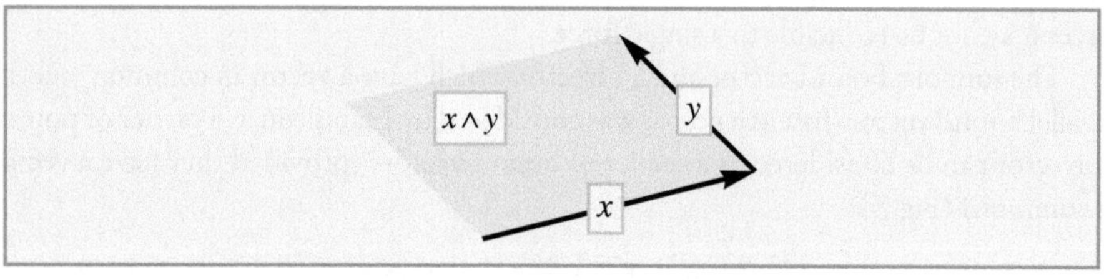

Fig. 2. Depiction of a bivector

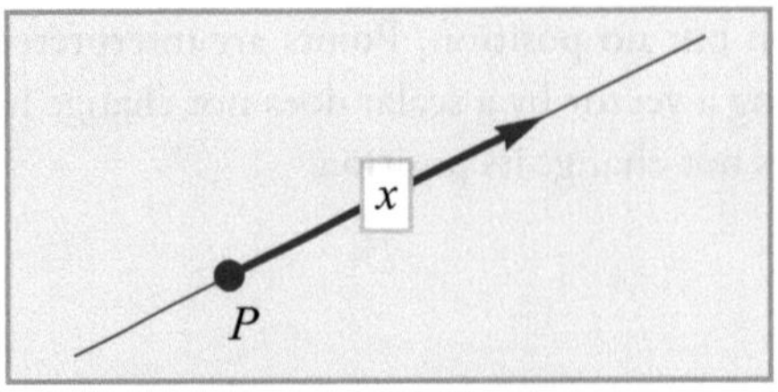

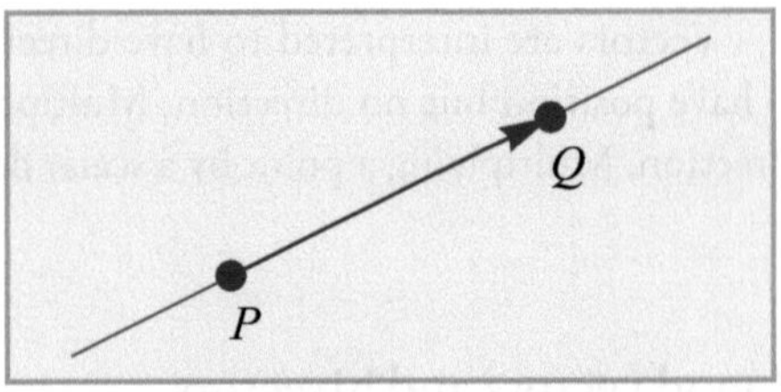

Fig. 3. Two depictions of a bound vector

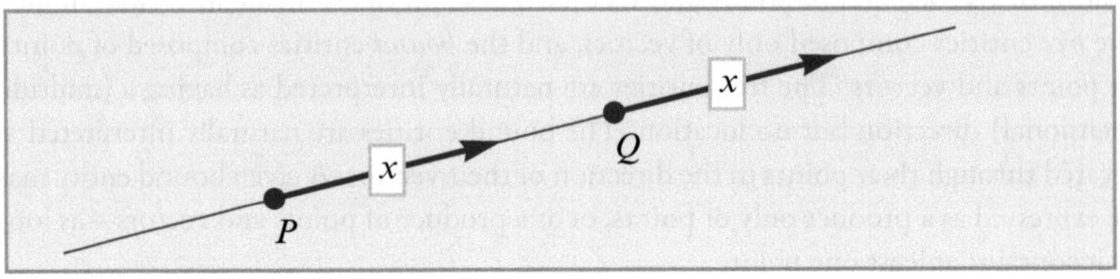

Fig. 4. A bound vector is located anywhere in its line

The power of Grassmann's algebra for doing geometry is that we can represent m-planes with bound m-vectors: we can compute with lines as if they were bound vectors, planes as if they were bound bivectors. Thus, it is often convenient to refer to an m-plane by any of the congruent bound m-vectors which define it. (Two entities are congruent if one is a scalar multiple of the other.)

Sums of bound vectors and bivectors

A sum of bound vectors may not be reducible (factorizable) to a bound vector. Bound vectors are precisely the type of entity we require to represent forces in mechanics. All mechanics textbooks tell us that a physical force has a line of action, and that a sum of forces may not be reducible to a single force.

The sum of a bound vector and a bivector which have a vector in common yields a parallel bound vector. Just as a vector was considered by Hamilton as a carrier of points, a bivector can be considered as a carrier of bound vectors (provided they have a vector in common) (Fig. 5),

$$P \wedge x + y \wedge x = (P + y) \wedge x.$$

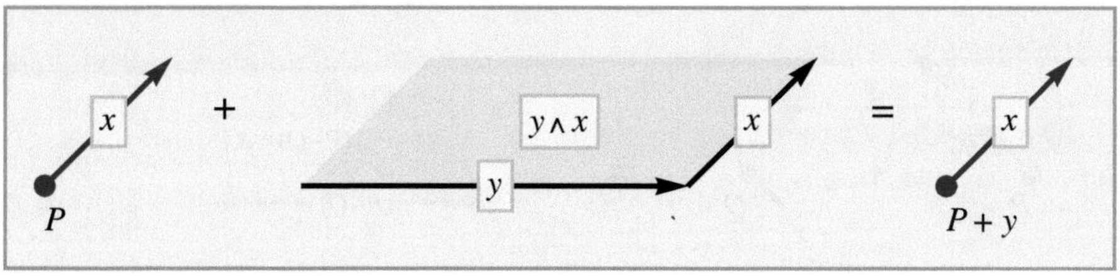

Fig. 5. A bivector as a carrier of bound vectors

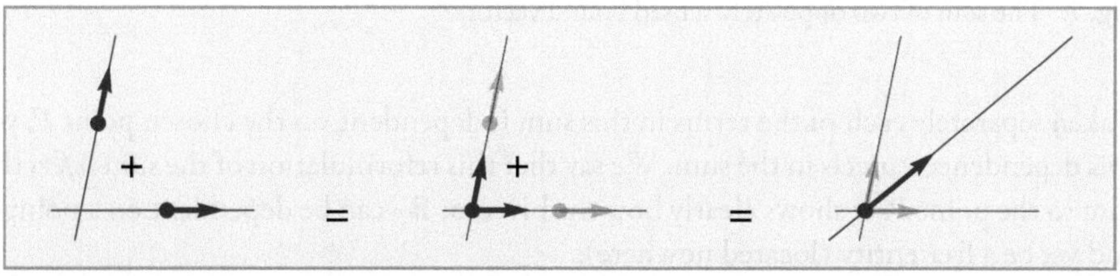

Fig. 6. The sum of intersecting bound vectors

The sum of two intersecting bound vectors is a bound vector through their point of intersection. To see this, slide the bound vectors along their lines so that they are expressed with a common point (their point of intersection), then factor out the point (Fig. 6). In mechanics this translates to: *forces acting through the same point can be resolved to a single force*,

$$P \wedge x + Q \wedge y = R \wedge x + R \wedge y = R \wedge (x + y).$$

The sum of two oppositely sensed bound vectors (that is, where the vectors differ only by a sign) is a bivector. Thus, the sum of the two bound entities (bound along their lines) reduces to a free entity (located nowhere) (Fig. 7). In mechanics this translates to: *two oppositely sensed forces result in a couple*,

$$P \wedge x + Q \wedge (-x) = (P - Q) \wedge x$$

In general, a sum of bound vectors may always be reduced to the sum of a bound vector through an arbitrarily chosen point, and a bivector. In special cases, as we see in Figs. 6 and 7, the bivector or the bound vector may be zero,

$$\sum_i P_i \wedge x_i = P \wedge \sum_i x_i + \sum_i (P_i - P) \wedge x_i = P \wedge X + \mathbb{B}_P.$$

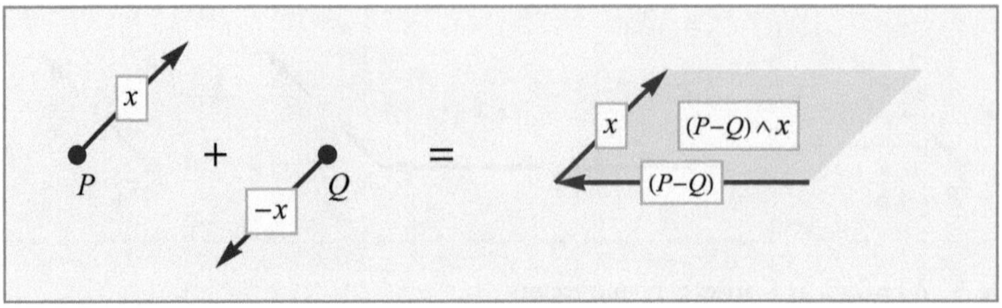

Fig. 7. The sum of two oppositely sensed bound vectors

Taken separately each of the terms in this sum is dependent on the chosen point P, yet this dependence cancels in the sum. We say that this reformulation of the sum *refers* the sum to the point P. It shows clearly how the bivector $\mathbb{B}_P$ can be dependent on a point P and yet be a free entity (located nowhere).

The equilibrium of a rigid body

A force may be faithfully represented by a *bound vector* $P \wedge f$ as the exterior product of *any point* P *on its line of action* with its *force vector* f.

A sum of forces may always be reduced to the sum of a single force *bound vector* through an arbitrarily chosen point, and a moment *bivector* (which, although a free entity, is dependent on the chosen point). In special cases, either the single force bound vector or the moment bivector may be zero. Thus, a sum of forces in space is not generally reducible to a single force,

$$\mathbb{F} = \sum_i P_i \wedge f_i = P \wedge \sum_i f_i + \sum_i (P_i - P) \wedge f_i = P \wedge F + \mathbb{M}_P.$$

Thus, the *single* condition for the equilibrium of a rigid body under a system of forces $\mathbb{F}$ in Grassmannian terms is

$$\mathbb{F} = P \wedge F + \mathbb{M}_P = 0.$$

On the other hand, the three-dimensional vector algebra approach requires *two* conditions: the sum of the force *vectors* must be zero *and* the sum of the moment *vectors* about any point must be zero. In a three-dimensional space all bivectors are simple, that is, they can be expressed as the exterior product of two vectors. In a metric space of any number of dimensions the Grassmannian complement (*Ergänzung*) of a simple bivector is a multivector orthogonal to the bivector and thus to any of its component vectors. In three dimensions, this multivector is a vector, and may be computed as a cross product. It was this fortuitous situation which allowed the Grassmannian bivector condition to be replaced by the Gibbsian vector condition in a metric 3-space. The single Grassmannian condition, however, does not require a metric for its specification and is valid in any dimension. For instance, it is valid in the plane (where the cross product is not defined).

Momentum

Textbooks rarely mention that *the momentum of a particle has a line of action*, just like a force. One author who does emphasize this point is E.A. Milne in his *Vectorial Mechanics* [Milne 1948], although he does not use a Grassmannian approach.

The *momentum of a particle* is a bound vector through the particle. By representing the particle as a mass point, and its velocity as the time-derivative of the point (or, since the origin is constant, the time-derivative of its position vector), its momentum can be written in any of the following forms:

$$P \wedge l = P \wedge \left(\overset{\circ}{mP} \right) = (mP) \wedge \overset{\circ}{P} = mP \wedge \overset{\circ}{P},$$

where P is the particle position,
l is the particle momentum vector,
m is the particle mass,
mP is the particle point mass,
$\overset{\circ}{P}$ is the particle velocity,
$\overset{\circ}{mP}$ is an alternative expression for the particle momentum vector.

The development of the expression for the momentum of a *system* of particles follows *mutatis mutandis* that for sums of bound vectors and systems of forces. The momentum of a system may always be reduced to the sum of a single linear momentum *bound vector* through an arbitrarily chosen point, and an angular momentum *bivector* (which, although a free entity, is dependent on the chosen point). In special cases, the single linear momentum bound vector or the angular momentum bivector may be zero,

$$\mathbb{L} = \sum_i P_i \wedge l_i = P \wedge \sum_i l_i + \sum_i (P_i - P) \wedge l_i = P \wedge L + \mathbb{H}_P.$$

A particularly simple result is obtained if we choose to refer the momentum of the system to the center of mass of the system, and use the results on sums of weighted points discussed in the above section "Points and weighted points" to express the momentum of the system as

$$\mathbb{L} = P_G \wedge \left(M\overset{\circ}{P}_G \right) + H_{P_G}.$$

Differentiating this with respect to time, and noting that the first term obtained is zero due to the repeated factor $\overset{\circ}{P}_G$, gives

$$\overset{\circ}{\mathbb{L}} = P_G \wedge \left(M\overset{\circ\circ}{P}_G \right) + \overset{\circ}{H}_{P_G}.$$

Thus, the time-rate of change of momentum of a system of particles may be written as the sum of

- The rate of change of linear momentum of the total mass at the center of mass.
- The rate of change of angular momentum about the center of mass.

Newton's Second Law

In Grassmannian terms, Newton's Second Law can now be stated without reference to separate linear and angular components, since the 2-entity representation includes both:

> *The sum of forces acting on a system of particles is equal to the time-rate of change of momentum of the system,* $\mathbb{F} = \overset{\circ}{\mathbb{L}}$.

We have seen that each of these can be written as a 2-entity referred to the center of gravity of the system,

$$P_G \wedge F + M_{P_G} = P_G \wedge \left(M\overset{\circ\circ}{P}_G \right) + \overset{\circ}{H}_{P_G}.$$

This one equation can be seen to imply two conditions: that the sum of the force vectors is equal to the mass of the system multiplied by the acceleration of the center of mass, and that the sum of the moment bivectors about the center of mass is equal to the time-rate of change of angular momentum about the center of mass,

$$F = M\overset{\circ\circ}{P}_G \qquad \text{and} \qquad M_{P_G} = \overset{\circ}{H}_{P_G}.$$

But the bivector form here allows validity in *any* number of dimensions (including the plane). And the *derivation* does not rely on a metric on the space, as the use of the cross product in three dimensions does. (However, a metric would appear to be necessary in order to express any given physical force in terms of a chosen basis.)

The regressive product

We have explored the nature of the bound and free entities resulting from distinguishing between points and vectors, and applied the results to a simple physical application. We now turn to some consequences for geometry.

In order to do geometry with the algebraic entities we can now construct, we need a way to find intersections, or *common factors*. The key is Grassmann's *regressive product*. In his *Ausdehnungslehre* of 1862, translated by Lloyd Kannenberg [$A2_K$, Sect. 95], he states (my parentheses):

> "If q and r are the orders (grades) of two magnitudes A and B, and n that of the principal domain (base linear space), then the order (grade) of the product $[AB]$ is *first* equal to $q+r$ if $q+r$ is smaller than n, and *second* equal to $q+r-n$ if $q+r$ is greater than or equal to n."

In terms of modern notation for the exterior and regressive products we can summarize this with

$$[AB] \iff A \wedge B, \qquad q+r<n \qquad \text{(exterior product)},$$

Projective Geometry	Grassmann Geometry	Grade
Point	Point	1
Point at infinity	Vector	1
Line	Bound vector	2
Line at infinity	Bivector	2
Plane	Bound bivector	3
Plane at infinity	Trivector	3
Multiplane	Bound $(m$-$1)$ – vector	m
Multiplane at infinity	m – vector	m

Fig. 8. Projective geometry from a Grassmannian point of view

$$[AB] \iff A \vee B, \qquad q + r \geq n \qquad \text{(regressive product).}$$

Thus Grassmann's $[AB]$ product *was actually two products*, the regressive product hidden in the exterior product! This was a beautiful and powerful notational device, although leaving the product type to be contingent on the grades of its factors rather than by being explicitly denoted, may well be the reason why Grassmann's regressive product is still not well known today.

It should be noted that Grassmann defines the result to be a scalar if $q + r$ is equal to n. In modern terms this would actually be a pseudo-scalar. His definition is not difficult to fix once we permit ourselves different notations for the different product operations. This can be done by making the regressive and exterior product axiom sets true duals of each other: that is, one can be transformed into the other by a few simple rules. It is this elegant algebraic duality which underpins the beautiful geometric duality which we are about to explore.

Projective geometry

A Grassmann algebra on a bound vector space equipped with both exterior and regressive products is all that we need for exploring projective geometry.

The key, as shown in Fig. 8, is to identify m-planes at infinity with $(m + 1)$-vectors, for example lines-at-infinity with bivectors. This is because *the regressive product (intersection) of parallel bound entities is their common vectorial direction*. Congruent entities represent the same geometric object.

Geometric constructions

Exploring Grassmann non-metric geometry often simply involves expanding and simplifying geometric formulas composed of products (exterior and regressive) of points, lines (bound vectors), planes (bound bivectors), ... For example, Grassmann showed that some geometric formulas can be viewed as a *prescription to construct*. Here is a recipe for constructing a conic in the plane (Fig. 9):

1. Construct the fixed points and lines: P_0, P_1, P_2, L_1, and L_2.
2. Construct a line L_c through P_2 to intersect line L_2 in point Q_2.
3. Construct a line L_b through Q_2 and P_0 to intersect line L_1 in point Q_1.
4. Construct a line L_a through Q_1 and P_1 to intersect line L_c in point P.
5. Repeat the construction by drawing a new line L_c through P_2 to form a new point P.

And here is Grassmann's formula encoding this construction. The variable point P occurs twice, leading to a curve of the second degree,

$$P \wedge P_1 \wedge (L_1 \vee (P_0 \wedge (L_2 \vee (P \wedge P_2)))) = 0.$$

This is the equation of the conic. Substituting symbolic or numeric coordinates reduces it to the expected algebraic relation. The ellipse in Fig. 10 is plotted by substituting numeric coordinates in the expressions for the fixed points and lines.

There is no restriction on composing geometric formulas in higher dimensions, though they are unlikely to be constructible. Here is a three-dimensional version of the planar conic where P is a variable point, P_0, P_1 are fixed points, L is a fixed line, and Π_1, Π_2 are fixed planes,

$$P \wedge P_1 \wedge (\Pi_1 \vee (P_0 \wedge (\Pi_2 \vee (P \wedge L)))) = 0.$$

Geometric theorems

Any theorem that can be reduced to a geometric formula can be proved by adopting symbolic coordinates for the geometric entities and simplifying the formula. For example, here is *Pappus' Theorem* (Fig. 10):

> Given two sets of collinear points P, Q, R, and U, V, W, then the intersection points X, Y, Z of line pairs PV and QU, PW and RU, QW and RV, are collinear.

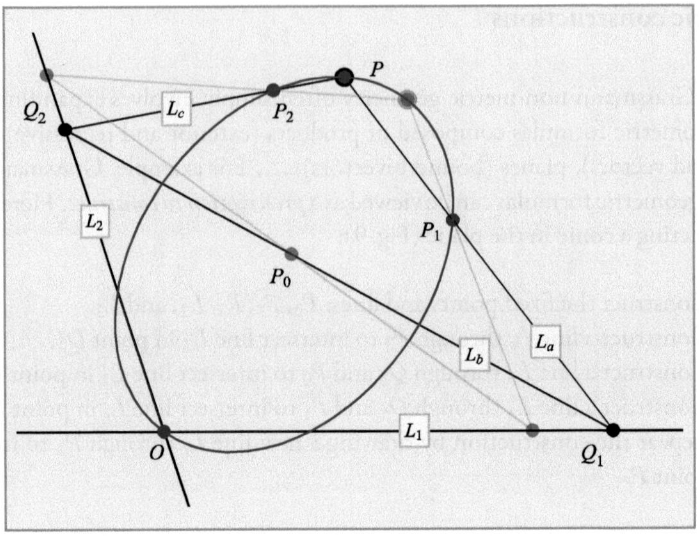

Fig. 9. An ellipse constructed from a Grassmann formula

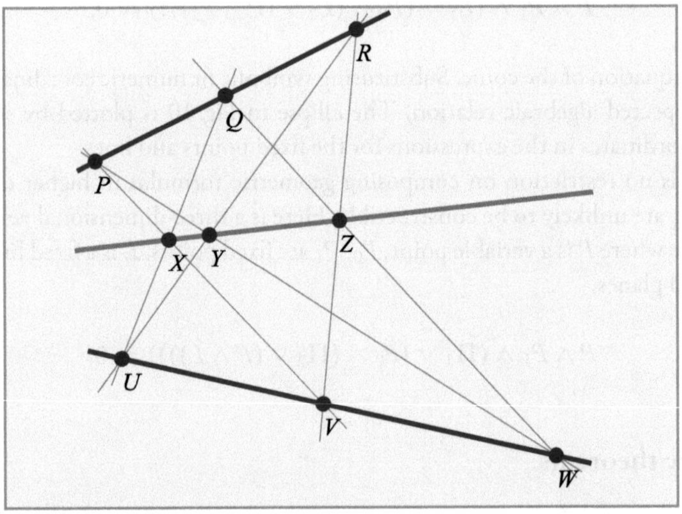

Fig. 10. A graphical depiction of Pappus' Theorem

The lines may be encoded as bound vectors by taking the exterior product of two of their points. The intersection of two lines is then computed as the point congruent to their regressive product,

$$X \equiv (P \wedge V) \vee (Q \wedge U), \qquad Y \equiv (P \wedge W) \vee (R \wedge U), \qquad Z \equiv (Q \wedge W) \vee (R \wedge V).$$

To prove the theorem using the Grassmannian approach, one only has to choose symbolic values for the points in an arbitrary basis, and compute the collinearity condition $X \wedge Y \wedge Z = 0$. In what follows we use the *GrassmannAlgebra* package [Browne 2009] to do the computations.

We begin by declaring that we wish to work in the plane, and then define the points P, Q, R, and U, V, W by means of two vectors u and v and four scalars a, b, c, d in such a way as to ensure the triplets of points are collinear,

```
{P = 𝕆+v, Q = 𝕆+av, R = 𝕆+bv, U = 𝕆+u, V = 𝕆+cu, W = 𝕆+du}.
```

The next stage is to compute the intersection points X, Y, and Z. We do this by using the *GrassmannAlgebra* functions `ToCommonFactor` and `ToPointForm`. `ToCommonFactor` computes the common factor as a weighted point, while `ToPointForm` determines the point congruent to it. The inputs and their results are as follows:

```
X = ToPointForm[ToCommonFactor[(P∧V)∨(Q∧U)]]
```

$$\longrightarrow \quad \mathtt{X} = \mathbb{O} + \frac{c(a-1)}{ac-1}\,\mathtt{u} + \frac{a(c-1)}{ac-1}\,\mathtt{v}$$

```
Y = ToPointForm[ToCommonFactor[(P∧W)∨(R∧U)]]
```

$$\longrightarrow \quad \mathtt{Y} = \mathbb{O} + \frac{d(b-1)}{bd-1}\,\mathtt{u} + \frac{b(d-1)}{bd-1}\,\mathtt{v}$$

```
Z = ToPointForm[ToCommonFactor[(Q∧W)∨(R∧V)]]
```

$$\longrightarrow \quad \mathtt{Z} = \mathbb{O} + \frac{cd(a-b)}{ac-bd}\,\mathtt{u} + \frac{ab(c-d)}{ac-bd}\,\mathtt{v}$$

Finally, we expand and simplify the exterior product of the three points. A value of zero proves that they are collinear, and hence proves the theorem:

```
Simplify[GrassmannExpandAndSimplify[X∧Y∧Z]]   ⟶   0
```

Conclusions

This paper has discussed the ability of Grassmann's algebra to construct higher order entities from points and vectors and to compute their intersections with the regressive product. Applications explored demonstrate the inherent beauty, simplicity, and power of Grassmann's approach. In mechanics they led to multiple insights on the relationship between force and moment, the generality of dimension, and the relevance of metric. In geometry they led to powerful methods for representing geometric objects, computing intersections, and proving theorems.

Representations of spinor groups using Grassmann exterior algebra

Nikolay Marchuk

An exterior algebra, invented by Grassmann in the year 1844 [A1$_K$], has many applications in different fields of mathematics and physics. Here we present a new application of the Grassmann algebra to the theory of spinor groups.

Clifford algebras. Let p, q, n be non-negative integers with $n = p + q$, and let $\mathcal{C}\ell(p, q)$ be a real Clifford algebra [Marchuk 2009] of the signature (p, q) with generators $e^1, \ldots, e^n$, such that

$$e^a e^b + e^b e^a = 2\eta^{ab} e, \quad a, b = 1, \ldots, n,$$

where e is the identity element of the Clifford algebra and η^{ab} are elements of the diagonal matrix of dimension n,

$$\eta = \mathrm{diag}(1, \ldots, 1, -1, \ldots, -1),$$

This work was partially supported by a grant of the President of the Russian Federation (project NSh-7675.2010.1) and by the division of mathematics of RAS (project "Modern problems in theoretical mathematics").

H.-J. Petsche et al. (eds.), *From Past to Future: Graßmann's Work in Context*,
DOI 10.1007/978-3-0346-0405-5_26,

with p pieces of 1 and q pieces of -1 on the diagonal. The Clifford algebra $\mathcal{C}\ell(p,q)$ can be considered a 2^n-dimensional vector space with basis elements

$$e, e^a, e^{a_1a_2}, \ldots, e^{a_1\ldots a_{n-1}}, e^{12\ldots n}, \quad 0 \le a_1 < \ldots < a_k \le n, \tag{1}$$

numbered by ordered multi-indices of lengths from 0 to n. Any element of the Clifford algebra $\mathcal{C}\ell(p,q)$ can be written in the form of a decomposition with respect to the basis (1),

$$U = ue + u_a e^a + \sum_{a_1<a_2} u_{a_1a_2} e^{a_1a_2} + \ldots + u_{1\ldots n} e^{1\ldots n}, \tag{2}$$

with real coefficients $u, u_a, u_{a_1a_2}, \ldots, u_{1\ldots n}$. Elements of the form

$$U = \sum_{a_1<\ldots<a_k} u_{a_1\ldots a_k} e^{a_1\ldots a_k}$$

are called *elements of rank k*. Denote by $\mathcal{C}\ell_k(p,q)$ the subspace of rank k elements. We have

$$\mathcal{C}\ell(p,q) = \mathcal{C}\ell_0(p,q) \oplus \ldots \oplus \mathcal{C}\ell_n(p,q).$$

An element $U \in \mathcal{C}\ell(p,q)$ is called *even (odd)* if this element is a sum of elements of even (odd) ranks. Hence

$$\begin{aligned} \mathcal{C}\ell_{\text{Even}}(p,q) &= \mathcal{C}\ell_0(p,q) \oplus \mathcal{C}\ell_2(p,q) \oplus \ldots, \\ \mathcal{C}\ell_{\text{Odd}}(p,q) &= \mathcal{C}\ell_1(p,q) \oplus \mathcal{C}\ell_3(p,q) \oplus \ldots, \\ \mathcal{C}\ell(p,q) &= \mathcal{C}\ell_{\text{Even}}(p,q) \oplus \mathcal{C}\ell_{\text{Odd}}(p,q). \end{aligned}$$

Reverse operation. Let us define a linear *reverse* operation $\sim: \mathcal{C}\ell(p,q) \to \mathcal{C}\ell(p,q)$ with the aid of the following rules:

$$e^\sim = e, \quad (e^a)^\sim = e^a, \quad (UV)^\sim = V^\sim U^\sim, \quad \forall\, U, V \in \mathcal{C}\ell(p,q).$$

In particular,

$$(e^{a_1} \ldots e^{a_k})^\sim = e^{a_k} \ldots e^{a_1}.$$

Spinor groups. Let $n = p + q \leq 5$. Consider the following set of even elements of the Clifford algebra $C\ell(p,q)$:

$$\mathrm{Spin}_+(p,q) = \{S \in C\ell_{\mathrm{Even}}(p,q) : S^\sim S = e\}.$$

This set is closed with respect to the Clifford product and contains the identity element e. Elements of this set are invertible. Therefore $\mathrm{Spin}_+(p,q)$ can be considered as a group (Lie group) with respect to the Clifford product. This group is called the *spinor group*.

The set of rank 2 elements $C\ell_2(p,q)$ with the commutator $[A,B] = AB - BA$ is the Lie algebra of the Lie group $\mathrm{Spin}_+(p,q)$ [Lounesto 1997].

Let us define an *exponent* of Clifford algebra elements $\exp : C\ell(p,q) \to C\ell(p,q)$ by the formula

$$\exp A = e + A + \frac{1}{2!}A^2 + \frac{1}{3!}A^3 + \ldots.$$

Spinor groups in cases $p + q = 4$. The sign $\simeq$ denotes group isomorphism. It is known [Lounesto 1997] that:

- $\mathrm{Spin}_+(4,0) \simeq \mathrm{Spin}_+(0,4) \simeq \mathrm{Spin}(4)$.
- $\mathrm{Spin}_+(1,3) \simeq \mathrm{Spin}_+(3,1)$.
- $\forall S \in \mathrm{Spin}(4)$ there exists $B \in C\ell_2(4,0)$ such that $S = \exp B$.
- $\forall S \in \mathrm{Spin}_+(1,3)$ there exists $B \in C\ell_2(1,3)$ such that $S = \exp B$, or $S = -\exp B$. Note that the set of exponents of rank 2 elements does not form a group.
- There exist elements $S \in \mathrm{Spin}_+(2,2)$ such that these elements cannot be represented in the form $\pm\exp B$, where $B \in C\ell_2(2,2)$.

Exterior (Grassmann) multiplication of Clifford algebra elements. Let us define the associative and distributive operation of exterior multiplication of Clifford algebra elements (denoted by $\wedge$)

$$e^{a_1} \wedge e^{a_2} \wedge \ldots \wedge e^{a_k} = e^{[a_1} e^{a_2} \ldots e^{a_k]},$$

where square brackets denote the operation of alternation of indices. The Clifford algebra $C\ell(p,q)$, considered with the exterior product, can be identified with the Grassmann algebra of dimension n.

Exterior exponent. Consider the exterior exponent $\widehat{\exp} : C\ell(p,q) \to C\ell(p,q)$,

$$\widehat{\exp}(B) = e + B + \frac{1}{2!}B \wedge B + \frac{1}{3!}B \wedge B \wedge B + \ldots, \tag{3}$$

where $B \in C\ell(p,q)$. We are interested in the exterior exponent of rank 2 elements. In this case on the right hand side of (3) there are finitely many non-zero summands. In particular, for the case $n = 4$ there are only three summands,

$$\widehat{\exp}(B) = e + B + \frac{1}{2} B \wedge B,$$

where $B \in C\ell_2(p,q), p + q = 4$. We see that

$$(\widehat{\exp}(B))^{\sim} = \widehat{\exp}(-B) = e - B + \frac{1}{2} B \wedge B.$$

It is not hard to prove that

$$(\widehat{\exp}(B))^{\sim} \widehat{\exp}(B) = \lambda e,$$

the coefficients of the element B.

Main theorem. If $U \in C\ell(p,q)$ is written in the form (2), then we denote $\operatorname{Tr} U = u$. Also denote $\ell = e^1 e^2 e^3 e^4$ and $\epsilon = \operatorname{Tr}(\ell^2)$.

Theorem 1. *Let $n = p + q = 4$. Any element S of the group $Spin_+(p,q)$ can be represented in one of the two following forms:*

- *If $\operatorname{Tr} S \neq 0$, then there exists $B \in C\ell_2(p,q)$ such that $\lambda = \lambda(B) > 0$ and*

$$S = \pm \frac{1}{\sqrt{\lambda}} \widehat{\exp} B. \tag{4}$$

The sign (plus or minus) on the right hand side is equal to the sign of the number $\operatorname{Tr} S$.

- *If $\operatorname{Tr} S = 0$, then there exists $B \in C\ell_2(p,q)$ such that $B \wedge B = 0$, $\epsilon(1 + \beta) \geq 0$, where $\beta = \operatorname{Tr}(B^2)$ and*

$$S = B \pm \ell \sqrt{\epsilon(1 + \beta)}. \tag{5}$$

The sign (plus or minus) on the right hand side is equal to the sign of the number $\operatorname{Tr}(\ell^{-1} S)$.

Proof. Let $S \in \mathrm{Spin}_+(p,q)$ be such that $\mathrm{Tr}\, S = \alpha \neq 0$. We write S and $S^\sim$ in the form

$$S = \alpha e + U + F, \quad S^\sim = \alpha e - U + F,$$

where $U \in \mathcal{C}\ell_2(p,q), F \in \mathcal{C}\ell_4(p,q)$. For $n = 4$ it follows that

$$U^2 = U \wedge U + \gamma e, \quad \alpha^2 e - F^2 - \gamma^2 e \in \mathcal{C}\ell_0(p,q), \quad 2\alpha F - U \wedge U \in \mathcal{C}\ell_4(p,q).$$

The identity

$$S^\sim S = (\alpha^2 e - F^2 - \gamma^2 e) + (2\alpha F - U \wedge U) = e$$

gives us $2\alpha F - U \wedge U = 0$, and hence $F = \frac{1}{2\alpha} U \wedge U$, which means

$$S = \alpha e + U + \frac{1}{2\alpha} U \wedge U = \alpha \,\widehat{\exp}\left(\frac{1}{\alpha} U\right).$$

Denoting

$$B = \frac{1}{\alpha} U, \quad \alpha = \pm \frac{1}{\sqrt{\lambda}},$$

we get

$$S = \pm \frac{1}{\sqrt{\lambda}} \widehat{\exp}\, B,$$

where

$$\lambda e = \widehat{\exp}(B)\, \widehat{\exp}(-B). \tag{6}$$

It is easy to prove that if an element $S \in \mathrm{Spin}_+(p,q)$ is such that $\mathrm{Tr}\, S = 0$, then S can be represented in the form (5). This completes the proof.

If $S \in \mathrm{Spin}_+(p,q)$ has the form (4), then we say that S is given with the aid of a semi-polynomial parametrization. That means coefficients of S in the basis (1) are polynomials of second degree of coefficients b_{ij} multiplied on one and the same factor $1/\sqrt{\lambda}$, where $\lambda = \lambda(B)$ is the polynomial of second degree with coefficients b_{ij}.

If $S \in \mathrm{Spin}_+(p,q)$ has the form (5), then we say that S is given with the aid of an adjoint semi-polynomial parametrization.

Let us write down the explicit form of elements $S \in \mathrm{Spin}_+(p,q), p+q=4$, using real coefficients of $b_{ij} \in B \in \mathcal{C}\ell_2(p,q)$,

$$B = b_{12}\,e^{12} + b_{13}\,e^{13} + b_{14}\,e^{14} + b_{23}\,e^{23} + b_{24}\,e^{24} + b_{34}\,e^{34}.$$

We have

$$\widehat{\exp}(B) = e + b_{12}\,e^{12} + b_{13}\,e^{13} + b_{14}\,e^{14} + b_{23}\,e^{23} \\ +b_{24}\,e^{24} + b_{34}\,e^{34} + (b_{14}\,b_{23} - b_{13}\,b_{24} + b_{12}\,b_{34})\,e^{1234}$$

Expressions for the scalar $\lambda = \lambda(B)$ that satisfy (6) and for $\epsilon(1+\beta)$ depend on the signature (p,q).

For $(p,q) = (0,4), (4,0)$:

$$\lambda = 1 + b_{12}{}^2 + b_{13}{}^2 + b_{14}{}^2 + b_{23}{}^2 + b_{14}{}^2\,b_{23}{}^2 - 2\,b_{13}\,b_{14}\,b_{23}\,b_{24} + b_{24}{}^2 \\ +b_{13}{}^2\,b_{24}{}^2 + 2\,b_{12}\,b_{14}\,b_{23}\,b_{34} - 2\,b_{12}\,b_{13}\,b_{24}\,b_{34} + b_{34}{}^2 + b_{12}{}^2\,b_{34}{}^2,$$
$$\epsilon(1+\beta) = 1 - b_{12}{}^2 - b_{13}{}^2 - b_{14}{}^2 - b_{23}{}^2 - b_{24}{}^2 - b_{34}{}^2.$$

For $(p,q) = (1,3)$:

$$\lambda = 1 - b_{12}{}^2 - b_{13}{}^2 - b_{14}{}^2 + b_{23}{}^2 - b_{14}{}^2\,b_{23}{}^2 + 2\,b_{13}\,b_{14}\,b_{23}\,b_{24} + b_{24}{}^2 \\ -b_{13}{}^2\,b_{24}{}^2 - 2\,b_{12}\,b_{14}\,b_{23}\,b_{34} + 2\,b_{12}\,b_{13}\,b_{24}\,b_{34} + b_{34}{}^2 - b_{12}{}^2\,b_{34}{}^2,$$
$$\epsilon(1+\beta) = -1 - b_{12}{}^2 - b_{13}{}^2 - b_{14}{}^2 + b_{23}{}^2 + b_{24}{}^2 + b_{34}{}^2.$$

For $(p,q) = (2,2)$:

$$\lambda = 1 + b_{12}{}^2 - b_{13}{}^2 - b_{14}{}^2 - b_{23}{}^2 + b_{14}{}^2\,b_{23}{}^2 - 2\,b_{13}\,b_{14}\,b_{23}\,b_{24} - b_{24}{}^2 \\ +b_{13}{}^2\,b_{24}{}^2 + 2\,b_{12}\,b_{14}\,b_{23}\,b_{34} - 2\,b_{12}\,b_{13}\,b_{24}\,b_{34} + b_{34}{}^2 + b_{12}{}^2\,b_{34}{}^2,$$
$$\epsilon(1+\beta) = 1 - b_{12}{}^2 + b_{13}{}^2 + b_{14}{}^2 + b_{23}{}^2 + b_{24}{}^2 - b_{34}{}^2.$$

For $(p,q) = (3,1)$:

$$\lambda = 1 + b_{12}{}^2 + b_{13}{}^2 - b_{14}{}^2 + b_{23}{}^2 - b_{14}{}^2\,b_{23}{}^2 + 2\,b_{13}\,b_{14}\,b_{23}\,b_{24} - b_{24}{}^2 \\ -b_{13}{}^2\,b_{24}{}^2 - 2\,b_{12}\,b_{14}\,b_{23}\,b_{34} + 2\,b_{12}\,b_{13}\,b_{24}\,b_{34} - b_{34}{}^2 - b_{12}{}^2\,b_{34}{}^2,$$
$$\epsilon(1+\beta) = -1 + b_{12}{}^2 + b_{13}{}^2 - b_{14}{}^2 + b_{23}{}^2 - b_{24}{}^2 - b_{34}{}^2.$$

With the aid of these formulas we get the general form (4)–(5) of elements $S \in \mathrm{Spin}_+(p,q)$.

The proof of the following known theorem is straightforward.

Theorem 2. *Let $n = p + q = 2, 3$. Any element $S \in Spin_+(p,q)$ can be represented in the form*

$$S = \pm e\sqrt{1+\beta} + B, \qquad (7)$$

where $B \in C\ell_2(p,q)$ is such that $\beta = \mathrm{Tr}(B^2) \geq -1$. The sign (plus or minus) on the right hand side of (7) is equal to the sign of the number $\mathrm{Tr}\, S$.

Hermann Grassmann's theory of linear transformations

Jörg Liesen

Introduction

In mid-1854 Hermann Grassmann started working on a second version of his *Ausdehnungslehre*. The first version of 1844, A_1 [A1], had failed to attract the attention of his contemporaries. But Grassmann was convinced that his theory was important, and his renewed effort was encouraged by Möbius,[1] to whom he sent the finished A_2 [A2] in October 1861. The new version featured a radically different style from A_1. Grassmann also had added a wealth of new material. The most notable addition was

This work was supported by the Heisenberg-Programm of the Deutsche Forschungsgemeinschaft.

[1] Möbius wrote to Grassmann on June 9, 1853: "Finally, I earnestly wish that you will be able to find time and leisure to prepare your *Ausdehnungslehre* in the new, simpler form and (as you wrote to me) in a form more suited to the real mathematician. In addition I wish that you can include the further investigations promised in the preface of the first part of this work. For I am sincerely convinced that this would greatly benefit the science of space." ("Schließlich wünsche ich Ihnen von Herzen, daß Sie bald Zeit und Muße finden mögen, um Ihre Ausdehnungslehre in der neuen, einfacheren und (wie Sie mir schreiben) dem eigentlichen Mathematiker mehr zusagenden Form bearbeiten und nebst den in der Vorrede zum ersten Teile des Werkes versprochenen weiteren Untersuchungen erscheinen lassen zu können. Denn ich bin innigst überzeugt, daß die Wissenschaft des Raumes dadurch einen großen Gewinn erhalten würde.") [Cited in Engel 1911, 163 - My translation.]

H.-J. Petsche et al. (eds.), *From Past to Future: Graßmann's Work in Context*,
DOI 10.1007/978-3-0346-0405-5_27, © Springer Basel AG 2011

the "Theory of Functions" (*Funktionenlehre*), which covers roughly the last third of the approximately 400 pages of A_2 (Nos. 348–527)[2].

This paper is concerned with a particular section of the *Funktionenlehre*, namely Nos. 377–391[3]. In this part Grassmann deals with a mathematical object he calls a "fraction" or "quotient" (*Bruch* or *Quotient*). Today this "fraction" would be called a linear transformation on a finite dimensional vector space. Since such transformations can be identified with matrices, Grassmann's section may also be considered a contribution to the theory of matrices. The goal in the following is to describe Grassmann's main definitions and results for the "fraction", and to relate them to contemporary developments and modern notation.

Definition of the fraction

The Nos. 377–391 in Grassmann's A_2 form a section entitled
"Complete functions of first degree. Quotient".[4]
It starts with the following definition:[5]

377. DEFINITION. *If* $a_1, a_2, \ldots, a_n$ *are magnitudes of the first (or* $(n-1)$*th) order in a principal domain of nth order, and* stand in no numerical relation to one another, *I mean by the* fraction (quotient)

$$Q = \frac{(b_1, b_2, \ldots, b_n)}{(a_1, a_2, \ldots, a_n)}$$

the expression which, multiplied by $a_1, a_2, \ldots, a_n$*, yields the values* $b_1, b_2, \ldots, b_n$ *respectively, so that*

$$\frac{(b_1, b_2, \ldots, b_n)}{(a_1, a_2, \ldots, a_n)} a_r = b_r.$$

[2] A detailed comparison of the contents of A_1 and A_2 is given in Grassmann's foreword to the second edition of A_1, which was written in 1877 [A1, 17–21]; see [$A1_K$, 19–22] for an English translation. In particular Grassmann writes that "[t]he second section (no. 348–527) is a completely new addition that treats function theory and the algebraic multiplication basic to it, along with differential calculus, infinite series, and integral calculus, and examines especially deeply the applied areas of analysis and both modern geometry and algebra" [$A1_K$, 22].

[3] This section (*Abschnitt II, Kapitel 1, Sect. 4*) is approximately 24 pages long; pp. 241–265 in the original A_2, and pp. 240–263 in the edited version of A_2 that appeared in Grassmann's *Gesammelte Werke* [HGW12].

[4] German: *Ganze Funktionen ersten Grades. Quotient.* The translation is taken from [$A2_K$, 207].

[5] The definition is quoted from [$A2_K$, 207], which is a translation of the version of A_2 that appeared in the *Gesammelte Werke* [HGW12]. The expression in curly brackets { } was added to Grassmann's original text by the editors of [HGW12], and this typesetting was adopted in [$A2_K$].

I call $(a_1, a_2, \ldots, a_n)$ *the* denominators *of the fraction,* $(b_1, b_2, \ldots, b_n)$ *their corresponding* numerators, *and set two fractions, or two expressions numerically derived from fractions, equal to one another if and only if they yield equals when they are multiplied by every magnitude of first {or* $(n-1)$*th} order. If in addition the numerators are magnitudes of first (or* $(n-1)$*th) order,* and stand in no numerical relation to one another, *I call the fraction* invertible, *and in this case, if*

$$Q = \frac{(b_1, b_2, \ldots, b_n)}{(a_1, a_2, \ldots, a_n)},$$

I denote the inverted fraction by $\frac{1}{Q}$*, that is, I set*

$$\frac{1}{Q} = \frac{(a_1, a_2, \ldots, a_n)}{(b_1, b_2, \ldots, b_n)}.$$

Let me briefly translate the key terms into modern language: Grassmann's "magnitudes of the first order"[6] may be considered *vectors*, and the "principal domain of nth order" is an *n-dimensional vector space*. The assumption that $a_1, a_2, \ldots, a_n$ "stand in no numerical relation to one another" means that these vectors are *linearly independent* (hence they form a basis). In this paper I will not discuss Grassmann's notion of "magnitudes of $(n-1)$th order", so this detail of the definition will not be considered.

Grassmann's definition only specifies the result of the "multiplication" of Q by each of the denominators a_r. It does not explicitly say what happens when Q is multiplied by a linear combination of the denominators. The key to understanding this lies in the term "multiplied by". A large part of A_2 (Nos. 37–215) is devoted to different types of multiplication of extensive magnitudes. Grassmann considers rather general products, and he studies non-commutative multiplication. All products considered in A_2 are linear in each factor, i.e., multilinear. In particular, in No. 44 Grassmann states that if $P_{qa+rb+\cdots}$ is a product of extensive magnitudes containing the factor $qa + rb + \cdots$, where $a, b, \ldots$ are extensive magnitudes and $q, r, \ldots$ are (real) numbers, then $P_{qa+rb+\cdots} = qP_a + rP_b + \cdots$.

Functions are considered extensive magnitudes by Grassmann, and thus Qa_r is a product of two of these, namely of a "complete function of first degree" and an

[6] Grassmann often uses the term "extensive magnitudes" (*extensive Größen*) instead of just "magnitudes".

"extensive magnitude of first order". Since this product is linear, the fraction Q is a *linear transformation*, i.e., for any $x = x_1a_1 + \cdots + x_na_n$ it satisfies

$$Qx = Q(x_1a_1 + \cdots + x_na_n) = x_1Qa_1 + \cdots + x_nQa_n. \tag{1}$$

Linearity of the fraction is implied by the previous development in A_2, which apparently was so obvious for Grassmann that he felt no need to mention it in No. 377. The first explicit occurrence of linearity is in the following No. 378, where he refers to No. 44 and applies (rather than defines) the identity (1).

The second part of Grassmann's No. 377 starts with the definition of the equality of fractions: If $Q_1x = Q_2x$ for every extensive magnitude x, then $Q_1 = Q_2$. Here Q_1 and Q_2 may be fractions, or expressions that are "numerically derived from fractions". "Numerical derivation" means *linear combination* in A_2, and this fundamental concept of Linear Algebra is introduced in No. 1 of the book. Since fractions are extensive magnitudes they may in principle be combined linearly, but at this point no further details are given. In the later No. 379 Grassmann shows how to linearly combine fractions having the same denominators:

$$\beta\frac{(b_1,\ldots,b_n)}{(a_1,\ldots,a_n)} + \gamma\frac{(c_1,\ldots,c_n)}{(a_1,\ldots,a_n)} = \frac{(\beta b_1 + \gamma c_1,\ldots,\beta b_n + \gamma c_n)}{(a_1,\ldots,a_n)}.$$

The definition of the fraction continues with the concept of invertibility. This is based on the observation that if the numerators $b_1,\ldots,b_n$ of a fraction Q stand in no numerical relation to one another, then they "qualify" as denominators of a fraction. Taking the original denominators $a_1,\ldots,a_n$ as numerators gives the "inverted fraction" $\frac{1}{Q}$. It may have been clear to Grassmann that in this case

$$\frac{1}{Q}(Qa_j) = \frac{1}{Q}b_j = a_j, \quad \text{and, similarly,} \quad Q\left(\frac{1}{Q}b_j\right) = Qa_j = b_j, \quad j = 1,\ldots,n.$$

In modern terms, the composition $\frac{1}{Q}Q = Q\frac{1}{Q}$ is the identity map on the given n-dimensional space. Grassmann does not discuss the composition of fractions at this point, and he does not consider it explicitly later in A_2. However, the composition of fractions is implicit in a number of places in the *Funktionenlehre* because all extensive magnitudes (and thus the fractions) can be multiplied (see the section "Exchanging the denominators" below for further comments on the composition of fractions).

In a remark following his definition Grassmann comments on his line of thought for the notion of the fraction. He writes that in Algebra the fraction $a : b$ of two numbers a and $b \neq 0$ is an expression that yields the number a when multiplied with b. In a domain of order n, he continues, it is not sufficient to specify the result of the multiplication of a fraction Q by a single extensive magnitude. Rather, the results of the multiplication of Q by n magnitudes standing in no numerical relation to one another (i.e., a basis) must be specified to completely determine Q. This again must have been so obvious to him that he left it at this brief remark; no formal proof of the validity of this construction is given. Thorough contemporary readers therefore may have wondered whether the concept of the fraction is well defined.

Peano's and Whitehead's takes on the fraction

Unfortunately for Grassmann, contemporary readers were rare. In the years following its publication, A_2 was even less successful[7] than the widely ignored A_1. Fifteen years after the publication of A_2, Grassmann wrote that "in the treatment of the quotient (Nos. 377–391) ... are contained the germs of developments which still await future processing."[8] In other words, he believed that his concept of the fraction had yet to be discovered by others.

Among the few nineteenth century mathematicians who carefully studied Grassmann's work was Peano. It is interesting to read Chap. 9 of his book of 1888 [Peano 1888] in light of Grassmann's definition of the fraction. In No. 75 Peano gives a rigorous definition of a linear transformation R via the two (nowadays common) rules

$$R(a + a') = Ra + Ra', \qquad R(ma) = m(Ra),$$

for all vectors[9] a, a' [Peano 1888, 145].

In No. 76 he defines, just like Grassmann, that the linear transformations R and S are equal if $Ra = Sa$ holds for each a. Unlike Grassmann, however, he explicitly defines the composition SR by the equation $SRa = S(Ra)$ (if R maps from A to B and S maps from B to C). For the composition Peano states (without proof) that $T(SR) = (TS)R$ (*associativity*), $S(R + R') = SR + SR'$ (*distributivity*), and he defines the concept of

---

[7] Engel writes: "Das betrübende Ergebnis war, daß die neue Ausdehnungslehre ebensowenig oder eigentlich noch weniger Erfolg hatte als die alte." [Engel 1911, 231].

[8] "... in der Behandlung des Quotienten (Nos. 377–391) ... Keime zu Entwickelungen enthält, welche noch zukünftiger Bearbeitung harren." [A1, 21] (My translation.)

[9] Peano uses the term "enti del systema", which may be translated "entities of the system".

commutativity, $SR = RS$. He also proves the following result (quoted from [Peano 2000, 124]):

77. Theorem. *If* $a_1, \ldots, a_n$ *are independent entities of a linear system A of n dimensions, and* $b_1, \ldots, b_n$ *are also entities of a linear system, one and only one transformation R of the system A is determined that satisfies the conditions*

$$Ra_1 = b_1, \ldots, Ra_n = b_n.$$

This theorem shows that Grassmann's concept of the fraction is indeed well defined: A linear transformation is uniquely determined when the images $b_1, \ldots, b_n$ for a given basis $a_1, \ldots, a_n$ of the space are specified.

Next, Peano writes the linear transformation R in the form

$$R = \begin{pmatrix} b_1, & b_2, & \ldots & b_n \\ a_1, & a_2, & \ldots & a_n \end{pmatrix},$$

which is very similar to Grassmann's notation. He states (among other identities), that if a, b, c are numbers, then

$$\begin{pmatrix} c \\ b \end{pmatrix} a = \frac{c}{b} a,$$

so he relates linear transformations to the quotient in Algebra (cf. Grassmann's line of thought). Peano continues with defining the inverse

$$R^{-1} = \begin{pmatrix} a_1 & \ldots & a_n \\ b_1 & \ldots & b_n \end{pmatrix},$$

if $b_1, \ldots, b_n$ are linearly independent. Because he has introduced the composition, he is able to write the (nowadays common) formulas

$$R^{-1}R = 1, \quad RR^{-1} = 1, \quad (R^{-1})^{-1} = R.$$

Peano's book [Peano 1888] is often characterized as a reworking of Grassmann's ideas in A_1; see, for example, [Kennedy 2002, 201], [Reich 1996b]. The above shows that Peano also dealt with material that goes well beyond A_1, particularly in his treatment of linear transformations. From a modern standpoint, Peano's text clarifies all ambiguities in Grassmann's definition, and it ties up some loose ends, for example, in the composition of transformations. Peano, however, does not cite A_2 or any other source for his treatment. It needs to be stressed that Peano by no means wanted to plagiarize Grassmann. As already indicated by the title of [Peano 1888], he did not claim originality for the ideas in his book; also cf. [Kennedy 1980].

In his "Universal Algebra" [Whitehead 1898], published a decade after [Peano 1888], Whitehead gave another "free translation of Grassmann's own writing" (p. 248) on the fraction. In the chapter "Matrices" [Whitehead 1898, Book IV, Chap. 6], Whitehead defines a matrix as a linear transformation (he uses the term "distributive in reference to addition") and he adopts Grassmann's notation by writing it in the form

$$\phi = \frac{a_1, a_2 \ldots a_\nu}{e_1, e_2 \ldots e_\nu}.$$

Besides Grassmann's A_2, Whitehead's chapter is based on a paper by Buchheim[10] which connected Grassmann's fraction with the theory of Hamilton, Cayley, and Sylvester, and attempted "to treat the theory of matrices from Grassmann's point of view" [Buchheim 1884, 70]. Whitehead suspects that Grassmann's A_2 had been written "in ignorance of Cayley's classical memoir on Matrices" [Whitehead 1898, 248].[11]

Remarkably, Whitehead introduces "matrices" as Grassmann's fractions. Only afterwards he proves that "[t]he operator ϕ – called by Grassmann a quotient – may be identified with Cayley's matrix" [Whitehead 1898, 249]. "Cayley's matrix" here means the nowadays common definition of a matrix. Whitehead's treatment of the subject goes far beyond Peano's and even a bit beyond Grassmann's, but this is out of the scope of this paper.

Exchanging the denominators

After defining the fraction in No. 377 and proving a few basic properties in Nos. 378 and 379, Grassmann considers the exchange of the denominators and its effect on the fraction. For example, if $\hat{a}_1 = x_1 a_1 + \cdots + x_n a_n$ with $x_1 \neq 0$, then $\hat{a}_1, a_2, \ldots, a_n$ stand

[10] Arthur Buchheim (1859–1888) studied at Oxford and also "for some time under Prof. Klein at Leipzig" [Sylvester 1888].

[11] Whitehead refers to [Cayley 1858].

in no numerical relation to one another[12]. Moreover,

$$Q\hat{a}_1 = x_1 Q a_1 + \cdots + x_n Q a_n = x_1 b_1 + \cdots + x_n b_n,$$

and thus

$$Q = \frac{(b_1, b_2, \ldots, b_n)}{(a_1, a_2, \ldots, a_n)} = \frac{(x_1 b_1 + \cdots + x_n b_n, b_2, \ldots, b_n)}{(\hat{a}_1, a_2, \ldots, a_n)}.$$

While here we have exchanged one of the denominators (a_1 by $\hat{a}_1$), Grassmann shows in No. 380 that every fraction can be written with n arbitrary denominators (standing in no numerical relation to one another) that are numerically derived from $a_1, \ldots, a_n$.

Today we typically identify a basis exchange with a *similarity transformation*[13], and we write this as $S^{-1}QS$ for an invertible S. Here the composition of linear transformations, or, equivalently, the multiplication of matrices, appears naturally. Matrix algebra, and in particular matrix multiplication as we use it today was introduced by Cayley [Cayley 1858]. Hence this was a very recent development when Grassmann wrote A_2, and he may have been unaware of it; cf. the above quote of Whitehead. Moreover, because with each change of a denominator the effect on the corresponding numerator is reflected by Grassmann's notation, it is simply not necessary to write a similarity transformation in terms of the composition of fractions. In the light of this it is not surprising that Grassmann did not formally introduce compositions of his fractions.

In No. 381 Grassmann shows that any fraction Q can be written in the form

$$Q = \sum_{r,s}^{n} \alpha_{r,s} E_{r,s},$$

where the n^2 fractions $E_{r,s}$, $r, s = 1, \ldots, n$, are defined by

$$E_{r,s} e_r = e_s, \quad \text{and} \quad E_{r,s} e_t = 0, \text{for all } t \neq r.$$

Moreover, with respect to the n^2 fractions $E_{r,s}$, Grassmann shows that "all fractions whose numerators and denominators are magnitudes of first order of the principal domain can be numerically derived from them, while they themselves stand in no numerical relation to one another" [$A2_K$, 209]. This means that the fractions $E_{r,s}$ form a basis of the vector space of all fractions (with n denominators). In addition, every

[12] Today this result is called the *Exchange Lemma*. It is proven in No. 19 of A_2, and inductively applied in the proof of the *Exchange Theorem* in No. 20. The latter result is often attributed to [Steinitz 1910].

[13] Similarity (*Ähnlichkeit*) of matrices (or bilinear forms) was first defined in [Frobenius 1878, 21].

fraction is represented in this basis by the corresponding $n \times n$ matrix containing the coefficients $\alpha_{r,s}$.

In No. 386 Grassmann proves that if m is the order of the domain numerically derived from the numerators of Q, then Q can be written in the form

$$\frac{(b_1, \ldots, b_m, 0, 0, \ldots, 0)}{(a_1, \ldots, a_m, c_{m+1}, c_{m+2}, \ldots, c_n)},$$

where $b_1, \ldots, b_m$ stand in no numerical relation to one another[14]. In modern terms m is the dimension of the image of Q, and Grassmann's result implies the *Rank-Nullity Theorem*: The dimension of the vector space on which Q operates is equal to its "rank"[15] (dimension of the image) plus its "nullity" (dimension of the kernel); here $n = m + (n - m)$.

Spectral theory

In No. 384 Grassmann defines the "power" (*Potenzwerth*) of a fraction $Q = \frac{(b_1, b_2, \ldots, b_n)}{(a_1, a_2, \ldots, a_n)}$ by

$$[Q^n] = \frac{[b_1, b_2, \ldots, b_n]}{[a_1, a_2, \ldots, a_n]}.$$

The square brackets on the right-hand side denote the "combinatorial product" of the respective extensive magnitudes. The power $[Q^n]$ can be considered the *determinant* of the linear transformation.[16]

The combinatorial product is the key tool in Grassmann's spectral theory, which starts with the following definition (quoted from [$A2_K$, 214]):

---

[14] I have slightly changed Grassmann's notation for consistency.

[15] The term "rank" (*Rang*) (of a determinant or matrix) was coined by Frobenius [Frobenius 1879, 1]. The concept of the "rank" had been implicit in earlier work of other authors, e.g. in [Sylvester 1850].

[16] The combinatorial product is studied in detail in Nos. 52–136 of A_2. Today this product would be called an *alternating multi-linear form*. This means that the product is multilinear, and the exchange of two factors leads to a sign change of the product, i.e., $[\cdots a \cdots b \cdots] = -[\cdots b \cdots a \cdots]$. In Nos. 62–63 Grassmann shows the relation between the combinatorial product and the determinant of "n rows of n numbers each", i.e. a square matrix.

387. DEFINITION. *If a fraction Q multiplied by a non-zero magnitude of first order yields a multiple of this magnitude, say the ρ-fold of it, so that*

$$Qx = \rho x,$$

I call the coefficient ρ (whether it is real or imaginary) a principal number *of the fraction Q, and the domain to which all those magnitudes x belong that satisfy that equation the* principal domain *of that principal number.*

No further explanation is necessary: This reads like a modern definition of eigenvalues and corresponding eigenvectors and eigenspaces. To find the principal numbers and principal domains of a fraction Grassmann derives in No. 388 the (characteristic) equation

$$[(\rho e_1 - Qe_1)(\rho e_2 - Qe_2)\cdots(\rho e_n - Qe_n)] = 0,$$

and, after evaluating the combinatorial product,

$$\alpha_0\rho^n - \alpha_1\rho^{n-1} + \alpha_2\rho^{n-2} - \cdots + (-1)^n\alpha_n = 0. \qquad (2)$$

The left-hand side of (2) is a polynomial of degree n, from which Grassmann concludes that each fraction Q has n principal numbers, which are the n roots of this equation. Moreover, he mentions that the number $\alpha_n : \alpha_0$ is equal to the power $[Q^n]$ (i.e., the determinant of the linear transformation).

Grassmann only considers the multiplicity of a principal number as a root of (2), i.e., the *algebraic multiplicity*. The concept of *geometric multiplicity*, i.e. the dimension of the corresponding principal domain, is implicit in his text but it is not explicitly defined or explored.

He shows in No. 389 that if the n principal numbers of Q are pairwise distinct, then the corresponding principal domains are all of first order and stand in no numerical relation to one another. In modern terms, Q is *diagonalizable* when its eigenvalues are pairwise distinct.

In No. 390 Grassmann tackles the tricky case of equal principal numbers (*multiple eigenvalues*). After several pages of algebraic manipulations he arrives at a result that today is called the *Primary Decomposition Theorem*[17]. His derivation is rather constructive. In fact, he gives a method for generating the primary decomposition by systematically exchanging the denominators of the given fraction. Using Grassmann's

[17] Fearnley–Sander writes that "it appears that Grassmann was the first to prove [this] result" [Fearnley-Sander 1979, 814]; according to Hawkins "[t]his result may have been generally known at the time" [Hawkins 1977b, 156].

wording and modern notation, the final result can be written as follows (cf. [A2$_K$, 220] for a direct translation of the German original):

Suppose that $\rho_1, \ldots, \rho_m$ *are the distinct principal numbers of the fraction* Q*, with respective (algebraic) multiplicities* $d_1, \ldots, d_m$, $\sum_{j=1}^{m} d_j = n$*, then there exist* n *extensive magnitudes* $a_1^{(j)}, \ldots, a_{d_j}^{(j)}, j = 1, \ldots, m$*, standing in no numerical relation to one another, so that*

$$Qa_\ell^{(j)} = \rho_j a_\ell^{(j)} + r_\ell^{(j)}, \quad j = 1, \ldots, m, \quad \ell = 1, \ldots, d_j,$$

where $r_\ell^{(j)}$ *is numerically derived from* $a_1^{(j)}, \ldots, a_{\ell-1}^{(j)}$.

This is a *reduction to block upper triangular form*, with one block for each of the distinct eigenvalues. Hence the size of each block is equal to the algebraic multiplicity of the corresponding eigenvalue. Each block itself is upper triangular and has the eigenvalue on its diagonal. For example, if $n = 3$ and Q has just one principal number ρ of algebraic multiplicity 3 (so that $m = 1$), then there exist 3 extensive magnitudes a_1, a_2, a_3 standing in no numerical relation to one another so that

$$Q = \frac{(\rho a_1, \rho a_2 + \beta_{12} a_1, \rho a_3 + \beta_{23} a_2 + \beta_{13} a_1)}{(a_1, a_2, a_3)},$$

for some numbers β_{ij} that are not further specified. Writing this as

$$[a_1, a_2, a_3] \xrightarrow{Q} [a_1, a_2, a_3] \begin{bmatrix} \rho & \beta_{12} & \beta_{13} \\ & \rho & \beta_{23} \\ & & \rho \end{bmatrix}$$

visualizes the reduction to block upper triangular form.

The Primary Decomposition Theorem is a first step in the proof of the *Jordan Canonical Form*[18], which is one of the most important results of Linear Algebra. To derive this result from the primary decomposition it is necessary to further decompose the diagonal blocks. This requires a distinction between algebraic and geometric multiplicities of the eigenvalues. In particular, if the algebraic and geometric multiplicities of an eigenvalue ρ are $m_a(\rho)$ and $m_g(\rho)$, respectively, then $m_a(\rho) \geq m_g(\rho)$, and the upper triangular diagonal block of size $(m_a \times m_a)$ corresponding to ρ in the primary decomposition further decomposes into $m_g(\rho)$ separate blocks.

[18] The first proofs of the Jordan Canonical Form were obtained independently in [Weierstraß 1868] and [Jordan 1870].

In a remark following No. 390 Grassmann discusses the special geometrical meaning of his theory for fractions having four denominators. The resulting cases of "collinearities in space" are treated in great detail in the *Anmerkungen* in [HGW12, 438–464] (written by Grassmann's son Hermann d. J.).

In the final No. 391 of the section Grassmann considers fractions Q that satisfy

$$[Qa|b] = [Qb|a] \quad \text{and} \quad [Qa|a] \neq 0 \tag{3}$$

for all non-zero extensive magnitudes a, b, where $[\cdot|\cdot]$ denotes the "inner product"[19]. In modern terms the first equation in (3) means that Q is *symmetric* (or *self-adjoint*) with respect to the given inner product. The second equation implies that Q is either *positive* or *negative definite*. Grassmann shows that such a symmetric Q always has n real eigenvalues[20]. Furthermore, he shows that there exists an orthogonal basis $e_1, \ldots, e_n$ consisting of eigenvectors, i.e., $Qe_r = \rho_r e_r$, $r = 1, \ldots, n$, and $[e_r|e_s] = 0$ for $r \neq s$. The proof of No. 391 contains an error; see the "Supplementary Notes" in [$A2_K$, 397] for details.

In a remark at the very end Grassmann relates his result to the Law of Inertia of quadratic forms (proven in [Sylvester 1852]) and to a theorem of Sturm on the roots of algebraic equations (most likely Grassmann refers to [Sturm 1829]). These are the only links to related results of others made in the section on the fraction.

Concluding remarks

In Nos. 377–391 of the A_2 Grassmann studied what today would be called linear transformations on finite dimensional vector spaces. He used an interesting notation, which explicitly reveals the effect of basis exchanges on the image of the transformation. He defined a number of important concepts (including the determinant, the characteristic polynomial, eigenvalues and eigenspaces), proved the Rank-Nullity Theorem, derived a sufficient condition for diagonalizability, and gave a constructive proof of the Primary Decomposition Theorem (reduction to block upper triangular form). These results, likely developed between the mid 1850s and October 1861, were very up to date. It

[19] The inner product is studied in detail in Nos. 137–215 of A_2, immediately following the treatment of the combinatorial product. The inner product allows a definition of the concept of *orthogonality*, which Grassmann introduces in No. 152 under the name "normal to one another" (*normal zu einander*).

[20] The fact that (real) symmetric matrices have real eigenvalues was first shown in [Cauchy 1829]. Cauchy used the term "Tableau" for matrices, but he did not define or use matrix algebra as such.

appears that Grassmann was not fully aware of the work of his contemporaries. They, on the other hand, to a large extent ignored A_2.

Consequently, Grassmann's theory of the fraction played little or no role in the mainstream research on linear transformations (linear substitutions, bilinear forms, or matrices) in the nineteenth century[21]. Among the few who carefully studied Grassmann's work on the fraction were Peano and Whitehead, but they did not influence work in this area either.

Reasons for the (initial) neglect of Grassmann's A_1 and A_2 have been analyzed by many authors; see, for example, [Crowe 1967; Dieudonné 1979; Petsche 2006; Rowe 1996; Schubring 1996b]. Felix Klein, who recognized Grassmann's work and cited him several times in his famous Erlanger Programm of 1872 [Klein 1872], elucidated the main problem: "Both works are extremely difficult to access, in fact, almost impossible to read."[22] In part this was due to the fact that Grassmann rarely related his results to concepts his contemporaries were familiar with. In the case of the fraction he preferred to give his own motivation for the definition, namely that the fraction generalizes the algebraic quotient, over describing its relation to linear substitutions, matrices, or bilinear forms.

Acknowledgments

I would like to thank Ivor Grattan-Guinness and Alvin Swimmer for helpful discussions during the *Grassmann Bicentennial Conference*. Thanks also to Robert Luce, Olivier Séte, Ivan Slapnicar, and Alvin Swimmer for carefully reading the manuscript and many helpful suggestions.

[21] Readers who are interested in the history of matrix theory should consult Hawkins' excellent series of articles [Hawkins 1975; 1977a, b] and the more recent [Hawkins 2008].

[22] "Beide Werke sind äußerst schwer zugänglich, ja fast unlesbar." [Klein 1979, 175] (My translation.)

The Golden Gemini Spiral

A. Swimmer

Introduction

In [Hestenes 1996, 247] David Hestenes eloquently and elegantly states "The insight that all *multiplicative* aspects of extension theory can be reduced to the single central product leads to the conclusion that Clifford Algebra is the proper completion of Grassmann's vision for a universal calculus." In this paper (at the risk of being accused of spreading the MV/G virus) we provide an example to suggest that the *additive* aspects of extension theory must also play an important role in the completion of Grassmann's vision.

Because geometric points do not directly lend themselves to algebraic manipulation, while mass points do, we follow Möbius and Grassmann and use mass points as the fundamental entities that we subject to algebraic manipulation. The fundamental properties of mass point algebra are discussed in several works among which are [Forder 1941; Swimmer 1990, 1996; Whitehead 1898]. Of necessity, our presentation of the basic facts is limited to the basic definitions and theorems.

H.-J. Petsche et al. (eds.), *From Past to Future: Graßmann's Work in Context*,
DOI 10.1007/978-3-0346-0405-5_28, © Springer Basel AG 2011

Notation

We use $\doteq$ instead of $=$ when using an equation to make a definition.

Capital letters designate mass points, which are characterized by their mass and location. The mass of the mass point A, which is a real number, is denoted $[A]$. Lower-case letters designate masses and other real numbers. The location of A, i.e. its associated geometric point, is denoted $\overline{A}$. To designate that mass points A and B have the same location (even though they have different masses) we write $\overline{A} \equiv \overline{B}$.

The oriented interval beginning at A and terminating at B is denoted $[A, B]$, and its oriented length is denoted $[AB]$. The fact that $[A, B]$ is oriented means that $[BA] = -[AB]$, and if C is *any* point on the line $\overline{AB}$, then $[AC] + [CB] = [AB]$.

The oriented triangle with vertices A, B, C is denoted $[A, B, C]$. Its oriented area is denoted $[ABC]$. The fact that it is oriented means that

$$[ABC] = -[ACB] = [CAB] = -[CBA] = [BCA] = -[BAC],$$

and if D represents any point in the plane $\overline{ABC}$, then $[DBC] + [ADC] + [ABD] = [ABC]$.

Two mass points are equal, $A = B$, iff $[A] = [B]$ and $\overline{A} \equiv \overline{B}$. The equation $A = B + C$ means that A is the center of mass (henceforth abbreviated *cm*) of B and C. Mathematically this means

$$A \in \overline{BC}, \quad [A] = [B] + [C], \quad \text{and} \quad \frac{[BA]}{[AC]} = \frac{[C]}{[B]}.$$

The reason we use the $+$ sign in this context is that the commutative and associative laws are valid, i.e., $A + B = B + A$ and $A + (B + C) = (A + B) + C$. In the proof of the associative law [Swimmer 1996], we find that if $P \doteq P_0 + P_1 + P_2$, then

$$[P_0] : [P_1] : [P_2] = [PP_1P_2] : [P_0PP_2] : [P_0P_1P].$$

It is also true that $[A + A] = 2[A]$ and that $\overline{A + A} = \overline{A}$. Generalizing this circumstance we define, for any real number x and any mass point A, xA to be the mass point satisfying $[xA] \doteq x[A]$ and $\overline{xA} \equiv \overline{A}$. Then for any two real numbers x, y and any two mass points A, B the following relations hold:

1. $(x + y)A = xA + yA$,
2. $x(A + B) = xA + xB$,
3. $x(yA) = (xy)A$,
4. $1A = A$.

With the usual definition of subtraction (i.e., $A - B = C \Leftrightarrow A = B + C$), it follows that $A - B = A + (-1)B$. When $[A] = [B]$, the difference $C \doteq A - B$ has mass 0 and divides $[A, B]$ in the ratio -1. This means that C is a point at infinity on $\overline{AB}$ [Swimmer 1990, 1996]. Since all points at infinity have mass 0, we distinguish between different points at infinity by their *measure*. By definition, $[[A-B]]$ designates the *measure* of $A - B$ and it is equal to $[[A - B]] \doteq [A][BA]$. Two points at infinity are equal if and only if they have the same measure and are located on the same or parallel lines.

Castor and Pollux, the Gemini Twins

Associated with the construction of a regular decagon and a regular pentagon inscribed in a circle, is a certain type of isosceles triangle with a vertex angle measuring $\frac{\pi}{5}$ radians. In the decagon this occurs as the triangle consisting of a side and the two radii which terminate at that side's endpoints. In the pentagon it occurs as the triangle consisting of a side and the two concurrent diagonals that terminate at that side's endpoints. Following the ancient Greek mathematicians and [Plummer 1970, 186] we call these triangles *Castor Triangles*.

In a Castor Triangle, the ratio $\phi \doteq \frac{\text{long side}}{\text{short side}}$ is the famous golden ratio with defining relation

$$\phi = 1 + \frac{1}{\phi}.$$

This relation yields the numerical value $\phi = \frac{\sqrt{5}+1}{2}$. Multiplying the relation by ϕ^n for any real number n yields the *Fibonacci Power Formula*

$$\phi^{n+1} = \phi^n + \phi^{n-1}.$$

Since each Castor Triangle is isosceles, each base angle measures $\frac{2\pi}{5}$ radians, exactly twice as much as the measure of the vertex angle.

Constructing the Golden Gemini Spiral

Let $C_0 \doteq [T_0, T_1, T_2]$ be a Castor Triangle. We choose $[T_1 T_2] \doteq 1$ for our unit of length and then $[T_2 T_0] = [T_0 T_1] = \phi$. For $i = 0, 1, 2$ let

$$[T_i] \doteq \phi^i, \tag{1}$$

and let $\overline{T_3}$ be the foot of the angle bisector emanating from the middle vertex T_1. The Angle Bisector Theorem then tells us

$$\frac{[T_0 T_3]}{[T_3 T_2]} = \phi.$$

Thus, we can define the mass point $T_3 \doteq \phi T_0 + T_2$. Note that $[T_3] = \phi + \phi^2 = \phi^3$ so that (1) continues to hold for T_3.

The angle bisector $[T_1, T_3]$ subdivides $\mathcal{C}_0$ into

$$\mathcal{C}_1 \doteq [T_1, T_2, T_3] \quad \text{and} \quad \mathcal{P}_1 \doteq [T_3, T_0, T_1].$$

Here $\mathcal{C}_1$ is another Castor Triangle, and $\mathcal{P}_1$ is also an isosceles triangle whose vertex angle measures $\frac{3\pi}{5}$.

Following [Plummer 1970] again, we call all triangles similar to $\mathcal{P}_1$ *Pollux Triangles*. A pair of triangles, one a Castor and the other a Pollux that subdivide a "Mother Castor" as described above are called *Gemini Twins*.

Using T_3 as the center we can swing a $\frac{3\pi}{5}$ radian arc circumscribing $[T_0, T_1]$. This arc is the beginning of the Gemini Spiral associated with $\mathcal{C}_0$.

We now repeat the process using $\mathcal{C}_1$ as the mother triangle. In detail we bisect the middle angle of $\mathcal{C}_1$ and call its foot on $[T_1, T_3]$,

$$T_4 \doteq \phi T_1 + T_3 = \phi T_0 + (\phi T_1 + T_2).$$

The point $\phi T_1 + T_2$ divides $[T_1, T_2]$ in the ratio 1, and hence it is the midpoint of that interval. The interval is simultaneously the first side of $\mathcal{C}_0$ and the last side of $\mathcal{C}_1$ (being opposite the first and last vertices of the respective Castor Triangles). In the first case we call the midpoint F_0 and in the latter case we call it $L_3 = F_0 = \phi T_1 + T_2$. Thus, $[T_0, F_0]$ is the first median of $\mathcal{C}_0$, while $[T_1, T_3]$ is its middle angle bisector and T_4 is their intersection. The new Pollux Triangle is $\mathcal{P}_2 \doteq [T_4, T_1, T_2]$. We draw the arc with center T_4 circumscribing $[T_1, T_2]$. T_1 is the transition point of the two arcs being the endpoint of the arc centered at T_3 and the beginning point of the arc centered at T_4. Since T_1 is on $\overline{T_3 T_4}$, the same line serves as tangent at T_1 for both arcs. Thus, the spline curve consisting of the two arcs is *tangent smooth* at their common point T_1.

(It is worth noting that for $\mathcal{C}_0$, $[T_0, F_0]$ is simultaneously the first angle bisector, the first median, the first altitude and the first perpendicular bisector. Consequently it contains the incenter, the centroid, the orthocenter and the circumcenter, and is the Euler Line. The incenter is T_4 since it is the intersection of the angle bisectors emanating from T_0 and T_2. The centroid of $\mathcal{C}_0$ is $G_0 \doteq \phi^2 T_0 + \phi T_1 + T_2 = \phi^2 T_0 + F_0 =$

$\phi T_1 + M_1 = T_2 + \phi L_2$, where $M_1 \doteq T_2 + \phi^2 T_0$ and $L_2 \doteq (\phi T_0 + T_1)$. $[F_0, \phi M_1, \phi^3 L_2]$ forms the (properly weighted) *medial triangle* for $\mathcal{C}_0$. Its vertices are respectively the (properly weighted) first, middle and last midpoints of $\mathcal{C}_0$. The orthocenter of $\mathcal{C}_0$ is $O_0 \doteq \phi T_0 + \phi^3 T_1 + \phi^2 T_2 = \phi T_0 + \phi^2 F_0 = \phi^3 T_1 + F_1$, and its mass is $[O_0] = (\phi + 2\phi^4)$. The circumcenter of $\mathcal{C}_0$ is the orthocenter of its medial triangle. So it is (using $G \doteq G_0$)

$$\begin{aligned} C_0 &\doteq \phi F_0 + \phi^3(\phi M_1) + \phi^2(\phi^2 L_2) \\ &= \phi(G - \phi^2 T_0) + \phi^4(G - \phi T_1) + \phi^4(G - T_2) \\ &= (\phi + 2\phi^4)G - (\phi^3 T_0 + \phi^5 T_1 + \phi^4 T_2) \\ &= (\phi + 2\phi^4)G - \phi^2 O_0. \end{aligned}$$

Thus, $(\phi + 2\phi^4)G = \phi^2 O_0 + C_0$ and $\frac{[O_0 G]}{[G C_0]} = \frac{[C_0]}{\phi^2 [O_0]} = 2 : 1$. This corroborates the Euler Line Theorem for Castor Triangles.)

By iterating this process we get the Golden Gemini Spiral, a spline curve consisting of an infinite number of circular arcs (joined at tangent smooth transition points), each of which circumscribes the last side of the appropriate Castor Triangle. Given $\mathcal{C}_n \doteq [T_n, T_{n+1}, T_{n+2}]$ we first bisect the middle angle with vertex T_{n+1} to get the foot $T_{n+3} = \phi T_n + T_{n+2}$ and a new pair of Gemini Twins,

$$\mathcal{C}_{n+1} \doteq [T_{n+1}, T_{n+2}, T_{n+3}] \quad \text{and} \quad \mathcal{P}_{n+1} \doteq [T_{n+3}, T_n, T_{n+1}].$$

Then we swing the arc with center T_{n+3} connecting T_n to T_{n+1}. Note that $[T_{n+3}] = \phi^{n+1} + \phi^{n+2} = \phi^{n+3}$ so that (1) continues to hold for all the T_n.

The eye of the Gemini Spiral

We now consider the mapping Φ_0 which transforms $\mathcal{C}_0$ into $\mathcal{C}_1$. In kinematic terms, we must rotate $\mathcal{C}_0$ by $\frac{3\pi}{5}$ radians and simultaneously shrink its size by a factor of ϕ. In this way the vertices of $\mathcal{C}_0$ map into those of $\mathcal{C}_1$ and every point $X_0 \doteq x^0 T_0 + x^1 T_1 + x^2 T_2$ (the superscripts are indices, not powers) of the plane $\overline{T_0 T_1 T_2}$ maps into $X_1 \doteq x^0 T_1 + x^1 T_2 + x^2 T_3 = (\phi x^2) T_0 + x^0 T_1 + (x^1 + x^2) T_2$.

In order that $\overline{X}_0$ be a fixed point of this mapping, X_0 (because of the increase in mass when T_i maps into T_{i+1}) must map into ϕX_0. Thus for a fixed point $\phi X_0 = X_1$, which implies

$$\phi x^0 = \phi x^2, \quad \phi x^1 = x^0, \quad \text{and} \quad \phi x^2 = x^1 + x^2.$$

The first two equations imply $x^2 = x^0$ and $x^1 = \frac{1}{\phi}x^0$. The last equation corroborates that the system is consistent. We choose $x^0 \doteq 1$ and call the fixed point $E \doteq T_0 + \frac{1}{\phi}T_1 + T_2$. It now follows that $\overline{E}$ is the fixed point for all the similar transformations Φ_n mapping $\mathcal{C}_n$ into $\mathcal{C}_{n+1}$ for all n.

Intertwining Gemini Spirals

The centroid of

$$\begin{aligned} &\mathcal{C}_0 \text{ is } G_0 \doteq \phi^2 T_0 + \phi T_1 + T_2, \\ &\mathcal{C}_1 \text{ is } G_1 \doteq \phi^2 T_1 + \phi T_2 + T_3, \\ &\mathcal{C}_2 \text{ is } G_2 \doteq \phi^2 T_2 + \phi T_3 + T_4. \end{aligned}$$

Hence $$\begin{aligned} G_1 - \phi G_0 = T_3 - \phi^3 T_0 &\Longrightarrow [G_0 G_1] = \tfrac{1}{3}\ [G_0, G_1] \parallel [T_0, T_3], \\ G_2 - \phi G_1 = T_4 - \phi^3 T_1 &\Longrightarrow [G_1 G_2] = \tfrac{1}{3\phi}\ [G_1, G_2] \parallel [T_1, T_4]. \end{aligned}$$

To get the third side of the 3-centroid triangle, we simply add the second equation to ϕ times the first one, yielding

$$G_2 - \phi^2 G_0 = (T_4 - \phi^3 T_1) + (\phi T_3 - \phi^4 T_0) = L_4 - \phi^2 L_2.$$

Reexamining the previous two equations we find that

$$G_1 - \phi G_0 = L_3 - \phi L_2 \quad \text{and} \quad G_2 - \phi G_1 = L_4 - \phi L_3.$$

(Notice that, just as $T_4 \doteq \phi T_1 + T_3$ follows simply by increasing the indices from the defining equation for T_3, the middle equation above follows from the first by simply increasing the index values by 1. We call this the *Increasing Index Principle*. It is valid for midpoints, angle bisectors, and all the other points associated with the Golden Gemini Spiral.)

Thus, $\mathcal{G}_0 \doteq [G_0, G_1, G_2]$ is a Castor Triangle and it is homothetic to $[L_2, L_3, L_4]$. This means that we can construct another Gemini Spiral starting from $\mathcal{G}_0$. The eye of this new Gemini Spiral is $G_0 + \frac{1}{\phi}G_1 + G_2$. By collecting terms in accordance with powers of ϕ, we find that the eye equation becomes $\phi^2 E + \phi E + E = 2\phi^2 E$. Thus, $\overline{E}$ is the eye of the new spiral also.

More surprisingly, given an arbitrary point X_0 as previously defined, along with its counterparts X_1 and $X_2 \doteq x^0 T_2 + x^1 T_3 + x^2 T_4$, which have the same relative position in $\mathcal{C}_1$ and $\mathcal{C}_2$ as X_0 does in $\mathcal{C}_0$, $[X_0, X_1, X_2]$ is a Castor Triangle. The Gemini Spiral that it generates has the same eye as those previously discussed. Thus, as soon as we have a single Castor Triangle, we can generate an infinite number of Golden Gemini Spirals, all having the same eye. (Proofs of these assertions are available from the author.) For each spiral there are an infinite number of Gemini Twins. For all of them together we have an infinity of infinities of Gemini Twins.

A short note on Grassmann manifolds with a view to noncommutative geometry

Jochen Pfalzgraf

Dedicated to Professor Arno Zaddach who wrote the extraordinary book
GRASSMANN ALGEBRA IN DER GEOMETRIE

Introduction

This is a very brief essay giving some comments on the basic importance of the notion of the Grassmann manifold, also called Grassmann variety or Grassmannian, for short. In the second part a new idea is sketched, how a generalized notion of the Grassmann manifold can be introduced in a noncommutative geometric space.

I dedicate this contribution to Arno Zaddach, professor emeritus, Universidad de Tarapacá, in Arica, Chile. With his book *Grassmann Algebra in der Geometrie. Mit Seitenblicken auf verwandte Strukturen* [Zaddach 1994] he pays great homage to Grassmann. Zaddach's book is a new work on Grassmann's famous *Lineale Ausdehnungslehre*, even a kind of revival, using notions and notation of Bourbaki. In 1984/1985 I was a professor of mathematics at Universidad de Talca, in Talca, Chile. This was the starting point of my involvement in Arno Zaddach's book project. For me it was an extraordinary experience. In short, during my time in the faculty of the Research Institute for Symbolic Computation, University of Linz, Austria (RISC-Linz), from 1990 to 1996, I was responsible for Arno Zaddach's original manuscript.

H.-J. Petsche et al. (eds.), *From Past to Future: Graßmann's Work in Context*,
DOI 10.1007/978-3-0346-0405-5_29, © Springer Basel AG 2011

He asked me to write a chapter for his book, which became Kapitel XXV, *Graßmann Varietäten*. As soon as the manuscript was complete I had to send it safely to Hermann Engesser, who was the editor at the BI Wissenschaftsverlag in Mannheim responsible for the printing of the book. The aim and Zaddach's wish was that the book had to be available in time for the Sesquicentennial Conference (150 years Lineale Ausdehnungslehre) taking place at Lieschow, Rügen, Germany, May 23–28, 1994.

A special task given to me by Zaddach was to pick up 250 copies of the book in Mannheim. I did this with my old car and drove to the conference location. Every participant received a copy. Besides that I sent further copies to people on a list given to me by Arno Zaddach, and I had a set of copies for my own purposes. At the Rügen conference I gave a talk entitled *An application of Grassmann geometry to a problem in robotics* [Pfalzgraf 1996]. Most contributions were published in [Schubring 1996a]. With these remarks I wanted to explain how I came into contact with the Grassmann conferences. I would also like to point out the following book by Arno Zaddach: *Algebra de Grassmann y Geometría Proyectiva* [Zaddach 1988].

The title of my talk submitted to the Grassmann Bicentennial Conference in Potsdam in 2009 was *Grassmann geometry in robotics*. Unfortunately, I was not able to attend the conference. For the following considerations I have chosen a different topic which, in some respect, continues my contribution on Grassmann manifolds in [Zaddach 1994].

On Grassmann manifolds

In this section we give a short introduction to the notions of Stiefel manifold and Grassmann manifold following Chap. XXV in [Zaddach 1994] (my contribution). For details we refer to the literature used therein, and the references [Harris 1992; Warner 1983; Fulton and Harris 1991; Eisenbud and Harris 2000; Husemoller 1966; Milnor and Stasheff 1974; Tom Dieck 2000].

We consider only the case of a finite-dimensional real vector space V, the Euclidean space $V = \mathbb{R}^n$, for short. $\{e_1, \ldots, e_n\}$ denotes the standard basis of the vector space, an orthonormal n-frame. Let $(v_1, \ldots, v_k)$ denote a k-frame, i.e., a k-tuple of vectors from V forming a k-dimensional orthonormal basis. The set of all k-frames $\mathsf{S}_{\mathbf{k},\mathbf{n}}$ is called a Stiefel manifold. For $n = 1$ we obtain the unit sphere S^{n-1}. A k-dimensional subspace of V is often called a k-plane. Obviously, a k-frame generates a k-plane. The set of all k-planes is the famous Grassmann manifold $\mathsf{G}_{\mathbf{k},\mathbf{n}}$. For $n = 1$ we obtain the projective space $\mathbb{P}(V)$.

The notion of an operation of a group on a set is of basic importance, we recall it here briefly. Let (G, X) denote a group operation (also called group action), i.e., a multiplicatively written group G with unit element 1 acting on a set X, described by a mapping

$$G \times X \to X, \quad (g,x) \mapsto g \cdot x \quad (= gx, \text{for short}),$$

subject to the conditions

$$\forall g,h \in G, x \in X : (gh) \cdot x = g \cdot (h \cdot x) \quad \text{and} \quad 1 \cdot x = x.$$

$G_x := \{g \in G | gx = x\}$ is called a stabilizer subgroup of G with respect to $x \in X$ (or stabilizer of x, for short), and $Gx := \{gx | g \in G\}$ denotes the orbit of x. The natural 1-1 correspondence $Gx \leftrightarrow G/G_x$, given by the bijection $gx \mapsto gG_x$, describes the orbit completely by group theoretic data. There is a decomposition of X into disjoint orbits, $X = \coprod_{i\in I} Gx_i$. We denote the set of all orbits by $G\backslash X$ (quotient set or quotient space). G acts transitively on X, if $X = Gx$ holds for every $x \in X$.

$GL_n(\mathbb{R})$ denotes the group of all invertible $n \times n$ matrices, the group of linear automorphisms of the space V. The subgroups $O_n(\mathbb{R})$ and $SO_n(\mathbb{R})$ are the orthogonal group and the special orthogonal group, respectively. These groups are classical Lie groups, they act naturally on the Euclidean space (matrix operation).

There are the following representations of manifolds as homogeneous spaces: The natural operation $O_n(\mathbb{R}) \times S^{n-1} \to S^{n-1}$, $(\sigma, v) \mapsto \sigma(v)$, leads to

$$S^{n-1} \cong O_n(\mathbb{R})/O_{n-1}(\mathbb{R}) \quad \text{and} \quad S^{n-1} \cong SO_n(\mathbb{R})/SO_{n-1}(\mathbb{R}).$$

We note that the operation is transitive, hence $S^{n-1} = O_n(\mathbb{R}) \cdot v$, and with $v = e_n$ we obtain $(O_n(\mathbb{R}))_{e_n} = O_{n-1}(\mathbb{R})$. The Stiefel manifold and the Grassmann manifold have the representations

$$\mathsf{S}_{\mathbf{k,n}} \cong O_n(\mathbb{R})/O_{n-k}(\mathbb{R}), \quad \text{and also} \quad \mathsf{S}_{\mathbf{k,n}} \cong SO_n(\mathbb{R})/SO_{n-k}(\mathbb{R}),$$
$$\mathsf{G}_{\mathbf{k,n}} \cong O_n(\mathbb{R})/(O_k(\mathbb{R}) \times O_{n-k}(\mathbb{R})).$$

$\pi_1 : \mathsf{S}_{\mathbf{k,n}} \to \mathsf{G}_{\mathbf{k,n}}$ is the natural mapping $(v_1, \ldots, v_k) \mapsto <v_1, \ldots, v_k>$ of a k-frame to the corresponding k-plane generated by the k-frame as its basis.

We also point out the Plücker embedding of the Grassmann manifold into the projective space of the exterior product, $\phi : \mathsf{G}_{\mathbf{k,n}} \to \mathbb{P}(\bigwedge^k(V))$. This and some further considerations are summarized in [Zaddach 1994, Chap. XXV].

Concluding remarks: Grassmann manifolds and Grassmann algebras are of basic importance in inter-disciplinary interactions of geometric, algebraic, topological and combinatorial structures. Their fundamental importance can be found in various disciplines like algebra, geometry, topology, physics, control theory, systems theory, robotics, combinatorics. Some quotations and references concerning the inter-disciplinary aspects can be found in my contribution to Zaddach's book in Chap. XXV.

Additionally, the basic importance of Grassmannians in the theory of fiber bundles is pointed out, where they play a fundamental role in the classification of vector bundles. D. Husemoller [Husemoller 1966, Sects. 3, 7 and 7.3] remarks that the problem of classifying vector bundles can be reduced to the calculation of sets of homotopy classes of maps. In this classification Grassmannians play an essential role. We mention the canonical bundle (tautological bundle) γ_k^n on the Grassmannian, where the total space $E(\gamma_k^n)$ is the set of all pairs $(U, v) \in \mathsf{G}_{\mathbf{k,n}} \times \mathbb{R}^n$ with $v \in U$, the base space is $\mathsf{G}_{\mathbf{k,n}}$, and the bundle map $p : E(\gamma_k^n) \longrightarrow \mathsf{G}_{\mathbf{k,n}}$ is the projection onto the first component of a pair. The fiber over the element $U \in \mathsf{G}_{\mathbf{k,n}}$ is given by the subspace (k-plane) U itself [Husemoller 1966, 2 (2.5); Tom Dieck 2000, IX (2.2)]. Furthermore, we mention that $\pi_1 : \mathsf{S}_{\mathbf{k,n}} \rightarrow \mathsf{G}_{\mathbf{k,n}}$ is an $O(k)$−principal bundle [Tom Dieck 2000, IX (1.7)].

The importance of Grassmannians in the theory of schemes is pointed out in [Eisenbud and Harris 2000]. We refer, for example, to the following remark in Sect. III.2.7: *Grassmannians exist in the category of schemes, and behave very much like Grassmannians in classical algebraic geometry.* There exists a functorial construction; in Sect. VI.2 the Grassmannian functor is introduced.

The previous remarks indicate that the literature is full of interesting work and many fascinating aspects dealing with Grassmannians.

A view to noncommutative geometric spaces

This section is again very short. A new idea is sketched dealing with a generalization of the notion of Grassmann manifold – or let me say Grassmannian, this sounds a little more general – in noncommutative geometric spaces. We present some notions and notation necessary for our considerations and refer to suitable references.

Noncommutative Geometry (NCG) was introduced by Johannes André nearly 40 years ago as a generalization of classical affine geometry [see, e.g., André 1974, 1983, 1988, 1992; André and Ney 1991]. Therefore, NCG is sometimes called noncommutative affine geometry. Elementary geometric configurations are the basis of the axioms which determine the structure of a geometry.

In the beginning of my own work on NCG, I introduced a new model to represent a geometric space by a so-called parallel map $<,> : X^2 \rightarrow R$. Such an object is denoted by $(X, <,>, R)$. This led to the introduction of the category of Noncommutative Geometric Spaces **NCG** [Pfalzgraf 1985a,b; 1986]. A brief summary of the main notions is contained in [Gehrke and Pfalzgraf 2001]; for a summary with greater detail see [Pfalzgraf 1998].

A line joining two points $x, y \in X$ is defined by $x \Box y := x \sqcup y \cup \{<x,y>\}$, where $<x,y>$ is called the direction of the line $x \Box y$ (also ideal point, or color), and $x \sqcup y := \{x\} \cup \{\zeta | <x,\zeta> = <x,y>\}$, is called the set of proper points of the line. This is the variety of solutions of the equation $<x,\zeta> = <x,y>$ together with the base point x. Lines are parallel if they have the same direction. We can consider $x \Box y$ as the projective version of the line $x \sqcup y$, where $<x,y>$ is the point at infinity.

Noncommutativity of the join operation $\sqcup$ means that in general $x \sqcup y \neq y \sqcup x$ holds for lines. $x \sqcup y$ is a straight line if $x \sqcup y = y \sqcup x$ (commutativity of the join).

Geometric axioms can be expressed by corresponding equations in the $<,>$-model, thus leading to a certain algebraization. The verification of an axiom amounts to verifying the solvability of the corresponding system of equations, and that means to verify the validity of the corresponding geometric constraints.

A morphism $f : (X, <,>_X, R) \to (Y, <,>_Y, S)$ is a map $f : X \to Y$ that respects the parallel structure (directions), i.e.,

$$\forall x, y, x', y' \in X : <x,y>_X = <x',y'>_X \quad \Rightarrow \quad <fx, fy>_Y = <fx', fy'>_Y .$$

Every group operation (G, X) induces a geometric space in the following way: G acts pairwise on X^2, i.e., $g \cdot (x,y) = (gx, gy)$. Then we can define the parallel map $<,>: X^2 \to G \backslash X^2$, $<x,y> := G(x,y)$, and for lines $x \sqcup y$ we obtain $x \sqcup y = \{x\} \cup G_x \cdot y$. Such a geometric space is called a group space, and it is denoted by $\underline{V}(G,X)$. Group spaces form an interesting and important class of geometric spaces.

Standard examples of noncommutative geometric spaces in a real vector space are the ray space and the sphere space in $\mathbb{R}^n$.

Ray space: For a natural number $n \geq 2$, let $X := \mathbb{R}^n$, $R := S^{n-1} \cup \{0\}$, where S^{n-1} denotes the $(n-1)$-dimensional unit sphere in $\mathbb{R}^n$. Then the n-dimensional ray space is defined by the parallel map $<,>: X^2 \to R$, $<x,y> := \frac{y-x}{\|y-x\|}$, if $x \neq y$, and $<x,x> := 0$. The set of proper points of the line joining x and y is the ray $x \sqcup y = x + \mathbb{R}_+ \cdot (y - x)$. Note that the union of sets $(x \sqcup y) \cup (y \sqcup x)$ is the (commutative) straight line of the underlying classical affine space. The ray space is a group space given by the standard action of the semi-direct product $\mathbb{R}^n \times \mathbb{R}^*_+$ on $\mathbb{R}^n$, where $\mathbb{R}^* := \mathbb{R} \backslash \{0\}$.

Sphere space: The example of a circle space in the real plane, where a noncommutative line $x \sqcup y$ (joining the points x and y) is a circle with center x and radius $\|x - y\|$, is a special case $(n = 2)$ of the more general sphere space. Let $X := \mathbb{R}^n$ and $R := \mathbb{R}_+$. Then the $(n-1)$-sphere space is defined by the polynomial parallel map $<x,y> := \sum_{i=1}^n (x_i - y_i)^2$, for $x = (x_1, \ldots, x_n), y = (y_1, \ldots, y_n) \in X$. The sphere space is a group space given by the standard action of the semi-direct product $\mathbb{R}^n \times O_n(\mathbb{R})$ on $\mathbb{R}^n$, and also by the action of $\mathbb{R}^n \times SO_n(\mathbb{R})$ on $\mathbb{R}^n$.

The classical real affine space of dimension n can be obtained from the ray space by identifying opposite directions: $<x,y> \equiv <y,x>$ (note: $<x,y> = -<y,x>$) for $x \neq y$. Thus, the parallel map for the affine space can be defined (for $x \neq y$) by

$$\begin{aligned} < x,y >_{\text{affine}} &:= [\pm < x,y >] \in S^{n-1}/\{\pm 1\} \\ &= (n-1)\text{–dimensional projective space.} \end{aligned}$$

Here $[\pm<x,y>]$ denotes the corresponding equivalence class of the pair of antipodal points. If we now apply the definition of a line to calculate $x \sqcup y$, then we obtain just the straight (commutative) line through x, y ($x \sqcup y = y \sqcup x$), namely $x \sqcup y = x + \mathbb{R}(y-x)$.

Our aim is to introduce a generalization of the notion of a Grassmannian. To this end we define the notions of a (generalized) k-frame and a (generalized) k-plane in a geometric space $(X, <,>, R)$. For the definition of a k-frame we look at the classical definition of a k-frame $(v_1, \ldots, v_k)$ in $\mathbb{R}^n$, which is a k-tuple of orthonormal vectors. For our purposes it is only of interest that such vectors are linearly independent. Therefore, we are looking for a suitable notion of independence of k elements in a geometric space X leading to the definition of a k-frame $(v_1, \ldots, v_k)$, a k-tupel in the cartesian product X^k. This leads to the definition of a (generalized) Stiefel manifold. We will call it Stiefel space, and denote it by $\mathsf{S}_k X$.

The next step is the generalization of the notion of a k-plane. In the classical case there exists the canonical mapping $\pi_1 : \mathsf{S}_{\mathbf{k},\mathbf{n}} \to \mathsf{G}_{\mathbf{k},\mathbf{n}}, (v_1, \ldots, v_k) \mapsto < v_1, \ldots, v_k >$. Since we want to generalize this too, we need to resort to the natural subspace notion in NCG and to the definition of a subspace in $(X, <,>, R)$ that is generated by a set of points. This will be considered subsequently.

We need to fix the following notation:
We define $u<x,y> = \{z \mid <u,z> = <x,y>\}$, for $u, x, y \in X$. The union of this set with the point u is the set of proper points of the line $u \sqcup y = \{u\} \cup u<x,y>$. In particular, it holds $x \sqcup y = \{x\} \cup x<x,y>$. Let U be a subset of X, then U is called subspace of X ($U \leq X$) if the following holds: $\forall u, x, y \in U : u<x,y> \subseteq U$. In other words, for $x, y \in U$ the whole line $x \sqcup y$ is contained in U, and if $u \in U$ and $x \sqcup y$ is a line in U, then the line parallel to $x \sqcup y$ with base point u is also contained in U.

If $M \subseteq X$ is a set of points, then the subspace of X generated by M, denoted by $sub(M)$, is the intersection of all subspaces containing M. Note that we do not use the usual notation $< M >$ for it, because we need the brackets $<,>$ to denote the parallel map of the geometric space. The set of directions of U is defined as the set $DU = \{<u,v> \mid u, v \in U\}$. If $Q \subseteq R$ is a set of directions of the space, then we define the set $xQ = \{y \mid \exists r \in Q :< x, y >= r\}$. For a subspace $U \leq X$ and $x_0 \in U$ the following holds: $U = \{x_0\} \cup x_0DU$.

We fix a base point $x_0 \in X$ and consider the pointed space X_{x0} (this is used in algebraic topology, especially in homotopy theory). Note that the base point x_0 plays a similar role as the zero vector (origin) in $\mathbb{R}^n$ in our general considerations. We make the following definition: $v_1, \ldots, v_k \in X \setminus \{x_0\}$ are called independent, if the element v_{i+1} is not contained in $x_0D\{< x_0, x_1 >, \ldots, < x_0, x_i >\}$, for $i = 1, \ldots, k-1$. This means that $< x_0, v_{i+1} > \neq < x_0, v_j >$, for $j = 1, \ldots, i$, and therefore the line $x_0 \sqcup v_{i+1}$ is not contained in this set.

Now we come to the definition of a k-frame. A k-tuple of points in X different from x_0 is called a k-frame (in X_{x0}), if the elements $v_1, \ldots, v_k$ are independent. It is important to note that while the definition is with respect to a base point, it does not depend on the selection of a particular base point. $\mathsf{S_k X}$ denotes the set of all k-frames in $(X, <,>, R)$ (we omit the index x_0).

A k-plane with respect to a base point x_0, also called k-plane in X_{x0}, is then defined as a subspace of X generated by a k-frame of $\mathsf{S_k X}$. More precisely, if $(v_1, \ldots, v_k)$ is a k-frame in X_{x0}, then the corresponding k-plane is given by the subspace $sub(x_0, v_1, \ldots, v_k)$ in X_{x0}. The set of all such k-planes will be denoted by $\mathsf{G_k X}$; we call it Grassmannian. Note that a more precise notation would be $\mathsf{G_k X}_{x0}$ and $\mathsf{S_k X}_{x0}$, but we do not need it when it is clear which base point is fixed.

Thus, we come to the generalization of the map

$$\pi_1 : \mathsf{S}_{\mathbf{k,n}} \to \mathsf{G}_{\mathbf{k,n}}, \quad (v_1, \ldots, v_k) \mapsto < v_1, \ldots, v_k >$$

by defining

$$p_1 : \mathsf{S_k X} \longrightarrow \mathsf{G_k X}, \quad (v_1, \ldots, v_k) \mapsto sub(x_0, v_1, \ldots, v_k).$$

As a simple example, for illustration, we now choose the ray space in real 3-space. This comes very close to the classical case of 3-dimensional Euclidean space, where the lines are commutative, i.e., straight lines. Let us take $x_0 = (0, 0, 0)$ as the base point of $X = \mathbb{R}^3$. A 2-frame in this ray space is given by two elements $v_1, v_2 \in X$, being independent. They generate the corresponding 2-plane, $sub(x_0, v_1, v_2)$, and it is easy to see that this generalized 2-plane in the ray space is the 2-dimensional linear subspace of $\mathbb{R}^3$ with basis given by v_1 and v_2. Considering more complex examples would be very interesting. Of particular interest will be the investigation of group spaces.

We conclude with a remark on the dilatation group $Dil(X)$ of a geometric space $(X, <,>, R)$. An automorphism of a space is a bijective morphism $f : X \to X$, where the inverse map f^{-1} is also a morphism. $Aut(X)$ denotes the group of automorphisms,

$Dil(X)$ is the normal subgroup of it consisting of all automorphisms δ with the property $< \delta(x), \delta(y) >=< x, y >$, i.e., a dilatation respects the directions, and therefore a dilatation maps a line into a parallel line (for more detail cf. [Pfalzgraf 1986]) and, more generally, a dilatation maps a k-frame into a parallel k-frame. In a group space every element of the group G acts as a dilatation.

Another interesting aspect concerns the simplicial structure of a space. Simplices are basic building blocks in topology and geometry. In [Pfalzgraf 1987a] I introduced the notion of a (geometric) simplex in NCG, and defined simplex configurations and simplex conditions (axioms). Actually, in a group space all simplex conditions hold. A k-frame $(v_1, \ldots, v_k)$ in X_{x_0} leads to a k-simplex $[x_0, v_1, \ldots, v_k]$ in a natural way. We cannot go into detail here, but wanted to indicate with these remarks that there are many interesting aspects and topics for further investigation.

Conclusion

This short essay deals with the notions of Grassmann manifold and Stiefel manifold and remarks on the basic importance of Grassmann manifolds in various disciplines. The second part of the article presents a new idea of how to introduce a generalization of these notions in noncommutative geometric spaces in a natural way. For this purpose generalized noncommutative versions of k-frames and k-planes are defined. We point out that these considerations were only a first step. As already indicated above, many interesting questions and aspects for future work arise.

For example, the theory of fiber bundles is very powerful with many deep results and applications. A fiber bundle integrates different structures, like topology, geometry, and algebra in one concept. For example, in a vector bundle over each point of the base space (topology) there is a vector space as fiber (algebra).

In my own work, fiber bundles and vector bundles gave me much inspiration. For example, in [Pfalzgraf 1987b] I introduced geometric fiberings, i.e., fibered geometric spaces, especially group fiberings. They helped me to solve some older problems (open questions) in NCG.

In the introduction we mention the tautological bundle on a Grassmannian in the theory of vector bundles. A generalization of the notion of a canonical (tautological) bundle in NCG will be part of our future program of work. At first sight, a natural definition can be as follows. We consider the pointed space X_{x_0} (for simplicity we omit the index x_0 in the subsequent considerations). If U is a k-plane in X, then $[U]$ denotes the corresponding element in $\mathsf{G_k X}$. Now we define the bundle γ_k (the generalization of the canonical bundle γ_k^n) with total space $E(\gamma_k) = \{([U], v) \mid [U] \in \mathsf{G_k X}, v \in U\}$

and base space $\mathsf{G}_k\mathsf{X}$. The bundle map $p_k : E(\gamma_k) \to G_kX$ is the projection onto the first component of a pair. Thus, the fiber $p_k^{-1}([U])$ over $[U] \in \mathsf{G}_k\mathsf{X}$ is determined by the k-plane U.

In [Pfalzgraf 1991] I introduced the new notion of logical fibering. The original idea was to replace the typical fiber in a vector bundle by a logical space (typically a 2-valued logic). In this way it is possible to combine logic with geometric and topological structures. This opens the way to a new theory in mathematical logic based on the powerful language of fiber bundles [Tom Dieck 2000]. In this context there is also the interesting question whether a generalized canonical logical bundle will arise in a natural way.

There remain many interesting aspects and topics for further work.

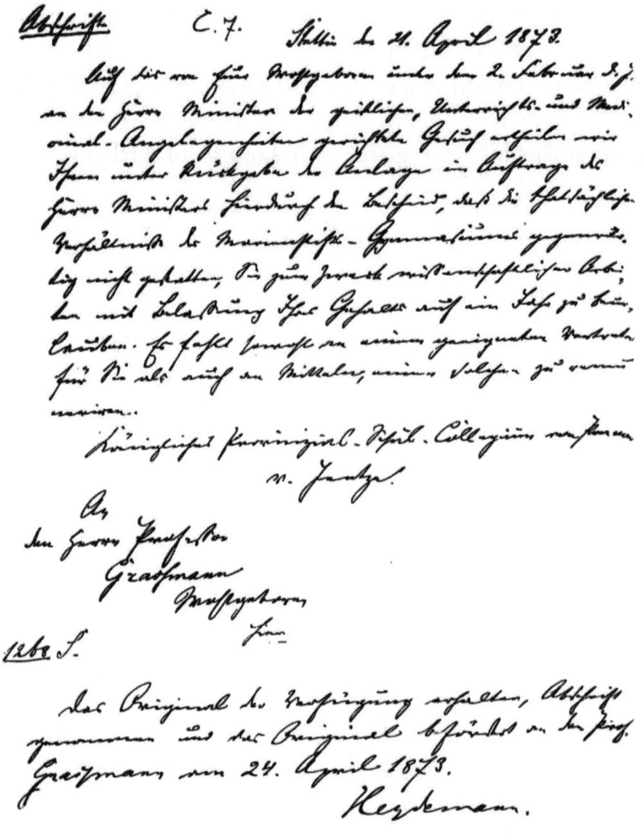

Abschrift C. 7. Stettin den 21. April 1873.

Auf das von Ihnen Wohlgeboren unter dem 2. Februar d. J. an den Herrn Minister der geistlichen, Unterrichts- und Medicinal-Angelegenheiten gerichtete Gesuch eröffnen wir Ihnen unter Rückgabe der Anlage im Auftrage des Herrn Ministers hierdurch den Bescheid, daß die Schulbedürfnisse des Marienstifts-Gymnasiums gegenwärtig nicht gestatten, Sie zum Zweck wissenschaftlicher Arbeiten mit Belassung Ihres Gehaltes auf ein Jahr zu beurlauben. Es fehlt bereits an einem geeigneten Vertreter für Sie als auch an Mitteln, einen solchen zu remuneriren.

Königliches Provinzial-Schul-Collegium von Pommern
v. Jagow.

An
den Herrn Professor
Grassmann
Wohlgeboren
hier.

12b S.

Das Original der Verfügung erhalten, Abschrift genommen und das Original befördert an den Prof. Grassmann am 24. April 1873.
Heydemann.

Illustration 13. Reproduction of a recently discovered negative reply to a petition by Hermann Grassmann, dated February 1873. He had asked for a paid sabbatical in order to complete scientific works. Further information on p. 501
[Archiwum Państwowe w Szczecine, Gimnazjum Mariackie w Szczecinie]

Present and future of Hermann Grassmann's ideas in philology

Hermann Grassmann: his contributions to historical linguistics and speech acoustics

John J. Ohala

Introduction

Hermann Grassmann's achievements in linguistics are well known and justly celebrated. His work in acoustic phonetics is less well known but has been noted. What I propose to do in this paper is to provide for the non-linguist reader some background to Grassmann's work in both areas and then offer a brief speculation on how he was able to accomplish so much in these and other domains.

Grassmann's work in historical linguistics

By the mid-nineteenth century comparative linguistics analysis of what we now call the Indo-European (IE) language family had made significant progress such that the major linguistic families that where part of IE had been identified: Sanskrit, Persian, Germanic, Latin, Greek, Celtic, Slavic and, of course, their dialectal descendants, (e.g., in the case of Latin: French, Italian, Portuguese, Spanish, Romanian, etc., and in the case of Germanic: Gothic, German, Danish, English, Dutch, Icelandic, etc.).

H.-J. Petsche et al. (eds.), *From Past to Future: Graßmann's Work in Context*,
DOI 10.1007/978-3-0346-0405-5_30, © Springer Basel AG 2011

Table 1. Examples of the correspondences informing Rask–Grimm's law

Sanskrit	Latin	Greek	Gothic	English
pitar	pater	pater	fadar	father
pada	pedem	poda	fotu	foot
tanu-	tenuis	tanaos		thin
tri	tres	treis		three
naktam	noctis	nuktos		night (OE: niht)
	centum	hekaton		hundred
daśan	decem	deka	taihun	ten
dant-	dentis	odontos		tooth
bhratar	frater	phrater	brothar	brother
bharami	fero	phero	baira	bear (verb)

Equally important, it had established the *nonmembership* in the IE family of various other languages that had previously been proposed as related: for example, Hebrew, Basque, various languages indigenous to North America, Polynesia, etc. It developed reliable criteria that identified and eliminated from consideration as evidence for family relationship such processes as borrowing, onomatopoeia (words imitative of naturally occurring sounds), and chance cognation (e.g., English *name* and Japanese *namae*). It did this by developing the concept of regular correspondences between the sounds of one dialect/language and another, for example, Latin, Greek, Sanskrit *p* corresponds to Germanic *f*, for example, Latin and Greek *pater*[1], Sanskrit *pitar* and German *fadar* and English *father*. Such correspondences were labeled "laws" (although today completely regular correspondences are not insisted upon). The principal pioneers in formulating these phonological relations were the Dane Rasmus Rask [Rask 1811, 1818] and the German Jacob Grimm [Grimm 1822]; thus these correspondences are commonly known as Rask–Grimm's Law. Table 1 gives further examples and Table 2 gives the sound correspondences (*Lautgesetz*) in terms of the manner or type of the sounds.

So, extracting just the stop manners, this yields the correspondences in Table 2.

An important contribution which facilitated Grassmann's contribution to historical phonology was a paper by [Lottner 1862] which classified and listed the exceptions to Rask–Grimm's Law. After noting similarities of vocabulary in different IE dialects due to false cognates, onomatopoeia, and borrowing, he notes cases where Sanskrit voiced stop corresponds to Germanic voiced stop, where, as indicated in Table 2, it should show a voiceless stop. For example, Sanskrit *gardh* "be greedy" and Gothic *grēdus*

[1] In most of these examples I ignore differences in accent and vowel length.

Table 2. The correspondences of stop manners within four branches of Indoeuropean

Sanskrit	Latin	Greek	Germanic
(simple) voiceless stop	(simple) voiceless stop	(simple) voiceless stop	voiceless fricative
(simple) voiced stop	(simple) voiced stop	(simple) voiced stop	(simple) voiceless stop
voiced aspirate stop	voiceless fricative	voiceless aspirated stop	(simple) voiced stop

"hunger, greed" and also Sanskrit *bandh* and Old High German *bindan*. (Significantly, all his examples also contain a voiced aspirated stop that corresponds to the expected simple voiced stop in Germanic.)

A further puzzle that Grassmann eventually solved was a curious behavior of inflected forms of roots that contained aspirated stops. In Greek, for example, voiceless aspirated stops never appear preceding the voiceless fricative *s*. But in some inflections when a stem beginning with a voiceless aspirate and ending with a simple voiceless stop is inflected with a suffix beginning with a vowel, the final stop becomes aspirated and the initial stop loses its aspiration, for example, *thrík-s* "hair" *tríkh-es* "hairs." Moreover, in both Greek and Sanskrit, the reduplicated forms of verbs starting with aspirated stops show the reduplicated prefix without aspiration [whereas with any other initial consonant, the reduplicated prefix is faithfully copied, e.g., Sanskrit: reduplication of a stem without an initial aspirate (to form the perfect of the verb), *tri* "to cross" *tatāra* (perfect of *tri*) vs. *babhūva* "became" (reduplication of root *bhū*); Greek: *péphúka* "converted" (reduplication of *phú-* "to engender")].

Further factors involved in Grassmann's work on the aspirated stops of Sanskrit and Greek was the belief by many of his contemporaries in IE historical phonology that Sanskrit was the parent language of the whole IE family. The Sanskrit texts were the oldest (or among the oldest) in the IE family, Vedic Sanskrit being dated to approximately the 2nd millennium BC. Certainly not everyone in Grassmann's time believed this. The alternate view was (and still is) that Sanskrit was a "sister" dialect among the other members of the family: Hellenic (Greek), Latin, Armenian, Germanic, Slavic, Celtic, etc. Still, in order to make the analysis that he arrived at he had to address this issue.

So, how did Grassmann solve the problems of the exceptions to the Rask–Grimm's Law and the odd behavior within Sanskrit and Greek of the aspirated stop in inflected forms? His argument involved command of and citation of evidence from the various branches of IE: Indic, Latin, Hellenic, Germanic, Slavic and the dialects within them, specifically Sanskrit, Avestan, Gothic, Classical Greek, Oscan, Latin, Lithuanian, Old English, Old Saxon, Old High German, New High German, Prakrit, English, Russian, Old Slavic, Old Norse, Zend (Persian). It is a feat something like Sherlock Holmes having to gather evidence from a multiplicity of sources and – like Holmes – arriving at a solution that brings order into what was previously chaos.

His solution is, in essence, very simple: He posited a parent language which, as mentioned above, was already accepted by some of his contemporaries – we now call this Proto-Indoeuropean – which did allow two aspirates within a root, for example, *bhudh- "to awake" (the asterisk signifies that the form is hypothetical or reconstructed – at least not attested in any text). Then there was a sound change in Sanskrit and in Greek whereby, given two aspirates separated by a vowel or other sonorant (nonobstruent), the first aspirate became deaspirated. This is known as Grassmann's Law[2] (GL). Apparently the same sound change arose independently in both languages. GL applied to the inflected forms within these two languages and gave rise to the otherwise puzzling behavior of aspirates such as its apparent shifting from one stop in a root to another. Thus to recall a Greek example given above, the root for "hair" must have been (at least in Proto-Greek) something like **thrikhos*; when the suffix *–s* was added, the second aspirate was deaspirated (by a separate rule) yielding ***thriks*** but in the form where the suffix showed an initial vowel, GL applied to yield ***trikhes***. Similarly, the "deviant" behavior of the reduplicated aspirates is accounted for by GL applying: given a root like *bhū,* this must have reduplicated in the normal way at an earlier stage, *bhabhu-*; then with the application of GL the first aspirate became deaspirated.

GL is an example of dissimilation – given two similar sounds in a word, one becomes different or disappears, for example, English *sword*, now pronounced [sɔɹd], i.e., without the [w], which, with its lip rounding, was too much like the rounded vowel [ɔ] and thus omitted or the not uncommon pronunciation of *library* as [ˈlɑjbɛɹi], with the first of the two "r"s (rhotics) neglected. Dissimilation of various kinds is found in many languages. I have presented an account of dissimilation which attributes it to an error of the listener and have likened it to the process of camouflage. The white arctic hare is camouflaged against the background snow because to a viewer its patch of white could be mistakenly attributed to the snow and thus not seen. Similarly a listener hearing two aspirated sounds or rhotics or rounded sounds might neglect or not detect its distinctive function on one of them – the feature present on one sound effectively camouflaging it on the other [Ohala 1993]. This occurs when the features involved are those known to spread onto adjacent sounds; aspiration is such a feature and this helps

---

[2] As often happens in science, this solution was anticipated a quarter century earlier. Rudolf von Raumer, another historical phonologist but one keenly interested in trying to accurately identify the phonetic character of the sounds involved in, among others, Rask–Grimm's Law, wrote in 1837: "Denn da das Skr. nie zwei Sylben hintereinander aspirirt, so kann man als ursprüngliche Form von bud' [in modern transliteration *budh*] ebensoght b'ud' annehmen. Dann aber wäre biudan nur der gewöhniche Uebergang von b' in b" [von Raumer 1837, 74]. [Lehman 1967] suggests that this idea was not noted at the time, perhaps because, unlike Grassmann, he did not work through all the relevant examples.

to give us some insight into the phonetic character of such speech sounds in the ancient languages.

Grassmann's work in this area of IE phonology thus solved some major linguistic problems, clinched the case for there being a parent language earlier than the IE languages known from textual material, and, by documenting this instance of dissimilation, contributed data that eventually led to recognition of the role of listeners in sound change.

Grassmann's contribution to the acoustic phonetics of vowels

It had been established at least since the time of [Ferrein 1741] that physically speech consists of a sound source (typically, the "hum" or "buzz" created by the vibrating vocal cords), plus an airspace, the vocal tract, that serves as a resonator, between this sound source and the point where the sound radiates to the atmosphere (thus reaching the ears of the listener). Such an arrangement was long familiar to musicians in which brass and wind instruments achieved their distinctive sound in the same way: in the bugle the sound source is the vibration induced in the bugler's pursed lips (by the air flowing through them) and the long metallic tube constitutes the resonator. In the bugle different resonant frequencies or "notes" are achieved by changing characteristics of the tube via valves.[3] In speech different sounds are made by changing the shape of the vocal tract. Different vowels are made by different tongue and lip configurations. In music there was a long tradition in labeling the notes in a standardized way, calibrated by tuning forks. This, however, was not easily applied to vowels. Between 1781 and 1803 there were some attempts at creating mechanical devices that could mimic speech sounds [Kratzenstein 1781; von Kempelen 1791; Darwin 1803[4]]. There is no question that these devices could successfully imitate speech sounds – replicas have been built and demonstrated – but they did not yield a theory that could give an account of how the vowel sounds, in particular, were made and what their acoustic structure was. In 1830 Robert Willis published an account of research he had done which attempted to remedy this. He made a cylindrical telescoping tube whose length could be varied and connected one end to a vibratory source (a reed), leaving the other end open. He claimed that he could produce all the variety of vowel sounds just by varying the length of the resonating tube. Significantly, he claimed that by gradually lengthening the tube the vowel series i e

[3] This comparison between speech production and wind instruments is made just for the sake of exposition; there are important differences, too, which I neglect.

[4] This was the grandfather of Charles Darwin.

a o u could be heard and then by lengthening it further the same vowels in inverse order. Whether this means he discovered multiple resonances for vowels is open to question.

In any case this was the state of knowledge when [Grassmann 1854] published his analysis of vowels in a small work apparently intended to accompany his teaching of physical acoustics at the Gymnasium in Stettin (Szczecin). I give here a long quote from that work:

> *The human voice* is produced by a device that is similar to the reed-pipe. Namely, two elastic pieces of skin, the vocal cords, are drawn over the upper part of the windpipe and can be stretched at will. As a result of the tension only a narrow channel, the glottis, can be made between them so that the air from the windpipe can only push through the glottis. While the vocal cords are not in tension the air goes in and out of the windpipe without making a sound. But as soon as they are put in tension vibrations of the vocal cords arise from the exhaled air and these [vibrations] follow one another all the more quickly as the vocal cords are tensioned more strongly, and therefore produce higher notes. With the falsetto notes only the edges of the vocal cords are vibrating at the glottis.
>
> The vocal cords simultaneously put the air in the mouth cavity into vibrations from which arise low overtones which, according to the shape given to the mouth cavity, turn out differently and which are related to the series of harmonic notes which have the note of the vocal cords as their "base" note. It is in this way that *vowels* arise. An attentive ear can easily hear a series of low harmonic overtones in the transition from *u* through *ü* to *i*, which can progress from the fourth octave *c* to the seventh octave *c* and can be produced with the same mouth positions. With the vowel *a* there is a whole series of harmonic overtones which the ear can perceive up to the fourth octave from the basis note, so therefore with the *a* a complete harmony of overtones sounds. Thus the transition from *a* through *o* to *u*, as well as that from *a*through *e* to *i*, is also explained.
>
> Among the *consonants* the semivowels are voiced. In the case of plosives there is no voice and the overtones do not occur purely any more but are mixed with a collection of unharmonic notes hard to distinguish from one another. The overtones of *a* occur with guttural sounds, those of *e* and *i* with the series of palatal sounds, those of *u* and *ü* with labial sounds, while with the dental sounds the highest overtones occur which do not belong to any vowels.[5]

[5] "*Die menschliche Stimme* wird durch eine der Zungenpfeife ähnliche Vorrichtung hervorgebracht. Nämlich über den oberen Theil der Luftröhre, den Kehlkopf, sind zwei elastische Häute, die Stimmbänder, gezogen, welche willkürlich gespannt werden können, und im Zustande der Spannung nur eine schmale Ritze, die Stimmritze, zwischen sich lassen, sodass die Luft aus der Luftröhre dann nur durch die verengte Stimmritze hindurchdringen kann. So lange die Stimmbänder nicht gespannt sind, geht die Luft ohne Tonbildung zur Luftröhre ein und aus; sobald sie aber gespannt sind, entstehen durch die ausströmende

The most impressive aspect of his analytical technique is that he came to this conclusion purely by auditory means. As he correctly indicates, the vocal cord sound consists of a large number of harmonics and by shifting the shape of the vocal tract different harmonics are emphasized, i.e., resonate selectively. I reconstruct what he must have done thus: intoned a given note, i.e., produced it on a monotone, he then moved in a continuous way from one vowel sound to another and was able to detect abrupt shifts in the harmonics that dominated the composite sound. This is, in essence, the technique used in so-called "throat singing" (more properly called "harmonic singing" or "overtone singing") that is well known from the Tuvan culture in southern Siberia: the sound source is a monotone and the different "notes" are the narrowly tuned harmonics.

In his discovery of multiple resonances for vowels Grassmann anticipated the same discovery by [Helmholtz 1863] by 9 years. Helmholtz, of course, made his discovery using what we now call "Helmholtz resonators," spheres of different dimensions, each tuned primarily to one acoustic frequency.

Conclusion

As is well known Grassmann did pioneering work in mathematics, linguistics, acoustic phonetics, physics (especially in optics) and was active in politics and other social and philosophical domains. There has been considerable discussion and speculation as to what character traits could have produced such a polymath. I add my speculation on

Luft Schwingungen der Stimmbänder, die um so schneller auf einander folgen, und also um so höhere Töne erzeugen, je stärker die Stimmbänder gespannt sind. Bei den Falsettönen schwingen nur die an die Stimmritze gränzenden Ränder der Stimmbänder.

Die Stimmbänder setzen zugleich die in der Mundhöhle befindliche Luft in Schwingungen, es entstehen dadurch leise Nebentöne, welche je nach der Form, die man der Mundhöhle giebt, verschieden ausfallen, und welche der Reihe der harmonischen Töne angehören, die den Ton der Stimmbänder zum Grundton hat. Auf diese Weise entstehen die *Vokale.* Ein aufmerksames Ohr hört leicht beim Uebergange von *u* durch *ü* zum *i* eine Reihe leiser harmonischer Nebentöne, welche vom zweigestrichenen *c* bis zum fünfgestrichenen *c* fortschreiten können, und welche man bei denselben Mundstellungen auch für sich hervorbringen kann. Beim Vokale *a* klingt eine ganze Reihe der harmonischen Nebentöne mit, welche das Ohr in der Regel noch bis zur vierten Oktave vom Grundton aus wahrnehmen kann, so dass also bei dem *a* ein voller Akkord von Nebentönen mitklingt. Hierdurch ist zugleich der Uebergang von *a* durch *o* zu *u*, sowie der von *a* durch *e* zu *i*, oder durch *ö* zu *ü* erklärt.

Unter den *Konsonanten* sind die semivocales noch von einem Stimmtone begleitet; bei den mutis fehlt der Stimmton, und die Nebentöne treten nicht mehr rein, sondern mit einer Menge unharmonischer, schwer von einander unterscheidbarer Töne vermischt hervor, und zwar die Nebentöne des *a* bei den Kehllauten, die des *e* und *i* bei der Reihe der Gaumenlaute, die des *u* und *ü* bei den Lippenlauten, während bei den Zungenlauten die höchsten (zischenden) Nebentöne, die keinem Vokale mehr angehören, hervortreten." [HGW22, 188f.]

this: he was gifted technically, to be sure, but so were a great many of his contemporaries who didn't achieve a fraction of what Grassmann did. I suggest that the difference is to be found in his fertile imagination, his boundless curiosity, and an approach to given problems that was unfettered by conventional wisdom.

Acknowledgements

I have relied heavily on Winfred Lehman's 1967 translations and on articles in Wikipedia.

Grassmann's "Wörterbuch des Rig-Veda" (Dictionary of Rig-Veda): a milestone in the study of Vedic Sanskrit

Maria Kozianka

Remarks on Ṛgveda (RV)

The RV, translated literally as "the knowledge of the hymns of praise", is the oldest literary monument of the Indian language. It contains 1,028 hymns (*sūktās*), that are distributed among ten books or song cycles (*maṇḍalās*) according to clearly recognisable principles. The division into ten books is not coincidental. Books II–VII, the so-called family books, stem from important families of poets. *maṇḍalā* IX contains mainly songs to Soma.

The composition of these hymns of praise was carried out over several centuries until approximately 1,000 BC. One has to note that the Vedic texts were handed down orally. Not until much later, around the middle of the first century BC, were the texts set out in writing. Up to the present day the exact recitation of the RV is cherished in some areas of India.

H.-J. Petsche et al. (eds.), *From Past to Future: Graßmann's Work in Context*,
DOI 10.1007/978-3-0346-0405-5_31,

Accomplishments of the Old Indic grammarians

Not only the literary works of ancient India cast a spell on us nowadays however. The achievements of the Old Indic grammarians also command our respect. To take an example, the systematic description of the vowel gradation (ablaut) – cf. Modern High German *singen, sang, gesungen* – with the aid of a three-step model, basic stage for example *kr̥-tá-* part.perf.pass. "made, done", Guṇa *kár-tar-* masc. "maker", Vr̥ddhi *brahma-kār-á-* adjective "offering prayers". Apart from this system of comparison, that captures the basic phenomena of the Indo-European ablaut, one has to point out the order of the Vedic alphabet. While the order of the letters of our Latin alphabet, or of the Greek or Cyrillic, is based on conventions, the Old Indic grammarians already arranged their alphabet strictly according to phonetic criteria, an astonishing scientific accomplishment by the standards of the time.

Vedic alphabet:

Vowels					
	short	long			
	a	*ā*			
	i	*ī*			
	u	*ū*			
	r̥	*r̥̄*			
	l̥	*l̥̄*			
Diphthongs					
	e	*o*	*ai*	*au*	
Consonants					
Gutturals	*k*	*kh*	*g*	*gh*	*ṅ*
Palatals	*c*	*ch*	*j*	*jh*	*ñ*
Cerebrals	*ṭ*	*ṭh*	*ḍ*	*ḍh*	*ṇ*
Dentals	*t*	*th*	*d*	*dh*	*n*
Labials	*p*	*ph*	*b*	*bh*	*m*
Semivowels	*y*	*v*			
Liquids	*r*	*l*			
Sibilants	*ś*	*ṣ*	*S*		
Aspirate		*h*			

The short and long vowels–syllabic liquids *r̥* and *l̥* are counted among these because they are syllable-bearers (cf. above *kr̥-tá-* "made, done")–follow diphthongs and consonants that are ordered according to their place of articulation. The manner of articulation is still distinguished into tenues, aspirated tenues, mediae and aspirated mediae. The last row contains the nasals, depending on the phonetical environment *n* occurs as guttural, palatal, cerebral or dental *n*. Moreover, there are also the Visarga *ḥ*, the Anusvāra *ṁ* and the Anunāsika *m̐*.

All Vedic Sanskrit and Sanskrit dictionaries follow this alphabetical order, also Grassmann's Rig-Veda dictionary.

Entries in Vedic dictionaries

Even the lemmata in Vedic dictionaries differ from others. If one takes a German dictionary such as, for example the Grimm dictionary, the verbs' entries will be given in the infinitive, for nouns the nominative singular, in Latin or Greek dictionaries it is the first person singular present tense for verbs, for example Lat. *ferō* "I carry, bring", Gr. *φέρω* [*pʰérō*] "I carry". It is different in Vedic Sanskrit. For nouns and adjectives the cardinal form is usually stated: *agní-* "fire, the god of fire", the nominative singular is *agníḥ*. For verbs the root is stated, because this is the starting point for all other forms of the verb: *bhṛ-* "carry, bring"; the following forms are derived from this, for example the third person singular present tense *bhárati* (class 1), *bíbharti* (class 3), the infinitives *bhártum* and *bhártave*.

Grammatical features of Vedic Sanskrit

Because of its old age, its abundance of forms and its extensive tradition, Vedic plays an important role for the comparison of languages. Thus, we distinguish between eight cases and three numeri for nouns and adjectives. Apart from the singular and plural, which are familiar to us, there is a dual for expressing two of the entities. These numeri are also present in verbs. After forming the present stem verbs are differentiated as thematic and athematic and are assumed to occur in ten verb classes respectively.

Grassmann's qualifications for such a dictionary

Not until he reached the age of forty did Grassmann start to devote himself to Vedic Sanskrit. He keenly studied Bopp's *Sanskritgrammatik* (Sanskrit grammar) and his *Vergleichende Grammatik der indogermanischen Sprachen* (Comparative grammar of Indo–Germanic languages). He had previously undertaken an intensive study of ancient languages. Old Persian, an old Iranian language, the Old Germanic Gothic, the Baltic Lithuanian and Old Prussian as well as Russian and Old Church Slavonic as Slavonic languages were a later concern of his. Being well prepared, Grassmann started the translation of the RV at the beginning of the 1860s. It was his intention to create a glossary as exhaustive as possible as a resource for his translation. Although it was at first only meant for his own use, it developed into a lexicographical masterpiece compared to others at the time, his *Wörterbuch des Rig-Veda* (Dictionary of Rig-Veda).

After ten years of hard work alongside school teaching he was able to conclude his dictionary at the end of 1872. It was published in six parts through 1875 by Brockhaus, Leipzig. The two-volume Ṛgveda-translation with commentaries was released in 1877.

The aids and resources for a translation and a dictionary of Ṛgveda were scarce at the time. As he commenced his work only the first volume of the Vedic text by Theodor Aufrecht was available in transcription. Volumes 1–3 of the Petersburg dictionary, that incorporates the entire Vedic literature, had been published, which only included the vowels, diphthongs and consonants *k* – *dh*. Furthermore, the centres of Sanskrit studies were a considerable distance from Stettin (today Szczecin).

Grassmann's Dictionary of Rig-Veda

1. Lemma entries

Let us take a closer look at individual lexicon entries. First of all, a noun. The lemma in bold print, here **sénā** [Grassmann 1996, 1578], is followed by the grammatical classification as female gender f. The derivational basis of the noun is given in square brackets [*si* "bind"], the meanings (1) *projectile* (*Geschoss*), (2) *army* (*Heer*), pl. *hosts* (*Heerscharen*) follow. In the subsequent form section the documented case endings[1] of *sénā* are given and – subdivided according to the meanings (1) and (2) – the corresponding RV-passages ascribed, so for example it says (2) *anavadyásya* 33, 6 beside the ending *-ām*. I.e.: The acc.sg. *sénām* with the meaning "army" can be found in RV I 33, 6. Attribute of *sénām* is the gen.sg. *anavadyásya* "the irreproachable".

In the case of endings that sound the same the corresponding case is stated; cf. in our example the ending *-ās*, that can be nominative plural or accusative plural depending on the context.

Concerning graded stems Grassmann systematically notes the low grade as lemma; cf. the nomina agentis with the vowel suffix full-grade *-tar-*, low-grade *-tṛ-*. The lemma

[1] The order of the Vedic Sanskrit case in paradigm is nominative, vocative, accusative, instrumental, dative, ablative, genitive, locative. In Grassmann's dictionary the vocative stands in first position because it is the least marked case, then the nominative.

hótr̥ masc. "sacrificer" (Opferer) is stated here; for adjectives with possessive suffix *-vant-/ -vat-* for example *kéśavat-* a. "having a mane (mähnig)". This low-grade approach is probably due to the Vedic Sanskrit grammarians, who have a basic grade in their ablaut system, which equates to the low grade. The Petersburg and subsequent dictionaries choose a full-grade lemma approach, *hótar-* and *kéśavant-*.

As previously mentioned, Vedic verbs are characterised by an extreme wealth of forms. To be able to handle this variety of forms Grassmann classifies the verbal forms according to the tense stems. He starts with the present stem, followed by perfect and aorist. If there is more than one present stem form of one verbal root these are subdivided into stem I, II, etc. The documented flectional forms of a verb are allocated to the individual stems including the passage. An example for illustrative purposes [Grassmann 1996, 596f.]:

The ascertained meanings of the root *dāś* are stated in the first section. Preverbs that occur in combination with *dāś* and the descriptive meanings of the composed verbs follow. The form section is next:

Stem I: *dāś*, it is equivalent to the second athematic class. *dāṣṭi* is documented in meaning (1) "worship a god" (einem Gott huldigen), god stands in dative: *agnáye*, so the *Agni*, and with the instrumental of means, here in plural *aráṇibhis*, "with churning-sticks" (mit Reibhölzern). *dāṣṭi* can be found in this connection in RV I 127, 4. It follows stem II: stressed *dā́śa-*, unstressed *dāśa-*. The stem equals the thematic first class. Stem III *dāśnu-* is formed with the suffix undergoing ablaut *-no/nu-* and is ascribed to the fifth athematical class. The stem IV that Grassmann assumed belongs to the perfect.

Apart from very frequent words like *ca* "and", *utá* "and, also" or *índra* "god Indra", the dictionary contains all references in an exemplary order. The instances are quoted according to the first edition of Aufrecht's Vedic text, which still counts the hymns from 1 to 1,028. The arrangement into books and strophes, which is common today, took place in the second edition. This makes the use of Grassmann's dictionary a little difficult but these shortcomings can be easily remedied with a concordance.

What does this brief glimpse of Grassmann's dictionary prove? The systematic layout and the neatly arranged material already reveal the hand of the mathematician. First he gives a consecutively numbered overview of the meanings that were gained from the R̥gvedic word material: it is ordered from the basic meaning to specific meanings. Then an entire scheme of the grammatical forms with corresponding meaning number and respective references is drawn up. The sections on meaning and form can, at all times, be related to each other by using the prefixed numbers. Thus, the dictionary user can find every word that occurs in the RV in a short space of time. Additionally, Grassmann's indications are characterised by a high degree of reliability.

A systematic backwards-sorted word register follows the article section and constitutes a good basis for the examination of Vedic word formation.

The usage of the transcription and not the Nāgarī-script, which is common for Vedic, made sure that the dictionary had a favourable reception at the time as well as today.

2. Concerning the illustration of compounds and derivations

Grassmann illustrates the individual members of compounds with a hyphen. In his translation he also tries to do justice to the different compositional types of Vedic. This can be observed among other things in the case of determinative compounds like the Tatpuruṣas, actually "his servant" (dessen Diener). In the case of this type the composition expresses a case relationship: cf. Modern High German *Goldschmied*, *Himmelstür*, *Königssohn* (gold smith, gates of heaven, king's son).

a) Vedic *rāja-putrá* masc. "son of a king (Sohn eines Königs) [*rā́jan*]" (RV X 40, 3),
b) *aśva-yūpá* masc. "the post, to which the sacrificial horse is bound" (der Pfosten, an den das Opferross angebunden wird), the first member, *áśva,* means "horse", the second, *yūpá,* "post" (RV I 162, 6).

In RV I 162, 6, where the compound occurs, [Geldner 1951, 1, 222] translates "steed post" (Rosspfosten), [Witzel/Gotō 2007, 290] and [Grassmann 1877a, 453] "horse post". The first member stands for a genitive here.

Or consider the equally determinative Karmadhārayas. These are compounds in which the last member is attributively determined by the first member, for example:

c) *mahā-vīrá* masc. "great hero" (großer Held) (RV I 32,6 akk.sg. *mahāvīrám*),
d) *su-devá* masc. "good god" (guter Gott) (RV VIII 69, 12 *sudevó asi varuṇa* "You are a good god, o Varuṇa" [Du bist ein guter Gott, o Varuṇa]).

Grassmann's aim, to simultaneously give an account of the corresponding compositional type with his translation, is very obvious in the case of the possessive compounds, the so-called Bahuvrīhis, as in "possessing or having a lot of rice" (viel Reis besitzend, habend). These exocentric compounds are primarily adjectival and are to be comprehended by "having a certain characteristic" (eine bestimmte Eigenschaft habend) and "being in a certain condition" (in einem bestimmten Zustand befindlich). They receive their gender from the term whose attribute they are. In Modern High German such compounds are for example *Dickkopf*, *Langfinger* or *Großmaul* (bullhead, pickpocket, bigmouth).

e) Vedic *rā́ja-putra* adj. "having kings [*rā́jan*] as sons [*putrá*]" (Könige [*rā́jan*] zu Söhnen [*putrá*] habend) (RV II 27, 7). Antecedent is the goddess *Áditi*, the mother of the kings. The word only differs formally due to the accent from the determinative compound above (a) *rāja-putrá* masc. "king's son" (Königssohn).
f) Vedic *áśva-peśas* adj. "having steeds as decoration [*péśas*], provided with beautiful steeds" (Rosse als Schmuck [*péśas*] habend, mit schönen Rossen versehen) (RV II 1, 16). As always Grassmann states the antecedent, which is the acc.sg.f. *rātím* "gift" (Geschenk).

Also syntactic information frequently appears in Grassmann's dictionary. A few examples will illustrate this. In the case of the neutral *s*-stem (f) *péśas* "decoration, adornment" Grassmann points out that the substantive occurs with the genitive of the decorated:

g) RV VII 34, 11 *péśo nadī́nām* (genitive plural) "adornment of the rivers" (Schmuck der Flüsse),
h) RV VII 42, 1 *adhvarásya* (genitive singular) *péśaḥ* "adornment of the sacrifice" (Zierde des Opfers).

The adjective *priyá-* "dear, beloved" (lieb, geliebt) demands the locative in the meaning "popular with" (beliebt bei), cf.

i) RV II 41, 18 *priyā́ deveṣu* (locative plural) *júhvati* "to sacrifice as beloved for the gods" (als Beliebte bei den Göttern opfern).

Exemplary comparison of Grassmann's dictionary with the Petersburg dictionary by Otto Böhtlingk and Rudolph Roth, pt. 2. (1856–1858)

In conclusion I will compare Grassmann's dictionary with the Petersburg dictionary (PD) on the basis of entries that start with *k*.

As already mentioned, Grassmann tries to illustrate the word formation when stating the meaning. This feature cannot be found as developed in the PD and its *k*-lemmata:

For the adjective *karambhín,* a derivation with the suffix *-in* as an expression of the "being provided with" (Versehenseins mit), [Grassmann 1996, 315] gives us "provide with grout" (mit Grütze versehen) with the reference to the sacrificial drink of Indra.

The PD [Böhtlingk and Roth 1856–1858, 112] provides us with the inappropriate meaning "accompanied by grout" (von Grütze begleitet). Another example is the adjective *kumārín*, a derivation of the masculine *kumārá-* "child, lad, teenager, son" (Kind, Knabe, Jüngling, Sohn), that [Grassmann 1996, 329] translates adequately as "provided with children" (mit Kindern versehen) and the PD [Böhtlingk and Roth 1856–1858, 338] as "provided with girls" (mit Mädchen versehen).

In distinction from Böhtlingk/Roth [ibid., 539] [Grassmann 1996, 365] also assumes the meaning "earth, soil" (Erde, Erdboden) for the feminine *kṣā́-* and not only "dwelling" (Wohnstätte), which is accurate for RV I 95, 10. 133, 6. 158, 4 and more.

Concerning a range of words, especially when they have only been documented once, the PD merely states the passage where the word occurs besides the Nāgarī-script. In contrast, Grassmann principally attempts a translation that he extracts from the context, the basis of the derivation, from the individual components of the compounds or from etymological consideration. An example is *kā́payā* (RV X 40, 3), which [Grassmann 1996, 322] quite justly determines to be an instrumental singular of a feminine *kā́pā-* perhaps "chariot" (Wagen). For the masculine *kī́ja-* (RV VIII 66, 3) [Grassmann 1996, 327] assumes the meaning "spur" (Sporn). *kī́jas*, which has the attribute *hiraṇyáyas* "golden", is connected with *mr̥kṣás áśvyas* "horse comb" (Pferdestriegel) by a coordinating *vā* "or, and".

Then the adjective *kārpāṇá-* (RV X 22, 10) "fought with the sword" (mit dem Schwerte [*kr̥pāṇa-*] gekämpft), that he quotes as "sword fight" (Schwerterkampf) [H. Grassmann 1996, 324], is similar to [Geldner 1951, 3, 158] "sword fight" (Schwertkampf).

The meaning of *kr̥ṣṇá-yoni-* (RV II 20, 7) "having black loins" (schwarzen Schoß [*yóni*] habend) is not given until the shorter version of the PD [Böhtlingk and Roth 1881, 96c].

And finally the adjective *kravaṇá-* (RV V 44, 9), that [Grassmann 1996, 358] assumes to have the meaning "terrible" (furchtbar), is actually "solidifying" (erstarren machend), because he puts the adjective into an etymological context with *kravíṣ-* "raw meat" (rohes Fleisch) and the basic meaning "coagulated blood" (geronnenes Blut). In his critical and exegetical notes to the RV, Hermann Oldenberg [Oldenberg 1909, 342] also considers a relationship to *kravíṣ-* and translates it as "slaughterer of the offering" (Opferschlächter).

kaśa-plaká- masc. maybe "buttock" (Hinterbacke) (RV VIII 33, 19) and the Bahuvrīhi *kṣipréṣu* "having fast arrows" (schnelle Pfeile habend) consisting of *kṣiprá-* "fast" (schnell) and *íṣu-* "arrow" (Pfeil) are not listed in the PD of 1856–1858, they were only listed in 1881 in the PD [Böhtlingk and Roth 1881, 40a, 126b].

Virtually all listed examples up to the present have not had a satisfactory explanation. Among many other reasons this can be ascribed to having only one documented formation that cannot be etymologised reliably.

Recognition of the linguistic accomplishments

From the very beginning, the linguistic works, especially the dictionary and the translation for the RV, as well as his dissimilatory law (Grassmann's law, Hauchdissimilationsgesetz), have attracted great interest among experts, in contrast to his mathematical works. The significance of the dictionary shows itself in the five new editions up to 1996. The first Indian edition, a reprint of the fifth edition, was published in Delhi in 1999. The assessment of the distinguished Indologist Albrecht Weber at the Stettin philological conference in 1880 is still accurate: "Sein Wörterbuch zum Rig-Veda gehört zu den household-works der Veda-Forscher" ("His dictionary of Rig-Veda belongs to the household works of Veda researchers") [cited in Engel 1911, 310].

The Rigveda Dictionary from a modern viewpoint

Toshifumi Gotō

The *Dictionary of the Rigveda* by Hermann Grassmann is the only standard Rigveda [RV] dictionary today.[1] The 5th reprint followed in 1976, and there exists a 6th edition, revised and supplemented by M. Kozianka [H. Grassmann 1996]. After its appearance, research has developed radically in the fields of philology (understanding of vocabulary and meaning) and grammar. The progress achieved in Indo-European linguistics in the twentieth century has been remarkable. The study of morphology including the functional analysis of grammatical forms has developed towards totally new dimensions. The *Dictionary*, nevertheless, is an essential tool even today. Its layout plays an important role in this regard. It seems that Grassmann could anticipate how the gaps should be bridged in the future, and how an entry would be more or less mechanically reinterpreted. I would like to report how this phenomenon was possible, foremost from my own three-year experience with the new translation of the Rigveda [Witzel and Gotō 2007], after the translations of Grassmann [RV1; RV2], [Ludwig 1876–1888], and [Geldner 1951] (completed in the 1920s, but published in 1951).

[1] Leipzig: F. A. Brockhaus 1873. Preface: Stettin, 18.8.1872. Afterword: 8.8.1875. For more information on the publication, cf. Preface, p. V and [Windisch 1920, 365].

H.-J. Petsche et al. (eds.), *From Past to Future: Graßmann's Work in Context*,
DOI 10.1007/978-3-0346-0405-5_32,

Lemmas, forms and meaning

The *R̥gveda* is a collection of more than 10,000 *ŕ̥c*-s "praising stanzas", codified as a sacred text (*veda*) presumably about 1,200 B.C. It is not only the oldest text in India, but also one of the most important documents written in the Indo-European languages. Grassmann's dictionary is an indispensable tool for RV scholars and for Indo-European studies in general.

Grassmann records all forms in the RV with almost all the occurrences and essential contexts. The occurrence is cited, however, inconveniently according to 2-figure numbers (sequential hymn and stanza), and not by the 3-figure numbers (book, hymn, and stanza) which are common today. The 1st edition of the RV by [Aufrecht 1861; 1863], which Grassmann used as a basis, gives us also the continuous numbering besides the 3-figure system. Unfortunately, in today's standard edition [Aufrecht 1877] the collateral data has been left out. The songs have been classified and codified according to the poets' families; within a family book, the number of the songs and stanzas as well as the length of the metrical line was standard for the arrangement of the songs. All these facts became obvious in full detail only after Grassmann's *Dictionary* had appeared, through investigations by [Bergaigne 1886; 1887] and [Oldenberg 1888]. But Grassmann had already essentially recognized this arrangement. He writes about this editorial principle in his translation of the RV [RV1, 2]. His translation consists of two parts, reflecting his understanding of the editorial background as well: 1. "The family books of the Rig-Veda": II–VIII [RV1], 2. "Collected books": I, IX, X [RV2]. The information showing in which of the ten books a certain word form is attested is important. It is true that the 3-figure parameter causes substantially more difficulty for presentation and correction. Did the mathematician Grassmann confuse the philologist Grassmann here? Anyway, one cannot easily understand his decision. Maybe he had noticed the importance of this fact only gradually, during his translation, and did not have enough time for the rearrangement.

At first, let a group of verbal roots be taken up to illustrate Grassmann's treatment, and to compare them with today's state of research ("xxx" suggests omission; "583,10 [VII 67,10]" means: 583,10 in the *Dictionary* corresponds to Book 7, 67th Hymn, Stanza 10; this reference is omitted in Book I; [] is my addition):

column 479

^{3}***jar***, originally **gvar*, substantially identical with *jur*, perhaps also with *jvar* ["be in fever"]. Basic meaning: "grind down, wear out, consume" (cf. Kuhn, in his *Zeitschrift für Vergleichende Sprachforschung...*, Curtius, *Grundzüge der griechischen Etymologie*, ^{2}1866). (1) "grind down, make fragile, make old"; (2) "let become old", i.e. "let reach an old age"; (3) "become fragile, become old, be consumed"; (4) "be ground down"; (5) causative "grind out, wear out, make old".

stem ***jára-***: *-anti* (1) *ná yám* ~ *śarádas* 465,7 [VI 24,7]. *-atam* (2nd person dual imperative) (2) *sūrī́n* 583,10 [VII 67,10].

jara-: *-anta* (3) *pūrvī́s uṣásas* 857,7 [X 31,7].

aorist ***jāriṣ-***: *-ṣus* (3) *mā́ dyumnā́ni* 139,8.

stem of causative ***jaráya-***, ***jāráya-*** see under participle.

participle ***járat-*** ((3) fragile, old, aged): *-antam* (3) xxx. *-antā* (dual) (3) xxx. *-atos* (genitive dual) (3) xxx. *-atas* (genitive singular) (3) xxx. *-atībhis* (4) xxx.

participle of causative ***jaráyat-***: *-an* (5) xxx. *-antam* (5) xxx. *-antī* (5) xxx. *-antīs* (accusative plural) xxx.

participle of causative ***jāráyat-*** (Padapāṭha *jaráyat-*): *-antī* 124,10 (*uṣā́s*).

column 493

jur, another form of ^{3}*jar*. (1) "become fragile, become old", of living beings, further (2) of the god's deeds; (3) "make frail, make old"; especially (4) participle II *jūrṇá-* "friable, brittle" (of a net 180,5), "frail" (of the casted-off snake's skin 798,44 [IX 86,44]); (5) *jūrṇá-* "very old", without nuance of fragility. – Cf. *jūrv* und *ni-júr-*.

stem I. ***jurá-***: *-átam* (2nd dual imperative) *paṇés ásum* 182,3.

stem II. ***jūrya-***: *-ati* (1) *sá (agnís) ayā́ kr̥pā́ ná* ~ 128,2. *-anti* (2) *ná vām* ~ *pūrviā́ kr̥tā́ni* 117,4.

imperfect ***ajūrya-***: *-an* (1) *devanídas ha prathamā́s* ~ 152,2.

participle ***jurát-*** (stem I): *-até* (1) *cyávānāya* 584,6 [VII 68,6]. *-atā́m* (1) 225,10 [II 34,10] (*járāya*).

jū́ryat- (stem II) (see *á-jūryat*): *-atsu* (1) *vánesu* (metaphorical) 257,1 [III 23,1], opposed to *ajáras* (*agnís*). *-antai* (dative feminine) (1) *pátim* ~ *adatta* 117,7.

participle perfect ***jujurvás-***, weak ***jujurús-***: (1) xxx

participle II ***jūrṇá-***: xxx (4) xxx. xxx (5) xxx. xxx (4) xxx. xxx (5) xxx.

verbal ***júr-***, "becoming old, aged", included in *a-*, *r̥ta-*, *dhiyā-*, *sanā-júr-*.

column 499

jūrv, basic meaning presumably "wear out, consume", and therefore related to *jur*, and came from this (cf. *nijur-*); but limited in consuming by heat (of fire, lightening, sun), thus: "scorch, consume through the heat (of the sun)".

With ***ní***, "burn down, consume through the heat (of the fire, lightening)". ***sám***, "burn, consume through the heat (of the fire)".

stem ***jū́rva-***: *-asi* ***sam*** 669,7 [VIII 60,7] *atasám*. *-ati* ***ni*** 1013,3 [X 187,3] *rákṣāṁsi*. *-athas* ***ni***: *rákṣas* 620,4 [VII 104,4]. *-ās* ***ni***: *śátrum* 221,5 [II 30,5].

stem ***jūrva-***: *-a* ***ní***: *spŕ̥dhas* 447,6 [VI 6,6].

aorist ***jū́rvī-***: *-īt* ***ni***: *ámānuṣam* 202,10 [II 11,10].

participle ***jū́rvat-***: *-an* (read *jūruan*) *víśvāni* 191,9. – ***ni*** 303,11 [IV 7,11].

Note: The more basic form *júr* is found in *ni-júr-* "scorching, consuming".

1. Analysis of the entry

Here we encounter various appearances of the one Proto-Indo-European [PIE] verbal root *ǵerh$_2$ "make old, infirm". Its strong form (accented full-grade) is *ǵérh$_2$ which is developed in Old Indo-Āryan [OIA] to *jár^{i}*, for example in the present stem *jár^{i}-a-^{ti}* "make old, make achieve old age". The superscript *i* is only a sign for a reflex of one of three PIE laryngeals for linguists' convenience; it appears as *i* in some circumstances, for example between consonants. "-^{ti}" is the ending of the 3rd singular active as representation. Greek γέρων (*gérōn*) "old" shows the same grade (cf. below).[2]

The verbal adjective from the root in the weak form (zero grade, with accent outside of the root or stem part) *ǵr̥h$_2$-nó- "ground" appears in OIA in *jūrṇá-* in the RV, and its younger form *jīrṇá-* "rotten, aged, frail", Latin *grānum*, German *Korn*, Old-Church-Slavic *zrьno* "grain". The phonological laws involved in these developments are: PIE *ǵ > OIA *j* (OIran. *z*); *r̥h > OIA *ur/ir* before a vowel, *ūr/īr* before a consonant (OIran > *ar*), Latin *rā*, Germanic *ur*, Slavic *r̥* (written *rь* or *rъ*). Grassmann, of course, had a profound knowledge of comparative Indo-European linguistics; the best illustration of this is the famous Grassmann's law about aspirate dissimilation in Indo–Iranian and Greek. But the existence of three laryngeals and the consequences they have left in Indo-European languages were not yet known to his generation. A strange reconstruction *gvar was the result of the efforts of contemporary scholars who recognized the relationship between the words but could not find phonological regularities underlying their connection.

An autonomous adjective *ǵérh$_2$-ont- "old" is inflected in the singular: nominative *ǵérh$_2$-ōn(t) > gr. γέρων (*gérōn*) "old, old man", accusative *ǵérh$_2$-ont-m̥ > OIA *járantam* (cf. *Dictionary*, under participle), dative *ǵr̥h$_2$-n̥t-éi̯> OIA *jurate* (cf. *Dictionary*, under *jur*). *járatas* (genitive singular) and *járatos* (genitive dual) are results of analogical levelling after strong stem-form. The feminine stem is *ǵérh$_2$-n̥t-ih$_2$- > OIA *járatī-*, attested in the instrumental plural *járatībhis*. [Gotō 1987, 153 n. 238].

The causative *jar-áya-^{ti}* goes back to PIE *ǵorh$_2$-éi̯e-^{ti}, formed with the causative suffix *-éi̯e- added to the root in the *-o-grade. It serves as the explicit substitute of *jár^{i}-a-^{ti}* in the meaning of 'make old', or as factitive to *jū́r-ya-^{ti}* 'become old'. Short *a*

[2] "Laryngeals" are laryngeal consonants postulated in Proto-Indo-European; three sorts are assumed, h_1 (with non-colouring effect), h_2 (with *a*-colouring effect on the basic vowel *e), and h_3 (with *o*-colouring effect on *e). They correspond to a large extent to "coefficient sonantique" theoretically assumed by [de Saussure 1879], and have been made common property through articles by É. Benveniste, especially J. Kryłowicz 1927–1928 (cf. [Kryłowicz 1935; 2004]), and others. The new impulse for the theory has been given from the discovery of Hittite in 1915, where some remainders are found in the sound *h*(corresponding to h_2 in today's theory). Cf. [Mayrhofer 1981].

in *jar* is due to Brugmann's law, i.e. PIE **o* becomes *ā* in Proto-Indo-Iranian in the open syllable, but to *a* in the closed syllable (syllable ending in a consonant). *jār-áya-*^{ti} with long *ā* is an analogic form after the type, for example *yātáyati* 'settle someone in his position' to *yátati*.

jū́r-ya-^{ti} in the RV with its younger variation *jī́r-ya-*^{ti} (Atharvaveda+) is a present stem characterized through the suffix **-i̯e-* as a "fientive" formation, i.e. a stem denoting the change of the subject's state or condition as opposed to "doing something" (facientive), in this case, thus, 'become old, frail'.

jū́r-va-^{ti} 'wear, exhaust' is a **-u̯é-* present (PIE **ǵr̥h₂-u̯é-*) with a secondary accent. This present formation is rare and its accurate function is yet to be discovered. Grassmann properly explains the form as belonging to the group of *jar*^{i}, *jur*, *jūr* (and *jī̆r*). Mayrhofer, who once judged this present stem differently [Mayrhofer 1956, 443], writes today [Mayrhofer EWAia, vol. I, 597], referring to [Gotō 1987, 153 n. 237], "correctly already Grassmann 499 ('to wear out'), who, however, arrives at the meaning 'to scorch, consume through the heat' because of repeated combination of *jūrv* with Agni [the fire god] (or the sun...); from there the false connection with *jvar*^{i} (*jval*^{i}) ['to scorch'] follows (Altindische Grammatik I 24...)." Grassmann owes his right judgement to the detailed treatment of all forms and occurrences as well as rational sorting out of the forms and meanings. He operates over the total figure of a word which was achieved in this way, towards its primary or central meaning. Thus, he could point out the history of a word group and possibilities of its development.

The interpretation of *ni-júr-* played a decisive role in this case: RV II 29,6 *trā́dhvaṃ no devā nijúro vŕ̥kasya trā́dhvaṃ kartā́d avapā́do yajatrāḥ*. Grassmann translates [RV1, 38]: "Save us that a wolf does not devour us, that, ye reverends, we do not fall into a pit! [Errettet uns, dass nicht der Wolf uns fresse, dass, Hehre, nicht wir in die Grube fallen]"; my translation: Save us, ye gods, from the wearing down [the herd of sheep] by a wolf! Save [us] from falling down into a pit, ye revered ones, cf. [Gotō 1987, 153 n.237]! Grassmann's translation observes the syllables' number in the original and uses iambic and trochaic rhythm imitating the original metre to some extent. Here he shows his faithful approach to the original in a way which is different to the analytical approach he takes in his *Dictionary*.

It is in itself a quite natural process to control all the forms and usages, to analyse them, and construct them into a transparent complex. Grassmann could perform this procedure ideally and rationally. He represents these operations in one format, i.e. in a synchronic list of the forms and meanings. We can observe every step of the operation in his list, which helps us to go deeper into the history of words and forms.

2. *Meaning entries*

Among the meanings Grassmann gives under *jar* (column 479), (1) 'grind down, make fragile, make old', and (2) 'let become old, i.e. let reach an old age' are facientive (or agentive)-transitive, i.e. the subject does some action or influences someone/something. The meanings (3) 'become fragile, become old, be consumed' and (4) 'be ground down' are fientive-intransitive, i.e. a change of state or condition happens in the subject. (5) "causative" 'grind out, wear out, make old' is factitive: the subject makes someone/something do something or something happen. With "*jur* (1) become fragile, become old (of living beings)", and "(2) the same (of the god's deeds)" a fientive-intransitive meaning of the verb is meant.

The root-form *jur* (column 493) is observed from a facientive-transitive realization: "(3) make frail, make old, especially (4) in participle *jūrṇá-* friable, brittle (of a net), frail (of the cast off snake's skin), (5) *jūrṇá-* very old without nuance of fragility". For (4) and (5), also a fientive or essive (statutive) interpretation (the subject is situated in some state) may be possible. But the interpretation of Grassmann is correct, because the root starts from a facientive meaning. That is why a simple present stem *jár-a-*^{ti} can mean 'make old', and a characterized stem *jū̂r-ya-*^{ti} is formed specially for the fientive meaning 'become old'.

Morphological means are used to vary the lexical content of a verbal root into various meanings and realizations. There are also analogical or levelled secondary formations. Today we could represent the forms and meanings in a combination of morphological and noematical (i.e. relating to lexical components) entries. But such a process is in some parts still the subject of monographic studies. Perhaps we will have to wait until a dictionary is able to incorporate the results of the research (a scholarly dictionary, not a translation dictionary). Thus, currently one will have to hope that the user is well informed about morphology.

The meanings are listed and classified well, based on external (syntactical) and internal criteria. This makes it possible to verify primary and developed meanings or uses. Grassmann's interpretations have been in some ways and in some places improved or varied upon. Still, on the whole, we remain within the framework which he set. If scholars do not arrive at a common solution, they still resort to "according to Grassmann, so-and-so".

3. *Form entries*

Grassmann takes up indicative, "conjunctive", optative and imperative for the moods of the verb. His "Conjunktiv" means subjunctive and injunctive ("the second conjunctive", p.VI) today. However, it scarcely disturbs us. The forms and functions of the injunctive were exactly ascertained only by Karl Hoffmann [Hoffmann 1967; first manuscript-version in 1950]. For a precise interpretation, whether a form is subjunctive or injunctive, an investigation is still necessary. Otherwise, the issue remains partly undeterminable. Thus, it was not a task of the dictionary at that time to make it clear whether a form is subjunctive or injunctive. The field of the stem formation and its ratio has made great progress up to the present day. Grassmann's judgement as aorist or present (including the imperfect, i.e. the preterit of the present stem which expresses a verb's action in its progressive aspect) cannot be accepted in all instances, especially in the case of the aorist stems which are used for the aspect without time-expansion. It is important, still, that the stems are sorted out separately. We can surely start our research from the *Dictionary*.

In frames and details of the nouns, pronouns, and adverbs, the views Grassmann presents are barely in need of revision today. Only some lemmas are to be put together, as is the case with the above-mentioned *járant-* and *jurat-*.[3] To divide one lemma into several is much more difficult than to fulfil the procedure in the opposite direction.

There may be cases, in which some forms are to be distinguished. In the case of verbs, for example: The forms of the secondary root *dad*, *dáda-*^{te} "hold" [Wackernagel 1942, 163f.] are registered under *dā* "give". We distinguish two homonymous roots ^{1}*yā* "travel, drive" and ^{2}*yā* "beg" [Schmid 1956, 222ff.] among the forms Grassmann gives under "*yā* go, wander … drive; arrive …; approach begging …; attack, pursue … ". Grassmann's '*cit* perceive, recognize', contains two separate roots ^{1}*cit* 'recognize, pay attention to' and ^{2}*cit* 'shine, distinguish' [Gotō 1987, 137–141; Mayrhofer EWAia I, 547f., 548f.]. The preset stem *cáy-a-*^{te} 'punish' under '^{1}*ci* place together, pile up" belongs to another root *ci*, which goes back to PIE **k*^{w}*ei̯* 'receive sanction' together with Gr. τίνω 'pay, atone for', τίνομαι 'make atone', ποινή 'penitence' [Gotō 1987, 132; Mayrhofer EWAia I, 532f.]. *math* 'steal' (present *mathnā́ti*) and *manth* 'twirl (in order to make fire, etc.)' (*mánthati*) were distinguished by [Narten 1960, 121–135]. Grassmann's inventory of forms and meanings serves, naturally, always as basic material for such renovation.

We know that there are three separate roots among the forms under "*ar*, *r̥* set in motion". They go back to PIE *h_1*er* 'come into, reach at', *h_2*ar* 'fit' und *h_3*or* 'set oneself into a (quick) motion', cf. [Kümmel in Rix et al. 2001, 238, 269f., 299f.]. They are, however, not yet duly distinguished, so every time we go back to Grassmann.

[3] Belong or go back to one and the same paradigm, e.g.: *ráyi-* 1149f. and *rái-*, *rā́-* 1182ff. 'wealth'; *dā́ru-* 595f. and *drú-* 647f. 'wood'; *sā́nu-* 1511 and *snú-* 1606 'back'; *hŕ̥d-* 1678 and *hā́rdi-* 1661 'heart'; *aśan-* 137 and *aśman-* 139 'rock, stone'; *dvā́r-* 651 and *dúr-* 613 'door'; *kṣám-* 363, *kṣā́* 365, *kṣmā́* 371, *kṣā́man-* 365, *gmā́-* 415 and *jmā́-* 502 'the earth'. Under *star-* 'star' one misses *tā́ras* 'stars' (nominative plural) VIII 55,2. *dyáv-*, *dyu-* and *div-* are duly registered together under *div-* 'heaven' (601ff.).

The treatment of nouns for 'great, big' *máh-*, *mahá-*, *mahā́nt-* (Grassmann "*mahát*"), and *mahā́-* is very impressive. The occurrences are very numerous and the interpretation is difficult. Indeed, we may meet with cases which should be better understood in another way, but still Grassmann's data remain essential for such precise study. *mahnā́*, *mah*$_a$*nā́*, and *mahábhis*, for which the *Dictionary* gives a lemma "*mahán-* neuter (from *mah-*) greatness, fullness, power, magnificence", are interpreted today as instrumentals of the masculine *mahimán-*.[4]

Metrical analysis

Grassmann makes efforts throughout to give, beside the forms transmitted in the text, also the forms which are to be assumed for metrical reasons in the original poetry. On the whole, we can rely on his assumptions. But each case has its own difficulty. Among others, Oldenberg has performed philological examinations in this field [Oldenberg 1888, 1909, 1912].

The striving of the generations following Grassmann and Oldenberg consisted of explaining such phenomena linguistically: exacter versions of the rules of syllabic structure,[5] identification of different consequences of disappeared sounds (three laryngeals), also in morphological elements.[6] The poets, in addition, had their "poetic freedom, poetic licence" to use older and newer forms of different times. *jána-*, for example, is a common word for "man, people". The stem is scanned as short-short or long-short. The latter is explained from the older constellation **jóh*$_2$|*no-* consisting of a closed (thus long) and a short syllable.[7]

---

[4] Cf. [Mayrhofer 1986, 159; Schindler 1977, 57; Tichy 1986, 99 n.7; Gotō 2005, 209]. The appearance and disappearance of *i* in *mahimán-* and *mahmn-_'* is explained from the "loss of a laryngeal" today. *mahā́m* (column 1016b under *mahát-*) belongs to "*mahā́* adjective 'great' (from *mah-*)" (column 1018).

[5] Sievers' law, now enriched by a more precise version of Schindler's syllabification law [Schindler 1977, 56]; some individual laws such as those of [Stang 1965, 292ff.; Lindeman 1965, 38ff., 69ff.]. Cf. [Mayrhofer 1986].

[6] E.g. nominal suffix *-*h*$_3$*on-*/-*h*$_3$*n-* (Hoffmann's possessive-suffix), *-*i̯h*$_2$*o-*, etc., the optative suffix *-*i̯eh*$_1$-/-*i̯h*$_1$-/-*ih*$_1$-. The OIA roots are divided traditionally into two groups *seṭ* und *aniṭ*. They are interpreted today, respectively, as the roots ending with a laryngeal and otherwise.

[7] *svàr-* 'sun's light' is delivered in the tradition only in this form. Grassmann registers it as "(*sv́ar*), *súar*, *sū́r*", it means: attested as *svàr-*, but to be read as *súar* (better *súvàr*, we write *súvàr* with a subscript character as the sign to be supplied). The word is reconstructed as *-*l*/*n-* heteroclite neuter of substance: nominative (A) **sáh*$_2$-*u̯l̥* (> Proto-Celtic **sāu̯al* > **hōu̯al* > **hōu̯l* > Middle-Breton *heol*), changed into (B) **sáh*$_2$*u̯el* after other *-*l*/*n-* or -*r*/*n-* heteroclites, or into (C) **sh*$_2$*uu̯el* (> OIA *svàr-*, *súvar-*, OAv. *huuarə*) after genitive **sh*$_2$-*u̯én-s* with Lindeman's variant *sh*$_2$-*uu̯én-s* (> Old-Avestan **huu̯əŋ*^{h} > *x*^{v}*ə̄ŋ*, Young-Avestan *hū*). OIA *súvàr* represents a nominative form (C). Only in RV II 35,6 the word *svàr* is scanned really as

Prepositions, particles, etc.

An important advantage of the *Dictionary* is its careful and thought-out treatment, as well as its rational presentation of modal words such as prepositions and particles which are decisive for understanding the "logic" of the poets. In order to illustrate how Grassmann treats such small words, let us cite the item about *ā́* 'towards something/someone', used when the directional sense is "towards here", i.e. seen from the terminal:

> column168–171 *ā́*:
> … The basic notion is "on", related first of all to a limit or surface, to the immediate proximity. Therefore it conveys the meaning of the adverb of the direction "up this way, near to; hither" (i.e. moving to the immediate proximity of the speaker), or "upwards, up to; towards" (moving to the immediate proximity of the imagined object). Therefore it functions as a preposition with a following ablative "up to, as far as", while it only emphasizes the meaning of the cases if it is used as a preposition with preceding accusative, ablative, locative. However, this word order is not unchangeably fixed in both cases. It connects words as a connecting particle in the meaning of "and", i.e. "attaching oneself to the expressed one arriving at this, stepping hereto". Finally, it serves to emphasize the preceding accented word in the sense that its notion is valid in its full contents, in its just proper meaning. Thus it is realized as:
> Word of direction with the verbs: xxx xxx xxx xxx xxxxxx xxx xxx. ["xxx" suggests omission by the article's author, as above in 1; here, the verbal roots are listed to which *ā́* is added.]

- Prepositions with preceding accusative: (1) 'up to' (of space) xxx xxx xxx xxx [the occurrences are omitted by the author]; (2) with an abstract noun: 'towards, aiming at (expressing the goal)' xxx xxx xxx; (3) connected with *ánu* 'along, after', see under *ánu*.
- Preposition with following ablative: (1) 'up to, as far as' (of space) xxx; (2) (of time) xxx xxx; (3) 'for, for the sake of' xx; (4) 'hither from …' xx.
- Preposition with preceding ablative: (1) 'hither from …' (of space) xxx xxx xxx; (2) 'from' (of space) xxx xxx; (3) 'originated from …' xxx; (4) '(to make) of …'

monosyllabic, and used as a genitive. There is an echo of the lost genitive form *sh_2-u̯én-s* behind the tradition (cf. Gotō in [Witzel and Gotō 1987, 814]). The form "*sū̂r*" Grassmann cites is related, e.g., to the genitive *sū̂r-as*, and besides this dative *sūr-é* with legitimate accent. They are mixed formations of *súu̯r-* as an artificial weak form of *súvar*, and a normal genitive ending *-ás*.

xxx; (5) 'from (the heart)' xxx; (6) 'away from ...' xxx; (7) 'before, above (others)', i.e. 'in higher, superior grade to others' xxx xxx; (8) 'towards ...' against the expected word-order xxx; (9) *ádhi ā́*'from ... hither' see under *ádhi*.

- Preposition with preceding locative: (1) denoting the place in which something exists, or at which something reaches (in case of verbs of motion), thus 'on, at, in, by, to' xxx xxx xxx xxx xxx; (2) 'in the case of' (some action), or 'in' (some state) xxx xxx; (3) of time: 'on, in' xxx; (4) *antár ā́* ['mid in ...'], *sácā ā́* ['together with ...'], see under *antár* and *sácā*.
- Preposition with adverbs that stand for case-forms: xxx. Combinations with other prepositions, see under the cases.
- Conjunction 'and' (1) between two jointed words: xxx xxx; (2) after the last element in case of multiple enumeration: xxx; (3) after the word (or words) in the jointed part which is identical with that in the first part according to wording or meaning, for example xxx xxx; (4) at the end or before the closing verb of the jointed part: xxx xxx.
- Emphasizing particle (1) after words of number or grade in order to express that the number or grade is reached completely: xxx; especially in the formula *trís ā́ divas* 'three times each a day' xxx xxx; (2) thus after adjective or adverb which can be emphasized in order to express a really high grade, 'really': xxx xxx; (3) beside substantive and adjective in the sense of 'representing oneself really as such', 'really as, really like': xxx xxx; after verbs: xxx xxx.

For elaborating categories such as 'preposition, particle, adverb, conjunction", further considerations and precisions might be necessary. For a dictionary, still, Grassmann's arrangement is absolutely substantial: preposition (governing the case), conjunction (small words for junction of words, sentences, or parts of a sentence, etc.), and particle (small words which express the speaker's attitude to the whole contents of a sentence), in addition, preverb and adverb. His verification and representation of the internal (meaning) and external criteria (case government, word order) are rational and compact. Here we have what remains as probably one of the best descriptions in this field even today. The works of Delbrück (1842–1922) are an irreplaceable basis in descriptive and comparative syntax, i.e. the philological, comparative and historical syntax in Indo-European languages. Delbrück [1888] often summarizes Grassmann's descriptions about prepositions in the RV, saying "I have nothing substantial to add to what Grassmann writes", and describes the findings from the prose literature of the Veda. Delbrück tries to treat the prepositions in connection with verbs as much as possible. In this sense, the basic meaning emerges more clearly in Grassmann's *Dictionary*.

One weaves the logical parts of a sentence with the help of prepositions and connecting particles (conjunctions). With the help of the modal particles the speaker adds his own attitude to the contents expressed in his speech, or "wraps" his sentence's substance. The same is applicable to the modal categories of the verbs. Grassmann is engaged in this domain with special intensity and shows his strength for providing rational and compact layout. This is due to his efforts to understand and reproduce an old foreign literature. It is very impressive that we feel as if there were nothing foreign in Grassmann's interpretation. It is as if Grassmann tried to understand the thinking and expressions of old Indian folk as his own language. He wanted to understand the RV in its entirety. For this, he did not use a poetic intuition or veil, but analyzed the expressions into their elements on various levels and constructed the thus-gained results in a clear structure. Grammar, which is not contained in school books, but in our brains, plays the decisive role. It reminds me of the method of Aristotle, but in the modern world. He has, in short, utilized all the methods of philology in an ideal way. Grassmann was conscious of the importance of providing accurate translations first, before attempting expositions of the contents and its background. His *Dictionary* is still an exemplary work on a closed text. His arrangement and the whole edifice of the work have a universal value.[8]

Abstract language and German

I do not know to what extent our logic itself depends on our mother tongue, and in some cases, also on other languages one can control if necessary. What Grassmann uses in his analysis seems to be an abstract language, as if, for example, a stranger would formulate something precisely with the most extreme intensity, but at the same time, completely. He does not avoid problems with German expression. His *Dictionary* is his message to the following generations showing how our brain should struggle and understand an appearance, an object in the aggregate. I am completely persuaded that Grassmann thought the human brain to be equipped commonly and universally since the time of the Rigveda, at the latest.

However, I must immediately add that the enterprise was made possible by German in a direct way. The fact that German preserves an old structure of the Indo-European languages and its word formation (root-suffix-ending, composition, preverbs

[8] Bartholomae was one of the best Indo-European linguists of all time and one of the best specialists in the field of Iranian studies. His *Dictionary of Old-Iranian* [Bartholomae 1904] registers the form under each meaning with its occurrence. Grassmann's principle, first the meaning with numerals, then the occurrences cited in the frame of forms with the figure of the meaning, is far more expedient. Bartholomae decided on his system probably because of the relatively small corpus of Old-Iranian texts. The inventory of the forms is important and makes things clearer. It is a desideratum to remodel the dictionary of Bartholomae after Grassmann's principle.

and prepositions, which can also be laid on each other), the genus of nouns, relatively free word order, and the usage of modal verbs – all of this favours the literal reproduction of the original text of the RV. In extreme cases, it is possible to transport the wording of the Rigveda into German and only then try to understand it. This possibility favoured the *Dictionary*. The philology of the RV is still today strongly impressed by the German language and its way of logical representation. Also, Grassmann played an important role here, alongside the *Petersburger Dictionary* by Böhtlingk and Roth (1855–1875), which was also a basis for Grassmann. With this reservation, one can say that the *Dictionary* is a most excellent exemplar in the analysis and representation of a closed corpus. It remains for us an active and irreplaceable instrument.

The decisive year of 1875

I would like to add some closing comments on the epoch we have been dealing with. I have not checked, nor am I informed about the question how much Grassmann owed to the *Petersburger Dictionary* [PW] (cf. contribution of Ms. Kozianka in this volume). The 7th and last volume (*ṣ, ś, s, h*) of the PW was printed on 7 August 1875, and has a preface dated 4 August 1875; the preceding volume (*y, r, l, v*) was printed on 8 December 1871. Grassmann's *Dictionary* has a preface dated 18 August 1872 and afterword, 8 August 1875. One may, however, safely assume from the friendship between Roth and Grassmann, which is attested by the afterword of the *Dictionary* and the letters [Petsche 2009, 167–185], that Grassmann was substantially informed of the whole material which the *Petersburger Dictionary* includes for the RV. Grassmann consulted the PW thoroughly, but he decided always for himself, as he himself says in the preface (p.V). He trusted only in his own reasoning. He did not rely on the Padapāṭha ('reading in words', i.e. segmentation into words of the RV-text, which is transmitted in contentious reading for recitation not separated word by word), by Śākalya, maybe about the sixth century B.C., nor the commentary of Sāyaṇa (fourteenth century A.D.). Grassmann profited, it is safe to assume, extensively from these traditional Indian contributions, but judged always only for himself.

The year of 1875, in which the PW and the *Dictionary* were completed, is thus an important date for the history of Indology. The next year, 1876, in which the 1st volume of Grassmann's translation of the Rigveda was published, is a revolutionary year which opened the way for the modern comparative Indo-European grammar with the monumental works of Leskien (the sound-change operates mechanically), Osthoff (the existence of PIE sonant **r̥*), Brugmann (the existence of PIE nasal sonants **n̥*, *m̥, and Brugmann's law, cf. 1.1), Verner (Verner's law in Germanic), Sievers (foundation of Phonology), Winteler (the description of Swiss German dialect), etc. (cf. "The *annus mirabilis* 1876 and posterity" by [Hoenigswald 1978], and [Schmitt 2009, 15ff.]). Grassmann's efforts in philology and linguistics seem to have been somewhat independent of the very active scholarly mainstream. How would it have been if Grassmann had written a grammar? But he lived perhaps a little too early to take on this task with his customary precision.

Grassmann's contribution to lexicography and the living-on of his ideas in the Salzburg Dictionary to the Rig-Veda

Thomas Krisch, Christina Katsikadeli, and Stefan Niederreiter

Introduction

The Rig-Veda is one of the most important religious texts of India; actually it is supposed to be the most important one. (In the year 2007 manuscripts of the Rig-Veda were added to the UNESCO Memory of the World Register.) As the oldest text of Vedic Sanskrit it is of invaluable significance for various fields of research as it is also the oldest attested text of Indo-Iranian and one of the oldest texts of all the Indo-European languages.

Some 134 years ago, Hermann Grassmann [Grassmann 1875[1], 1976[5], 1996[6] revised by Maria Kozianka] published his monumental dictionary to this important text. It still is a very useful work and one can consider it an innovative and revolutionary concept of a dictionary even from today's point of view.

H.-J. Petsche et al. (eds.), *From Past to Future: Graßmann's Work in Context*, DOI 10.1007/978-3-0346-0405-5_33, © Springer Basel AG 2011

The main part of this paper intends to show how Grassmann's dictionary has inspired us to compile a new one, which we have abbreviated as RIVELEX[1] and which adapts Grassmann's central ideas and combines them with the present state-of-the-art in Indology and Indo–European Studies.

Comparing Grassmann and RIVELEX from a modern lexicographical point of view

As modern research in lexicography underlines [Atkins 2008, passim], the compilation of a dictionary consists of three phases: (1) pre-lexicography, (2) elaboration of a macrostructure and (3) working out a microstructure. We will deal with these three stages, comparing Grassmann's concept with our own approach.

Pre-Lexicography

In the pre-lexicographical phase, one of the main points is the *motivation* for making a dictionary. Grassmann's motivation can be reconstructed from the situation of indological studies and linguistics of his time and from the hints he gives us in the preface of his monumental dictionary. Sanskrit and Indo–European studies were young but booming fields of research at that time and the study of the Rig-Veda played a central role in both of them. In the preface and the epilogue to his dictionary, Grassmann mentions another motivation: he wanted to create a dictionary as a tool for his own translation of the Rig-Veda. Our motivation at the beginning of the twenty-first century is manifold: Indo–European and indological research has flourished enormously, especially in recent years and decades. A great number of dissertations, monographs and articles have been published on Vedic Sanskrit, especially on the language of the Rig-Veda, and on Indo–European etymology and grammar, which also includes important research on the Vedic language. The bibliography of the published first volume of

[1] RIVELEX is a bundle of projects funded by the Austrian Science Fund (FWF = Fonds zur Förderung der wissenschaftlichen Forschung; Austrian Science Fund; FWF projects P14411–SPR (October 2000 – September 2002), P15833–G03 (December 2002 – November 2004), P17799–G03 (December 2004 – November 2006, prolonged to December 2007) and the still running Projects P20311–G02 (January 2008–December 2009), P 20310–G02 (November 2007 – October 2010) and P 20615–G12 (May 2008–April 2010). The editor and supervisor of RIVELEX is Thomas Krisch, contributors to the project are: Christina Katsikadeli, Stefan Niederreiter, Sabine Ziegler, Konstantinos Sampanis, Thomas Kaltenbacher (English translations).

[RIVELEX] cites a large amount of them. Another motivation is equally important for us, namely the compilation of a "multiple user dictionary" of this important text. We try to compile a dictionary which covers the interests of comparative philologists, linguists, Indologists, theologists and other philologists and classical scholars interested in the Vedic Sanskrit language and culture. In the ideal case, our dictionary should enable someone with intermediate linguistic background knowledge to access the Rig-Veda.

The second aspect to be accounted for in the pre-lexicographical phase is the definition of the *sources of evidence*. As H. Grassmann writes in his preface, his word list was taken from the great Petersburg dictionary by Otto Böhtlingk and Rudolph Roth [Böhtlingk and Roth 1855–1875] and from the glossary of Theodor Benfey's *Die Hymnen des Sâma-Veda* [Benfey 1848]. The text version of the Rig-Veda his dictionary was based on was Theodor Aufrecht's transcription [Aufrecht 1861–1863] with corrections by Max Müller. As far as we know, Grassmann did not have a complete translation of the Rig-Veda at his disposal. RIVELEX is based on the new *Rig-Vedic Word Concordance* by Alexander Lubotsky [Lubotsky 1997]. The text version of the Rig-Veda which we use is the metrically restored text by [van Nooten and Holland 1994], which also exists in an electronic version and is based on Aufrecht's edition, as in Grassmann's text. The translations we consult at present are: [Elizarenkova 1999; Witzel and Gotō 2007; Geldner 1951; Renou 1955–1969; Grassmann RV1; RV2].

Elaboration of a macrostructure

After this pre-lexicographical phase, the lexicographer has to make decisions about the macrostructure of the lexical entries, also called "lemmata". The macrostructure of a lexical entry is mainly its lay-out. As an example of Grassmann's layout cf. Sample 1.

(á-pūrvya), á-pūrvia, a., der nichts vorhergehendes oder vorzüglicheres [pūrviá] hat, 1) der *erste*, 2) der *vorzüglichste*, *unvergleichliche*, daneben oft purutáma (410,5; 473,1; 849,6).

-a 2) (indra) 641,1; 698,5.	-am [n.] 2) sumnám 638,1.
-as 1) (vāyús) 134,6 [parallel prathamás].	-ā [n.] 2) vácāṅsi 473,1; bráhmāṇi 675,11.
-am [m.] 2) agním 247,5; sárgam 410,5; stómam 849,6.	-ā [f.] 1) uṣâs 46,1.

Sample 1.

Grassmann starts with the "headword" of the entry *á-pūrvya-* with some morphological information. After that he offers a set of translations, which is followed by a block of morphological information about the attested forms. The translations and the attested forms are combined by using numbers. These are called "section marker" and "subsection markers" in lexicography. This type of arrangement is already an important innovation for Grassmann's time. The RIVELEX entry follows the same macrostructure as Grassmann in general, cf. Sample 2.

ápūrv$_{i}$ya- (adj.), m. f.(- ā-), n. „NICHT FRÜHER DAGEWESEN"[1] – "NOT HAVING BEEN THERE EARLIER"[1]

adj. Verwendung – adj. use
1a attr.
1b präd. – pred.
subst. Verwendung – nom. use
2 m. der Vorzüglichste, der Erste
▪ the most excellent, the first

Nom.Sg.m. *ápūrv$_{i}$yas* **1b** 1,134,6
Vok.Sg.m. *apūrv$_{i}$ya* **2** 8,21,1; 8,89,5
Akk.Sg.m. *ápūrv$_{i}$yam* **2** 3,13,5 (oder 1b); 5,56,5 (oder 1a)[2] ; 10,23,6
Nom.Sg.f. *ápūrv$_{i}$yā* **1b** 1,46,1
Akk.Sg.n. *ápūrv$_{i}$yam* **2** 8,18,1
Akk.Pl.n. *ápūrv$_{i}$yā* **1a** 8,66,11 (oder 2); **2** 6,32,1

Determinativkompositum aus *a-* (Privativpartikel) + *pūrv$_{i}$yá-* „vormalig, früher dagewesen, frühest" (s.d.) mit der Bedeutung „nicht früher dagewesen".

(C.K.)

[1] Auch im Sinn der implikativen Bedeutung „hervorragend" gebraucht.
▪ Also used in the sense of the implicative meaning "excellent/outstanding".
[2]Falls man *sárgam* als apokoinu zu *gávām* und zu *marútām* gehörig interpretiert.

Sample 2.

The lemma starts with the headword (**ápūrv$_{i}$ya-**) containing basic morphological information and a general translation. If there are more meanings than the one covered by the general translation, this headword is followed by a set of translations (as in Grassmann's dictionary). We chose the form of highlighting this set with grey colour. Such a type of marking is a late-comer in lexicography in the 1990s. The third section is a morphological block.

But there are also differences in the macrostructure between Grassmann and RIVELEX. The most obvious one is that we have a fourth section in addition to Grassmann. The fourth section (written in smaller characters) contains information about the linguistic structure of the lexical entry and its etymology. One can find such information also in Grassmann, but not systematically offered within each lexical entry. He puts this information into his headword section.

There is another feature in the macrostructure of RIVELEX which has no counterpart in Grassmann's dictionary. We offer endnotes below the respective lexical item which deal with the philological interpretation of difficult passages, linguistic analysis and other remarks which we consider important to the user. Finally, every lexical item is signed by the respective main contributor.

In general, we try to follow a more rigid "style guide" than Grassmann. This is in accordance with modern dictionary making. Applied to the information about the linguistic structure of the lexical entry and its etymology, this means that we include this feature in every entry, also in cases where the facts may be completely clear to specialists. RIVELEX also provides negative information in the etymological section, i.e. it is stated explicitly when an etymology is still unclear. Another difference occurs within the morphological section: RIVELEX offers all attestations of every entry in the text, while Grassmann reduces the number of attestations in extensive entries. In short entries, as here in ***ápūrv$_i$ya-***, he lists all the attestations like RIVELEX. Nevertheless Grassmann offers a large number of attestations also for extensive entries, like pronouns or proper names, for example names for gods. This also has to be considered an innovation in Grassmann's times. Furthermore, inside the morphological section Grassmann abbreviates the forms by putting a hyphen plus word-ending, while he mentions the word stem as a headword, cf. Sample 1. RIVELEX, on the other hand, cites the full form of the words including a morphological analysis of every attestation, cf. Sample 2.

Working out a microstructure

After the pre-lexicographical phase and the macrostructure, the lexicographer has to make decisions about the *microstructure* of the lexical entries. Since the microstructure fills up the macrostructure, we will discuss it according to the sections of the macrostructure.

Inside the headword, cf. Sample 1, Grassmann obeys the following order: lexical entry in the form of a stem followed by grammatical information, here "adjective". In the case of *ápūrv$_i$ya-* the headword also contains a short analysis of the word. Grassmann does not offer such information for every lexical entry. RIVELEX has a strict order inside the headword: the lexeme is presented in the form of a stem which is indicated by a hyphen.[2] This lexeme is followed by an indication of the part of speech (here

[2] In the case of verbs, the lemma is the verbal root as it appears in Mayrhofer's etymological dictionary [EWAia, 1986–2001].

"adjective") and of the gender – in this case all *three* genders, namely masculine, feminine (including declension type) and neuter. The grammatical information is followed by a general "core meaning" (according to the tradition in semantics, written in capital letters). This accounts for the claim in modern lexicography, that one of the most important tasks for the lexicographer is to "[m]ake sure that TL [target language] words given as direct translations are general enough to suit most contexts" [Atkins and Rundell 2008, 212]. Another important feature of the direct translation in RIVELEX is that it appears in German and in English in order to provide broad usability, ideally for users worldwide. Inside the "grey box" (the so-called "menu" in lexicography) RIVELEX offers semantic and syntactic information in German and in English. In the case of an adjective, we distinguish between adjectival and nominal use. Inside the section of the adjectival use, we distinguish between attributive and predicative use. In Vedic it is not as easy to distinguish between these uses as it is in German and in English. We give our criteria for this distinction in Vedic in the preface to RIVELEX. Grassmann also has a similar section where the headword is divided into senses, but this section is less rigid in structure than ours and less theoretically oriented, for example Grassmann classifies *ápūrv*$_{i}$*ya-* as an adjective but translates nominally "der erste" ("the first one"), "der vorzüglichste" ("the eminent one").

In RIVELEX, the morphological section below the grey box contains paradigmatic and syntagmatic information, cf. Sample 2. The paradigmatic data in RIVELEX consist of all the attestations of the word in the Rig-Veda, ordered by morphological categories like case and number. This information is cross-classified – by numbers and letters – with the classifications contained in the grey box. Thus, the nom.sg.m. *ápūrv*$_{i}$*yas* is classified as 1b (predicative use). The three numbers, separated by commata, indicate the commonly accepted numbering of a passage in the Rig-Veda.

Such a cross-classifying structure of the morphological section was Grassmann's innovation, cf. Sample 1. RIVELEX has adopted this notable advantage of Grassmann's concept because it increases the clarity and conciseness of the presentation.

But there is a difference concerning the presentation of the morphological data between Grassmann and RIVELEX: Grassmann, cf. Sample 1, presents morphophonological information first, for example *-a, -as, -am,* etc., whereas RIVELEX analyses the forms: Nom.Sg.m. etc. Moreover, RIVELEX is also produced in electronic form, and so one can search for grammatical information as well. Hence, in addition to a dictionary one also obtains a minimal grammar of the Rig-Veda. Grassmann sometimes also adds grammatical information, but not systematically, cf. Sample 1: in only four of the six instances Grassmann indicates the gender. He does not add any information about case and number.

Concerning syntagmatic information, there is another fundamental difference between Grassmann and RIVELEX: the latter does not provide explicit microcontexts of the respective attestations. Grassmann mentions such collocations, but without theoretical background. This information is very useful for the researcher, but it does not spare the scholar from looking up the original text[3]. In fact we do provide some syntagmatic information about the context, for example in verbal lexical entries, but we do so in a more theoretical way and not inside the morphological section.

Another difference between RIVELEX and Grassmann is due to technical innovations: RIVELEX also provides an electronic version on CD-Rom (Word-Format and PDF file), especially useful for the search of attestations and morphological and syntactic categories.

The section below the morphological component has the following microstructure. First, information about word formation is provided, in difficult cases with more detailed argumentation. Second, the etymology is dealt with tersely; RIVELEX offers a more extensive discussion with additional comparative data in unclear cases only. The lemma in Sample 2, again, can serve as an example: *ápūrv$_i$ya-* is characterized as a determinative compound formed by the negative particle *a-* "not" plus the adjective *ápūrv$_i$ya-* "earlier". In this case one obtains no information about etymology, because this may be found inside the lexical entries *a-* and *pūrv$_i$yá-* respectively.

The last section of a lexical entry in RIVELEX contains the "endnotes". Their microstructure consists of a discussion of specific philological problems, translation and interpretation of difficult or unclear passages and linguistic remarks. In the case of *ápūrv$_i$ya-*, cf. Sample 2, endnote 1 deals with semantic information in addition to the translation. Since this endnote is added to the headline, it is also translated into English: "Also used in the sense of the implicative meaning 'excellent/outstanding'". Endnote 2 refers to a specific passage. This type of endnote appears in German, only. In this case, the English endnote reads: "if one interprets *sárgam* as apo koinou to *gávām* and to *marútām*". *Apo koinou* means that one part of speech is shared by two clauses. This is a device normally used in poetry. An English example of this is: "There were three crows sat on the tree" [Webster's 1981, 101]. In Sample 3, the part of speech which is common to both clauses (*sárgam*, the Vedic word for "herd/swarm/flock") is written in italics.

[3] If we had included this feature in RIVELEX, this would have stretched the dictionary considerably. The collection and analysis of rigvedic collocations are topics of research of other projects, like the on-going project "Family-grammar of the Rgveda" at the University of Zürich under supervision of Prof. Dunkel, http://www.research-projects.uzh.ch/p7734.htm seen: Nov. 23 2009).

RV 5,56,5 :
[út tiṣṭha nūnám eṣāṃ stómaiḥ sámukṣitānām]|
marútām purutámam ápūrvyaṃ gávāṃ *sárgam* iva hvaye ||
[first line: "Now stand up with (songs of) praise for them who grew up together. (the poet speaks to himself)]

Sample 3.

The difficult construction appears in the second line: Without assuming an *apo koinou*-construction the interpretation of the second line would be: "I call the manifoldest one of the Marut, the excellent one, like a herd (*sárgam*) of cows". With *apo koinou* the interpretation would be: "I call the very numerous excellent [swarm, *sárgam*], of the Maruts, like a herd (*sárgam*) of cows".[4]

Adjectives normally are quite unproblematic in comparison to other parts of speech. Verbs especially are very difficult, because there are so many grammatical categories to deal with, cf. Sample 4.

arc- (v.) **facientiv-transitiv/intransitiv „SINGEN, PREISEN" – "SING, PRAISE"; ánu** (etw. nachsingen, verkünden – repeating s.th. in speech song, proclaim); **abhí, prá, abhí prá** (besingen, preisen – sing of, praise); **sám** (gemeinsam, allgemein besingen – sing of s.th./s.o. in common, in general)
Tiefenkasus-Schemata (semantische Rollen):
• Deep Case Frames (Semantic Roles/Theta Roles):
ACTOR (– EXPERIENCER/BENEFACTIVE) (– THEME) **1** ACTOR (- THEME); **2** ACTOR– EXPERIENCER / BENEFACTIVE; **3** ACTOR – BENEFACTIVE / EXPERIENCER – THEME
1a ACTOR – THEME „ein (Lob-)Lied singen" ACTOR = Nom. +bel., -abstr.; THEME = Akk. -bel., -abstr. (**Simplex; abhí; prá**) Aktiv
• ACTOR – THEME "sing a song (of praise)" ACTOR = Nom. +bel., -abstr.; THEME = Akk. -bel., -abstr. (**Simplex; abhí; prá**) active ... etc.
2 ACTOR – EXPERIENCER/BENEFACTIVE „jmdn./etw. besingen, preisen"
• ACTOR – EXPERIENCER /BENEFACTIVE "praise s.o./s.th. in song, praise"
... etc.

Präsensstamm themat. (I. Kl.) (*arc-a-*):
**Aktiv
*Indikativ Präs. 1.Sg. *árcāmi* **2aa** 1,138,1; **2b** 4,4,8[1]; 10,12,4; etc. ...
*Imperfekt 3.Sg. *árcat* **1b** 5,33,9; **1c** 1,120,3 (prá); 3.Pl. *árcan* **1c** 1,52,15; 5,45,7; **2aa** 5,29,2 (ánu, Tm.)
*Injunktiv 3.Sg. *árcat* **1c** 1,173,2; *arcat* **3a** 1,165,14; **3a** 1,62,1; *arcāma* **2aa** 9,97,4 (abhí); etc. ...
*Partizip Nom.Sg.m. *árcan* **1c** 3,31,8; 10,149,5; **4** 3,31,7; Dat.Sg.m. *árcate* **4** 1,87,2; etc ...
**Medium
*Indikativ Präs. 1.Sg. *arce* **2aa** 5,41,8 (abhí, Tm.); **3c** 6,22,1 (abhí)
*Prospektiv 1.Sg. *arcase*[2] **3c** 10,64,3 (abhi)

Perfektstamm (*ānṛc-*[3]):
**Aktiv
*Indikativ Perfekt 3.Pl. *ānṛcúr* **1a** 1,19,4; **2aa** od. **2ac**[4] 5,6,8; etc. ...

Vgl. VIA: 166f.; EWAia s.v.; MAYRHOFER 2005: 13. Iir., vgl. khot. *āljs-* „singen"; idg. *(h₁)erk*- „singen, preisen", „strahlen"(?). Der Befund könnte zum Ansatz zweier ursprungsverschiedener Wurzeln (*arc*¹-, *arc*²-) verführen, da sonst Nomina wie *arci(s)-* „Flamme, Strahl" schlecht erklärbar wären. Etc. ...

(S.N.)

[1] *árcāmi te sumatím* „ich lobe deine Güte"; *te* wird hier als Gen. aufgefasst; Ge. übersetzt dativisch.
[2] vgl. RASMUSSEN 1985: 390; vgl. auch GOTŌ 1987: 98f.
[3] Lokalanalogie nach den Perfektformen von *añj-* „salben, verherrlichen", vgl. KRISCH 1996: 40f.
[4] zu ergänzen *tvā* (Akk.) oder *te* (Dat.).

arc, ṛc. Erweiterung aus ar. Die Grundbedeutung „in Bewegung setzen, abschiessen" findet sich in AV. 1, 2, 3, wo es vom Abschiessen des schwirrenden Pfeiles gebraucht wird, und wol auch AV. 12, 1, 39, wo es mit

mit sám (ganz ähnlich wie ar mit sam 4 u. 5) „feststellen" (die beiden Welten durch Stützen) bedeutet. Also 1) *abschiessen*, AV.; 2) *strahlen, glänzen*; 3) *singen, lobsingen*; 4) *singen*, einem Gotte oder göttlichen Wesen [Dat.]; 5) *singen*, ein Lied [Acc.]; 6) *singen*, einem

ánu, jemandem [A.] *zujauchzen*.
abhí 1) *singen* (ohne Cas.); 2) *singen*, ein

pra 1) *vorleuchten* (265, 8); 2) *anheben zu singen*; 3) besonders: einem Gotte [D.];

Stamm **árca**:
-āmi 7) (pūṣáṇam) 138, 1. — 8) sumatím 300, 8; ápas 838,4.
-ati 6) te práçastim 538,3.
-athas 5) gātúm 151,6* (mitra varuṇa).
-āmasi 7) tuā 462,6.
-anti 1) çarám anusphu-

arkám 10,1; 166,7*. — 6) túbhya arkám 384,6*. — 7) tvā 383, 1*; índram 6*; vājínam 973,3.
-at 2) vṛ́ṣā (índras) 173, 2*.
-āma 5) sáma 173,1. — 6) arkám náre 62,1.

Sample 4. *Left*: RIVELEX, excerpts. *Right*: Grassmann, excerpts.

[4] Toshifumi Gotō draws our attention to the fact that normally only one word of a constituent is put in front of the particle *iva* and thus one would expect * *gávām* iva *sárgam* instead of *gávām sárgam* iva. This claim accounts for almost all the attestations of *iva* in constructions like that, but there are exceptions where *iva* occurs after two words belonging to the same phrase (e.g. 1,117,18; 1,130,9g; 8,1,17). [Witzel/Gotō 2007] translate the passages 1,117,18 and 1,130,9g differently.

If one compares Grassmann with RIVELEX one may notice that RIVELEX has adapted Grassmann's idea of cross-classification not only for semantics and morphology but also for the combination of the verbum simplex with preverbs, like *abhí.* A feature in addition to Grassmann is the use of semantic theta-roles (deep cases) utilized by the most prominent linguistic paradigms both in Europe and in America (e.g. Fillmore, Chomsky, Radford). We apply the following roles and assign them to grammatical case-forms.[5]

- ACTOR: animate or inanimate performer of an action. The semantic role incorporates the classical role of AGENT, who is animate and performs an action deliberately with the metaphorical use of AGENT for inanimate forces.
- THEME: person or thing that is affected or created by an action or an event/state. If there are two THEMES (i.e. with causative verbs or with the verb "to be") then this is described as "THEME 1" and "THEME 2" (in predicative function), which indicate different roles. The verb "to be" links rather than assigns semantic roles. We have refrained from introducing another semantic case frame label for cases like that.
- EXPERIENCER: bearer of a state of mind. A semantic role experiencing a psychological state leading to an experience.
- BENEFACTIVE: animate beneficiary/"maleficiary" of an event, an action or a state.
- LOCATIVE: place of an event/state.
- SOURCE: source of a movement/action.
- GOAL: goal of a movement/action.
- PATH: the way along which a movement takes place.
- INSTRUMENT: means for fulfilling an action.

RIVELEX also includes modern insights into the structure of the Vedic language, for example the so-called "prospective" (category introduced by [Rasmussen 1985]), in our Sample 4.

[5] There are evident semantic distinctions between semantic roles and grammatical case, e.g. the semantic role of ACTOR and EXPERIENCER both appear in nominative case, cf. the following examples: *the boy* (nom., ACTOR) *hits the ball* (Acc., THEME) vs. *the boy* (nom., EXPERIENCER) *sees the ball* (Acc., THEME).

Final remarks

We want to stress how fortunate a lexicographer of Vedic can be to have Grassmann's dictionary as a great example which combines mathematical systematicity with a user-friendly interface. Other philologies of ancient languages do not have this advantage, and therefore dictionaries to ancient texts still follow the traditional style of compilation: they did not have such an innovator as was Grassmann. Grassmann's concept corresponds to concepts discussed in recent lexicographical literature and some of his features gradually find their way into modern bilingual dictionaries and computerized thesauri. Grassmann's merits can be summarized in one sentence: His dictionary combines almost all the features a corpus dictionary can provide in macrostructure and microstructure. Finally, we want to emphasize another practical aspect of his achievements: Grassmann managed to complete his dictionary in an incredibly short time. Sue Atkins, a famous lexicographer of our times, remarks "the scholarly and historical dictionary is [...] a work often with few length constraints, and sometimes little pressure to complete within a specific time period – but also with a tendency to run out of money around letter C, or take 50 years to get there." [Atkins 2008, 31].

We hope that RIVELEX, the Salzburg Dictionary to the Rig-Veda, will prove to be nearer to Grassmann's brilliant example than to the numerous cases of dictionary projects which Sue Atkins describes.

Illustration 14. Hermann Grassmann was a collector of Pomeranian folksongs. For "singing in the family", he arranged them for three voices. The song reproduced here was one of Grassmann's favourites. See also on p. 502

Hermann Grassmann's impact on music, computing and education

Calculation and emotion: Hermann Grassmann and Gustav Jacobsthal's musicology

Peter Sühring

It is quite remarkable that for many generations and in many different genealogical lines of the Jacobsthal's family clan of academics, scholars dedicated their energy to three specific disciplines: medicine, mathematics, and music.[1] This essay aims to analyze the connections between mathematics and music as they appear in the intellectual history and biography of a Stettin pupil of Hermann Grassmann: Gustav Jacobsthal. Jacobsthal's son Walther was noted in his day for being a competent mathematician and headmaster of a school in Berlin, while his nephew Ernst Jacobsthal became a very well-known mathematician. Ernst Jacobsthal received international acclaim while living as an emigrant in the Norwegian town of Trondheim. His achievements lie in the field of number theory. Before he fled the Nazis, he taught at the Berlin Polytechnic University. If Grassmann had had his way, Ernst's uncle Gustav Jacobsthal would also have become a mathematician. But Gustav followed the advice of his music teacher, Carl Loewe, who in this case was competing with Hermann Grassmann for the young man's attention. Without giving up his interest in mathematics and in its potential for applications in musicology, Gustav Jacobsthal chose a career in musical research. He became one of the

[1] See in this context [Sühring forthc.].

H.-J. Petsche et al. (eds.), *From Past to Future: Graßmann's Work in Context*,
DOI 10.1007/978-3-0346-0405-5_34, © Springer Basel AG 2011

German Empire's most prolific musicologists and historians of music – even though much of his work was never published.

From 1872 to 1905, Jacobsthal was a professor of music in Strasbourg. His teaching covered many aspects of the history of music, ranging from Antiquity to Beethoven. But his specialty was research into the music of the Middle Ages. In 1897, he became Germany's first full professor of musicology, an academic discipline which had only been recently established. Besides his dissertation, two essays from the early days of his academic career and a carefully phrased edition of motet song-lines in Old French and Latin, he only published one book. As a consequence, a large segment of his insights remained confined to his unpublished papers, which consisted of handwritten notes and sketches of his lectures. Therefore, on the one hand, Gustav Jacobsthal's name fell into oblivion after his death. On the other hand, some of his pupils illegitimately claimed to be his intellectual successors and distorted his initial intentions and concepts.[2]

In 1858, at age 13, Gustav Jacobsthal came to Stettin in order to attend the Gymnasium. Five years later, in spring of 1863, he had completed his secondary-school education. In Stettin, Jacobsthal lived at an uncle's house. It must have been a quite crowded environment, since Jacobsthal usually did his homework sitting on the staircase, surrounded by children. One might imagine the serious and diligent secondary-school pupil poring over Grassmann's *Textbook of Arithmetic* [LA], trying to concentrate despite the noise or the occasional stray ball: a tough exercise in persistence, which would prove helpful in later years.

Jacobsthal's son Erwin was a medical doctor, and it was in biographical notes of Erwin's that the first clues hinting towards Jacobsthal's teacher from Stettin appeared. Exiled in Guatemala, Erwin Jacobsthal composed an autobiographical essay in 1946, which he entitled *Ups and downs of a life*:

> "When he was still quite young, my father had to leave his parents' house in order to attend the famous Marienstifts-Gymnasium in Stettin. Apparently, he possessed a special gift for two subjects: mathematics and music. In both, his teachers urged him to consider a career. In mathematics, this had been Gassner [sic!], who was so far ahead of his time that his books were destroyed because nobody could understand them; later on, the few remaining copies aroused attention and were republished by the Berlin Academy of Sciences [sic!]. And who had been the musician? The great Carl Loewe, a composer of immortal ballads. But there was no way the boys in school could have fathomed the greatness of their music teacher." [E. Jacobsthal 1945–1946].

[2] See in this context the reproductions of Jacobsthal's notes and sketches in [G. Jacobsthal 2010].

And we find the following passage in a second text from 1947:

> "It was no coincidence that these two men played such an influential role. Both were trying to persuade my father to get involved in their own academic fields. But, later on, my father actually combined the two disciplines in his own peculiar way. He was a follower of Dove and Helmholtz. Later, he also felt close to Stumpf's line of thinking. Therefore, he used mathematics and physics as a foundation for the physiology of the sense-perception of music and dedicated many hours of his personal studies to this approach." [E. Jacobsthal 1946–1947].

These two documents show some degree of imprecision, given the fact that the author's memory had faded (Grassmann had been a member of the Göttingen Academy of Sciences and his works had been published by the Academy of Saxony), which is also why Grassmann's name is not rendered correctly. But they nevertheless clearly indicate that Grassmann was a strong influence in Jacobsthal's intellectual development. Even though much work remains to be done, this paper will try to highlight and describe some aspects of this influence.

In 1858, Hermann Grassmann had already spent six years as his father's successor at the Stettin Gymnasium, teaching mathematics and physics. Another two years would pass before he published his first textbook of mathematics [LA]. It was in this period of time that Grassmann supposedly wrote the words to a song for piano and choir. The music was by his friend and colleague Carl Loewe. 13 years his senior, Carl Loewe had taught the gifted young man to play piano and bass in the Grassmann family home. The song is entitled "Das 'Dolce far niente' ". The song's second stanza alludes to Newton, which prompted the editor of Loewe's songs, Maximilian Runze, to surmise that Grassmann might not only have performed the song with his choir – the local men's choir, also known as the *Holy man's Club* (*Pater-Verein*) – but that he might have in fact written the words [Runze 1899, XVI]. It is clear that Grassmann admired Newton. In 1852, he had written a contribution to Newton's theory of color, which he had first presented to the Stettin Physics Society and then, after some modifications, published a year later in volume 89 of the journal *Poggendorffs Annalen der Physik und Chemie* [H. Grassmann 1853]. The article is a masterpiece of scientific diplomacy and persuasive arguing: On the one hand, Grassmann protected Newton against criticism brought forward by Helmholtz by proving that, in the end, both theorems of color-mixing led to the same conclusion. Therefore, Grassmann wrote, Helmholtz's "observations, instead of weakening [Newton's] theory, could actually serve to support or complement it" [H. Grassmann 1853, 69]. But, on the other hand, Grassmann urged the reader to rethink Newton's theory completely. Helmholtz, Grassmann believed, had taken the first steps in that direction. Grassmann wished to revise the way in which Newton distributed homogeneous colors on the circumference of his circle. We cannot elaborate on this. In any case, Grassmann concluded:

> "Only when this question has been sufficiently discussed can we dare to try and find the law according to which the oscillations of the ether, which correspond to the different colors, come together in the nerves or sensory organs to form simple color-impressions. The theory of different colors and colorless light depends on how we solve this problem." [H. Grassmann 1853, 84].

It is remarkable that, at an earlier point, Goethe had already discussed that question. But there can also be no doubt that in 1852 Newton's and Goethe's controversy concerning the theory of color was no longer a pressing issue among physicists. Goethe had also worked on a theory of tonality while dealing with acoustics, an interest Grassmann and Goethe shared. Needless to say, Grassmann's efforts were much more relevant to Jacobsthal than Goethe's theory of color.

One will have to wonder whether or not the combination of empirical laws and mathematical calculations, which in this case referred to colors, might also hold in the realm of musical tone and sound. One could ask whether the multidimensionality which Grassmann developed for spatial extensions might also be applied to the temporal extensions of multi-dimensional music: polyphonic music, that is. In 1876, Wilhelm Preyer was attempting to apply Grassmann's extension theory to sensation, mostly to the perception of color. This was an effort Grassmann was aware of and which he commented upon. But in that same year, in a tract entitled *The limits of audible perception* [Preyer 1876b], Preyer also undertook calculations concerning oscillations and the limits of their audibility. He found the lower limit to be 14–24, the upper limit to be 40,000 oscillations per second. In 1877, in his *Elements of a pure theory of sensation* [Preyer 1877], Preyer noted that the perception of contrasts required the subject to perform an act of multiplication. This insight could also be applied to the perception of sound. Grassmann wrote an appendix to the book [H. Grassmann 1877b].

If we wanted to place ourselves in the position of the young Jacobsthal and search for the first elements of acoustics, we would quickly encounter a pedagogical outline of acoustics which Grassmann wrote for the Stettin Gymnasium. Jacobsthal learned how important it was to avoid unwanted elements in vocal sounds when singing, namely the hissing and aspirated sounds so common in German. In this text, Grassmann explicitly declared "how much time and effort" music teachers had to "invest in order to eliminate them." [H. Grassmann 1877d, 609]. One can imagine Grassmann toiling away as the leader of the school's choir, of which young Jacobsthal was a member and thus also the object of Grassmann's pedagogical efforts. In later years, when Jacobsthal was a member of the Berlin Singakademie and the leader of the academic choir of Strasbourg

University, he must have dealt with the same kind of problems. This theory of singing was based on what Grassmann had established as early as 1854 on the last pages of his *Outline of acoustics and basic optics* [H. Grassmann 1854]. It was here that he had made first steps towards a theory of vocal tones, to which he proudly referred in 1877 in one of his last essays as a "pedagogical guideline of acoustics written mainly for my pupils" [H. Grassmann 1877, 606]. He also remarked that this theory of 1854 had not been called into question by the work of Helmholtz. The introductory words to this theory were the following:

> "The vocal chords cause the air within the mouth cavity to oscillate, which also creates softer secondary tones. The variety of these secondary tones depends on the make-up of the mouth cavity. They belong to the harmonic series of tones from the musical scale which has the tone generated by the vocal chords as its keynote. This is how vowels come into being."[3]

When in 1877 – the year of his death – Grassmann humbly stated that his theory had gone completely unnoticed, he must have limited his judgment to the scientific community. Jacobsthal was on the way to becoming a member of this community, but even though he never discussed Grassmann's theory in any kind of publication, Jacobsthal nevertheless was aware of its practical applications and used it whenever he needed to. From the earliest days of his secondary-school education in Stettin, Jacobsthal had known about an advanced physical and physiological theory concerning the generation of vocal sounds, which was repeatedly put to the test in the school's choir and in singing lessons. Jacobsthal always considered singing the basis of music, even though he also learned to appreciate the historical evolution of instrumental music. His teaching obligations at the university included singing lessons. Grassmann and Loewe attributed a dominant role to singing as well. After arriving in Berlin, Jacobsthal took lessons with Heinrich Bellermann and Eduard Grell (at the time the director of the Berlin Singakademie). This influence strengthened his views on singing, which led him to the following statement in the second thesis to his 1870 dissertation: "It is impossible to attain a deeper understanding of music merely by mastering an instrument. Rather, this understanding arises from the theory and practice of vocal music." [G. Jacobsthal 1870]. And even in a memorandum for the Prussian Ministry of Culture dealing with the question of how the teaching of music at university level could be improved, written in 1883 by Jacobsthal in Berlin, his advice is reminiscent of these initial influences:

> "The artful laws of its own development, the musical forms, have arisen from a concept of beauty that we cannot find in nature. Rather, with great rigor, music has

[3] He quotes himself [H. Grassmann 1854] in [H. Grassmann 1877d, 606f.].

> followed its own laws. But it is also true that in earlier days of music, since up until the sixteenth century it was mostly vocal music, it did parallel the development of language and poetry. For both, speech and singing, rely on the same medium to express themselves, namely the human voice, the vocal sounds. Therefore the musical repertoire of forms owes a lot to language and poetry, and poetry owes a lot to music." [G. Jacobsthal 2002, 304].

Grassmann does not directly concern himself with how the ear and the sensation of sound influence the artistic process of structuring music. But when he aimed to lose the unwanted sounds in singing, he was nevertheless following an ideal of beauty that ran against the grain of nature and which Jacobsthal would have considered the consequence of a historical evolution. In fact, the influence of Goethe's approach to a theory of tonality ("Tonlehre") on Jacobsthal and probably even Grassmann has been massively underestimated so far. Goethe had tackled his "Tonlehre" in 1812, immediately after having completed his theory of color. Riemer published it in 1834 along with the correspondence between Goethe and Zelter. Grassmann must have read Goethe's text and appreciated it, for he chose the same point of departure, the human ear and its capacity for perceiving sound. Today, it reads like an apologetic of Goethe's approach when, in his 1877 *On the physical nature of the sounds of speech*, Grassmann wrote:

> "The human ear shows an extraordinary capacity to perceive these simple tones and, after some well-executed training, to differentiate them clearly. In this respect, if the cacophony of sound is not too great, it is far superior to all man-made instruments we know to date. Therefore, my theory is grounded in the immediate perceptions of the ear. These perceptions are by no means purely subjective impressions, but they can be objectively communicated to a larger audience without the aid of man-made instruments." [H. Grassmann 1877d, 609].

Goethe also emphasized that his theory was "grounded in experience" and that it could be subdivided into three scientific categories, each of which represented a type of musical sensation: (1) organic (subjective), (2) mechanical (a mixture), and (3) mathematical (objective) [Goethe 1970]. According to Goethe, the world of sound begins with the voice (point of departure of his theory of singing), reaches us through the ear (acoustics) and affects the entire body (the theory of rhythm). This is how Goethe described the ear and its susceptibility:

> "The ear being a higher organic entity, we must say that it follows its own laws of susceptibility, which alone enables it to perceive and structure that which reaches it from the outside." [Goethe 1970, 135].

Even though Goethe admitted that the objective dimension of the world of sound could be grasped mathematically, he nevertheless was less open to mathematics as such and often quite hostile towards the mathematicians of his time. As far as I know, we find Goethe's most violent attack in a letter to Zelter, dated 28 February 1811. The context is the controversy aroused by his theory of color:

> "Mathematicians are a silly lot. They are so far removed from even the slightest notion of what is essential here that one cannot even condemn them for their prejudices. It will be interesting to see who among them will be the first to understand the matter, and do so gracefully: not all of them are blind to the facts, and not all of them have bad intentions. In any case, this just tells me something I have known for a long time, namely that the entire mathematical approach to the world is one-sided and uninspiring." [Goethe 1991, 249].

Goethe must have felt that mathematicians were incapable of appreciating the subjective aspects of the world of sound. He might have changed his mind if he had had the chance to learn about Hermann Grassmann and his specific views on acoustics.

But Jacobsthal remained true to Goethe's views in yet another point. Jacobsthal's Berlin teachers had led him to disregard instrumental music, but Jacobsthal quickly changed his mind when he realized how important it had been during the Middle Ages. Jacobsthal became aware of the aesthetical importance of instruments in music, especially opera, after Monteverdi's exquisite instrumental accompaniment to *L'Orfeo*. The thoughts expressed by Goethe in his sketches of the "Tonlehre" must have played a part in this. In section two, "the mechanical," Goethe explained the following things about instruments:

> "Their relationship to the mathematical: instruments arise from insights into the proportions of units and numbers. Through diversification, they contribute to this type of knowledge. Their relationship to the human voice: they replace it, they accompany it discretely, but, as with the voice, a knowledgeable and sensitive hand guides them to higher levels." [Goethe 1970, 136].

We can find a remarkable analogy (parallelism or identity would be saying too much) between Grassmann's work on extension in multi-dimensional space and Jacobsthal's research on early forms of polyphonic singing in the Middle Ages. Jacobsthal examined the various ways in which the voices in early motets could come together. He did so by taking the combinations of temporal durations into account as they were expressed in the individual voices and their notation. Therefore, Jacobsthal concentrated on a codex of motets from the twelfth and thirteenth centuries which represented the greatest stylistic and experimental variation concerning the interplay of the individual voices: the codex Montpellier H 196. Basically, a motet resembles an early version of what would later become the complex ensemble of Mozart's operas. Four singers in different voices sing, simultaneously, four different texts. Maybe Grassmann would have considered just this the kind of "cacophony" which, according to him, was too complex for the human ear to untangle. Unfortunately, Jacobsthal's extensive preliminary studies never led to a complete analysis of the Montpellier codex, see [G. Jacobsthal 2010, 11–235]. But Jacobsthal managed to show how the great rhythmic and harmonic variation in the interplay of the individual voices still creates an artful and unified structure. This did not necessarily require one voice to be rhythmically dominant, which traditionally would have been the tenor, with its melodic structure dating back to Gregorian singing. Nor did this structure arise from one of the typical rhythmic patterns of the time, which consisted of sequences of longer or shorter durative units (longa and breve). Occasionally the melody or cantus firmus was even complemented by a rhythm taken from the chorus of a – sometimes obscene – French popular song or from an anticlerical tune. The meters derived from these songs usually arose from the vernacular language.

Grassmann's calculations were linked to displacements and, at a later point in time, multi-dimensional space. By taking their directions into account (vectors), the multiplication of displacements of opposite orientation in space could produce a shorter, or even a negative displacement. Jacobsthal's approach to polyphonic motets followed a similar path. One will have to admit that the temporal development of a piece of music is irreversible. But Jacobsthal's calculations described the unfolding of polyphonic musical structures by keeping in mind the combinations and relations of their polyphonic architecture. This meant that, even though the unified whole of the piece remained intact, rubatos ("stolen" temporal units) could occur within the polyphonic structure of the piece. Rubatos were caused by inevitable deviations from tempo. These deviations were inevitable because there was no other way to create a well-structured piece and harmonious endings in vocal music, given the fact that the number and length of syllables and the rhythmic accent of the song's words were not the same in each individual voice.

Jacobsthal's scientific approach to music was characterized by an analytic mindset which usually permitted him to present more than one solution to a problem. He never shied away from the statistical side of his work, which led him to carefully establish comparisons and solutions. Jacobsthal did not prefer the emotional interpretation of music to its mathematical description, but this openness sometimes led to conflicts between the two points of view. A lecture on an opera which Mozart composed in his childhood provides a typical example. Jacobsthal criticized young Mozart and his very early opera in Latin, *Apollo et Hyacinthus*, for over-emphasizing the dramatic importance of one of its figures, the angry princess Melia. According to Jacobsthal, this exaggeration prompted Mozart to make a blunder in the rhythmical structure of the piece. This is how Jacobsthal justified his somewhat pedantic views:

> "When it comes to analyzing the relationship between words and music in a musical composition, I always find it very useful to write down just the words and, if the text is in verse, the individual lines. This is what the composer had in front of him when he was looking for something to work with. Our work becomes much more precise when, just like the composer, we have the words in front of us. Neither do I demand, nor do I wish to recommend a painstakingly precise philological approach, which cold-bloodedly aims to understand and dissect a work of art while it in fact rips it to pieces. Rather, it is a simple and necessary tool for us to observe the composer at his desk and follow his steps. [...] Does this mean that the composer, in planning or writing a section of his work, proceeded cautiously and thoughtfully, or that he should have done so? And does this mean that Mozart committed the blunder we uncovered because he did not proceed cautiously and thoughtfully?

One might be led to believe that I was going to answer 'Yes' to both questions. But then one could tell me: this is not how works of art come into being, and a man who propagates this view by answering these questions affirmatively knows nothing about art and nothing about how a work of art comes into being. His view of artistic creation stands for nothing but calculated experimentation, it resembles a mathematical exercise."

Jacobsthal then went on to say:

> "But my answer to the questions is 'No', and let me comment on and justify my torturously slow and maybe overly thoughtful way of analyzing the passage we are dealing with. Whoever wants to understand how a work of art came into existence, which forces helped bring it to light, whether its stylistic elements are well-chosen or not, will have no other choice but to reflect on, examine, experience and retrace the path which the artist took in creating the work of art in his artistic flight, that is, by letting artistic rationale domesticate his imagination." [G. Jacobsthal 1888].

But there were times when the clear-minded mathematician was confronted with the limits of rational understanding and had to rely on his imagination. His son Erwin tells the story of such an incident from Jacobsthal's life as a scholar:

> "He had experienced a strange psychological phenomenon. He spent about 15 years of his life trying in vain to find a solution to a problem in his field, constantly working away at it. [This must have been a problem which posed itself in the context of the original melodic structures of liturgical songs, which had only been inadequately transmitted to posterity and which Jacobsthal had rediscovered.] Suddenly, while he was asleep at night, the solution dawned on him. Half awake, he fumbled for a pencil and jotted down his idea. [The next morning,] much to his surprise, he discovered the solution to his problem on the cabinet

> next to his bed. This unsettling experience probably tells us that imagination and rational understanding are, in a certain sense, opposites. Systematical work on the problem had limited the powers of the imagination. But, on the other hand, without the preparatory work of rational understanding, the imagination could have done nothing to help." [E. Jacobsthal 1945–46].

A number of things we know about Jacobsthal's everyday life indicate that mathematics remained a constant companion during his days as a scholar. Bibliographical references to mathematical works and questions concerning mathematics keep turning up in his unpublished papers. Again, his son Erwin is a valuable source when he tells the story how Jacobsthal was very capable of dealing with mathematical problems when his children needed help with their schoolwork:

> "When my grandfather fell ill with gangrenous pneumonia my father left Strasbourg to join him by his deathbed. But despite his great worries, he only left at the very last minute because at the time I was in sixth grade and, since I was a very bad mathematician, my father and I were practicing cross-multiplication. I have always felt grateful for that" [ibid.].

We can safely assume that, as opposed to Erwin, Jacobsthal's son Walther did not need this kind of attention. Walther began his career by publishing on the didactics of mathematics, before becoming a teacher and the headmaster of a school in Berlin. In 1939, he emigrated from Germany to the United States and lived there under his wife's maiden name, Bruns. His Strasbourg dissertation [W. Jacobsthal 1899], *On the asymptotic representation of integrals in a specific linear differential equation of the second order* was published in 1899. In 1905, he collaborated with two professors from Strasbourg, Heinrich Weber and Josef Wellstein, on a fundamental academic textbook of mathematics, the *Encyclopedia of elementary mathematics. A handbook for students and teachers* [Weber and Wellstein 1905]. Jacobsthal wrote the significant sixth section of the book: Spheres and spherical trigonometry. Apparently, in the years before the First World War, he also taught mathematics at the University of Berlin: his name appears alongside Georg Jakob Wallenberg's, a specialist for differentials, in certain university records. We know of two other books by Walther Jacobsthal: (1) a textbook on differential equations, published with Andrew R. Forsyth [Forsyth 1912], following the model of the third edition of Forsyth's English original *A Treatise on Differential Equations*. (2) *Phases of the Moon, Computus and the Perpetual Calendar* [W. Jacobsthal 1917], published in Berlin in 1917.

Walther Jacobsthal became a professional mathematician, a career which Grassmann had already wanted Walther's father Gustav to choose. Paradoxically, this is how continuities in the history of ideas can hide behind biographical contingencies.

Classification of complex musical structures by Grassmann schemes

Guerino Mazzola

Global compositions

The precise understanding of complex musical structures has made considerable progress in the last 50 years. On the one hand, such structures have been described in a unified way by the use of modern mathematics, in particular by conceptual tools developed in the algebraic geometry of schemes and more generally in topos theory. Both fields were essentially created by Alexander Grothendieck and allow for a conceptual elegance that can capture rather subtle features in music-theoretic thinking. We refer to [Mazzola 2002] for a comprehensive treatise of these achievements.

On the other hand, the new techniques developed in algebraic geometry also enable the solution of difficult music-theoretic problems. This means that modern mathematical conceptualization not only can help music theorists to step out of amateurish concept work, as is typically the case with so-called American Set Theory, but it also provides them with technical tools that answer questions arising from musical situations, once their conceptualization has been turned into an acceptable state. That precise conceptualization is necessary, or even desirable, is far from self-evident. For example, Rudolph Reti in his book *The Thematic Process in Music* [Reti 1951, 12] makes the

H.-J. Petsche et al. (eds.), *From Past to Future: Graßmann's Work in Context*,
DOI 10.1007/978-3-0346-0405-5_35, © Springer Basel AG 2011

snobbish remark that "in general", he "does not believe in the possibility or even desirability of enforcing strict musical definitions." This sanctions fuzziness as a scientific quality.

Classification issues have always been some of the weakest parts in traditional music theory. Writing down complete lists of representatives of all possible instances of a given type of structure seemed a meaningless endeavor. The reasons were twofold: First, the fuzzy conceptualization often made it impossible to distinguish different instances of a concept. And it also looked like being way too complicated to write down all specimens without exception. For example, complete chord lists, or complete lists of all possible allowed counterpoint steps for a given species (even the first!), looked too long and boring to be tackled. Second, music theory as such had no need for complete lists, at least it behaved like it was not necessary to know about all cases since the "interesting" ones (also a ghost of a thought) were "clear" anyway. But it is for such reasons (among others) that harmony was abandoned, Riemann's program of a complete identification of *all* possible chords with respect to tonal functions (tonic, dominant, subdominant) for a given tonality reference remained a torso, as has been observed by the great German music theorist Carl Dahlhaus [Dahlhaus 1966].

In this paper we want to support the idea that a complete classification of a given type of structure in music can not only be achieved, but that this work also reveals essential properties of such structure. This is also a logical necessity. In fact, writing down a complete list of all representatives of a given type of structure does not mean writing down all instances of that structure, but all representatives thereof with respect to some classification aspect. This means that we may consider instances of that structure as being essentially the same, or equivalent, and then only choose one of the members of each equivalence class. For example, if chords are considered as being essentially the same if they are transpositions of each other, then one may stick to chord representatives having middle C as their deepest pitch. Now, this understanding means that we only focus on interval relations within those chords, and not on the absolute pitch of their notes. If a music theorist says "a fifth", he/she means exactly this relation, and not the absolute pitch position of the interval's members.

In other words, classification reveals a way of understanding the structural properties. And this is what one needs to construct the list of representatives of that classification. This construction needs a selection process that reveals one's understanding of the musical situation. The final list is not what classification means, this is just the output of a complex process. Such a list is useless except if one understands what type of identification underlies its construction.

In our context, we are dealing with global musical structures, coined *global compositions* [Mazzola 1985]. These are a modeling of the formal procedure in music (theory, performance, composition, analysis, no matter what), which consists of the covering of a set K of notes or tones by a quasi-geographic "atlas of charts", i.e., a collection of (nonempty) subsets K_i, $i = 1, 2, \ldots m$, of the given set. So we are mimicking the concept of a differentiable manifold, which is defined by an atlas of charts, which are a covering of a set M by charts M_i that correspond to open sets in $\mathbb{R}^n$, and such that their intersections, when retracted to the open sets in $\mathbb{R}^n$, are connected by diffeomorphisms. In music theory, the open sets are simulated by (usually finite) subsets $K_i \subset M_i$ of modules M_i over a given commutative ring R, the K_i being in bijection with subsets $G_i \subset G$ of a supporting set G, which cover G, such that on the non-empty intersections of any two such charts, the retracted intersections $K_{ij} \subset M_i$ and $K_{ji} \subset M_j$ are isomorphic. Such "local" isomorphisms are special cases of isomorphisms between local structures, much as differentiable morphisms between open subsets of some $\mathbb{R}^n$ are the local morphisms for manifolds. In the musical context, the local objects are the subsets $K \subset M$ of modules, while the morphisms $f : K \subset M \to L \subset N$ are set maps $f : K \to L$, which can be extended to affine morphisms $F : M \to N$ of the underlying modules. Refer to [Mazzola 2002, Chap. 13] for a rigorous definition.

Similar to mathematical manifolds, the category of global compositions can be defined by an evident concept of a morphism between two global compositions: This is a map between the supporting sets that restricts to morphism between charts. The typical example is the atlas of the seven degree triads $X \subset C, X = I, II, III, IV, V, VI, VII$ in the pitch class major scale $C = \{0, 2, 4, 5, 7, 9, 11\} \subset \mathbb{Z}_{12}$ in the chromatic octave $\mathbb{Z}_{12}$ of semitone pitch classes modulo octave. This global composition is denoted by $C^{(3)}$. Here the transition isomorphisms are the identities since all charts are living in one and the same $\mathbb{Z}$-module $\mathbb{Z}_{12}$.

But in general this is not required, and there are simple examples where such a situation is not possible, in the sense that there is no big module comprising all charts and yielding the global composition up to isomorphism of global compositions. Such global compositions are called *noninterpretable*, as opposed to the above example of the seven triadic degrees, which is called *interpretable*. See our example in Fig. 2 for such a composition. Non-interpretable global compositions correspond to differentiable manifolds that are not open sets of $\mathbb{R}^n$, for example, a sphere surface or a torus. Musically speaking, non-interpretable global compositions are constructions of musical structures that are thought from a strictly global perspective. This means that such a global composition can only be thought in terms of musical parameters in the charts (the parameters being given by the underlying module), but a global representation of the atlas within one and the same module is not possible.

One very important geometric shape associated with such a global composition is its nerve. It is the nerve $Nerve(I)$ of the covering $I = (G_i \subset G)_i$ of G by the atlas of the charts G_i. The nerve's zero-dimensional simplexes are the charts, the one-dimensional simplexes are the non-empty intersecting pairs of charts, etc. In the case of $C^{(3)}$, the

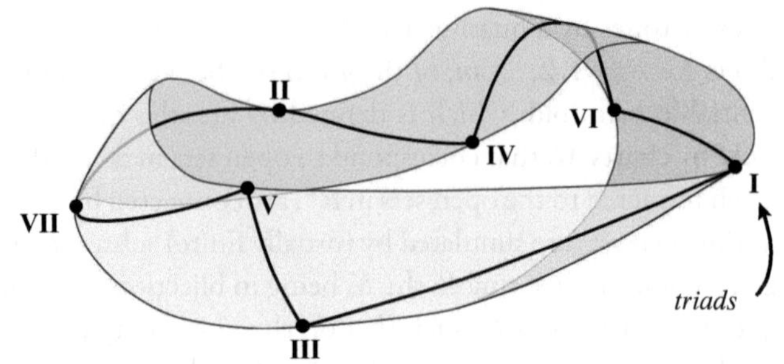

Fig. 1. The harmonic band is a Moebius strip

nerve $Nerve(C^{(3)})$ is called the harmonic band. It is a Moebius strip; see Fig. 1. On its boundary, the fifth sequence is visible. The absence of orientation is responsible for the failure of Hugo Riemann's original plan to construct a function-theoretic harmony.

This example makes clear that global compositions are central structures in the precise conceptualization of music. So we are asked to classify these structures, i.e., to establish the complete list (finite or not) of all possible isomorphism classes of global compositions.

Classification of global compositions

Classification of global compositions is described in great detail in [Mazzola 2002, Chap. 15]. It is obvious that this is not feasible for all possible mathematical parameter spaces, since the technical tools from algebraic geometry, which are necessary in this endeavor, are only effective for special rings and modules involved in the theory of global compositions. However, the classification theorem 18 in [Mazzola 2002, Sect. 15.3.2] is quite powerful and can be understood in an intuitive way. It states that the isomorphism classes of global compositions can be parameterized as points on an algebraic scheme. So they are represented as points in an algebraic "manifold". Of course this type of "listing" isomorphism classes is a geometric one, which shows that these classes are far more than isolated items: They are connected by a strong global structure (locally) defined by polynomial equations.

Let us describe the general procedure in order to access the algebraic-geometric yoga underlying this theorem. The situation for any given global composition has two components: the first is its nerve. This one is of purely combinatorial nature. It takes care of the intersection configuration of the composition's points, irrespective of their geometric position in the charts. This is the combinatorial frame of classification. We can fix this data and just work in a fixed nerve type.

The difficult component is the control of the points' positions in the chart modules, and also in the nerve's simplexes. The immediate idea is to continue the combinatorial perspective from the nerve to the points. But how can we reduce a global composition to a purely combinatorial configuration? In geometry, there is one situation, where the module-theoretic context degenerates to a combinatorial one in disguise, namely the situation, where the points are as free as possible from each other. What does this mean? They have to build a system of points, which are independent in the sense of linear algebra, which means that they are in *general position*. This means the following. Given a finite set $K \subset M$ of $k+1$ points $x_0, x_1, \ldots x_k$ in a R-module M, we want that the module generated by the k differences $x_i - x_0$, $i = 1, \ldots k$, is free of rank k. But we want more, we also want these points sitting in M in such a way that they can be mapped wherever we want in terms of morphisms of local compositions. This can be achieved if we let them be basis points of an affine basis of a free R-module.

This can be done if we have a free module of dimension k and our points are just the affine standard basis of the free module R^k: $e_0 = 0, e_1 = (1, 0, \ldots 0), \ldots,$ $e_k = (0, 0, \ldots, 0, 1)$. So we replace K by the standard $(k+1)$-point local composition $\Delta_k \subset R^k$. Evidently, there is exactly one morphism of local compositions $\Delta_k \to K$ that sends e_i to x_i, for all $i = 0, 1, \ldots, k$. This construction can be extended to global compositions, and we have a global composition ΔG^I for every global composition G^I given from a support set G and a covering $I = (K_i \xrightarrow{\sim} G_i)_i$ of G by charts K_i in modules M_i. We also have a projection $\Delta G^I \to G^I$ onto the original global composition G^I, which is the global extension of the above local morphism. This construction is called the *resolution of* G^I,

$$res_{G^I} : \Delta G^I \to G^I.$$

The left-hand global composition consists of charts, which all live as standard local compositions in their free modules over R. Their nerve is isomorphic with that of G^I, more precisely, the nerve map of res_{G^I} is an isomorphism of nerves.

For example, if each chart has four points, then the three differences generate a three-dimensional space, that is: the four points define a tetrahedron. The example in Fig. 2 shows such a global composition. It has six points $1, 2, \ldots, 6$, and three charts: $G_1 = \{1, 2, 3, 4\}$, $G_2 = \{1, 2, 5, 6\}$, $G_3 = \{3, 4, 5, 6\}$, which are all singular as their four points lie in a plane. The charts are glued on three pairs of points: $(1, 2)$, $(5, 6)$, and $(3, 4)$. The pathology here is the gluing of points $1, 2$, which is twisted, as shown in the figure. This global composition is not interpretable. But its resolution Γ is interpretable, as shown in the figure: its charts lie in a three-dimensional space and are nonsingular. In fact, each of the three charts $\Gamma_1, \Gamma_2, \Gamma_3$ generates a tetrahedron.

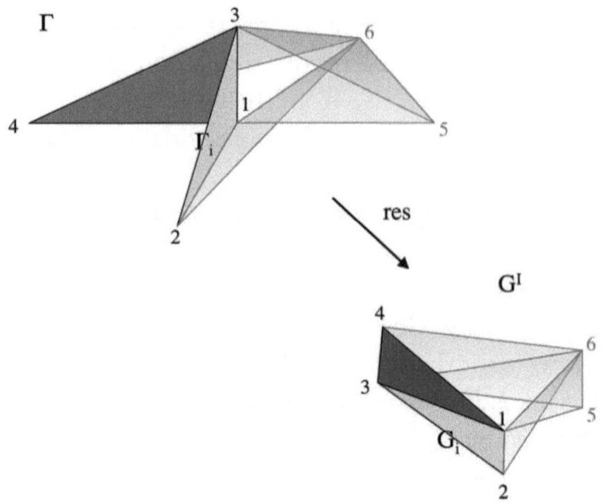

Fig. 2. The resolution Γ of a global composition G^I. It is mapped one-to-one onto G^I, and chart Γ_i corresponds to chart G_i

Now, every global composition has such a resolution, and morphisms between global compositions $f : G^I \rightarrow H^J$ induce canonical corresponding morphisms $\Delta f : \Delta G^I \rightarrow \Delta H^J$ of their resolutions, which are compatible with the resolution projections. In other words, the entire construction is natural in the sense of category theory.

Of course, this is not the solution of our problem, but it is a first step in a representation of the category of global compositions in terms of free objects and maps. The critical process comes now. It consists of lifting the special situation into data within the free context.

More precisely, we have to lift the special information about the given global composition in order to be able to *reconstruct the original data* from the lifted data. The idea is as follows: We have to use the given resolution $res_{G^I} : \Delta G^I \rightarrow G^I$ in order to lift the G^I data to the free data of ΔG^I. The crucial question is: What special data do we want to lift? The points have already been lifted to their general counterparts. What has not been lifted is the "environment" of the points, the space, where these points live.

How can we lift that environment? The idea is very musical here: The ambient module of a point is an abstract entity, it does not provide us with any musically useful information. What is needed in music is to know how such points may sound, when played. This means that we have to ask about the parameters that could be attributed to such points in order to play them on a musical instrument: loudness, pitch, overtone component, etc.

But are there such parameter associations? The only thing we have are all possible (locally affine) functions from the points to the underlying ring of coefficients R. We are therefore looking at all possible (locally) affine functions $f : G^I \rightarrow R$ on the global composition G^I. This means that on each chart of G^I, f is an affine morphism into R. Denote by ΓG^I the R-module of these functions and call it the *module of affine functions on* G^I. Evidently, the retract $\Delta\Gamma G^I = \Gamma G^I \circ res_{G^I}$ is a submodule of $\Gamma\Delta G^I$, the module of all affine functions on the resolution of G^I. This solves our problem of reconstructing G^I. It can be shown that under good mathematical conditions (which we omit here, but see [Mazzola 2002, Sect. 15.3.2]), the original global composition can be reconstructed by embedding the points k in general position in ΔG as linear operators $\dot{k}$ acting on the retracted modules of affine functions on the charts, where the k live.

So the special information on the resolutions ΔG^I is to consider all possible ways to "play" the resolved composition's points in musical parameter spaces, and to interpret points as operators on the modules of these parametrizations. The variable parts of this representation are no longer the original points, but modules of affine functions, entire spaces, which may eventually be used to reconstruct the original points. So we are confronted with the resolution $\Delta\Gamma G^I$ and on its nerve, we are given a classical "coefficient system" of modules of affine functions, namely for each simplex σ a module of functions $\Delta\Gamma G^I(\sigma)$ on the intersection of all charts of this simplex, including the canonical (surjective) restriction morphism $p_{\sigma,\tau} : \Delta\Gamma G^I(\sigma) \to \Delta\Gamma G^I(\tau)$ for each inclusion $\sigma \subset \tau$ of simplexes.

The situation now is that we have replaced the global compositions G^I by the coefficient systems $\Delta\Gamma G^I$ of their affine functions on the free resolution structures ΔG^I. This assignment is also functorially valid. In particular, if we have isomorphic global compositions, then their coefficient systems also correspond under automorphisms of their resolutions. And the latter are just permutations of the free points. In other words, classification of global compositions boils down to orbits of coefficient systems of function modules under the symmetric groups of automorphisms of resolutions.

Grassmann's technique

Instead of local compositions that are patchworks of points in chart modules, their classification by means of resolutions shifts to coefficient systems of modules of affine functions. This looks quite complex at first sight, because we now deal with coefficient systems, i.e., limits of nerve diagrams of spaces, whose coordinates are subspaces of big modules of affine functions. But this looks more complex than it is since now, the big modules are free modules since they live in resolutions, and then the subspaces are not really big structures. This is what Grassmann taught us.

What is the yoga of this insight? Let us look at the local situation, and let us stick to the essential situation, where R is a commutative field: One big vector space R^n, and in this module we look at subspaces $N \subset R^n$. In our context, we may even suppose that

N is a direct summand of R^n of given rank k. This is known as the *Grassmannian* space $Grass_{k,n}(R)$. Is there any chance to compactify this data such that N can be interpreted as being a point in some associated space? The answer is "yes", but we have to conceive that space as a patchwork of local charts. In other words: the answer is positive locally speaking. It works as follows: Take a subspace $N \subset R^n$ of rank k. Then there is a linear complement $C_{i_1,i_2,\ldots,i_{n-k}} = Re_{i_1} \oplus Re_{i_2} \oplus \ldots \oplus Re_{i_{n-k}}, i_1 < i_2 < \ldots < i_{n-k}$, i.e., $N \oplus C_{i_1,i_2,\ldots,i_{n-k}} = R^n$ (this is Steinitz' theorem). Here the basis $e_1, e_2, \ldots, e_n$ is the canonical basis of R^n. Denote by $C^{i_1,i_2,\ldots,i_{n-k}}$ the space spanned by the remaining canonical basis vectors. Then the canonical projection $N \to C^{i_1,i_2,\ldots,i_{n-k}}$ is an isomorphism, and therefore the other projection $N \to C_{i_1,i_2,\ldots,i_{n-k}}$ defines a linear map $\mu_N = C^{i_1,i_2,\ldots,i_{n-k}} \to C_{i_1,i_2,\ldots,i_{n-k}}$, which we identify with a $k \times (n-k)$-matrix. Therefore, the total space of all rank k subspaces of R^n is covered by spaces that are all isomorphic to the space $\mathbb{M}_{k\times(n-k)}(R)$ of $k \times (n-k)$-matrices with coefficients in R. This is a nice affine space, and therefore we have covered $Grass_{k,n}(R)$ by a set of affine spaces. The special case of one-dimensional subspaces is known as the projective space $\mathbb{P}_{n-1}(R) = Grass_{1,n}(R)$. Although this looks like a special case it can be shown that there is a closed immersion $Grass_{k,n}(R) \rightarrowtail \mathbb{P}_{\binom{n}{k}-1}(R)$, due to Julius Plücker, whose work had been generalized by Grassmann.

This setup enables the complete classification of global compositions, since the operation of the resolution's automorphism group, which is a finite group (essentially a symmetric group), admits good quotient schemes, so the classification is settled in the sense described above.

The musical meaning of Grassmann's approach

Although Grassmann's method is standard in technical terms, its meaning in the musical context is not automatic. In fact, among the scholars specialized in mathematical music theory, the classification theorem and its correlated method of resolutions and Grassmannian schemes have never been appreciated to this date. I am perhaps feeling a bit like Grassmann did: you are just not in the spirit of present science.

Therefore, it is necessary to rethink the musical interpretation of this approach. To begin with, the construction of the resolution is a musically dramatic process. What is it that this resolution $res_{G^I} : \Delta G^I \to G^I$ realizes? In its combinatorial and set-theoretic setup, it is trivial: We have a bijection between the points of ΔG^I and the points of the given global composition G^I, and we have a corresponding isomorphism of nerves $Nerve(res_{G^I}) : Nerve(\Delta G^I) \xrightarrow{\sim} Nerve(G^I)$. But the points in the resolution lie in a space that is dramatically different from the original spatial setup. In fact, in the resolution, all points lie in a big free module R^m, and the resolution is interpretable. Musically speaking this means that the coordinates of points in ΔG^I all have the same set of m coordinates in R. They can therefore be realized as points in a big shared musical parameter space.

Concretely, this means that in contrast to the local coordinates of points in G^I, we can play these points as sounds within in a big sound parameter space. For example, if the original composition was conceived as living in a piano score space, its points would have *local* coordinates onset, pitch, loudness, and duration. So on each chart, such parameters would be available. But this does not mean that the global composition as a whole could be played since it could be construed in a non-interpretable way. The above example in Fig. 2 illustrates this situation. One would only be capable of playing notes within the charts, not the entire piece. The resolution is achieved by first adding enough parameters in such a way that the totally available parameters enable the embedding of the resolution. Which means that we have to add so many dimensions that all the charts can be embedded in a big space, its points being some of the big set of canonical basis vectors in that big space. For example, we may add glissando parameters, crescendi, sound colors, envelopes etc., whatever could be used to generate sufficient dimensions.

This is to say that additional dimensions of musical parameters add a potential of expressivity. And conversely, if we abolish musical sound parameters, performing the piece might become impossible. This is particularly relevant regarding the question of instrumentation. The prototype of such a situation is string quartet theory [Mazzola 1985, Part XIV]. The expressivity enabled by the instruments of the violin family (violin, viola, violoncello) is a function of the many parameters that are provided by these instruments, and it serves the shaping and communication of complex musical structures, such as harmony and counterpoint of the late 18th and early nineteenth century. So the point here is not that one cannot play a piece in a reduced interpretation. The critical fact is that one would aspire to play it as it is understood in its global configuration.

The musical meaning of Grassmann's approach positions the classification process in the free environment of the resolution. We are given a big repertory of musical parameters and can play the resolved composition in this space. But the original composition did not have all those parameters, it had much less, and we have to understand to what extent this restriction can be understood in the free framework of the resolution. The idea is to look at all possible coordinate functions, i.e., the space $\mathrm{Aff}(R^n, R)$ of affine functions on the free module R^n. In the construction of the coefficient system, we were looking at submodules $N \subset \mathrm{Aff}(R^n, R)$ of such function modules, namely those induced from affine functions on the original composition. The reconstruction of the original composition uses the identification of a free point k with its linear operator $\dot{k}$ on N.

This means that instead of admitting all those fancy parameters necessary to play the piece's resolution in free space, we interpret points k as being operators on musical parameters stemming from the original composition. So a point is viewed as being an action upon a space N of musical parameterizations. To put it in more intuitive terms: a sound event is viewed via its potential to be played in musical parameters. The identification of such events is achieved by the difference in their performance, by some parameter value, where their performance differs and this is very natural: You distinguish two sound events by their differences in performance. One sounds louder, or is longer in duration, or has a different instrumental characterization, you name it. But the difference need not be a specific one, it is sufficient that there exists one parameter function separating two points.

Therefore, the reconstruction identifies a global composition as an atlas of general points, which are identified as operators on spaces of sound parameterizations. Or, to put it more radically: A sound is what it can sound like in a space of parameters. The abstract identity is replaced by the potential to sound. And this potential is provided by a point in a Grassmann scheme. So instead of looking at the abstract sound symbol, we look how it can be parametrized under all possible perspectives. The point in the Grassmann scheme is a variety of parametrizations, and the sound is operating on the variety of these parametrizations and yields parameter values for each of these parametrization functions. Which means that we have to understand the sound as the total of all parameterizations, not just one selected item. This means that we have to vary parametrizations and to look at the behavior of the operator under this system of variations. In other words, we are identifying the sound with the sum of all possible parameterizations, by their *integration*.

Varèse's interpretation

It is remarkable that Edgar Varèse had somehow anticipated these insights from his compositional work, when creating the work (1924–1925) with the spoken title "Intégrales" [Varèse 1960, 67]:

> "Die Intégrales wurden für eine räumliche Projektion entworfen. (...) Während wir in unserem musikalischen System Klänge anordnen, deren Werte festgelegt sind, suchte ich eine Verwirklichung, bei der die Werte fortwährend im Verhältnis zu einer Konstanten verändert werden. (...)
> Um dies besser zu begreifen, übertragen wir, da das Auge viel schneller und geübter ist als das Ohr, diese Vorstellungen ins Optische und betrachten die wechselnde Projektion einer geometrischen Figur auf eine Fläche, wobei Figur und Fläche sich

beide im Raum bewegen, aber jede nach ihren eigenen Geschwindigkeiten, die veränderlich und verschieden sind, die sich verschieben und rotieren. Die augenblickliche Form der Projektion ist durch die Relation zwischen Figur und Fläche in diesem Augenblick bestimmt. Aber wenn man erlaubt, daß die Figur und die Fläche ihre eigenen Bewegungen haben, ist es möglich, mit der Projektion ein äußerst komplexes und scheinbar unvorhersehbares Bild zu erhalten. Diese Qualitäten können noch vermehrt werden, wenn man die Form der geometrischen Figur ebenso wie ihre Geschwindigkeiten variiert. (...) Ich hoffe, innerhalb kurzer Zeit einen Apparat zur Verfügung zu haben, der es erlauben wird, ein räumliches Relief zu geben. Nur des Beweises wegen würde ich daran interessiert sein, die Intégrales einmal so zu realisieren, wie sie ursprünglich konzipiert worden sind."

In 1960 Maurizio Kagel transferred these principles to paper strips and discs of the score for "Transición II" for piano, percussion, and two tapes.

"Varèse's" idea basically is a remake of the *classical variation principle*. Bach's "Goldberg Variationen" (BWV988), Beethoven's "Diabelli Variationen" (op.120), or Webern's "Variationen für Klavier" (op.27) are compositions in this spirit. The subject is always an artistically interwoven change of perspectives of a theme: the variation in the parts of the theme and their relations.

Especially for Webern a composition is a cellular organism, a connected manifold (in the naive sense) of transformations, of ever changing perspectives, of metamorphoses of a single cellular germ (in the sense of Goethe), which in fact is Schönberg's dodecaphonic series.

In front of this historical background the classification problem of global compositions – together with its central process of resolution – appears as a *canonical program*. In particular, the nerve of a resolution, a concept related to that of a "cell complex" from algebraic topology, reminds us of the cellular organism alluded to by Webern. And the projections which Varèse describes in a visionary fashion show a surprisingly similar geometry to the projections of a resolution onto the original composition, projections which are distinguished in that they project a general position onto specializations. Finally, the variation of these projections corresponds to the variation of the modules of affine functions, i.e., the variation of the compositions which are distinguished from each other via their retracted function modules on one and the same resolution.

But the variational principle is not only a compositional strategy, it equally, or even more dramatically, applies to the performance level. Performance deals with a transformation from the mental score space to the physical space of the acoustic realization. But this transformation locally is a deformation of the 'rigid' parameter values set out on the score. Why should the artist deform a perfect opus? Wouldn't this be blasphemy or at least a tremendous lack of respect? No, the added value of such a deformation is not a destruction of given structure, it is a subtle change of parametric perspectives which let the auditory system still recognize the written relations, but on top of that puts configurations into general position such that their generic, or better: resolved, structure becomes "visible" on the auditory level – to restate it in the wording of Varèse.

New views of crystal symmetry guided by profound admiration of the extraordinary works of Grassmann and Clifford

Eckhard Hitzer

Soli Deo Gloria

Introduction

Already Hermann Grassmann's father Justus (1829, 1830) published two works on the geometrical description of crystals, influenced by the earlier works of C. S. Weiss (1780–1856) on three main crystal forces governing crystal formation. In his 1840 essay on the derivation of crystal shapes from the general law of crystal formation Hermann established the notion of a three-dimensional vectorial system of forces with rational coefficients, that represent the interior crystal structure, regulate its formation, its shape and physical behavior. In the *Ausdehnungslehre* of 1844 (Sect. 171) [A1] he finally writes: "I shall conclude this presentation by one of the most beautiful applications which can be made of the science treated, i.e., the application to crystal figures" [cited from Scholz 1996, 44]. The geometry of crystals thus certainly influenced the *Ausdehnungslehre*.

Grassmann's work in turn influenced W. K. Clifford [Clifford 1878] in England: "I propose to communicate in a brief form some applications of Grassmann's theory

DOI 10.1007/978-3-0346-0405-5_36,

...I may, perhaps, therefore be permitted to express my profound admiration of that extraordinary work, and my conviction that its principles will exercise a vast influence upon the future of mathematical science." Conformal Clifford (geometric) algebra has in turn led at the beginning of the twentieth century to a new fully geometric description of crystal symmetry in terms of so-called versors [Hestenes and Holt 2007]. Versors are simply (Clifford) geometric products of five-dimensional vectors conformally representing general planes in three-dimensional (3D) Euclidean space (by their 3D normal vector and the directed distance from the origin). Each plane's vector geometrically represents a reflection at the plane, the geometric products of several plane vectors represents the combination of reflections at the respective planes (Cartan–Dieudonné).

As expected, three crystal specific 3D vectors are enough to construct all symmetry versors of any type of crystal. With graphics software such as CLUCalc [Perwass 2001] that is capable of geometric algebra this concept can be implemented in every detail, such that the abstract beauty of the enormously rich symmetry of crystals can be fully visualized by state-of-the-art 3D computer graphics. The Space Group Visualizer (SGV) is a tailor-made CLUCalc Script [Hitzer and Perwass 2004, 2005a,b, 2006, 2009; Perwass and Hitzer 2005a,b]. To be precise, the SGV is thus capable of showing every plane of reflection and glide-reflection symmetry, all axes of rotations, screw-rotations and rotary inversions, and every center of inversion. It further allows dynamic visualization of the action of any symmetry operation on a general element (representing atoms, molecules or ions).

We thus have, 165 years after the *Ausdehnungslehre* of 1844, an explicit form of the beauty, which Grassmann may have had in mind, when he wrote eloquently: *one of the most beautiful applications.*

The next section uses the symmetries of diamond in order to demonstrate how the SGV visualizes space group symmetry. For mathematically interested readers the appendix introduces the Clifford geometric algebra description of crystallographic space groups.

Computer visualization of crystal symmetry

Geometrically a diamond cell lattice (type: face-centered cubic = fcc) is highly symmetric. That means there is an enormous variety of possible geometric transformations that leave the lattice as a whole invariant, including all lengths and angles. These symmetry operations include single cell transformations that leave a cell vertex point invariant: planes of reflections (through the vertex), rotations (with axis through the vertex, and inversions $\mathbf{x} \mapsto -\mathbf{x}$, centered at the vertex), and roto-inversions (inversions followed by a rotation). The 24 symmetry transformations of a diamond vertex point group create

Fig. 1. *Left*: Diamond cell in Space Group Visualizer. *Right*: 24 general elements in 3D showing diamond point symmetry of one vertex

24 symmetric copies of a general asymmetric element placed next to the invariant point, see Fig. 1 (left), or enlarged in Fig. 1 (right). In pure diamond one carbon atom is located at the center of this cluster (plus one at 1/4 distance away along a cubic space diagonal).

The inclusion of integer lattice translations (from fcc vertex to fcc vertex) can lead to new planes of reflection, see Fig. 2 (left). The combination of a plane of reflection with a lattice translation not perpendicular to the plane leads to a combined glide reflection, see Fig. 2 (right), where (red) vectors indicate the parallel glide motion. The perpendicular translation component displaces the reflection plane in normal direction, and the parallel translation component creates a glide motion parallel to the plane. Pairs of characteristic diamond glides are shown in Fig. 3 (left).

A sequence of two reflections at two planes results in a rotation, see Fig. 3 (right). This rotation has the intersection line of the two planes as its axes and twice the (dihedral) angle between the two planes is the resulting rotation angle. All the rotation axes seen in Fig. 3 (right) are lines of intersection of reflection planes of Fig. 2 (left). A lattice translation perpendicular to the rotation axis after a rotation, effectively creates another rotation also already contained in Fig. 3 (right). But if we perform a translation not normal to the rotation axis, with a translation component parallel to the rotation axis, we get a new transformation, a so-called screw. So a screw is a rotation followed by a translation along the screw axis, resulting in a directed helical motion around the screw axis, see Fig. 4 (left).

Combining an inversion with a subsequent lattice translation yields a new center of inversion, see Fig. 4 (right). The combination of an inversion with a rotation leads

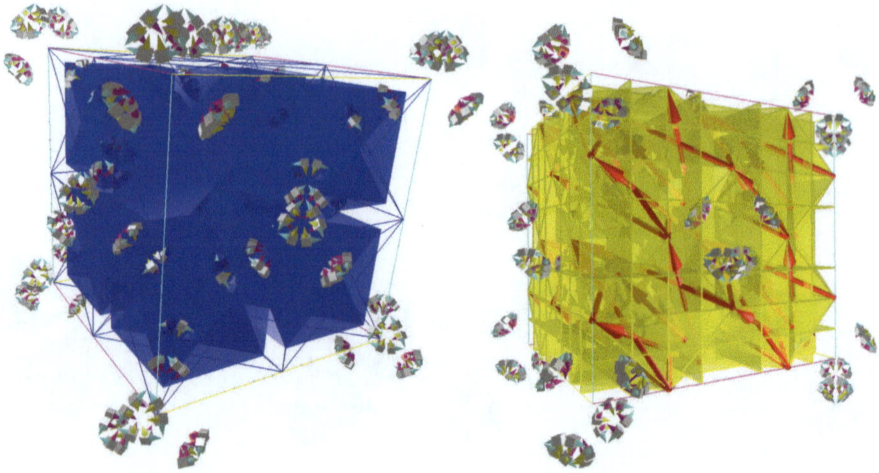

Fig. 2. *Left*: Diamond reflection planes. *Right*: All glide planes of a diamond cell

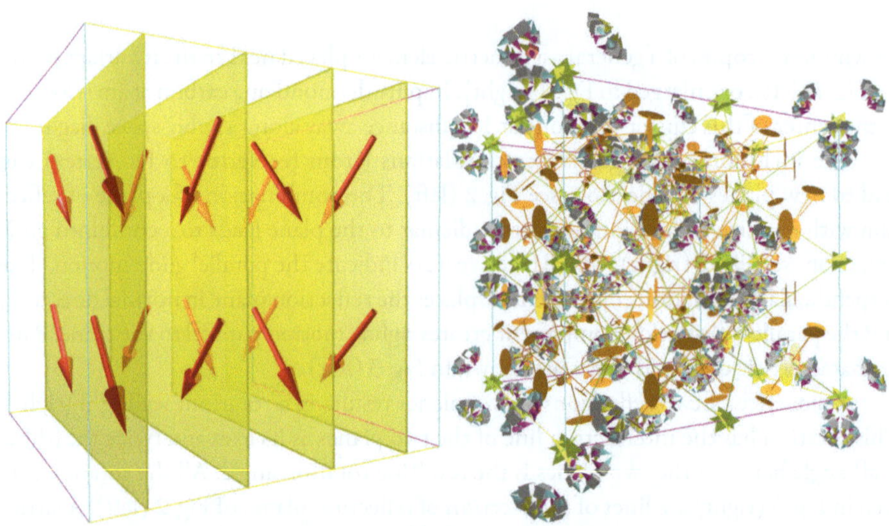

Fig. 3. *Left*: Pairs of diamond glide planes. *Right*: Symmetry rotation axes. Angles indicated by colors and arc segments

to a roto-inversion. Characteristic for the diamond lattice are the 90° roto-inversions depicted in Fig. 5 (left). The total graphical depiction of these symmetries in Fig. 5 (right) gives an idea of the intricate complexity of the symmetries possessed by the diamond lattice. The International Tables of Crystallography, Vol. A [Hahn 2005],

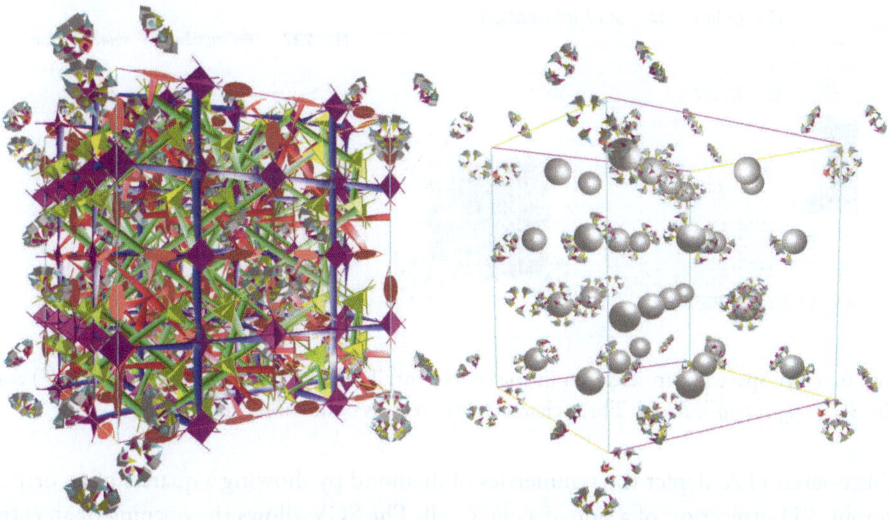

Fig. 4. *Left*: All screw symmetry axes of diamond. *Right*: All centers of inversion symmetry of diamond

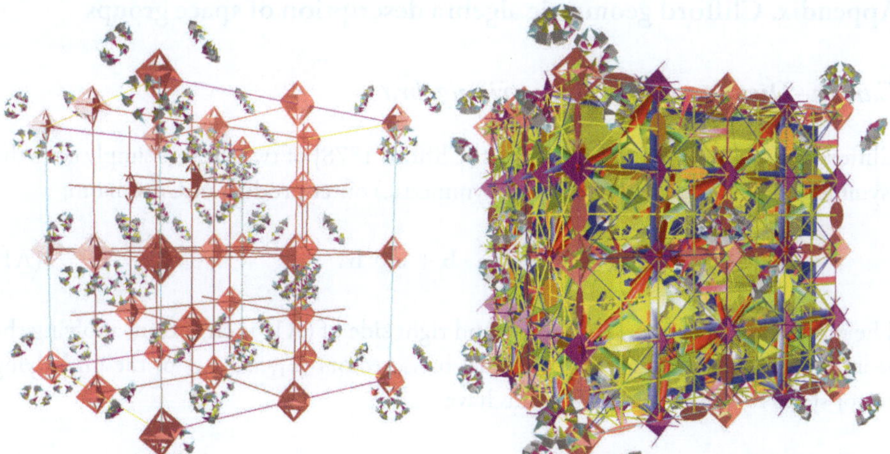

Fig. 5. *Left*: All axis (and centers) of roto-inversion symmetry of diamond. *Right*: Diamond cell in Space Group Visualizer, showing all symmetries

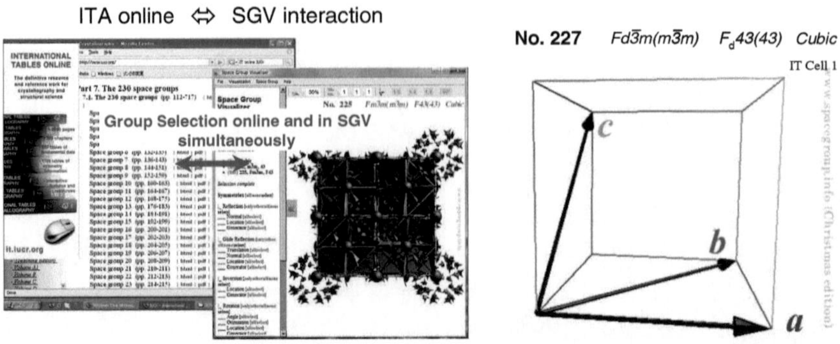

Fig. 6. *Left*: Space group selection in the online Int. Tables of Crystallography (*left half*) and the SGV (*right half*). *Right*: Three characteristic vectors of diamond cell

abbreviated ITA, depict the symmetries of diamond by showing a quarter of an orthographic 2D projection of a side of a cubic cell. The SGV allows the opening of an extra window with the ITA online and synchronous navigation in both, see Fig. 6 (left).

Appendix. Clifford geometric algebra description of space groups

Cartan–Dieudonné and geometric algebra

Clifford's associative geometric product [Clifford 1878] of two vectors simply adds the (symmetric) inner product to the (antisymmetric) outer product of Grassmann,

$$\mathbf{ab} = \mathbf{a} \cdot \mathbf{b} + \mathbf{a} \wedge \mathbf{b}\,. \tag{A1}$$

The mathematical meaning of the left and right side of (A1) is clear from applying the geometric product to the n orthonormal basis vectors $\{\mathbf{e}_1, \ldots, \mathbf{e}_n\}$ of the underlying vector space $\mathbb{R}^{p,q}$, $n = p + q$. We thus have

$$\mathbf{e}_k\mathbf{e}_k = \mathbf{e}_k \cdot \mathbf{e}_k = +1, \quad \mathbf{e}_k \wedge \mathbf{e}_k = 0, \quad 1 \leq k \leq p, \tag{A2}$$

$$\mathbf{e}_k\mathbf{e}_k = \mathbf{e}_k \cdot \mathbf{e}_k = -1, \quad \mathbf{e}_k \wedge \mathbf{e}_k = 0, \quad p+1 \leq k \leq n, \tag{A3}$$

$$\mathbf{e}_k\mathbf{e}_l = -\mathbf{e}_l\mathbf{e}_k = \mathbf{e}_k \wedge \mathbf{e}_l, \quad \mathbf{e}_k \cdot \mathbf{e}_l = 0, \quad l \neq k,\ 1 \leq k, l \leq n. \tag{A4}$$

Under this product parallel vectors commute and perpendicular vectors anticommute,

$$\mathbf{a}\mathbf{x}_{\parallel} = \mathbf{x}_{\parallel}\mathbf{a}\,, \qquad \mathbf{a}\mathbf{x}_{\perp} = -\mathbf{x}_{\perp}\mathbf{a}\,. \tag{A5}$$

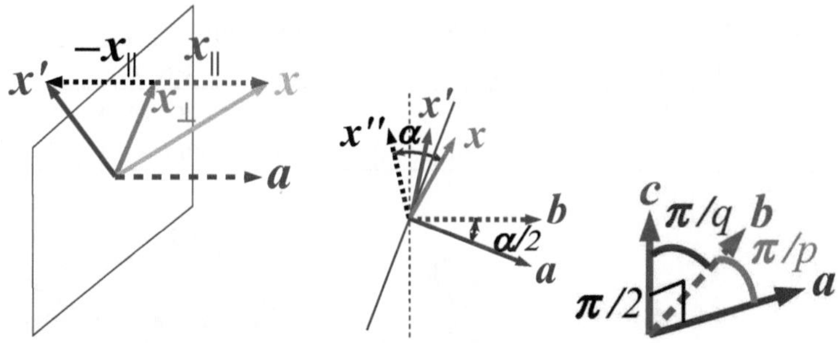

Fig. 7. *Left*: Reflection at a hyperplane normal to **a**. *Center*: Rotation generated by two successive reflections at hyperplanes normal to **a**, **b** by twice the angle $\angle(\mathbf{a},\mathbf{b})$. *Right*: Angular relations of pairs of geometric cell vectors of **a**, **b**, **c**: $\angle(\mathbf{a},\mathbf{b}) = \pi/p$, $\angle(\mathbf{b},\mathbf{c}) = \pi/q$, $\angle(\mathbf{a},\mathbf{c}) = \pi/2$, $p, q \in \{1,2,3,4,6\}$

This allows us to write the *reflection* of a vector **x** *at a hyperplane* through the origin with normal **a** as (see left side of Fig. 7)

$$\mathbf{x}' = -\mathbf{a}^{-1}\mathbf{x}\,\mathbf{a}, \qquad \mathbf{a}^{-1} = \frac{\mathbf{a}}{\mathbf{a}^2}. \tag{A6}$$

The composition of two reflections at hyperplanes, whose normal vectors **a**, **b** subtend the angle $\alpha/2$, yields a rotation around the intersection of the two hyperplanes (see center of Fig. 7) by α,

$$\mathbf{x}'' = (\mathbf{ab})^{-1}\mathbf{x}\,\mathbf{ab}, \qquad (\mathbf{ab})^{-1} = \mathbf{b}^{-1}\mathbf{a}^{-1}. \tag{A7}$$

Continuing with a third reflection at a hyperplane with normal **c** according to the Cartan–Dieudonné theorem yields rotary reflections (equivalent to rotary inversions with angle $\alpha - \pi$) and inversions

$$\mathbf{x}' = -(\mathbf{abc})^{-1}\mathbf{x}\,\mathbf{abc}, \qquad \mathbf{x}'' = -i^{-1}\mathbf{x}\,i, \qquad i \doteq \mathbf{a}\wedge\mathbf{b}\wedge\mathbf{c}, \tag{A8}$$

where $\doteq$ means equality up to non-zero scalar factors (which cancel out in (A9)). In general the geometric product of k normal vectors (the versor S) corresponds to the composition of reflections of all symmetry transformations [Hestenes and Holt 2007] of two-dimensional (2D) and 3D crystal cell point groups (also called crystal classes),

$$\mathbf{x}' = (-1)^k S^{-1}\,\mathbf{x}\,S. \tag{A9}$$

Table 1. Geometric and international notation for 2D point groups

Crystal	Oblique		Rectangular		Trigonal		Square		Hexagonal	
Geometric	$\bar{1}$	$\bar{2}$	1	2	3	$\bar{3}$	4	$\bar{4}$	6	$\bar{6}$
International	1	2	m	2mm	3m	3	4mm	4	6mm	6

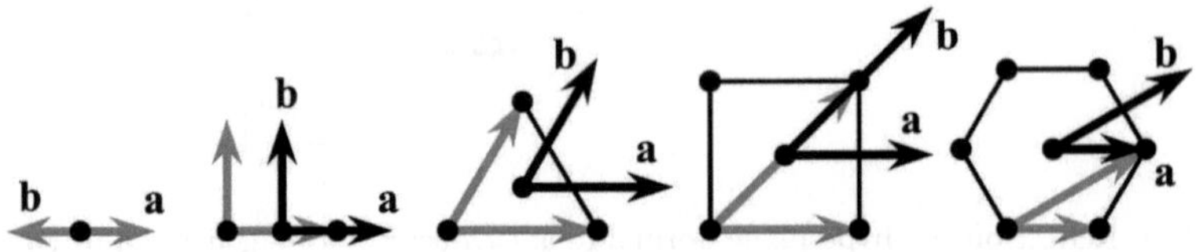

Fig. 8. Regular polygons ($p = 1, 2, 3, 4, 6$) and point group generating vectors **a**, **b** subtending angles π/p shifted to center

Two-dimensional point groups

2D point groups [Hestenes and Holt 2007] are generated by multiplying vectors selected [Hitzer and Perwass 2004, 2005a; Perwass and Hitzer 2005b] as in Fig. 8. The index p denotes these groups as in Table 1. For example the trigonal (square) point group (the symmetry group of the equilateral triangle (square), leaving the center point invariant) is given by multiplying its two generating vectors **a**, **b**,

$$3 = \{\mathbf{a}, \mathbf{b}, R = \mathbf{ab}, R^2, R^3 = -1, \mathbf{a}R^2\}. \tag{A10}$$

$$4 = \{\mathbf{a}, \mathbf{b}, R = \mathbf{ab}, R^2, R^3, R^4 = -1, \mathbf{a}R^2, \mathbf{b}R^2\}. \tag{A11}$$

In (A10) and (A11) the vectors **a**, **b** represent reflections (A6) at lines normal to **a**, **b** and passing through the center of the equilateral triangle (square) of Fig. 8. The rotor $R = \mathbf{ab}$ represents as in (A7) a double reflection at the two lines passing through the center and normal to **a** and **b**, respectively. Because $\angle(\mathbf{a}, \mathbf{b}) = 60^\circ$ (45°), the resulting rotation is by $2 \times 60^\circ = 120^\circ$ $(2 \times 45^\circ = 90^\circ)$ around the center. The cyclic rotation subgroups are denoted in Table 1 with bars, for example,

$$\bar{3} = \{R = \mathbf{ab}, R^2, R^3 = -1 \doteq 1\}, \quad \bar{4} = \{R = \mathbf{ab}, R^2, R^3, R^4 = -1 \doteq 1\}, \tag{A12}$$

containing the three (four) symmetry rotations of the equilateral triangle (square) of Fig. 8 around its invariant center by 120°, and the multiples 240° and 360° $(90^\circ, 180^\circ, 270^\circ$ and $360^\circ)$. The vectors $\mathbf{a}R^2$ in (A10) $(\mathbf{a}R^2, \mathbf{b}R^2$ in (A11)) are the normal directions of the remaining one (two) lines (passing through the center) of reflection symmetry.

Table 2. Geometric 3D point group symbols [Hestenes and Holt 2007] and generators with $\mathbf{a}, \mathbf{b}, \mathbf{c}$: $\angle(\mathbf{a},\mathbf{b}) = \pi/p$, $\angle(\mathbf{b},\mathbf{c}) = \pi/q$, $\angle(\mathbf{a},\mathbf{c}) = \pi/2$, $p, q \in \{1,2,3,4,6\}$

Symbol	$p = 1$	p	$\bar{p} = \bar{1}$	$\bar{p}$	pq	$\bar{p}q$	$p\bar{q}$	$\bar{p}\bar{q}$	$\overline{pq}$
Generators	$\mathbf{a}$	$\mathbf{a}, \mathbf{b}$	1	$\mathbf{ab}$	$\mathbf{a}, \mathbf{b}, \mathbf{c}$	$\mathbf{ab}, \mathbf{c}$	$\mathbf{a}, \mathbf{bc}$	$\mathbf{ab}, \mathbf{bc}$	$\mathbf{abc}$

Three-dimensional point groups

The selection of three characteristic vectors $\mathbf{a}, \mathbf{b}, \mathbf{c}$ (see Fig. 6 (right) for diamond) from each crystal cell is sufficient [Hestenes and Holt 2007; Hitzer and Perwass 2004, 2005a; Perwass and Hitzer 2005b] for generating all 3D point groups.

For the purpose of point groups keeping a single cell *as a whole invariant*, the vectors $\mathbf{a}, \mathbf{b}, \mathbf{c}$ have always to be attached to the invariant cell center. These three vectors are normals of characteristic planes passing through the cell center. The plane reflections which the vectors represent and their combinations as in (A6) to (A9) constitute all point symmetries of the 3D crystal cells. The point symmetry transformations are applied to every vertex of a cell and keep the cell as a whole invariant, transforming each vertex into another vertex.

Using $\angle(\mathbf{a},\mathbf{b})$ and $\angle(\mathbf{b},\mathbf{c})$ (right side of Fig. 7) we can denote all 32 3D point groups as in Table 2 ($pq = 43$ for diamond). Again the overbar notation, for example, $\bar{p}$, means that the two vectors concerned are only to be used in their fixed rotor combination, for example, $\mathbf{ab}$. If the closed overbar extends over both indexes $\overline{pq}$ all three vectors are only to be used in the fixed roto-inversion (alias rotary reflection) combination $\mathbf{abc}$ of (A8).

Note that the notation of Table 2 is fully isomorphic to the notation used in Table 2 of [Coxeter and Moser 1980], with p, q in the very same roles.

Space groups

The smooth composition with translations is best done in the conformal model [Lie 1871; Ahlfors 1986; Li 2008] of Euclidean space (in the GA of $\mathbb{R}^{4,1}$), which adds two null-vector dimensions for the origin e_0 and infinity e_∞,

$$X = \mathbf{x} + \frac{1}{2}\mathbf{x}^2 e_\infty + e_0, \quad e_0^2 = e_\infty^2 = X^2 = 0, \quad X \cdot e_\infty = -1. \qquad \text{(A13)}$$

The inner product of two conformal points gives their Euclidean distance and therefore a plane m equidistant from two points A, B as

$$X \cdot A = -\frac{1}{2}(\mathbf{x}-\mathbf{a})^2 \Rightarrow X \cdot (A-B) = 0, \quad m = A - B \propto \mathbf{n} - d\,\mathbf{e}_\infty, \tag{A14}$$

where $\mathbf{n}$ is a unit normal to the plane and d its signed scalar distance from the origin. Reflecting at two parallel planes m, m' with distance $\mathbf{t}/2$ we get the *transla*tion opera*tor* (by $\mathbf{t}$)

$$X' = m'mX\,mm' = T_{\mathbf{t}}^{-1}XT_{\mathbf{t}}, \quad T_{\mathbf{t}} = 1 + \frac{1}{2}\mathbf{t}\mathbf{e}_\infty. \tag{A15}$$

Reflection at two non-parallel planes m, m' yields the rotation around the m, m'-intersection by twice the angle subtended by m, m'.

Group theoretically (i.e. by the identical structure of multiplication tables) the conformal group $C(3)$ is isomorphic to $O(4,1)$ and the Euclidean group $E(3)$ is the subgroup of $O(4,1)$ leaving infinity $\mathbf{e}_\infty$ invariant. Now general translations and rotations are represented by geometric products of invertible vectors (called Clifford monomials, Lipschitz elements, or *versors*).

Applying these techniques one can compactly tabulate geometric space group symbols and generators [Hestenes and Holt 2007]. Diamond has space group F_d43 with generators $\{\mathbf{a}T_{\mathbf{c}/4}, \mathbf{b}, \mathbf{c}, T_{\mathbf{a}}, T_{\mathbf{b}/2}, T_{\mathbf{c}/2}\}$ (SGV correction to [Hestenes and Holt 2007]!).

Acknowledgments

I wish to thank God for his wonderful creation with the words of H. Grassmann: "Ich glaube also den in der Bibel geoffenbarten Wahrheiten nicht darum, weil sie in der Bibel stehen, sondern weil ich ihre seligmachende Kraft, ihre ewige, göttliche Wahrheit in meinem Bewußtsein erfahren habe." [H. Grassmann 1878] (Translation: I therefore believe the truths revealed in the Bible, not because they are written in the Bible, but because I have consciously experienced their power of blessing, their eternal, divine truth.) I thank my family for their loving support, and I thank O. Giering, D. Hestenes, C. Perwass, M. Aroyo, D. Litvin, and H.-J. Petsche.

From Grassmann's vision to geometric algebra computing

Dietmar Hildenbrand

Introduction

What mathematicians often call *Clifford algebra* is called *geometric algebra* if the focus is on the geometric meaning of the algebraic expressions and operators. Geometric algebra is a mathematical framework to easily describe geometric concepts and operations. It allows us to develop algorithms fast and in an intuitive way. It is based on the work of Hermann Grassmann [A2] and his vision of a general mathematical language for geometry. William Clifford combined Grassmann's exterior algebra and Hamilton's quaternions [Clifford 1882a, 1882b]. Pioneering work has been done by David Hestenes, who first applied geometric algebra to problems in mechanics and physics [Hestenes and Sobczyk 1984; Hestenes 1985]. His work culminated some years ago in the invention of conformal geometric algebra [Hestenes 2001]. Grassmann's outer product leads to high-dimensional multivectors. Currently, we can recognize a shift from single-processor platforms to parallel computing platforms which are able to treat these multivectors in an efficient way. During the past decades, especially from 1986 until 2002, processor performance has doubled every 18 months. Currently, this improvement law is no longer valid because of technical limitations. The new trend

This work was supported by the DFG (Deutsche Forschungsgemeinschaft) project HI 1440/1-1.

H.-J. Petsche et al. (eds.), *From Past to Future: Graßmann's Work in Context*,
DOI 10.1007/978-3-0346-0405-5_37,

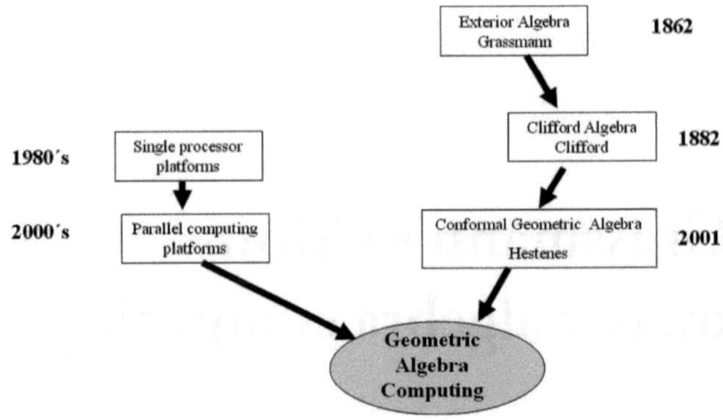

Fig. 1. Two trends towards what we call geometric algebra computing

towards parallel systems will most likely dominate the future. Thanks to multi-core architectures or powerful graphics boards for instance based on the CUDA technology from Nvidia, the ATI stream technology of AMD, or on the future Larrabee technology of Intel, one can expect impressive results using the powerful language of geometric algebra. We call this combination of applying geometric algebra on parallel platforms *geometric algebra computing* (Fig. 1)

In this paper we highlight some benefits of conformal geometric algebra and present some examples of *geometric algebra computing* on different parallel computing platforms.

Benefits of conformal geometric algebra

Geometric algebra as a general mathematical system unites many mathematical concepts such as vector algebra, quaternions, Plücker coordinates and projective geometry, and it easily deals with geometric objects, operations, and transformations. Many applications in computer graphics, computer vision, and other engineering areas can benefit from these properties. In a ray-tracing application, for instance, the intersection of a ray and a bounding sphere is needed. According to Fig. 2, this can be easily expressed with the help of the outer product of these two geometric entities.

In the following we highlight some of the properties of geometric algebra that make it advantageous for many engineering applications.

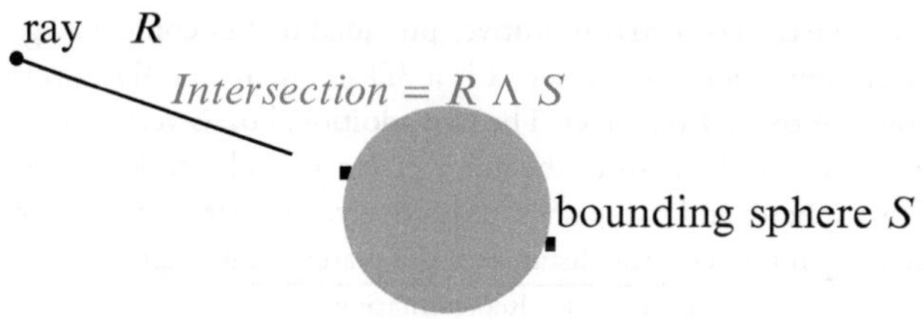

Fig. 2. Spheres and lines are basic entities of geometric algebra computation. Operations such as their intersection are easily expressed with the help of their outer product. In a ray-tracing application, for instance, the result of the intersection of a ray and a (bounding) sphere is another geometric entity: the point pair of the two points of the line intersecting the sphere. The sign of the square of the point pair easily indicates whether there is a real intersection or not

Table 1. Multiplication table of 2D geometric algebra. This algebra consists of basic algebraic objects of grade (dimension) 0, the scalar, of grade 1, the two basis vectors e_1 and e_2, and of grade 2, the bivector $e_1 \wedge e_2$, which can be identified with the imaginary number i squaring to -1

	1	e_1	e_2	$e_1 \wedge e_2$
1	1	e_1	e_2	$e_1 \wedge e_2$
e_1	e_1	1	$e_1 \wedge e_2$	e_2
e_2	e_2	$-e_1 \wedge e_2$	1	$-e_1$
$e_1 \wedge e_2$	$e_1 \wedge e_2$	$-e_2$	e_1	-1

Unification of mathematical systems

In the wide range of engineering applications many different mathematical systems are currently used. One notable advantage of geometric algebra is that it subsumes mathematical systems such as vector algebra, complex analysis, quaternions, or Plücker coordinates. Table 1, for instance, describes how complex numbers can be identified within 2D geometric algebra. This algebra not only contains the two basis vectors e_1 and e_2, but also basis elements of grade (dimension) 0 and 2. Grade 0 represents the scalar. With the help of the rules of geometric algebra it can be shown quite easily that the element $e_1 \wedge e_2$ (of grade 2) squares to -1 and can be identified with the imaginary unit i. The linear combination of these two elements describe all the complex numbers. This imaginary unit as well as all the imaginary units of quaternions, describing 3D rotations, can be identified in Fig. 4 as elements in conformal geometric algebra, the geometric algebra of conformal space.

Table 2. List of the basic geometric primitives provided by 5D conformal geometric algebra. The bold characters represent 3D entities ($\mathbf{x}$ is a 3D point, $\mathbf{n}$ is a 3D normal vector, and $\mathbf{x}^2$ is the scalar product of the 3D vector $\mathbf{x}$). The two additional basis vectors e_0 and e_∞ represent the origin and infinity. On the basis of the outer product, circles and lines can be described as intersections of two spheres and two planes, respectively. The parameter r represents the radius of the sphere and the parameter d the distance of the plane to the origin

Entity	Representation
Point	$P = \mathbf{x} + \frac{1}{2}\mathbf{x}^2 e_\infty + e_0$
Sphere	$S = P - \frac{1}{2}r^2 e_\infty$
Plane	$\pi = \mathbf{n} + d e_\infty$
Circle	$Z = S_1 \wedge S_2$
Line	$L = \pi_1 \wedge \pi_2$

Fig. 3. Spheres and circles are basic entities of geometric algebra. Operations such as the intersection of two spheres are easily expressed

Intuitive handling of geometric objects

Conformal geometric algebra is a 5D geometric algebra based on the 3D basis vectors e_1, e_2 and e_3 as well as on the two additional basis vectors e_0 representing the origin and e_∞ representing infinity. This algebra is able to easily treat different geometric objects. Table 2 shows the representation of points, spheres, planes, circles, and lines as the same entities algebraically. Consider the spheres of Fig. 3, for instance, which are simply represented by

$$S = P - \frac{1}{2}r^2 e_\infty, \tag{1}$$

based on their center point P, their radius r, and the basis vector e_∞. The circle of intersection of two spheres is then easily computed using the outer product to operate on the spheres as if they were vectors,

$$Z = S_1 \wedge S_2. \tag{2}$$

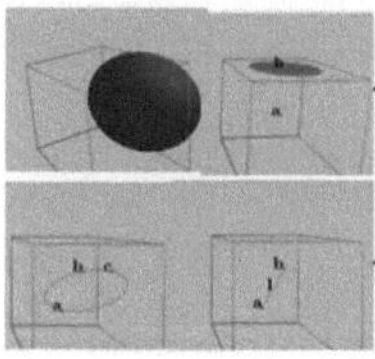

grade	term	blades	nr.
0	scalar	1	1
1	vector	$e_1, e_2, e_3, e_0, e_\infty$	5
2	bivector	$e_1 \wedge e_2, e_1 \wedge e_3, e_2 \wedge e_3, e_1 \wedge e_\infty, e_2 \wedge e_\infty, e_3 \wedge e_\infty, e_1 \wedge e_0, e_2 \wedge e_0, e_3 \wedge e_0, e_0 \wedge e_\infty$	10
3	trivector	...	10
4	quadvector	$e_1 \wedge e_2 \wedge e_3 \wedge e_\infty, e_1 \wedge e_2 \wedge e_3 \wedge e_0, e_1 \wedge e_2 \wedge e_0 \wedge e_\infty, e_1 \wedge e_3 \wedge e_0 \wedge e_\infty, e_2 \wedge e_3 \wedge e_0 \wedge e_\infty$	5
5	pseudoscalar	$e_1 \wedge e_2 \wedge e_3 \wedge e_0 \wedge e_\infty$	1

3.1416

i, j, k

Fig. 4. The blades of conformal geometric algebra. Spheres and planes, for instance, are vectors. Lines and circles can be represented as bivectors. Other mathematical systems like complex numbers or quaternions can be identified based on their imaginary units i, j, k. This is why transformations like rotations can also be handled within the algebra

This way of computing with conformal geometric algebra clearly benefits engineering applications.

Intuitive handling of geometric operations

Looking more closely at the details of conformal geometric algebra, it consists of blades with *grades* (dimension) 0, 1, 2, 3, 4, and 5, whereby a scalar is a *0-blade* (blade of grade 0). The element of grade 5 is called the pseudoscalar. A linear combination of blades is called a *k-vector*. So a bivector is a linear combination of blades with grade 2. Other *k*-vectors are vectors (grade 1), trivectors (grade 3), and quadvectors (grade 4). Furthermore, a linear combination of blades of different grades is called a *multivector*. Multivectors are the general elements of a geometric algebra. Figure 4 lists all the 32 blades of conformal geometric algebra. It shows the power of algebraic expressions of geometric algebra in describing different geometric objects as well as geometric operations. Operations such as rotations, translations [Hildenbrand et al. 2004], and reflections can be easily treated within the algebra. There is no need to change the way of describing them with other approaches (vector algebra, for instance, additionally needs matrices in order to describe transformations).

Robotics application example

Let us look, for instance, at an inverse kinematics application of the simple robot of Fig. 5. Assuming that we already know the locations of the joints P_0 and P_2, we have

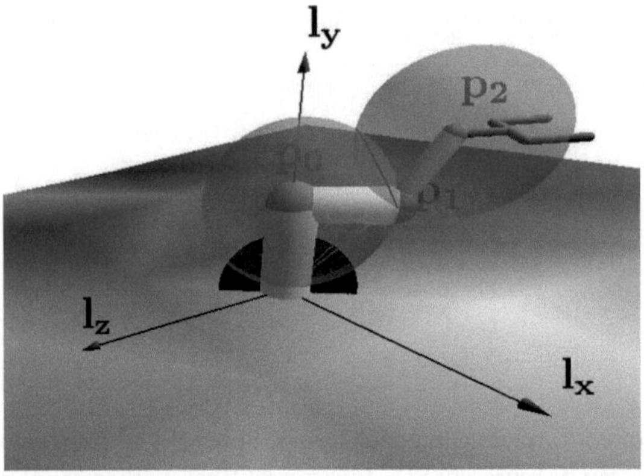

Fig. 5. Computation of P_1

to compute P_1 in the next step. Computing this point is usually a difficult task. However, using conformal geometric algebra we can determine it by intersecting the spheres S_1 and S_2,

$$S_1 = P_0 - \frac{1}{2}d_2^2 e_\infty, \tag{3}$$

$$S_2 = P_2 - \frac{1}{2}d_3^2 e_\infty, \tag{4}$$

with the plane π_1 (describing the plane the joints are acting in)

$$Pp_1 = S_1 \wedge S_2 \wedge \pi_1. \tag{5}$$

The result is a point pair Pp_1 and we have to choose one point from this point pair.

The complete algorithm may be found in [Hildenbrand 2005].

Geometric algebra computing technology

The power of geometric algebra as described in the previous section comes along with a complex algebraic structure of high-dimensional multivectors. Fortunately, this algebraic structure offers a high potential for optimization and parallelization that we can advantageously use for highly efficient implementations on current parallel processor platforms.

Multivectors of a n-dimensional geometric algebra are 2^n-dimensional. At first glance, this seems to be computationally very expensive. But there is a lot of potential for optimization and parallelization of multivectors by symbolically precomputing geometric algebra expressions. Since all of the coefficients of the multivectors can be computed in parallel, geometric algebra computations benefit significantly from parallel structures.

This is why we separate geometric algebra computing into two layers:

- Geometric algebra (GA) compilation layer.
- Platform layer.

At the GA compilation layer, geometric algebra operations such as geometric product, outer product, inner product, dual and reverse on multivectors are handled. These operations are compiled to the platform layer. On this layer only basic arithmetic operations on multivectors with a high potential for efficient computations on parallel platforms are available.

Our geometric algebra computing architecture is presented in Fig. 6. Algorithms (described by the geometric algebra programming language CLUCalc [Perwass 2005]) are compiled to an intermediate representation using an adequate compilation approach. On the basis of this representation implementations for different sequential and parallel platforms can be derived.

We are presently developing a compiler for reconfigurable hardware using the Maple-based compilation approach (see the following subsection). A proof-of-concept of our approach has been done with the help of a FPGA (field-programmable gate array) implementation of an inverse kinematics algorithm. Naively implemented, the inverse kinematics algorithm was initially slower than the conventional one. However, with our symbolic computation optimization approach the software implementation became three times faster [Hildenbrand et al. 2006] and with a hardware implementation about 300 times faster [Hildenbrand et al. 2008] (three times by software

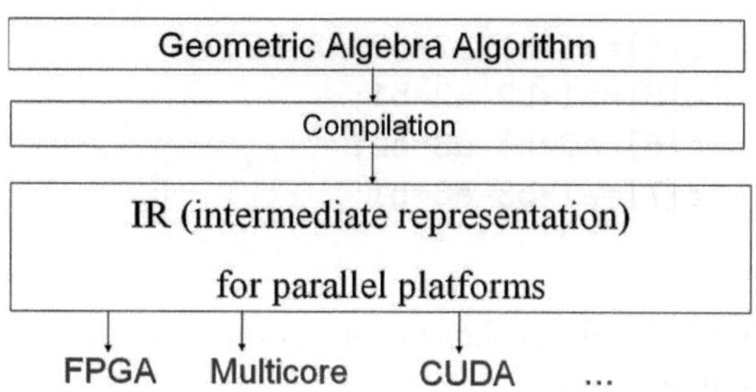

Fig. 6. Geometric algebra computing architecture. Algorithms are compiled to an intermediate representation for compilation to different computing platforms

optimization and 100 times by additional hardware optimization) than the conventional software implementation. Herewith, we could show for the first time that implementations of geometric algebra algorithms can be faster than solutions based on standard mathematics.

Some examples of geometric algebra computing based on current parallel architectures may be found below. They vary in different parallelization concepts such as multi-core or SIMD (single instruction multiple data).

Compilation

In order to achieve highly efficient implementations, geometric algebra algorithms have to be optimized first. Currently we use a Maple-based compilation approach. It needs the commercial Maple system [MAP 2009] as well as a specific geometric algebra package [Abłamowicz and Fauser 2009] and is restricted to geometric algebras with dimension at most 9. The Maple-based compilation uses the powerful symbolic computation features of Maple. Since all of the results of geometric algebra operations on multivectors are again multivectors we symbolically compute and simplify the resulting multivectors in order to determine which of the coefficients are actually needed and what is the simplest expression for each coefficient. The following short algorithm, for instance,

```
a=a1*e1+a2*e2+a3*e3;
b=b1*e1+b2*e2+b3*e3;
c=a*b;
```

computes the geometric product of two 3D vectors and assigns it to the multivector c. The compiler optimizes this algorithm in the following form:

```
c[1]=a1*b1+a2*b2+a3*b3;
c[5]=a1*b2-a2*b1;
c[6]=a2*b3-a3*b2;
c[7]=a1*b3-a3*b1;
```

with simple arithmetic operations for each coefficient which is needed for the multivector c.

Adaptation to different parallel processor platforms

One goal of this research is to adapt geometric algebra computing to the best suitable parallel processor platforms. Below are some examples of how this could work from the point of view of the programming languages of different example architectures.

OpenMP. OpenMP can be used to parallelize GA algorithms. The programming language C can be extended with OpenMP directives for an incremental approach to parallelizing code. For details on OpenMP, please refer to [Chapman, Jost and van der Pas 2008].

OpenMP supports task-parallel computations. The data of all the different threads is shared by default. This is why the coefficients of multivectors can be computed in parallel (as well as independent multivectors). Using OpenMP for C, parallel sections can be programmed as follows:

```
#pragma omp parallel {
  #pragma omp sections {
    #pragma omp section
    ... Section 1 ...
    #pragma omp section
    ... Section 2 ...

    #pragma omp section
    ... Section 3 ...

  }/*End of sections block */

} /*End of parallel region */
```

Each section consists of computations of different coefficients of a multivector as well as of computations of independent multivectors.

Intel's Ct. Intel researchers are developing Ct, or C/C++ for Throughput Computing [Rajagopalan and Ghuloum 2007] to support their new multi-core platforms.

Ct offers parallelism on so-called *indexed vectors* suitable for sparse multivectors. The first step of our example generates a multivector which can be described as the following indexed vector

```
c= [(1 -> a1*b1+a2*b2+a3*b3),
```

```
    (5 -> a1*b2-a2*b1),
    (6 -> a2*b3-a3*b2),
    (7 -> a1*b3-a3*b1),
    (_ -> 0)
   ]
```

Note that the underscore denotes a default value for empty coefficients.

All operators on indexed vectors are implicitly parallel. This is why the addition of multivectors

```
d=a+c;
```

can be done very easily in Ct.

ATI Stream. The ATI stream technology combines multiple thread computing with parallel computing within the threads. The following sample code computes the geometric product of the above example with the help of float4 vectors.

```
kernel void MV (float4 a<>,
                float4 b<>,
                out float4 c<>){
  float4 result;
  result.x=a.x*b.x+a.y*b.y+a.z*b.z;
  result.y=a.x*b2-a2*b.x;
  result.z=a.y*b.z-a.z*b.y;
  result.w=a.x*b.z-a.z*b.x;
  c=result;
}
```

The four computations for the coefficients x, y, z, w are computed in parallel.

CUDA. CUDA [NVIDIA 2009] is Nvidias technology for their parallel computing platforms. It supports multiple threads which are able to run the same code with different data on many parallel processors. This SIMD (single instruction multiple data) technology can be used – as usual – for geometric algebra algorithm tasks with the same code operating on different data.

For high-dimensional algebras, there is also another advantage. The results of products in n-dimensional geometric algebras are always 2^n-dimensional multivectors. Using multiplication tables, each of the 2^n coefficients can be computed as a sum of (signed or unsigned) products of coefficients of the multivectors to be multiplied. These computations can be distributed to 2^n threads, each computing one coefficient. The kernel code can be identical for all the threads, assuming that each thread knows its individual part of the multiplication table.

Conclusion

Geometric algebra and especially 5D conformal geometric algebra can be applied in a wide range of engineering applications with a geometric background. It is a very powerful mathematical framework in terms of geometric intuitiveness, compactness, and simplicity.

Who can benefit from the properties of geometric algebra? There is no need for students to learn different mathematical systems and the translations between them, because many other mathematical systems are already included in geometric algebra. Researchers can benefit when developing new solutions in their field of research. They are able to gain new insights based on a more general mathematical system. From the academic point of view, geometric algebra computing is a very inter-disciplinary topic between mathematics, computer science, and engineering. It is still a basic research topic, but with a high potential for engineering applications in industry. For companies, geometric algebra technology can lead to an enhancement of quality as well as to a reduction of costs for the development, documentation, and maintenance of their products and solutions.

Last but not least there is a longer term vision based on geometric algebra technology: For the hardware industry the need for suitable parallel architectures is quite evident. After adapting geometric algebra algorithms to current hardware architectures, future architectures can even be influenced and driven by geometric algebra computing technology.

Grassmann, Pauli, Dirac: special relativity in the schoolroom

Martin Erik Horn

Introduction

Hermann Günther Grassmann spent nearly all of his professional life as a teacher of mathematics and physics in Stettin. Unfortunately, it is not known to what extent he taught his new mathematical concepts at school and how intensively he tried to implement his theory of extensions in his way of teaching standard mathematics and standard physics in the schoolroom. No special traces of Grassmann's modern mathematical concepts can be found by analysing the recollections of his students. Their reports only seem to reflect difficulties with classroom discipline and the students' inability to grasp large parts of his mathematical teachings. Modern didactical concepts were rare and mathematics seemed to be taught in the same way that languages were taught. And, very often, the only task of the students was to repeat what the teacher had presented [Schwartze 1996].

But now, more than 175 years after Grassmann started his teaching career in 1834 in Berlin, two very important things have changed: Firstly, mathematics and physics education specialists have developed new teaching methods. And we also now know about the importance of self-regulated student activities and the possibility and necessity for students to solve mathematical problems on their own.

H.-J. Petsche et al. (eds.), *From Past to Future: Graßmann's Work in Context*,
DOI 10.1007/978-3-0346-0405-5_38,

Secondly, large parts of the mathematics and physics curriculum have been rearranged and reconstructed in a didactical manner. Mathematics and physics teachers now follow different, well-tested paths when teaching their subjects, which have been evaluated with respect to learning student outcomes [Kircher, Girwidz and Häußler 2007].

Therefore, I felt the time had come for a new attempt at implementing large parts of Grassmann's ideas when teaching physics or mathematics. Anti-commutative relations, which form the centre of Grassmann's way of reconstructing mathematical and physical phenomena, are nothing mysterious. On the contrary, they might be a valuable tool when applied in a convincing context. And the context I have chosen for my classroom experiment is – for a number of reasons – special relativity.

Grassmann's mathematical parenthood

In a newspaper article describing aims and intentions of the Grassmann Bicentennial Conference, Wienand characterised Grassmann as the forefather of linear algebra[1] and wrote: "Yet Grassmann investigated vectors in abstract spaces with arbitrarily many dimensions. Herewith he stands at the beginning of a branch of mathematics, without which Albert Einstein would not later have been able to develop special relativity." [Wienand 2009][2].

But Grassmann is not only the forefather of the usual vector and tensor algebra we apply when we solve present-day physics problems conceptually. Grassmann's ideas developed into the innumerable concepts we possess today to model our world mathematically. Even such strange, unwieldy (but nevertheless interesting) mathematical peculiarities like the pseudo-complex non-associative algebra Tsabary and Censor use to describe special relativity [Tsabary and Censor 2004] can be traced back to Grassmann.

At the core of all of these mathematical instruments we find the concept of non-commutativity, which is one of the essential pillars of modern mathematics Grassmann invented. Relativity cannot be thought about without it. But what is relativity?

---

[1] "Erzvater dieser Disziplin" [Wienand 2009].

[2] "Graßmann allerdings untersuchte Vektoren in abstrakten Räumen mit beliebig vielen Dimensionen. Damit steht er auch am Anfang eines Mathematikzweiges, ohne den später Albert Einstein seine Relativitätstheorie nicht hätte entwickeln können" [Wienand 2009].

Space and perception

The stage of physics is three-dimensional space, which Einstein widened into a four-dimensional spacetime. Thus special relativity is the interaction between space and time, resulting in this very powerful notion of spacetime.

Yet before we can understand spacetime we should try to understand what space is. That is not an easy task because our impression of the three-dimensional world is heavily distorted by our ability to see [Bohm 2006]. "The spatial intuition that sighted people have is based on the image of the world that is projected onto their retinas; thus it is a two- (and not three-) dimensional image that is analysed in the brain of a sighted person. A blind person's spatial intuition, on the other hand, is primarily the result of tactile and operational experience. It is also deeper – in the literal as well as the metaphorical sense." [Sossinsky 2002, 13]. Knot theorist Alexei Sossinsky therefore concluded that it is not surprising at all that "almost all blind mathematicians are (or were) geometers." [Sossinsky 2002, 13].

To motivate students to think about the nature of space, it is helpful to demonstrate some of the strange properties of space which we do not expect. The spin of $1/2$ is such a property which we normally assign to quantum mechanical objects like electrons or other fermions. In fact, a spin of $1/2$, or a rotation symmetry of 4π, is not a quantum mechanical quality, but a simple consequence of the fact that we live in a three-dimensional world. Stephen Hawking actually used this fact to explain Grassmann numbers to a wider audience [Hawking 2003, 56f.]. "[...] rotating one time around can be distinguished from doing nothing at all. In fact, it's rotating twice around that is about the same as doing nothing," Feynman noted, when he demonstrated Dirac's visualisation of a spin of $1/2$ [Feynman 1987, 29f.].

Following Misner, Thorne and Wheeler, my teaching unit about special relativity therefore starts with a demonstration of the disentanglement of an object which was rotated twice, that is about 4π [Misner, Thorne and Wheeler 1995, 1149]. This can be done with a magazine fixed with several strings and rotated about a horizontal axis. An analysis of this "Dirac belt trick" with some didactical hints can for example be found in [Hart, Francis and Kauffman 1994].

Mathematical models of space

"Physicists employ a miscellaneous assortment of mathematical tools in ways that contribute to a fragmentation of knowledge. We can do better!" These words can be found in the Oersted medal lecture of David Hestenes [Hestenes 2003a, 104], referring to

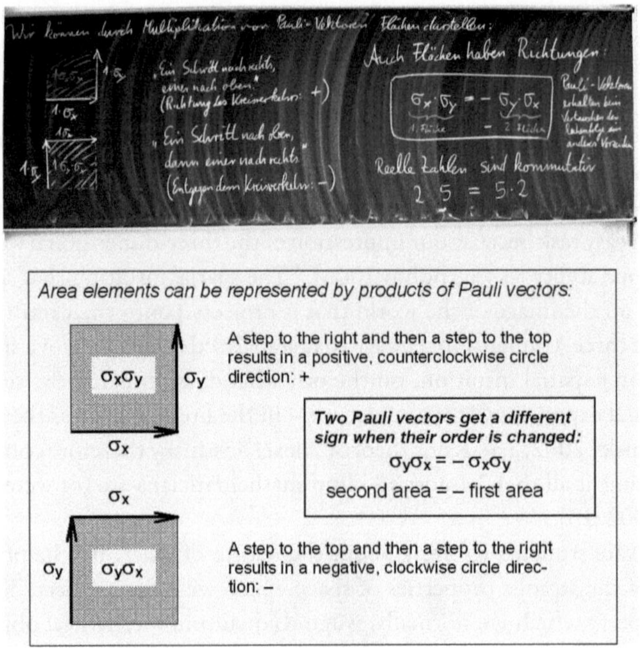

Fig. 1. Blackboard presentation of base vector multiplication

all the different mathematical concepts used by physicists (see for example Fig. 1 in [Hestenes 2003a, 106]). All of these tools are applied to physical situations which appear in three-dimensional space – hence they must include an intrinsic mathematical model of space. This "wasteful redundancy" [Hestenes 2003a, 106] can be resolved following the idea of Grassmann to stick to geometry when modelling space.

Therefore we should not ask how we can apply algebraic structures to space, but we should follow the opposite direction, like Grassmann did before us: How can we use basic elements of space to form an algebra?

Thus, the starting point is orthonormal base vectors which can be called $\boldsymbol{\sigma}_x$, $\boldsymbol{\sigma}_y$, and $\boldsymbol{\sigma}_z$. A multiplication of two of these base vectors results in a basic area element (see Fig. 1). If the order of multiplication is changed, the area element changes its sign. This anticommutativity is inherent in every spatial structure. We can hide it and confuse students, using vector algebra with two different multiplications, or we can show and state directly this structural feature of space, using the geometric product [Doran and Lasenby 2003]:

$$\boldsymbol{\sigma}_x\boldsymbol{\sigma}_y = -\boldsymbol{\sigma}_y\boldsymbol{\sigma}_x$$
$$\boldsymbol{\sigma}_y\boldsymbol{\sigma}_z = -\boldsymbol{\sigma}_z\boldsymbol{\sigma}_y$$
$$\boldsymbol{\sigma}_z\boldsymbol{\sigma}_x = -\boldsymbol{\sigma}_x\boldsymbol{\sigma}_z$$

Grassmann did nothing else when he formulated his algebra of extensions. The relation between this external product and the internal product of two vectors is "exceedingly interesting", as Charles Sanders Peirce commented in 1877 [Peirce 1877, 116], because "in truth, Grassmann has got hold (though he did not say so) of an eight-fold algebra" [Peirce 1877, 115].

The next step is to apply another multiplication to get the product of all three base vectors:

$$\boldsymbol{\sigma}_x\boldsymbol{\sigma}_y\boldsymbol{\sigma}_z = I$$

It "is a solid, and therefore a new kind of quantity" [Peirce 1877, 115]. This is the reason why we are able to represent space using Grassmann's ideas. Whatever geometric objects exist in three-dimensional space, they can be represented by a linear combination of these eight basic elements:

One scalar	**1**	representing numbers
Three base vectors	$\sigma_x, \sigma_y, \sigma_z$	representing lines
Three area elements	$\sigma_x\sigma_y, \sigma_y\sigma_z, \sigma_z\sigma_x$	representing planes
One volume element	$\sigma_x\sigma_y\sigma_z$	representing solids

Peirce used his own style of writing mathematical formulae ("an eight-fold algebra, which may be written in my system as follows: [...]" [Peirce 1877, p. 116]). The multivector $\underline{\mathbf{M}}$ he proposed can be written in our present-day mathematical language as

$$\underline{\mathbf{M}} = a + b_1\boldsymbol{\sigma}_x + b_2\boldsymbol{\sigma}_y + b_3\boldsymbol{\sigma}_z + c_1\boldsymbol{\sigma}_x \wedge \boldsymbol{\sigma}_y + c_2\boldsymbol{\sigma}_y \wedge \boldsymbol{\sigma}_x + c_3\boldsymbol{\sigma}_z \wedge \boldsymbol{\sigma}_x + d\boldsymbol{\sigma}_x \wedge \boldsymbol{\sigma}_y \wedge \boldsymbol{\sigma}_z$$

which of course can be identified with the multivector

$$\underline{\mathbf{M}} = a + b_1\boldsymbol{\sigma}_x + b_2\boldsymbol{\sigma}_y + b_3\boldsymbol{\sigma}_z + c_1\boldsymbol{\sigma}_x\boldsymbol{\sigma}_y + c_2\boldsymbol{\sigma}_y\boldsymbol{\sigma}_x + c_3\boldsymbol{\sigma}_z\boldsymbol{\sigma}_x + d\boldsymbol{\sigma}_x\boldsymbol{\sigma}_y\boldsymbol{\sigma}_z$$

$$a,\ b_i,\ c_i,\ d \in R$$

because the exterior product and the geometric product of different base vectors are equal:

$$\boldsymbol{\sigma}_x \wedge \boldsymbol{\sigma}_y = \boldsymbol{\sigma}_x \boldsymbol{\sigma}_y$$
$$\boldsymbol{\sigma}_y \wedge \boldsymbol{\sigma}_z = \boldsymbol{\sigma}_y \boldsymbol{\sigma}_z$$
$$\boldsymbol{\sigma}_z \wedge \boldsymbol{\sigma}_x = \boldsymbol{\sigma}_z \boldsymbol{\sigma}_x$$

and

$$\boldsymbol{\sigma}_x \wedge \boldsymbol{\sigma}_y \wedge \boldsymbol{\sigma}_z = \boldsymbol{\sigma}_x \boldsymbol{\sigma}_y \boldsymbol{\sigma}_z$$

The influence Peirce had on communicating Grassmann's ideas cannot be overestimated – and he was a friend of William Kingdon Clifford [Chisholm 2002, 2]. Until shortly before his death in March 1879, Clifford knew very little about Grassmann's theory of extensions, but then, one year before his death, he published a paper clarifying the relation between the theory of extensions and his biquaternions.

In this paper Clifford celebrated Grassmann for his deep and profound insight. Clifford, feeling his poor health, deeply regretted that he did not have more time to apply Grassmann's theory in further detail: "I propose to communicate in a brief form some applications of Grassmann's theory, which it seems unlikely that I shall find time to set forth at proper length, though I have waited long for it. Until recently I was unacquainted with the Ausdehnungslehre, and knew only so much of it as is contained in the author's geometrical papers in Crelle's Journal and in Hankel's Lectures on Complex Numbers. I may, perhaps, therefore be permitted to express my profound admiration of that extraordinary work, and my conviction that its principles will exercise a vast influence upon the future of mathematical science." [Clifford 1878, 350].

Although Chisholm states that "it is intriguing to note that Grassmann may well not have known of Clifford's 1876 abstract or Clifford of the Grassmann 1877 paper" [Chisholm 2002, 54f.], it is very likely that Clifford had learned a lot about Grassmann's concepts by reading notes by Peirce about Grassmann because Clifford, in his 1878 paper, refers to Peirce several times ("in the sense of Professor Peirce" or "in the language of Professor Peirce", see [Clifford 1878, 356]).

Didactical aspects of the geometric product

Didactically, the geometric product $\underline{a}\,\underline{b}$ of two vectors

$$\underline{a} = a_1 \boldsymbol{\sigma}_x + a_2 \boldsymbol{\sigma}_y + a_3 \boldsymbol{\sigma}_z$$

and

$$\underline{b} = b_1 \boldsymbol{\sigma}_x + b_2 \boldsymbol{\sigma}_y + b_3 \boldsymbol{\sigma}_z$$

can be treated in two different ways. It can be considered as an educational starting point, and then split into two different parts, namely a scalar part and a bivector part:

$$\underline{\mathbf{a}}\,\underline{\mathbf{b}} = <\underline{\mathbf{a}}\,\underline{\mathbf{b}}>_0 + <\underline{\mathbf{a}}\,\underline{\mathbf{b}}>_2$$

Hestenes follows this didactical path in his Oersted Medal Lecture [Hestenes 2003a, 107] when he defines the geometric product and then decomposes it into the symmetric inner product and the antisymmetric outer product:

$$<\underline{\mathbf{a}}\,\underline{\mathbf{b}}>_0 = \underline{\mathbf{a}}\cdot\underline{\mathbf{b}} = \frac{1}{2}(\underline{\mathbf{a}}\,\underline{\mathbf{b}} + \underline{\mathbf{b}}\,\underline{\mathbf{a}}) = \underline{\mathbf{b}}\cdot\underline{\mathbf{a}}$$

$$<\underline{\mathbf{a}}\,\underline{\mathbf{b}}>_2 = \underline{\mathbf{a}}\wedge\underline{\mathbf{b}} = \frac{1}{2}(\underline{\mathbf{a}}\,\underline{\mathbf{b}} - \underline{\mathbf{b}}\,\underline{\mathbf{a}}) = -\underline{\mathbf{b}}\wedge\underline{\mathbf{a}}$$

Alternatively, this didactical way can be reversed, like Hestenes does in his influential book *New Foundations for Classical Mechanics* [Hestenes 1990]. In this case, the geometric product is presented as a synthesis of inner and outer products:

$$\underline{\mathbf{a}}\,\underline{\mathbf{b}} = \underline{\mathbf{a}}\cdot\underline{\mathbf{b}} + \mathbf{a}\wedge\underline{\mathbf{b}}$$

Now we have two educational starting points: two different products which have totally different symmetry properties.

This synthesis is "a kind of mathematical taboo" [Hestenes 1990, 30], and this taboo may have "delayed Grassmann from considering it" [ibid.]. It delayed him, but it did not stop him. "Grassmann combined his interior and exterior products into a single, central product. [...] The central product is precisely Clifford's product of vectors, which Grassmann arrived at independently from (and slightly prior to) Clifford" [Doran 1994, 4].

It is of interest to note that Grassmann called his central product formula "Fundamentalgleichung" (fundamental or essential equation, see [H. Grassmann 1877c, 378]), emphasising the importance of this mathematical synthesis. Of course, Grassmann inserted the minus sign correctly, which indicates a change from a left-handed coordinate system into a right-handed coordinate system and normally is a "potential source of great confusion" [Doran and Lasenby 2003, 34]. And, in his paper [H. Grassmann 1877c], Grassmann already pointed to an important possible application in mathematics and physics, which was later presented in a didactically clear and modern formulation in [Hestenes 1990; Doran 1994; Calvet 2001; Doran and Lasenby 2003] and many other works.

The Quantum-mechanical misconception

I was struck by geometric algebra when I first read a paper by [Gull, Lasenby and Doran 1993] with the ambiguous but nevertheless unambiguously precise title *Imaginary numbers are not real.* The authors present, in an overwhelmingly clear, impressive, striking, and at the same time provoking way, the essential and relevant features of geometric algebra. I even remember the line which struck me: "We have now reached the point which is liable to cause the greatest intellectual shock." [Gull, Lasenby and Doran 1993]. The lines of this paper, noted below, concerning Pauli matrices, indeed deeply impressed me.

Wolfgang Pauli invented Pauli matrices in May 1927 to model spin without relativity [Pauli 1927; see also Pais 2002, 289]. In physics courses at universities, Pauli matrices were usually introduced as operators, nearly always in the same way Enrico Fermi had done in his quantum mechanics lecture in 1954 [Fermi 1995, 108/109, slides 25–1 and 25–2]. Thus, students learned and were told that Pauli matrices are something intrinsically quantum mechanical. This is commented on by Gull, Lasenby and Doran with the words: "...our present thinking about quantum mechanics is infested with the deepest misconceptions. We believe, with David Hestenes, that geometric algebra is an essential ingredient in unravelling these misconceptions." [Gull, Lasenby and Doran 1993].

This misconception can be resolved by comparing Pauli matrices with imaginary numbers. Imaginary numbers can be interpreted in two totally different ways: First, they can be considered as coordinates in the Argand or Gauss plane. Secondly, imaginary numbers can be interpreted as operators causing, for example, rotations of mathematical objects in that plane. This dual nature of imaginary numbers is an important property and one reason why imaginary numbers can be applied in such an amazing range of different mathematical situations. It is even more amazing and highly astonishing that mathematicians and physicists ignore the fact that Pauli matrices have been showing a similar dual nature for more than 70 years!

Ten years after the invention of Pauli matrices, Élie Cartan wrote: "In particular if we consider the matrices associated with the base vectors

$$\mathrm{H}_1 = \begin{pmatrix} 0 & 1 \\ 1 & 0 \end{pmatrix}, \quad \mathrm{H}_2 = \begin{pmatrix} 0 & -\mathrm{i} \\ \mathrm{i} & 0 \end{pmatrix}, \quad \mathrm{H}_3 = \begin{pmatrix} 1 & 0 \\ 0 & -1 \end{pmatrix},$$

the squares of these matrices equal 1, and the product of any two of these will change sign if the order of the factors is reversed." [Cartan 1981, 44]. Therefore the base vectors H_1, H_2, and H_3 can immediately be identified with the base vectors $\boldsymbol{\sigma}_x$, $\boldsymbol{\sigma}_y$, and

σ_z of three-dimensional space mentioned earlier. We know that, but it needed a charismatic scientist like David Hestenes who dared to take Cartan's idea literally and who said about Pauli matrices: "I claim that their physical significance is derived solely from their correspondence with orthogonal directions in space." [Hestenes 2003a, 115]. Pauli matrices can be interpreted as base vectors of three-dimensional Euclidean space. Cartan surely thought about that when he wrote: "We shall often refer to the 'vector X' instead of the 'vector which is associated with the matrix X." [Cartan 1981, 43].

Cartan presented the whole mathematics needed to understand Pauli matrices: "In the real domain, the matrices 1, i, H_1, H_2, H_3, iH_1, iH_2, iH_3 (which are linearly independent in this domain) form an algebra of order 8 over the field of real numbers. Any element of the algebra is uniquely the sum of a real scalar, a real vector, a real bivector and a real trivector." [Cartan 1981, 45]. That is the "eight-fold algebra" mentioned in [Peirce 1877], which "Grassmann has got hold of."

It is of great importance didactically that "the representation of σ_i by 2×2 matrices is irrelevant to physics" [Hestenes 2003a, 115]. Therefore "we can concentrate on the algebra and forget explicit representations" [Baylis 2002, 12]. Instead of speaking about Pauli matrices or linear combinations of Pauli matrices it therefore makes sense to speak about Pauli vectors or Pauli multivectors. To distinguish between vector algebra and Pauli algebra in the schoolroom, vectors in vector algebra can be written with an arrow, while Pauli vectors are indicated by underlined letters.

Students can be trained in this style of writing at the blackboard by comparing situations without mixed terms. If there are no bivector parts and only scalar multiplication is used, all formulae will show an identical structure. As an example, circular motion has been discussed with students in this way.

Didactical aspects of special relativity

Special relativity is the interaction between space and time: Space can be transformed into time and time can be transformed into space. But instead of emphasising this transformational nature of relativity, the discussion of special relativity in nearly all physics books used in schools, high schools or universities in Germany, for example [Grehn 1991, 333], start with the experiment of Michelson and Morley.

"What a complicated start it is!" exclaims Hermann Bondi and goes on to analyse some didactical consequences: "First you have got to explain the picture of ideas of the nineteenth century, which we now know to be inapplicable, out of which state of mind the desire to make this experiment grew. Then you have to say that it did not show the result that had been expected; therefore something must be wrong in what we have just taught you. And into all this goes a good deal of description of an awkward, difficult experimental technique, and a little reference, perhaps, to the doubts that have arisen from time to time about this experiment. All this hasn't helped the understanding of relativity very much." [Bondi 1967, 25]. Similar arguments can be found in [Beyvers and Krusch 2009], trying to implement special relativity in a school-oriented way. And even authors discussing the Michelson–Morley experiment in length acknowledge that it is of minor importance for understanding special relativity [Grehn 1993, 271].

Bondi's conclusion is simple: "What one surely wants to do is take Einstein's principle of relativity and try and deduce from this principle easily observable results – which is not difficult. And then you can go forward." [Bondi 1967, 25].

This principle of relativity directly leads to Minkowski's spacetime interval.

Spacetime algebra

Special relativity seems to be a perplexing and at the same time contradictory mental concept. On the one hand, there is the first and foremost claim that space-like and time-like directions should be treated in the same way. This claim suggests that there is no conceptual difference between space and time – they are both mere shadows of the higher entity of spacetime. One reason for this conceptual identification of space and time is the fact that space can be transformed into time, and time can be transformed into space. We observe that in nature: Two observers who have a constant relative velocity with respect to each other measure different distances and time intervals between the same two events.

On the other hand we know, of course, that space-like directions and time-like directions are not equal. We can go forwards and backwards in space, but we only can go forwards in time, and never backwards. Space and time are different. Of course we observe this simple fact as well.

This feigned self-contradiction is one reason for Max Planck to assess that "this new conception of the meaning of time demands of physicists the very highest ability to abstract; this conception exceeds by far everything in boldness which until now has been achieved in speculative natural sciences [...]. Compared to it, non-Euclidean geometry is a children's playground." [Biener 2005][3].

Mathematically, the obvious differences between space and time appear as different signs of space and time in the spacetime interval of Hermann Minkowski [Minkowski

[3] "diese neue Auffassung des Zeitbegriffs an die Abstraktionsfähigkeit des Physikers die allerhöchsten Anforderungen stellt; sie übertrifft an Kühnheit wohl alles, was bisher in der spekulativen Naturforschung (...) geleistet wurde. Die nichteuklidische Geometrie ist Kinderspiel dagegen" [Biener 2005].

1909; Corry 1997]:

$$r^2 = (ct)^2 - x^2 - y^2 - z^2$$

There are now three structural possibilities: We can either carry along openly these different signs in all our relativistic equations. This then results in really confusing equations showing no symmetry between space and time.

Or we can use imaginary coordinates. This then results in complex equations, and very often we will have problems interpreting complex numbers in a geometrically satisfactory way.

Or we can assign (and sort of hide) these different signs to (or in) the base vectors of spacetime. This then results in equations showing the desired symmetry between space and time. The great advantage of this strategy is that we then only have real coefficients, which can be interpreted in a straightforward and geometrically convincing way. This is what spacetime algebra does. Four-dimensional vectors are then defined as

$$\underline{\underline{r}} = ct\boldsymbol{\gamma}_t + x\boldsymbol{\gamma}_x + y\boldsymbol{\gamma}_y + z\boldsymbol{\gamma}_z$$

These vectors can be called Dirac vectors. To distinguish them from vectors of vector algebra and Pauli algebra, they will be underlined twice in this text. Dirac vectors are symmetric with respect to space and time.

The signs only reappear when we evaluate relativistic equations getting squares of the base vectors:

Basis vector of time: $\boldsymbol{\gamma}_t^2 = 1$
Basis vector of x-direction in space: $\boldsymbol{\gamma}_x^2 = -1$
Basis vector of y-direction in space: $\boldsymbol{\gamma}_y^2 = -1$
Basis vector of z-direction in space: $\boldsymbol{\gamma}_z^2 = -1$

Again, the geometric products of two base vectors are anticommutative [Doran and Lasenby 2003]:

$\boldsymbol{\gamma}_x\boldsymbol{\gamma}_y = -\boldsymbol{\gamma}_y\boldsymbol{\gamma}_x$	$\boldsymbol{\gamma}_x\boldsymbol{\gamma}_t = -\boldsymbol{\gamma}_t\boldsymbol{\gamma}_x$
$\boldsymbol{\gamma}_y\boldsymbol{\gamma}_z = -\boldsymbol{\gamma}_z\boldsymbol{\gamma}_y$	$\boldsymbol{\gamma}_y\boldsymbol{\gamma}_t = -\boldsymbol{\gamma}_t\boldsymbol{\gamma}_y$
$\boldsymbol{\gamma}_z\boldsymbol{\gamma}_x = -\boldsymbol{\gamma}_x\boldsymbol{\gamma}_z$	$\boldsymbol{\gamma}_z\boldsymbol{\gamma}_t = -\boldsymbol{\gamma}_t\boldsymbol{\gamma}_z$

With these structural features we now posses an algebra which really fulfils Minkowski's promise in a mathematically concise way: "Henceforth, space by itself, and time by itself, are doomed to fade away in the shadows, and only a kind of union of the two will preserve an independent reality." [Minkowski 1909][4].

The quantum-mechanical misconception revisited

When reading the paper [Gull, Lasenby and Doran 1993] some years ago, the next "greatest intellectual shock" awaited me: The basis vectors of spacetime can be identified with Dirac matrices. "The Dirac matrices are no more and no less than matrix representations of an orthonormal frame of spacetime vectors and thereby they characterise spacetime geometry." [Hestenes 2003b, 694]. Dirac matrices are not quantum mechanical objects. They are the basis vectors of the world we live in since Einstein discovered special relativity.

But it is a great riddle for me why "Dirac never said any such thing!" [Hestenes 2003b, 694]. Dirac, being one of the central figures of theoretical physics, "has done more than anyone this century, with the exception of Einstein, to advance physics and change our picture of the universe" [Hawking 1998, XV]. He surely had the intellectual capacity to understand geometric algebra and spacetime algebra. But he decided to remain silent about that. "In fact, Dirac was unaware of Clifford's much earlier work and, in effect, had to rediscover the necessary part of Clifford's theory for himself", Roger Penrose explains [Chisholm 2002, 182]. But I cannot believe that Dirac was unaware of Cartan and his straightforward connection between Dirac algebra and the algebra of the space of special relativity (see Chap. VII of [Cartan 1981]). Although presented in "a different (and more limited) point of view, by Élie Cartan in 1913" as Penrose [Chisholm 2002, 182] wrote, everything needed by a genius like Dirac to grasp the deeper geometrical meaning of his matrices can be found in Cartan's book.

I guess that there must be psychological reasons why Dirac never spoke about Dirac matrices as basis vectors of four-dimensional spacetime. It seems that there is a strong psychological barrier which our contemporary way of speaking about physics erects between the domains of the quantum world and the classical world. One can even find the reverse situation in literature. Roger Penrose presents the basis vectors $\boldsymbol{\gamma}_1, \boldsymbol{\gamma}_2, \boldsymbol{\gamma}_3, \ldots, \boldsymbol{\gamma}_n$ of n-dimensional space and clearly uses Dirac algebra with some slightly different applications of the plus and minus signs in his well-written book *The road to reality* [Penrose 2005, 208f.]. But Penrose identifies a product of two basis vectors with quaternions and, at the same time, he identifies quaternions with the basis vectors in ordinary Euclidean space [Penrose 2005, 203] – which is really astonishing, because it is well

[4] "Von Stund' an sollen Raum für sich und Zeit für sich völlig zu Schatten herabsinken und nur noch eine Art Union der beiden soll Selbständigkeit bewahren" [Minkowski 1909].

known that "Hamilton attempted to identify pure quaternions (null scalar part) with vectors, but we now see that they are actually bivectors." [Doran and Lasenby 2003, 35].

Penrose calls his basis vectors $\boldsymbol{\gamma}_1, \boldsymbol{\gamma}_2, \boldsymbol{\gamma}_3, \ldots, \boldsymbol{\gamma}_n$ basic reflections, but basic reflections are basis vectors in geometric algebra: A "basic reflection (...) $\boldsymbol{\gamma}_r$ reverses the rth coordinate axis, while leaving all the others alone" [Penrose 2005, 209]. In geometric algebra the basis vector $\boldsymbol{\gamma}_r$ just shows this behaviour, because the transformation

$$\underline{\underline{\mathbf{r}}}' = \boldsymbol{\gamma}_r \underline{\underline{\mathbf{r}}} \boldsymbol{\gamma}_r \quad \text{(when} \quad \boldsymbol{\gamma}_r^2 = -1)$$

results in a reflection at the plane perpendicular to the basis vector $\boldsymbol{\gamma}_r$. Basic reflections and basis vectors are modelled by the same mathematical objects $\boldsymbol{\gamma}_k$. They are mathematically identical – and, again, mathematicians and physicists have been ignoring this dual nature for more than 70 years.

Remark about the history of the interpretation of Dirac matrices

Paul Adrien Maurice Dirac invented Dirac matrices in January 1928 to model the relativistic electron [Dirac 1928; Pais 2002, 290]. In physics courses at universities, Dirac matrices were usually introduced as operators, nearly always in the same way Enrico Fermi had done in his quantum mechanics lecture 1954 [Fermi 1995, 150–152, slides 34–1 till 34–3]. Thus students unfortunately learned and were told that Dirac matrices are something intrinsically quantum mechanical.

But already in 1935 Walter Franz noted in a paper presented to the Bavarian Academy of Sciences by Arnold Sommerfeld, that "there are certainly great advantages not to write down γ matrices in explicit form, but to use only the simple relations existing between them, when general relations are derived from the Dirac equation" [Franz 1935, 380][5]. And he claims that it is of great didactical and practical importance to use methods of solving the Dirac equation which guarantee that "the assignment of the $\boldsymbol{\gamma}$ matrices to coordinates – which is inherent to the Dirac equation – does not get lost" [Franz 1935, 381][6]. This was written in the introduction. Franz then tries hard to follow his concept using quaternions, with the consequence that his presentations show

[5] "Freilich bietet es bei der Ableitung allgemeiner Beziehungen aus der Dirac Gleichung (...) große Vorteile, die γ-Matrizen nicht explizit hinzuschreiben, sondern nur die zwischen ihnen bestehenden einfachen Beziehungen zu benützen" [Franz 1935, 380].

[6] "ein Lösungsverfahren, welches die in der Dirac-Gleichung vorhandene Zuordnung der γ-s zu den Koordinaten in keinem Augenblick verloren gehen läßt" [Franz 1935, 381]. The German word "Zuordnung" can be translated as relation, as identification or as assignment.

the usual problematic features mentioned earlier. But on the last page of his paper, he writes clearly and precisely about the identification of Dirac matrices with an orthonormal frame of space: The isotropy of Clifford numbers C_n "allows the assignment of all n $\boldsymbol{\gamma}_k$ to an n-dimensional coordinate space in an isotropic way. Since if you assign $\boldsymbol{\gamma}_k$ to the $\boldsymbol{x}_k$ axis, the coordinate axes are in no way different, it is rather possible to assign every direction to the unit vector e in a totally isotropic way by $(\boldsymbol{\gamma} e) = \Sigma e_{\mathrm{k}} \boldsymbol{\gamma}_{\mathrm{k}}$." [Franz 1935, 435][7]. Thus, the idea of identifying Dirac matrices with base vectors has been around since 1935. And this is the idea Hestenes evaluated 30 years later in such a convincing manner.

Main focus at school

At an advanced physics course[8] of a Berlin comprehensive school,[9] I taught special relativity using Pauli and Dirac algebra. The following topics were discussed with the students:

- Pauli representation of vectors in three-dimensional space
- Reflexion and rotation of Pauli vectors
- Orthogonality of Pauli vectors
- Dirac representation of vectors in four-dimensional spacetime
- Minkowski diagrams
- Time dilation
- Twin paradox
- Orthogonality of Dirac vectors
- Length contraction
- Lorentz transformation

Since most of the students did not attend the advanced mathematics course and therefore only had minor knowledge of vector algebra, the students did not have to relearn how to deal with vectors. Instead it was something new for them, and therefore they quickly adopted the anti-commutative structure of Pauli algebra. These students of

[7] "Sie gestattet es, die n $\boldsymbol{\gamma}_{\mathrm{k}}$ einem n-dimensionalen Koordinatenraum in isotroper Weise zuzuordnen. Wenn man nämlich $\boldsymbol{\gamma}_{\mathrm{k}}$ der $\mathbf{x}_{\mathrm{k}}$-Achse zuordnet, so sind die Koordinatenachsen damit keineswegs ausgezeichnet, es kann vielmehr jeder beliebigen Richtung mit dem Einheitsvektor e in vollkommen isotroper Weise $(\boldsymbol{\gamma} e) = \Sigma e_{\mathrm{k}} \boldsymbol{\gamma}_{\mathrm{k}}$ zugeordnet werden." [Franz 1935, 435].

[8] In German, these classes with a special focus in grades 12 and 13 are called "Leistungskurs".

[9] In German: "Gesamtschule".

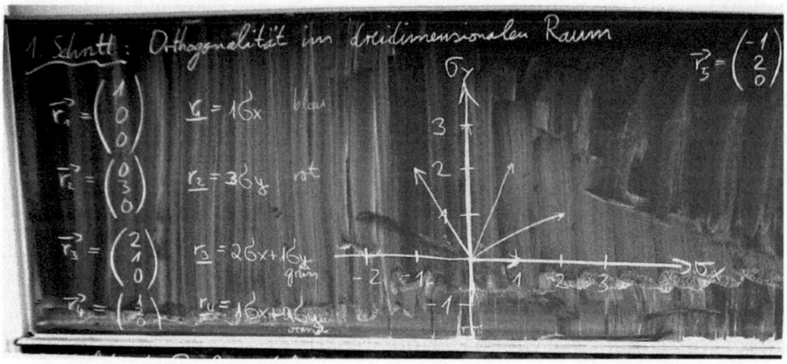

Fig. 2. Orthogonality in space and in spacetime: Which vectors are perpendicular to each other?

13th grade, shortly before their final examinations, handled the exercises and problems successfully. After the course, they were able to explain the relevant physical arguments concerning time dilation, length contraction and the twin paradox.

Of special interest is the way in which orthogonality can now be discussed in the classroom (see Fig. 2), because orthogonal directions in the Minkowski diagram are represented by vectors which lie symmetrically with respect to the world line of light.

In three-dimensional space, the Pauli vectors $\underline{\mathbf{a}}$ and $\underline{\mathbf{c}}$ are perpendicular because their geometric product possesses no scalar part:

$$\begin{aligned}\underline{\mathbf{a}}\,\underline{\mathbf{c}} &= (2\boldsymbol{\sigma}_x + 1\boldsymbol{\sigma}_y)\,(-1\boldsymbol{\sigma}_x + 2\boldsymbol{\sigma}_y) = -2\boldsymbol{\sigma}_x^2 + 4\boldsymbol{\sigma}_x\boldsymbol{\sigma}_y - 1\boldsymbol{\sigma}_y\boldsymbol{\sigma}_x + 2\boldsymbol{\sigma}_y^2 \\ &= -2 + 4\boldsymbol{\sigma}_x\boldsymbol{\sigma}_y + 1\boldsymbol{\sigma}_x\boldsymbol{\sigma}_y + 2 = 5\boldsymbol{\sigma}_x\boldsymbol{\sigma}_y\end{aligned}$$

Having the scalar part of 4, the two Pauli vectors $\underline{\mathbf{a}}$ and $\underline{\mathbf{b}}$ are not perpendicular:

$$\begin{aligned}\underline{\mathbf{a}}\,\underline{\mathbf{b}} &= (2\boldsymbol{\sigma}_x + 1\boldsymbol{\sigma}_y)(1\boldsymbol{\sigma}_x + 2\boldsymbol{\sigma}_y) = 2\boldsymbol{\sigma}_x^2 + 4\boldsymbol{\sigma}_x\boldsymbol{\sigma}_y + 1\boldsymbol{\sigma}_y\boldsymbol{\sigma}_x + 2\boldsymbol{\sigma}_y^2 \\ &= 2 + 4\sigma_x\sigma_y - 1\sigma_x\sigma_y + 2 = 4 + 3\sigma_x\sigma_y\end{aligned}$$

In four-dimensional spacetime, the mathematical analyses carried out by the students show that now the Dirac vectors $\underline{\underline{a}}$ and $\underline{\underline{c}}$ are no longer perpendicular. Their product possesses a scalar part of 4:

$$\begin{aligned}\underline{\underline{\mathbf{a}}}\,\underline{\underline{\mathbf{c}}} &= (2\boldsymbol{\gamma}_x + 1\boldsymbol{\gamma}_t)(-1\boldsymbol{\gamma}_x + 2\boldsymbol{\gamma}_t) = -2\boldsymbol{\gamma}_x^2 + 4\boldsymbol{\gamma}_x\boldsymbol{\gamma}_t - 1\boldsymbol{\gamma}_t\boldsymbol{\gamma}_x + 2\boldsymbol{\gamma}_t^2 \\ &= 2 + 4\boldsymbol{\gamma}_x\boldsymbol{\gamma}_t + 1\boldsymbol{\gamma}_x\boldsymbol{\gamma}_t + 2 = 4 + 5\boldsymbol{\gamma}_x\boldsymbol{\gamma}_t\end{aligned}$$

Thus, the different signs of squares of space base vectors and time base vectors produces a totally different behaviour in spacetime, compared to the situation in three-dimensional space. The product of Dirac vectors $\underline{\underline{a}}$ and $\underline{\underline{b}}$ is a pure bivector without a scalar part, showing that these vectors $\underline{\underline{a}}$ and $\underline{\underline{b}}$ are now perpendicular:

$$\begin{aligned}\underline{\underline{\mathbf{a}}}\,\underline{\underline{\mathbf{b}}} &= (2\boldsymbol{\gamma}_x + 1\boldsymbol{\gamma}_t)(1\boldsymbol{\gamma}_x + 2\boldsymbol{\gamma}_t) = 2\boldsymbol{\gamma}_x^2 + 4\boldsymbol{\gamma}_x\boldsymbol{\gamma}_t + 1\boldsymbol{\gamma}_t\boldsymbol{\gamma}_x + 2\boldsymbol{\gamma}_t^2 \\ &= -2 + 4\boldsymbol{\gamma}_x\boldsymbol{\gamma}_t - 1\boldsymbol{\gamma}_x\boldsymbol{\gamma}_t + 2 = 3\boldsymbol{\gamma}_x\boldsymbol{\gamma}_t\end{aligned}$$

This is a direct consequence of the spacetime structure of the world we live in. To understand Minkowski diagrams correctly, orthogonality has to be analysed carefully with the students. Similar simple exercises and worksheets of lessons used in class can be found in [Horn 2009a; 2009b].

The results of these lessons show that special relativity can be taught in a straightforward and successful way in secondary schools using geometric algebra and the ideas of Hermann Grassmann. As physics teachers, we should dare to do so because it makes sense didactically – and because the words of Sturm, Schröder and Sohncke in their memorial address for Hermann Grassmann are still true today: "The methodological interest of this mathematical approach, however, remains of highest importance, and no advanced learner and no teacher of mathematics should disregard it, for it is the unquestionable merit of the two Grassmanns to have given a new and powerful impetus to arousing and to satisfying the interest in mathematical rigor of the foundations of the arithmetic discipline." [Sturm, Schröder and Sohncke 1878, 32][10].

---

[10] "Sehr bedeutend aber bleibt das methodologische Interesse dieser Behandlungsweise, die keinem Vorgerückteren vorenthalten und keinem Lehrer der Mathematik fremd bleiben sollte, und unbestreitbar der beiden Grassmann Verdienst, für die Weckung und grossentheils auch Befriedigung des Interesses an mathematischer Strenge der Begründung der arithmetischen Disciplin einen neuen und mächtigen Anstoss gegeben zu haben." [Sturm, Schröder and Sohncke 1878, 32].

Illustration 15. [illegible]

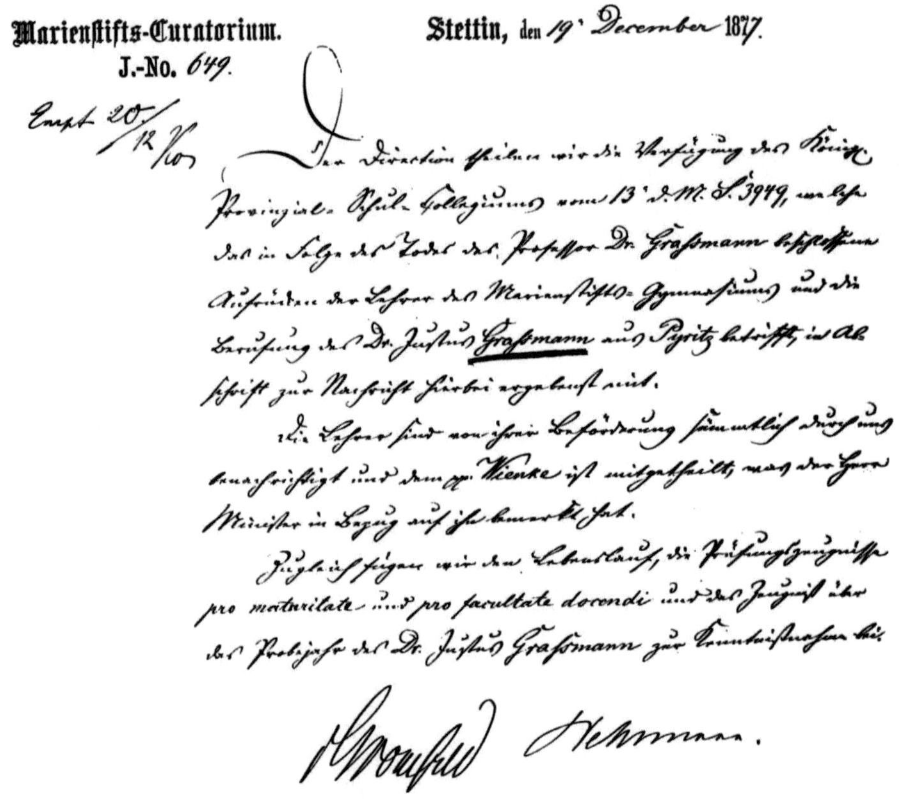
Marienstifts-Curatorium. J.-No. 649.
Stettin, den 19. December 1877.

Empf. 20/12 77

Der Direction theilen wir die Verfügung des Königl. Provinzial-Schul-Collegiums vom 13. d. M. S. 3949, welche das in Folge des Todes des Professor Dr. Grassmann beschlossene Aufrücken der Lehrer des Marienstifts-Gymnasiums und die Berufung des Dr. Justus Grassmann aus Pyritz betrifft, in Abschrift zur Nachricht hierbei ergebenst mit.

Die Lehrer sind von ihrer Beförderung sämmtlich durch uns benachrichtigt und dem p. Wienke ist mitgetheilt, was der Herr Minister in Bezug auf ihn bemerkt hat.

Zugleich fügen wir den Lebenslauf, die Prüfungszeugnisse pro maturitate und pro facultate docendi und das Zeugniß über das Probejahr des Dr. Justus Grassmann zur Kenntnißnahme bei.

Illustration 15. Hermann Grassmann's son Karl Justus (1851–1909) became his father's successor at the Stettin Marienstiftsgymnasium. The transcription and translation are given on p. 502
[Archiwum Państwowe w Szczecine, Gimnazjum Mariackie w Szczecinie]

Appendix

Zu

der öffentlichen Redeübung

welche Montags den 15. October 1827 Nachmittags um 3 Uhr

in dem großen Hörsaale

des Gymnasii zu Stettin

veranstaltet werden soll

ladet

die Beschützer Gönner und Freunde

unserer Schulanstalt

ehrerbietigst und ergebenst ein

Dr. **Friedrich Koch,**

Königl. Schul-Rath, Director und Professor des vereinigten Königl. und Stadt-Gymnasii, wie auch Director des mit demselben verbundenen Seminarii für Gelehrteschulen, Mitglied der Lateinischen Gesellschaft zu Jena und Ritter des rothen Adlerordens 3. Klasse.

Gedruckt bei Carl Wilhelm Struck.

Illustration 16. Title page of the school publication series in which Justus Grassmann's work on the theory of number was published in 1827. Further information on p. 503

Ueber
den Begriff und Umfang
der
reinen Zahlenlehre
von
Justus Günther Grassmann
Professor.

On the concept and extent of pure theory of number (1827)

Justus Günther Grassmann Professor

[Program des Königl. und Stadt-Gymnasiums zu Stettin, 1827]

The more science gains in breadth, the more necessary it is to organize the mass of material thus presented, not only in order to facilitate the entry of the beginner, but rather, and more importantly, thus to elevate the raw information to a genuinely ordered

H.-J. Petsche et al. (eds.), *From Past to Future: Graßmann's Work in Context*,
DOI 10.1007/978-3-0346-0405-5, © Springer Basel AG 2011

knowledge, in which the position, the connections, the function of every part is distinctly perceived relative to the whole, and so that this latter can appear as an organism, as a manifestation of an infinite intellect as it becomes clear to us in a particular sphere.

These remarks find their application not just practically – and particularly in natural science, in which the mass of material mainly accumulates these days; rather, they apply especially to pure science, and particularly to mathematics, by which the laws of outer nature are given to us in inner development as a rich dowry of the human intellect. The extent of mathematics has already increased so significantly, and expanded in the most varied directions to such a degree, that the intellect sinks under the weight of the presentations, and they themselves run the danger of becoming blind instruments that one can only apply, as it were coincidentally, to the appearances of nature, rather than their attending the explanation of nature in an orderly way, or rather, since they are independent of it, their being a model and standard for its presentation.[1]

This lack of organization obtains not only in the mutual relations of the mathematical disciplines however, but also in each of them individually; indeed it extends to the arrangement of the individual theorems in them, and it is a much too one-sided attitude, tenaciously maintained by mathematicians above all, that is *only* necessary to verify whether a theorem can be completely demonstrated from what has preceded it. Without a doubt it is the lack of communication and of instruction that has secured the victory of this attitude over all others; but it can safely be predicted that the methods for science must coincide with the methods for instruction as soon as the inner economy of a discipline has assumed a genuinely organic form, resulting from its inner essence, since that which in every sense belongs together must also be most fully understandable in its unity. All methodological endeavors intended to facilitate beginners in science therefore appear most problematical, and in any case of very subordinate value, if they do not at the same time, through a greater clarification of the elements,[2] advance the science

[1] The most striking example of how in our times the inner development of mathematics has fallen behind compared to natural science is crystallography, in which the natural scientists have displayed to mathematicians a series of structures scarcely guessed by them, so full of inner relations and having such perfect organisms, that it is indeed admirable how the simple observation of the syntheses of nature can have raced so far ahead of the intellectual syntheses of mathematics. It is true, to be sure, that in this case the mathematics has now been given, but the entire series of structures should have been presented long since, since it is just the solution of a simple problem of combination theory, as the author hopes to show most convincingly in a paper now in preparation on general and geometric combination theory. Astronomy took a completely different course from crystallography. The development of the intellectual synthesis in mathematics, in particular the theory of conic sections, long anticipated the observation of the synthesis of nature in the orbits of the planets. The astronomers hit upon it, and it was then a question of recognizing the correspondence, which *Kepler* expressed in the laws named for him, and *Newton* proved as a necessary consequence of the hypotheses of gravitation.

[2] The author, who himself has published a couple of methodological textbooks for elementary instruction (entitled *Raumlehre*), is far from believing that he has attained this goal, but is sincerely conscious of having

to which they are devoted, present clearly the inner connections of the operations and constructions, allowing each step occasioned by the nature of the subject to enter as such, and, granting the intellect a resting point in the results achieved, simultaneously use them as a starting point for higher developments. It appears to be preeminently the problem of the schoolman to improve the science by a good economy, to shape its structures, and to communicate the breath of life by which it attracts the intellect and thus captures it. Every development, however small its extent, whose purpose is to yield a contribution to this end can thus serve as an appropriate subject for a school-program.

Mathematics generates its first concepts by a synthesis characteristic of it (which we call a construction in the broader sense), in which it completely filters out the content of those to be conjoined. But its subject is not however the form of this synthesis, but its product, and thus it is distinguished from logic, which does indeed assume some content in general but abstracts from it, whereas in mathematical constructions a content emerges from the fact that one regards those to be conjoined as contentless.

This requires a little explanation. If we take any given sentence as a proposition, then in it we attribute or deny a predicate to a subject. Now insofar as we take notice, not just of the form, but of the content, we will attribute to the proposition a truth or validity if the predicate, as a consequence of its assertion, must really be conjoined to the subject, or not; so the grounds lie in the result or in something else. Now such a synthesis, the validity of which depends on the content of those to be conjoined, we will call a *synthesis according to inner relations.* Although logic does indeed consider only the form of the proposition, and leaves the matter, or its content, completely disregarded, still it always assumes in general that such content is at hand.

Now the mathematical synthesis attributes no truth in this sense, and precisely thus it is distinguished from a synthetic proposition. If for example we take that synthesis whereby a number results $(1 + 1)$, we do indeed find here the same constituents, subject, predicate, copula (except that here there is no essential distinction between subject and predicate), but nothing can be said here about the objective validity of this synthesis – that is, whether one unit really amounts to another – the conjunction can take place unconditionally, and the concept created thereby, the product of this synthesis, is the number two. But if one wishes to say that the proposition is actually this: $1 + 1$ is 2, then on the other hand one would with justice be reminded that here one has no longer introduced a concept complete in itself, but has simply labeled it by a word.

One can therefore assume a content for those to be conjoined or not. In the latter case they must then, precisely because they are to be regarded as contentless, be defined as either *merely equal* or *merely unequal.* Now since we have called that relation,

aspired to it, and to have subordinated every other consideration to it. On the subject discussed here, see the Foreword to Part Two of *Raumlehre*, Berlin 1824.

according to which those to be conjoined are regarded as amounting to one another, and thereby assuming a content, a synthesis according to inner relations, we must then call that in which they are asserted to be merely equal or merely unequal a *synthesis according to outer relations,* whereby it then follows that:

> *Mathematics is the science of the synthesis according to outer relations, that is as equal or unequal.*

The complete justification of this conception of mathematics would lead me too far from the goal previously stated, and must therefore remain withheld for another place; I will content myself here with its application to geometry. Thus, if we wish to construct the spatial magnitudes synthetically, then it is obvious that we must proceed from the absolute spatial limit, the point; the first geometric construction is then the drawing of a straight line, which geometry demands at once, by the progression of the point. Now it is immediately clear that, although this construction is of a completely characteristic type, precisely that on which the particular nature of geometry is based, all its characteristics are given at once in its beginnings, and that the whole construction of the line consists in a synthesis of similars, whereby it is precisely a subject of mathematics.

Now mathematics establishes theorems in which a subject is conjoined to a predicate, not in an arbitrary way, to be sure, but rather with respect to their content; but this content is precisely only such as is laid down by the synthesis as equal or unequal. Thus, these theorems are only expressions of the nature of that characteristic mathematical synthesis, and that which is given along with it.

The fact that, once we have removed the content of those to be conjoined, we can regard them, not just as equal, but also as unequal, and then conjoin them, seems clear at once, since inequality is expressed simultaneously with equality. But here there is not to be imagined any specific qualitative inequality, but only a general distinction, as one of the contentless entities. *But the conjunction when unequal yields combination theory.* Generally, where things are unequal, they can be used as elements of a combination, and can be bound into complexes; in pure combination theory one considers the elements as unequal yet without any definite content. Unit and element, units and elements, number and combination are therefore oppositions that are produced reciprocally. No equality can be imagined without an inequality. In the formation of the numbers this latter is also inescapable. One and one can therefore only be bound together as two by the fact that within their complete equality they are still distinct, since otherwise they would be identical and could only be regarded as one. Likewise an equality is necessarily imagined in inequality, as already follows since the inequality in the combinatorial elements is a common predicate, by virtue of which one can enumerate them. Their equality, that in which they agree, is then precisely this inequality. Equality and inequality are only factors with respect to those to be conjoined, and it is only a question of which of the two applies in the fundamental definition of the synthesis, although both are always at hand.

The fact that among the combinatorial elements equality can also obtain, as for example when the problem is posed of developing the combinations of the elements *aaabc*,does no harm to the concept of combination theory just formulated, since it is easy to show that these equal elements can only yield new combinations insofar as, either by their number or their position, a distinction among them can be occasioned. The complexes of the sum written above no longer belong to pure combination theory in isolation, but rather constitute an application of it to arithmetic, just as conversely the definition of number complexes is an application of arithmetic to combination theory.

It is a fortunate circumstance for mathematics that its content is completely independent of the expression of its concept, and that it is defined by its field itself. There is no definition of it known to me that encompasses combination theory in its complete purity, isolated from the admixture of other mathematical disciplines; but no mathematician doubts that it is a perfectly proper and integral part of pure mathematics, even though pure combination theory assumes nothing whatever of magnitude as such.

Much more detrimental than the lack of a satisfactory definition is the lack of precise distinctions among its various disciplines. By the many applications that are allowed between them their boundaries have become so intertwined that the appearance is produced of a continuous transition from one to another. But arithmetic, geometry, combination theory all have a completely definite part that is completely independent of all the rest; this part must be precisely definable, and it is necessary that it be defined in order that one have a clear general perception of which domain one finds oneself in, where its boundaries are, and where they are overstepped. Only with such a distinction is a proper ordering and classification of the great fields of mathematics possible; without it individual successful theories can indeed evolve, but no satisfactory overview of the whole domain.

Mathematics as the science of the synthesis according to outer relations, i.e. as equal or as unequal, decomposes according to this definition into the theory of magnitude {Grössenlehre} and combination theory. The synthesis of similars gives us magnitude; it is ***discrete*** *if with its generation that to be conjoined* (by which synthesis the magnitude results) *is regarded as given;* on the other hand it is ***continuous*** *if that to be conjoined is only generated by the synthesis itself.* The mathematics of continuous magnitudes is geometry, not only because spatial magnitudes can only be generated as continuous magnitudes, since it is clear that the same is true of temporal magnitudes and of intensive magnitudes as well; rather it is because it already includes all others, and extends far beyond them. There is no relation of these latter magnitudes that has not already been made completely clear about the straight line, and for that one does not even need the other two dimensions of space. On the other hand, new sciences, the pure theory of motion {kinematics} and dynamics, emerge upon the combined consideration of them with spatial magnitudes.

I will omit here those branches of mathematics for which the division of magnitudes into continuous or discrete is either indifferent or in which the one is considered in the form of the other, in order to proceed directly to my main subject. I must however note that combination does not decompose into discrete and continuous in the same way as magnitude, since it yields and can yield no synthesis of dissimilars, by which the elements of the combination would be generated in the synthesis itself and by it. Combinations are therefore essentially discrete, and the division according to the scheme

equal	unequal
discrete	continuous

is therefore not fully articulated. Precisely on this account it is arbitrary whether one divides mathematics into the theory of magnitude and combination theory, where the former decomposes into arithmetic and geometry, or into discrete and continuous mathematics, where combination theory then appears as a subdivision of discrete mathematics.

There are grounds for both divisions. But in either case combination theory stands on an equal footing with arithmetic, and deserves a different treatment than that it has previously had from mathematicians.[3]

[3] Combination theory is still in its infancy, as if in arithmetic one had proceeded no further than addition. Sad to say, it was immediately made the handmaid of analysis; this early servitude has impeded its growth and development, and even now the times are so inclined against it that only by this servitude can its existence be delayed. But the day will come when the stepchild will appear in her undimmed beauty, and be acknowledged, and while one asked nothing of her, assigned her no duties, she will in her harmless presence cast her rays upon all the sciences. Yet how completely different it was with the Greeks, where interest in pure speculation, suggestions within their own sphere, every relation to another, outweighed usefulness, thus cultivating and assisting the embryonic sprout so that it could be formed according to its own nature, and thus satisfy itself, instead of which we always lead it artificially to a predetermined goal. It is my conviction that someday combination theory will be the same for natural history and chemistry as the theory of magnitude is for physics. I would however have doubts about expressing this conviction if I saw a means of securing its recognition in another way, and if there did not inhere in its own character grounds that require the synthesis of nature to seek the counsel of this synthesis of the intellect. Thus, most combinatorial discussions lead, when one prosecutes them, to great prolixity, and I regard it as a central problem to find the most appropriate restrictions to eliminate this prolixity. But in so doing the regard of combinations in nature is as a rule of the most decisive utility, and at least restrains the arbitrariness in those restrictions, which only too easily is made into a biased opinion, stops with it, and this then holds for the individual.

Arithmetic textbooks usually include, along with the concepts and theorems related to the formation and combination of numbers, a more or less extended development of the general theory of magnitude, which one usually calls algebraic computation {Buchstabenrechnung}.

This custom, which at first glance seems unmethodical, can however, even absent the requirements of instruction with which those textbooks are principally charged, very well be justified. Thus, first of all I cannot keep the general theory of magnitude in the abstract from every particular type of magnitude, without losing all secure footing. Even the experienced thinker requires a resting point from time to time, where that hanging in abstraction is again gathered into a structured product, and thus provides satisfaction, with the one series regarded as concluded, and where the results achieved can serve as the starting point for a new series. Then again, and this seems to be the main point, the general theory of magnitude cannot dispense with number, without collapsing into nearly nothing. For there remains nothing of it but the general logical conjunction and its decomposition, which, as will be shown below, are addition and subtraction. The concept of multiplication is already an essentially numerical concept, and if geometry also had something analogous to exhibit, this would still depend on the characteristic nature of the spatial synthesis, and could in no way be applied to magnitudes of all types. A scheme signifying a magnitude generally – that is neither particularly continuous nor discrete, but only a magnitude, no matter whether continuous or discrete -, can therefore not be used as a multiplier, nor at the same time a root, much less an exponent – can therefore only stand for a multiplicand or as an entry (term), and be combined as such, and yet a product ab has no sense at all if neither a nor b is a number, but rather both represent magnitudes in general.[4]

Should however the general symbol not denote magnitudes generally, but only numbers, known and unknown, then the general theory of magnitudes can indeed still apply to magnitudes of all types, and likewise to number, but the so-called algebraic computation then appears as just a particular method of representing the rules of numerical conjunction schematically, and therefore belongs properly to arithmetic. Algebra and analysis also certainly assume arithmetic, since in their schemata each of the factors combined in a term, except for one, must denote numbers. It is therefore completely impossible to posit algebraic computation, and that which is connected to it, prior to arithmetic, as the general theory of magnitude.

A discrete magnitude is, according to the above, one for which those to be conjoined as similar are regarded as simply given, without having first to appear in the construction of the magnitude, or being assumed to have any particular nature. Discrete magnitudes are numerical magnitudes, however.

---

[4] Cf. *Klügels Wörterbuch*, "Buchstabenrechnung", 1 and 2.

The difficulty in providing a proper definition of that which is a number is often acknowledged. It lies principally in that this definition must be such that it admits both one and zero, which are obviously represented as numbers in the course of arithmetic, and are undoubtedly required by it, and so must be included within it. Precisely the same difficulty, and of exactly the same nature, is also found in combination theory. Here also unity as well as nullity, meaning its opposite, must be regarded as complexes. The science, in the course of its development, requires it thus; all sequences remain incomplete, all terms appear displaced from their natural positions, all formulas are more complicated and lose their simple form, if one omits the nullity from the complexes. Now such a demand of the science itself absolutely cannot remain disregarded, and it is by no means dependent on our free choice whether or not we reckon one and zero among the numbers, unity and nullity among the complexes, if they undoubtedly prove to be such in the science itself. On the contrary, we must preserve as such the concept of such objects, as required by the science, and either regard our definitions themselves as incorrect so long as they do not represent the true concept – for such a concept may by no means be regarded as arbitrary, one we can define thus or otherwise, and shape it arbitrarily by definition – or we must, if the concept is an elementary one, completely abandon attempting to specify it by a definition.

With respect to the numbers it must not be overlooked that we would have no notion of them if we could consider things only as one and zero. No definition of number can therefore be abstracted from them, and only the opposition between one and zero can awaken this concept in us. Thus, if we return to the *origin* of the number concept within us, then it is clear that only by considering a manifold, a multiplicity in our conceptions, i.e. insofar as it exceeds one here, that we can proceed to the notion of number. The genetic interpretation of number must therefore necessarily turn out so that it is no longer adapted to one and zero. This however by no means suffices to produce the sequence of numbers, since the way a concept originates within us by no means specifies the concept as such.

The concept of number is therefore awakened in us by the perception of a multiplicity in our conceptions, if one abstracts from them differing contents, and defines them all as equal. The concept of a number therefore comes to the point, that this specific multiplicity is conjoined in a unit of consciousness. Each of the conceptions imagined as a similar one calls the *unit* in relation to number.

The preceding interpretations are genetic, i.e. they establish the possession of the concept and incidental reasons for its origin. However, number as such must be regarded as the *specific quantity of the setting of unity,* and since one can set unity once or not at all, one and zero must also be reckoned among the numbers, although they result from no conjunction of a manifold. But why the genetic interpretation must turn out this way was just illustrated above.

Now a number is called concrete or "named" if the unit is a specific conception; if however it is only the conception of a general given, without specific content (the conception of a conception), then the number is called abstract or unnamed.

If arithmetic is to be a pure science in the strict sense, then not only must it be removed from all circumstances of outer experience, but generally from everything *alien to it.* It thereby follows, above all, that the concrete number, whose unit is a definite conception, does not belong in the pure theory of number, if in addition to number it still includes something else alien to this, taken perhaps from outer experience or perception, to wit the content of the given conception; yet this does not prevent the concrete number in general from being considered as such in pure arithmetic. Just as little can the unit be permitted any such character as arithmetic has not itself imposed upon it. One therefore already steps outside the domain of pure arithmetic in the strict sense if one conceives the unit as a continuous magnitude, i.e. if one supposes that it is in any respect divisible.

It will be necessary to tarry a bit upon this point, since the principal classification of arithmetic, and particularly the concept of the pure theory of number, depends on a correct understanding of it.

The intent is not that, in conceiving the unit as a continuous magnitude, one steps outside of the domain of pure understanding. We can internally construct continuous magnitudes, such as spatial magnitudes, temporal magnitudes, intensive magnitudes, set them equal, and enumerate them, without thereby being in need of exterior perceptions. Arithmetic would then always remain a science of pure understanding, even though one would conceive the unit as a continuous magnitude. The assertion is only that if one understands the unit as a continuous magnitude, one would import to arithmetic something originally alien to it, and that a rigorous method requires that those parts of it in which one assumes the unit without any special qualities be isolated from those in which one understands by the unit a continuous magnitude; and this isolation seems to me to be just as important for science as for instruction, insofar as scientific insight is aimed for thereby.

I will therefore call that part of pure mathematics in which the unit is given as the origin of all conjunction, and as the end or limit of every resolution, or briefly as the simple given, which has no further stipulation upon it, *pure isolated arithmetic, or pure theory of number,* and on the other hand that part in which the unit appears as a continuous magnitude; that is partly in that respect, or endowed with the relation of the subject, *pure non-isolated arithmetic.* In order to specify the concept and extent of the former it is necessary to give a brief survey of the relations and conjunctions that are found in it.

I do not fear on this account the objection that properly there is still nothing without magnitude and without parts, and that not just the atom of the physicists is an absurdity, but rather that conceptions themselves stand in temporal relations, and even have an intensive magnitude in a certain sense. The pure theory of number sets this problem completely to one side. There may be indivisible things or not, they may be conceivable or not, these questions are a matter of indifference for it, for if they can be conceived, they can be enumerated, and it is not implicit in the concept of the simple unit whether they are indivisible; so little that one can even regard the arithmetic of the physicists as an experiment of subjecting nature to whole numbers, thus to encompass its evolutions within the relations of the pure theory of number and so understand them. If it is implicit in the concept of the unit that it must be divisible, then one could not enumerate indivisible things. Indivisible things can however be conceived and are at hand as such, and one can enumerate them; one can speak of 3 geometric points, 3 abstract concepts, 3 prepositions, 3 persons (individuals), etc., and it is therefore clear that the concept of divisibility is not implicit in the concept of the arithmetic unit, this the more as that from which all developments in the pure theory of number begin, and upon which all resolution is based, must be regarded as the arithmetic atom, the absolute impenetrable of number, beyond which one may not proceed without introducing something originally alien to the pure theory of number. Pure isolated arithmetic is therefore concerned only with whole numbers.

The three orders of enumeration

1st order

The intellectual activity of the production of a number from the unit one calls *enumeration,* and this consists in the comprehension of given units in a unit of consciousness.[5]

The inverse of enumeration one calls *decomposition,* whereby that united in the consciousness is again separated, that comprehended again dispersed into its permanent parts.

Enumeration and decomposition are the two fundamental intellectual activities in numerical arithmetic; all arithmetic constructions must ultimately be reduced to them.

[5] The word "unit" appears here in a threefold sense. Thus it denotes union, amalgamation into a whole, and to that extent requires no arithmetic interpretation. Then it is used above all to signify the general concept, by means of which one wishes to regard the manifold as similar, in order to enumerate it; finally it also means the individualized concept (transferred to an individual, as it were), one of the constituent parts of the manifold, which is to be unified in the number. Only in this meaning can it be used for the most part.

2nd order

In the pure theory of number the unit is not to be an object given by outer perception, since number theory is an a priori science; nor is it to be a continuous magnitude, since this is alien to number: both of these conditions are met if the unit is a product of the arithmetic synthesis itself, since in this case we do not depart from the domain of pure comprehension, nor from that of number theory. But the only product of the arithmetic synthesis that has previously occurred to us is number itself; so we will take this again as the unit, and enumerate it.

Numbers can be enumerated in a twofold sense, either as numbers generally, or as named numbers. The former gives us nothing new, except perhaps the concept of a named number in a purely arithmetic way; so we simply set it aside.

Should specific numbers be enumerated, then they must be *equal.* This necessarily follows from the concepts of the unit and of enumeration, as we have interpreted them above. Only by setting them completely equal can they be units, and serve for an arithmetic structure. The product of this enumeration is a number of numbers, but of specific ones, e.g. 4 nines, the so-called multiplicative conjunction.

In this conjunction one distinguishes the passive or enumerated number, the number of first order, in the above example the number 9. In this union it is the substitute for the unit, the basis or substance of the enumeration, the *multiplicand.* One distinguishes from it the active or enumerating number, which first arises from this higher enumeration, thus the number of second order, in the above example the number 4. It results from the number of first order in exactly the same way as did that from the unit, and in this union is the agent or quantifier, the number proper, the *multiplier.*

One usually says that the multiplier is simply an unnamed number. From the above development however it follows that, to the contrary, it is the first named number we encounter in a purely arithmetic way. Thus, its name is the multiplicand, and precisely on this account it can have no other name, since it is already found here. For it to be examined it is unnamed, to be sure, since otherwise it would have two names in the union, which is clearly absurd.

Such a multiplicative conjunction can again be enumerated, whence there results a union of three factors, a multiplicand and two multipliers; this can again be enumerated, and so forth. But this is by no means one of the distinguished types of enumeration, whence no new types of numerical magnitude arise hereby. Every multiplicative conjunction must therefore consist of a multiplicand and an arbitrary number of multipliers.

3rd order

Since in this way a multiplicative conjunction can include an arbitrary number of factors, there once more arises the possibility of enumerating the factors not only generally as such, but in setting them equal as a specific value. But between the factors there is the essential difference that the one is a multiplicand, the others are multipliers, completely different in their significance. Thus, the equating required for enumeration cannot take place unless one transforms the multiplicand as well into a multiplier, but this is always possible, since one can append unity as a multiplicand, whence it follows most clearly, if one names the multiplicand. Thus $4 \cdot 4 \cdot (4\text{parts}) = 4 \cdot 4 \cdot 4 \cdot (1\text{part})$.

Now no further obstacle stands in the way of this new higher enumeration. The numbers are all equal in their significance, for they are all multipliers; if one also sets them equal as absolute numbers, they can then again serve as units for a higher enumeration, and thus arises a new conjunction, which we will call *dynamic.*

The number of third order arising from this enumeration is called the *exponent;* its unit, that of which it is composed, is called the *root,* but it does not enumerate this unit simply, but as a factor, or rather as a multiplier for the unit, which alone may be named.

In this order, what had previously been the agent or quantifier is demoted to the basis or measure, at least taking its place; the multiplier becomes the root; on the other hand a new agent emerges, a new quantifier in the number of third order or in the exponent, which one can also call a factor number. Thus, it denotes a number of enumerations, all of which are applied to the same absolute numerical magnitude. The relation of the dynamic conjunction to the unit (as the deeper basis of the conjunction) can certainly and under no circumstances be neglected. Thus, first, the dynamic conjunction makes no sense without this relation. Multipliers without something multiplied, without a multiplicand, are numbers without something numbered. $4^2 = 4$ times 4 times, makes no sense; but if one appends the unit as multiplicand, it becomes clear. But then, with this self-evidently necessary relation to the unit as multiplicand, the concept of the exponent is at once presented so clearly that from this concept the whole theory of powers, including both negative and fractional exponents, is developed immediately with the greatest clarity.[6]

[6] Since we will not return to this in the pure theory of number again, it may be indicated here. The exponent enumerates factors which are appended to the unit (as factors), or removed from it; the first if it is positive, and thus enumerates the factors in the sense of the positive enumeration, the second if it is negative. The fractional exponent, e.g. in $64^{2/3}$, indicates, since it enumerates factors, that to the unit 2 thirds of the factors from 64 are to be appended, (as factors, i.e. multiplicatively) or, if the exponent were negative, that they were to be deducted from it. The root exponent is equivalent to the denominator of the power exponent on the same grounds that division by a is equivalent to multiplication by $1/a$.

Now one can again enumerate a number of factors, i.e. raise a power to a power, this yet again to another, and so forth, and thus it apparently follows that, by setting the root and exponent equal, one could proceed to a still higher order of numerical structure. But, as will be proved later, there can be no such higher order of enumeration, and that, with numbers of the third order arithmetic reaches a limit beyond which it cannot proceed. Just as space has three dimensions and no more, so arithmetic has three orders of enumeration and no more, and the claim of many mathematicians and philosophers that one could proceed to higher numerical forms, whose development is however not attempted, since it has not previously been produced, is completely baseless.

Pure theory of number is therefore closed with the third order of enumeration, exponentiation, and is fully encompassed in this respect, and from these three orders of enumeration we have the following arithmetic results:

1st Order – simple numbers
2nd Order – multipliers
3rd Order – exponents.

In each of these three orders there is an active and a passive, a measure and a quantifier, and that which in one order was the quantifier is always the measure in the next higher order.[7]

[7] At this point I must return once more to combination theory. Since it conjoins the simple given as unequal, arithmetic as equal, it already follows that the two sciences correspond, and take a common route in opposite directions, as was already shown in the equal difficulty of viewing the unity and the nullity as complexes. The comparison of the two sciences would therefore be most instructive if combination theory were developed to the degree of arithmetic, which however will remain a distant goal for a long time. The necessity of a distinction between pure isolated and non-isolated combination theory thus becomes almost as urgent as that of the same distinction in arithmetic. Thus, as already noted above, the development of the complexes of the array written above does not belong in isolated number theory at all, but amounts to a characteristic application of it to arithmetic, therefore asserting a completely analogous relationship between combination theory and arithmetic as does trigonometry between arithmetic and geometry. But in addition the development of complexes with repetitions seems to me not to belong to this isolated part, since in it the distinction necessary for the combination will only result if an element simultaneously appears as a number; in fact the question can be asked, whether permutations belong in this part, since here the distinction consists in the sequence, and thus in a temporal relation (not spatial; spatial permutations are, as I well know, not yet attempted at all), whereby, as always, something originally alien to the science would be introduced.

On the other hand, it permits one to see that in combination theory there must be formed relations analogous to the 3 orders of enumeration. From the first combination of absolute elements there results a series of complexes or forms. A complex of forms would give the complex of second order (the analog of the multiplicative conjunction), and a complex of formations of the third order. That combination theory also attains its limit with this, and that there is no higher order, I prefer not to prove just now, but only to make it plausible. Combinations of second order are however given in what the crystallographers call a combination, most clearly perceived as soon as it is proved that what *Mohs* calls the elementary structure is the product of a combination, and can perhaps already be anticipated from *Haüy*'s law of decrescence. These

The general conjunction

In the above survey of the orders of enumeration it may seem surprising that although the synthetic or progressive conjunctions, namely multiplication and exponentiation, appeared, the first of them, the conjunction of addition, was completely missing.

The reason for this is that it has a more general character than the others. Thus, it is not just an arithmetic, but the general intellectual synthesis, or the general logical conjunction, applied to number. That is, as subject and predicate are conjoined in a sentence with respect to their contents, so also can one conjoin one number with another with respect to their contents, and this conjunction is not mathematical as such, but only insofar as these contents are introduced by the mathematical synthesis, i.e. here by the enumeration of the subjects of conjunction. The juxtaposition of two lines, two surfaces, or bodies, the addition of a condition to a concept that did not previously include it, is as such the same act as the conjunction of two numbers by addition, except that there it is applied with respect to concepts and spatial magnitudes, here to numerical magnitudes. On the other hand, the multiplicative and the dynamic conjunctions appear, the former as just a mathematical synthesis, the latter already completely defined only as an arithmetic synthesis, i.e. a synthesis of givens as similar. Geometry has yet another synthesis analogous to multiplication,[8] which however is essentially limited to three factors; on the other hand no conjunction analogous to exponentiation is allowed. Here then is the reason why the conjunction of addition cannot result from the orders of enumeration, since it is precisely a conjunction with respect to the content of those to be conjoined, so these are not set equal in it, and cannot be enumerated. Otherwise it is self-evident that it can be applied to numbers of each order; so we will have to place it ahead of the other conjunctions.

The types of calculation

We now have the following conjunctions of numbers: Conjunction by addition, conjunction by multiplication, and conjunction by exponentiation.

Each of these conjunctions, if the things conjoined are numbers, has its specific value, which must always be represented as a number of first order. Now in order not to confuse the value of the conjunction with the conjunction itself, we will, where

analogous relations of arithmetic and combination theory will excuse me if, in the footnotes, I have returned to combination theory more often than one would expect from the subject of the article.

[8] See my *Raumlehre,* Part Two, p. 194fn.

necessary, distinguish them, calling the latter a *composition,* the former a *unification.* Thus 7 + 8 is a composition, but 15 the unification by addition. This distinction is the more necessary as we will immediately encounter such compositions as could only very improperly be called conjunctions.

Thus, each of the above-named conjunctions can be decomposed once more, and thus there next results, when they have previously become unified, those compositions that we will call *analytic* or *regressive,* those of subtraction, division, and deponentiation. These also can have values that are expressed in numbers of first order, so that here as well there can be a unification from a composition. It may however also happen that the composition itself, cast into its elementary form, must stand in place of the unification, since this cannot be expressed as a number of first order.

Ascertaining the value of a given composition in a number of first order one calls a *calculation* in the narrow sense. *Thus, calculation means to make a unification from a composition,* and there are as many types of elementary calculation as there are different compositions, that is 6, or properly 7, since deponentiation divides into two completely different types of calculation, according as the root or the exponent is sought.

Now for each of the analytic types of calculation there arises the case where to resolve that which is imagined as conjoined into its constituents cannot be produced from them; for example the sum 3 is to be resolved into two terms or parcels, one of which is 5. Arithmetic is therefore required to admit new numerical magnitudes that we will call analytic numbers, since they arise from the analytic types of calculation. Their sense we can tentatively base on it.

The following small table comprises a survey of the elementary types of calculation, to which the analytic numbers are appended. The names mechanical and chemical for types of calculation and conjunction I regard as temporary makeshifts; the compounds of third order should then properly be called organic, but I have not chosen this name, but prefer to allow a nonuniformity in the sequence of names, especially since those chosen are supported by common usage.

Survey of the types of calculation

	Mechanical	Chemical	Dynamic
Synthetic	Addition	Multiplication	Exponentiation
Analytic	Subtraction	Division	Deponentiation
Analytic numbers	Negative numbers	Fractions	Irrational numbers and Impossible magnitudes

It will now be necessary to go over the individual compositions, as we have found them above, partly in order to show how one can make a unification from each composition by the fundamental operations of arithmetic, enumeration, and decomposition, i.e. to go over the different types of calculation, partly in order to understand better the significance of the analytic numbers, and to see the extent to which their treatment belongs in the pure theory of number.

Mechanical conjunction

The synthetic mechanical conjunction or addition is the general logical conjunction applied to number, i.e. with respect to the content of those conjoined. But this content is such as is introduced by the mathematical synthesis, here by enumeration. Now if, from the composition that the synthesis first demands, there is a unification in which the synthesis is completed, then these contents must be resolved and then unified, i.e. in order to find the sum in addition one must *decompose both terms* (summands) *into their units, and then enumerate all of them together.* Addition is therefore, as required, reduced to enumeration and decomposition.

From this it is immediately clear that the sequence or order of the terms has no influence on the sum, for precisely the complete equality makes the units what they are for it. Likewise the terms themselves must relate to the same unit, which then is also the basis for the sum.

The mechanical-synthetic conjunction is now decomposed again by the mechanical-analytic. If the synthetic conjunction is still just a composition, then the decomposition is accomplished by dropping the symbol of the unification of terms and considering them separated. If however the terms are already unified into the sum, and only this is given, those terms are no longer evident. If one decomposes the given sum into its units, then unless the sum is very small there are a multitude of ways one can partition it into two or more segments, and enumerate these individually. This procedure is called disassembly or disception. Since one only intends to formulate the possible *different* cases of disception, and this is achieved best by a combinatorial procedure, one treats this problem best in combination theory applied to arithmetic, where it appears under the name of the disception problem.

From the given sum the terms of which it consists cannot be specified, even if only two of them are present. If, however, one of the terms is given in addition to the sum, then the other term can be found by decomposing the given sum, the minuend, into its units, enumerating as many of them as indicated by the given term, the subtrahend, and isolating these from the rest. Enumerating the rest gives the term sought, the remainder. To find the second term from a given sum and one given term is called subtraction. The relation between addition and subtraction is clearest if one says: *To add means to append terms; to subtract means to remove terms.*

We pass over all the theorems concerning the evolutions and combinations of given magnitudes and their influence on what is sought, since our subject does not necessarily require them, and only consider the single theorem: If the minuend and subtrahend both increase or decrease by equally many units, the remainder always remains the same. The validity of this theorem is clear at once from the procedure for subtraction. Thus, if one resolves the minuend into its units, and isolates, for example from the units on the left those of the subtrahend on the right, then it is clear that units appended to or taken from the right increase or decrease the minuend and subtrahend equally, but that the remainder is completely unaffected by these evolutions. Thus, it is now possible to express this remainder as an infinite set of compositions of subtraction by increasing or decreasing the minuend and subtrahend by equal amounts.

Now the problem can arise that the subtraction is impossible, if the given term is greater than the given sum. In this case one can only indicate the problem, i.e. the composition, but not make the unification, e.g. $3 - 5$, but this composition can, without its value being changed, be expressed in an infinite number of different ways. Thus $3 - 5 = 2 - 4 = 1 - 3 = 0 - 2$. Among these expressions the last is distinguished in that it is the limit of a series, and is the simplest of all. All problems of a similar type can be cast into the same form, $0 - a$; one may therefore drop the 0 and symbolize the problem as simply -2, or generally $-a$.This is now a negative number, which here, in the pure theory of number, just appears as an insoluble subtraction problem cast into its simplest expression.

Chemical conjunction

The multiplicative composition results from a higher enumeration, in which the number already formed takes the place of the unit. The only enumeration that we have encountered so far, and which alone is valid in the pure theory of number, is now a conjunction of units in the sense of addition, i.e. of incrementing. (The synthesis of enumerations and addition are the same as such, i.e. syntheses in the proper sense, except that the former requires that those to be conjoined be set equal, the latter however considers their contents, and is therefore an application of the general synthesis to number). They must therefore be taken here in precisely the same sense as well, i.e. one must apply the enumerated numbers as many times as the multiplication indicates, and combine them by addition if one is to produce a unification from the factors of the product. The procedure of finding the product from the factors is thus reduced to the addition of enumerations and resolutions.

That the product is named the same as the multiplicand follows from the addition. That in order to interchange multiplicand and multiplier themselves, when both are named, one can transpose their names can be proved rigorously and generally by a spatial presentation,[9] and thus does not require such a proof as *Legendre* has given in the Introduction to his *Théorie des Nombres*. From this it easily follows that for more than two factors the product is also indifferent to their order or sequence. In addition the nature of the constituents of a composition are not distinguished at all, in multiplication they are distinguished by the concept, but as numerical values they are mutually indifferent; for exponentiation however the distinction is fully developed, root and exponent are distinguished, not only as concepts, but in addition one can no longer interchange them. Should it not then already be concluded from this that exponentiation is the highest order of numerical conjunction?

We pass over the disassembly of the multiplier and multiplicand. From the concept of the multiplier as a number of second order, i.e. a number of numbers, there follows immediately that:

If two or more numbers include the same factor, then their sum and difference also include it. The numbers 63 and 117, for example, both include the factor 9. One can always regard this as the multiplicand; then 63 consists of a certain number of nines, and117 likewise. Thus, if one adds or subtracts these, one adds or subtracts a number of nines from another number of nines, and thus a number of nines remain as sum or difference. The theorem therefore only expresses that the sum of several numbers must be named the same as the terms, since the multiplicand is in fact just the name of the multiplier. Thus, here again we have nothing but the general or mechanical conjunction of numbers of second order, i.e. applied to the multiplier. From this it follows that, if a number includes a factor, so also every multiple of it, as well as every sum of its multiples, etc., must include this factor.

In addition the most important concepts of prime numbers and compound numbers can be given immediately here. That every number includes 1 and itself as factors follows, since it can be enumerated by units, as only thereby can it be produced. One finds which numbers contain the factor 2 if one enumerates the natural number series of 2's; which numbers include the factor 3, if one enumerates the 3's, etc. That one can pass over the numbers recognized as compound, I mean, that one need not enumerate the numbers 4 and 6, for example, since these are already known to be compound from the enumeration of 2's, is clear, since it is only a question of finding the prime numbers that precede them. Those numbers, which one does not encounter upon enumeration by numbers other than 1 and the number itself, are prime numbers; every other is compounded from those factors by which one has encountered it. By this procedure one

[9] Cf. *Klügels* Wörterbuch, "Multiplication."

obtains the prime numbers and the compound numbers with all their factors by a complete induction. One remarks at once that this procedure agrees in essence with the contrivance that one calls the sieve of Eratosthenes.[10]

The chemical-synthetic conjunction is decomposed again by the chemical-analytic. If the synthetic conjunction is only a composition, then it is decomposed by dropping the symbol of multiplication and considering the combined factors as isolated. If however the composition has previously become unified, and only this, the product, is given, then in many cases this can be resolved into two or more factors in many ways. Thus, if one intends to investigate and display all possible cases of resolution into factors, then one attains one's goal here as well by combinatorial procedures.[11]

Thus, the factors themselves cannot be specified from the given product. If however there are only two present, one of which is given, then the other can be found. *To find one factor given a product and the other factor is called division.* Assuming the concept of the factors, the mutually opposed relation of multiplication and division can most clearly be stated thus: *Multiplication means to adjoin factors, division means to remove factors* (as factors, that is).

The direct concept of division is double, according as the divisor is the multiplier or the multiplicand. In the first case it is called *partition,* in the second *comparison.* Both follow easily from the concept of the multiplier as a number of second order. If for example the divisor (multiplier) is the pure number 4, the dividend (product) the named number 20 parts, then the number of parts is sought which, enumerated by 4's, or applied 4 times, gives 20 parts. This is however found by partitioning the 20 parts into 4 equal parts, since only those equal as such can be enumerated. In the other case the named number 20 parts is to be divided by the named 5 parts. I am therefore to find the factor which, multiplying 5 parts, gives the product, 20 parts. This must now be the multiplier that I am to discover, how many times 5 parts must be applied in order to obtain 20 parts, or how many times 5 parts is included in 20 parts. In order to discover this I must therefore compare 20 parts with 5 parts, for the investigation of how many times a magnitude is included in another means precisely to compare it with that magnitude.

The procedure of division, and how it can be reduced to enumeration and decomposition, is accomplished most easily if the divisor is the multiplicand, and thus the division is a comparison. Thus, from the concept of comparison it follows that one must remove or subtract the divisor as many times as needed to exhaust the dividend. The number of subtractions gives the quotient. This procedure is however not applicable if the divisor is not named, or the multiplier is. Indeed we know that, if only the

[10] Cf. *Klügels* Wörterbuch, "Eratosthenes."

[11] Cf. *Gauss, Disquisitiones Arithmeticae,* Sect. 17. *Legendre, Théorie des Nombres,* Introduction.

result is sought, one can always imagine the divisor as named, and only after the discovery of the quotient transfer the name to it. However, it is also advantageous for this case, completely different from the previous one, to give a direct procedure in which the division is reduced to terms to enumerate and resolve. Now this is nothing but what children use when they share out nuts. Each gets one, and another one in turn, until the supply is exhausted. Then each of them enumerates how many he has. In other words: One imagines as many receptacles (columns, groups) as the divisor has units, and places them serially in these receptacles until the dividend is exhausted. If one enumerates the units in each receptacle, then one has the named quotient.

The problem of reduction may be mentioned as an example of a division of the first type. Thus, for example 96 inches are to be reduced to feet. Here however the concept of comparison with 12 inches is applicable, and the quotient is the pure number 8. It is however not the quotient that is sought, but rather the multiplicative composition of the two factors, and thus $8 \cdot 12$ inches $= 8$ feet. One perceives how the assertion expressed on P. 465 above, that the multiplier in the composition is properly not unnamed, but that the multiplicand assumes its name, is confirmed here, and that the multiplicative conjunction is represented as a higher enumeration.

We also pass over here the disassembly of the dividend, for which the procedure in division is based on compound magnitudes. Since every division is a dropping of factors, a division with several divisors is a dropping of several factors, thus taking the place of a division by the product of them. Also conversely, instead of dividing by a product one can resolve it into factors, and remove them from the dividend one after the other. Now if one imagines an arbitrary division problem, and expresses it by the usual symbol, for example $30/6$, then this is a composition whose value is the quotient. If we remove a factor from the divisor and dividend of it, for example 2, then a part of the division is really completed, and $15/3$ expresses what is yet to be done. The value of the expression is therefore still the same, that is the quotient. The case is completely the same for the adjoining of factors to both. The value of a division problem, or the value of a quotient is therefore not altered if one adjoins equal factors to the divisor and dividend, or removes them.

Now the problem can arise that the given factor (divisor) is not included in the given product (dividend). In this case, provided the dividend is greater than the divisor, one can still isolate a term from it that includes the given divisor as a factor, and whose second term is smaller than the divisor. One calls it the division remainder, and says of this division that it doesn't go, the number cannot be divided without a remainder. Thus, the quotient of only one term has been found. One therefore reduces it to the second case, in which the given factor is greater than the given product. In this case it is however clear that the division is only indicated, but not carried out, i.e. that one can indeed form the composition, but not the unification. Such a composition, insofar as it is regarded as the representative of the unification (the quotient), one now calls a *fraction,* which one can indeed cast (reduce) into its smallest expression by dropping the common factors in the dividend and divisor (numerator and denominator), which however always retains the form of a complete division problem. It is valid here in the pure theory of number as a required but insoluble division problem, and gives us the second type of analytic number.

Here I will take up a few of the main theorems on prime numbers and compound numbers, which in textbooks on the elements are usually either not found at all, or are not proved rigorously, in order to show how the foregoing can be used for establishing simple proofs.

1 Since division must be regarded as the voiding of a previous multiplication, and dividing means nothing but dropping factors, the dividend is decomposed into factors by every completely soluble division, and every number can be divided by all its factors without remainder. Thus, a prime number can be divided by nothing but itself or 1, since it can be enumerated in no other way. Now if one resolves a number into any two factors, then those can be either prime numbers, i.e. not resolvable, or compound numbers, i.e. further resolvable into factors. If in this case one continues the resolution until all factors are prime numbers, then one says that one has resolved the number into its *elementary factors.* From this it is immediately clear that every compound number can be resolved into its elementary factors. Thus $35 = 5 \cdot 7$; $60 = 6 \cdot 10 = 2 \cdot 3 \cdot 2 \cdot 5 = 2 \cdot 2 \cdot 3 \cdot 5$.

2 If a prime number p does not go into any of the numbers a, b, c, d, etc., then it also does not go into their product. Thus, were it included in the product, then this elementary factor p must either have been in *one* of the factors $a, b, c, d, \ldots$, which is contrary to assumption, or in a product of *several* of them, which however is again impossible, since p is a prime number and is therefore not compounded of factors.

3 Every number can be resolved into elementary factors in only one way. Consider for example $360 = 2 \cdot 2 \cdot 2 \cdot 3 \cdot 5$. According to the previous theorem the number (360) can contain no elementary factors other than 2, 3, 5; but in addition none of them can be missing, since otherwise it would not be divisible by them. Thus, should there be another decomposition of 360 into elementary factors, it can only be distinguished from the first in that it includes one factor, for example 2, more often, and perhaps another less often. But if one removes

the one factor, for example 2, equally many times from both decompositions, one eventually discovers that one decomposition still includes it, but the other one no longer does, which, since the quotients must be equal, is impossible.[12]

4 If one can decompose two or more numbers into their factors, one finds their greatest common factor as the product of all their common factors. The numbers are relatively prime if they have no common elementary factors. A common multiple of several numbers, i.e. the number into which they all go without remainder, one finds by a composition in which all the elementary factors of each individual number are included in full. In order to obtain the least common multiple one applies each factor only as many times as it appears in that number in which it is most often repeated.

I pass over all further developments, since it is not my aim to provide a complete pure theory of number, but only to specify its concept and extent, and only that insofar as it belongs to elementary arithmetic. A complete treatment is found in the works of *Gauss* and *Legendre* cited above.

Dynamic conjunction

The concept of the dynamic conjunction is based on the concept of the exponent; this is the number of third order, and enumerates factors, or more properly multiples of the unit. The root is therefore a multiplier and both magnitudes combined in the dynamic composition are pure numbers. Their unification, the power, is therefore related to that unit which as multiplicand must necessarily be appended to the multiplier. From this it already follows how exponentiation can be reduced to enumeration and decomposition by multiplication and addition. In order to demonstrate this immediately with a simple example, let the symbol (•) denote a named point; then

$$\begin{aligned}4^3 = {} & [((\bullet\;\bullet\;\bullet\bullet) + (\bullet\;\bullet\;\bullet\bullet) + (\bullet\;\bullet\;\bullet\bullet) + (\bullet\;\bullet\;\bullet\bullet)) + ((\bullet\;\bullet\;\bullet\bullet) + (\bullet\;\bullet\;\bullet\bullet) \\ & + (\bullet\;\bullet\;\bullet\bullet) + (\bullet\;\bullet\;\bullet\bullet)) + ((\bullet\;\bullet\;\bullet\bullet) + (\bullet\;\bullet\;\bullet\bullet) + (\bullet\;\bullet\;\bullet\bullet) + (\bullet\;\bullet\;\bullet\bullet)) \\ & + ((\bullet\;\bullet\;\bullet\bullet) + (\bullet\;\bullet\;\bullet\bullet) + (\bullet\;\bullet\;\bullet\bullet) + (\bullet\;\bullet\;\bullet\bullet))].\end{aligned}$$

Here each parenthesis encloses a particular enumeration. Since one regards the unit as the number one, i.e. as a number of first order, the lowest parentheses already give a number of second order $= 4 \cdot 1$. This is again enumerated up to the same magnitude,

[12] Cf. *Gauss*, *Disquisitiones Arithmeticae*, Sect. 16, where the proof is given in full in general arithmetic form.

and thus the middle parentheses give $4 \cdot (4 \cdot 1) = 4 \cdot 4 \cdot 1$. This is once more enumerated just as far, and thus in the third parenthesis one obtains $4 \cdot (4 \cdot 4 \cdot 1) = 4 \cdot 4 \cdot 4 \cdot 1 = 4^3$. The exponent is the number of enclosing, or mutually subordinating, parentheses; the number of enumerations a^n says that the unit is to be multiplied n times by a, or enumerated n times up to a, in such a way that the product of each earlier enumeration acts as the basis of the next.

The disassembly of the root yields the binomial and polynomial theorems for whole positive exponents, which belong completely within the pure theory of number, but which, in order not to descend into too many details, we must pass over here. The resolution of the root into factors leads with great ease and evidence to the theorem that $a^n b^n = (ab)^n$.

Since the exponent enumerates factors, an addition to it is an adjoining of factors, i.e. a multiplication, a removing a dropping of factors, i.e. a division, and thus it is immediately clear from the concept of the exponent as factor enumerator that $a^{n+m} = a^n \cdot a^m$; on the other hand $a^{n-m} = a^n : a^m$; where in the pure theory of number it must be assumed that $n > m$. Likewise $(a^n)^m$ can say nothing but that the unit is to be multiplied m times by a^n, i.e. m times n times with a,or that $(a^n)^m = a^{nm}$. The proof of these theorems can easily be obtained from the general arithmetic scheme; it is however always advantageous, and raises knowledge to a higher order (since it lends it a greater immediacy), if one already discovers them in the original concept, and only thereby does one perceive its full power.

The synthetic dynamic conjunction is to be decomposed by the analytic. If it is only a composition then one has its elements, the root and the exponent, but isolated; if however the unification, i.e. the power, is given then in many cases it can be dynamically resolved in more than one way. Thus $64 = 2^6 = 4^3 = 8^2 = (2^2)^3 = (2^3)^2$. The problem of finding, from a given power, the root and exponent from which it is produced can also be indefinite in the pure theory of number. Thus, if it is the aim to find all possible cases of decomposition, then one can attain one's goal by a resolution of the power into its elementary factors and applying a combinatorial procedure.

If, however, either the root or the exponent is given in addition to the power, the problem becomes specific in the pure theory of number; it is to be decomposed by *deponentiation.* Since however the root and the exponent are numbers of completely different natures, i.e. the root can in no way be transformed into an exponent (as the multiplicand can be transformed into the multiplier), it follows that the problem of deponentiation divides into two completely different parts, the *extraction of the root* and the *extraction of the exponent,* according as the exponent or the root is given in addition to the power. One calls the given exponent in this combination the root exponent; by analogy one must call the given root the exponent root.

To find the root from a given power and a given exponent is called extraction of the root – Since the exponent has enumerated equal factors, which are unified with the unit in the power, but are to be separated again, the root exponent indicates into how many equal factors the power is to be resolved. Since all factors are multipliers, the unit remains completely unaffected by this operation. Thus, the expression $\sqrt[n]{A}$ requires that A be resolved into n equal factors, and that the unit is to be multiplied by one of them. From this it then follows that for the completion of the above theorem $\sqrt[m]{a^n} = a^{n/m}$, whence the pure theory of number must always assume that n is divisible by m.

I know of no direct procedure for reducing root extraction to enumeration and decomposition, and in general it seems impossible. The procedure, supplied by arithmetic, which is based on disassembly, always requires that the root of the first term of the power can be found, and finds this first term by raising the first term of the root to the desired power, is indirect. On the other hand there is no difficulty at all in the pure theory of number in reducing extraction of the exponent to enumeration and decomposition. The number of divisions – or the number of subordinating parentheses, which one can easily insert in the power decomposed into its units, gives it, as one perceives at once from the pattern on P. 476, if one imagines that in it, for the power 64 and root 4 the exponent is sought. Arithmetic usually postpones the problem until after the theory of logarithms, since with it one finds a very simple and general solution, but this does not belong in the pure theory of number.

Thus, the problem of deponentiation can only be fully solved in the pure theory of number if the number given as a power is really an exact power of the order sought. If in this case one can decompose it into its elementary factors, one finds the root sought by partitioning those elementary factors into the given number (defined by the exponent) of equal groups; the exponent sought, on the other hand, by partitioning into groups whose magnitude is given by the root, and whose number defines the exponent. One sees immediately that these are the two types of division (partition and comparison), here applied to the number of equal factors. Thus, since $\sqrt{144} = 2{\cdot}2{\cdot}3$ here the factors must be *partitioned* into two equal groups. On the other hand, should the exponent be sought from the equation $4^x = 64$, i.e. from the exponent for the root 4 be extracted, then one obtains it, since $64 = 2 \cdot 2 \cdot 2 \cdot 2 \cdot 2 \cdot 2$ and $4 = 2 \cdot 2$, by *comparing* the first group with the second, and thus finding how many times the latter is included in the former. That in the first case one resolves the exponent, in the second case the root into elementary factors, and can undertake the required operations one by one, easily follows from what was said above about the resolution of the divisor and dividend.

If the given number whose root or exponent one is to find is not an exact power of the given exponent, or of the given root, then the pure theory of number can only indicate the problem, but is insoluble in it. The expression $\sqrt[n]{a}$ is a composition which, insofar as it is a valid representative of the unification, one calls a root magnitude. For the exponential magnitude there is, as far as I know, no common expression by virtue of which one can indicate, from the expression $b^x = a$, by a combination of b and a, that the unknown exponent is sought. The characteristic property of the irrational magnitudes, in not being subject to a more precise consideration of fractions, cannot be made clear, but from the theorem in the previous section on prime numbers and compound numbers it can easily be shown that the root sought cannot be a division problem (fraction). It is self-evident that the root must always be assumed positive, and that the concept of the imaginary or impossible magnitude cannot be admitted here. The problem of root extraction can however often be cast into a simpler expression if the number given as a power can be resolved into factors, among which there are a sufficient set of factors, or if the given exponent can be resolved into factors. Thus $\sqrt[6]{576} = \sqrt[3]{24} = 2\sqrt[3]{3}$ etc., as easily follows from the concept of power and root exponentiation.

On the negative numbers

We have seen above that the analytic numbers result from the regressive types of calculation if one demands that a number be resolved into parts from which it cannot result. In the pure theory of number they are insoluble from their calculation problem reduced to their elementary expressions, cannot be realized for them, and have no independent, but only a combination value.

Before we go into this more deeply, we will try to answer a question about its systematic sequence. Thus, how is it that arithmetic usually takes up the theory of fractions, but leaves negative numbers unconsidered, and in addition that in systematic textbooks one also treats fractions before negative numbers, since it is clear that the place of the latter, as resulting from the mechanical conjunction, belongs ahead of the former in the system of arithmetic? The most obvious reason to be sure is that fractions, regarded independently, are grasped in a much simpler, easier sense than negative numbers. But this only pushes the question further back. Why do fractions have a more easily graspable sense than negative numbers? The answer to this question is closely connected with what was said above about the orders of enumeration and about the general conjunction; it appears the same as the way the orders of enumeration do indeed lead to the chemical and dynamic, but not to the mechanical conjunction. The chemical conjunction was only a higher order of enumeration, and thus just as simple as enumeration itself; it therefore stands nearer to the fundamental arithmetic constructions, enumeration and decomposition; one may say that it is more arithmetical than the mechanical conjunction, which is of a more logical type. The opposition of addition and subtraction is therefore the logical opposition, not of enumeration and decomposition, but of affirmation and denial, and where this opposition is situated in the number itself the negative number appears. Its comprehension is therefore the comprehension of the logical opposition, and requires a completely different and greater abstraction from the opposition of enumeration and decomposition, yet can be applied to these. Since in my opinion the nature of negative numbers can only be understood from this relation, I will, as far as space allows, make an attempt here to clarify it more precisely. First it must be remarked that, if we join the conjunction of addition with the conjunction of subject and predicate in a logical statement, then by this we always mean what the philosophers call a synthesis. The predicate to be conjoined with the subject does not inhere in its concept, but rather is accessory to it, and broadens our knowledge of the subject, not just indicating it. Likewise the negative sentence must be thought of as removing from the subject a predicate previously regarded as being at hand. Now just as in the logical sentence the denial, which belongs to the form of the statement, can be situated in the predicate so that the statement, which has the form of an affirmation, is still a denial in its significance, and conversely, so one can also in the general arithmetic conjunction situate the negation in the number to be conjoined itself, so that the conjunction remains in the form of an addition while it is still a subtraction in its significance, and conversely can have the form of a subtraction while from its significance it is still an addition. The subject is the same as such, just applied in one case to the concept, in the other to the number, as one easily sees if one compares a logical and an arithmetic statement; for example

Caius - is - learned;	Caius - is not - learned;	Caius - is - unlearned;
$a + b$	$a - b$	$a + (-b)$
	Caius - is not - unlearned.	
	$a - (-b)$	

From this a negative number is therefore one in which what originally belonged to the form of the conjunction is inserted into the number itself, and the arithmetic rule, according to which an odd number of minus signs referred to a number makes it subtractive, an even number additive, is the same as the logical rule, whereby an odd number of denials denies, an even number affirms. The minus sign of a negative number can therefore belong to the number itself, regarded as tightly bound to it, but otherwise has exactly the significance and effect of the subtraction symbol. In this connection one may judge the extent to which one can call the negative numbers subtractive, and whether it is necessary to introduce a special symbol for them.[13] The advantages are well known; it consists in that one replaces the oppositions of subordinate conjunctions with an otherwise higher consideration, and thereby a set of cases can be comprehended. It finds its application above all where there is a contrary (but not contradictory) opposition

[13] Cf. *Rothe, Arithmetik*, Part I., Foreword.

between the relations of magnitudes, and realizes arithmetic more as the negative predicate realizes speech. Which of two opposed magnitudes one calls positive and which negative is indeed arbitrary, as soon as the relations of space, in particular of direction, as well as of time, are spoken of; within the limits of arithmetic itself however it is by no means arbitrary, if it is to say more than that one could have chosen the symbol – instead of the symbol + to denote addition, which is admittedly self-evident. But within arithmetic there appears the opposition of multiplication and division, of exponentiation and deponentiation, which, on account of the significance of the multiplier and the exponent must be reduced to enumeration and decomposition. Now these oppositions do indeed have their characteristic symbols, but this does not prevent them from being expressed by those general ones, as is shown with exponents, where it is arbitrary whether one makes use of the one or the other, since $a^{-n} = 1/a^n$, where in the first expression the symbol is the general one, in the second the characteristic opposition of the chemical conjunction.

Since the synthesis of enumeration, as the characteristic arithmetic synthesis, can be subsumed under the synthesis of addition as the general logical synthesis (enumeration regarded as repeated addition), it must be denoted by the symbol for addition (+), whence it follows that decomposition is to be denoted by the other symbol. One must however carefully distinguish negative enumeration from the enumeration of negatives (negative units), which as enumeration can still always be positive. Since it is unimportant on which of the three orders of enumeration this is to be made clear, I will choose the second. In $3 \cdot 7$[14] let the number 3 be the multiplier. The expression then gives just $7 \cdot 7 \cdot 7$. But here the seven is regarded as produced by the synthetic enumeration of units, the number 3 by equal enumeration of seven. Now if we grant the possibility of reversal, then the 7 must be denoted +7, i.e. as an enumeration in the sense of addition, the 3 likewise as +3, the 7 enumerated in the sense of addition. On the other hand −3 as multiplier would indicate that the multiplicand is to be subtracted 3 times. Consequently

$$(+3) \cdot (+7) = +(+7) + (+7) + (+7) = +21,$$
$$(+3) \cdot (-7) = +(-7) + (-7) + (-7) = -21,$$
$$(-3) \cdot (+7) = -(+7) - (+7) - (+7) = -21,$$
$$(-3) \cdot (-7) = -(-7) - (-7) - (-7) = +21,$$

[14] The development of a general theorem on a composition consisting of specific numbers has the same generality as development in the general scheme, where letters replace numbers, provided one remains aware that the particular nature of the numbers used as no influence n the development. Thus a specific figure is the basis for developments in geometry, without its thereby lacking generality.

in which, according to the remark on P. 471, 0 must always be imagined as the subject of the conjunction if one wishes to take the sign as the subtraction symbol. Thus, the negative multiplier does not enumerate negative numbers, but rather enumerates negatively, i.e. it applies it as subtraction, and thus completely defines the form of the conjunction in accordance with its nature, while the multiplicand only gives its basis. Admittedly in the pure theory of number the limit of what is on hand is the limit of the solution; but it is the nature of the negative number that in its formation this limit must be overstepped; precisely on this account it no longer belongs in the pure theory of number, since it can add no independent value. This can only maintain it if the unit is subject to relations of the direction x, and this circumstance is decisive in order to refer it to the pure theory of number, and to distinguish this from the theory of whole numbers in the sense that this expression is used by *Legendre* and others.

Likewise the positive exponent enumerates factors in the sense of the synthetic, progressive conjunction, i.e. factors added as such, or multiplied, the negative such as are to be divided, and here as well it is not arbitrary whether one indicates multiplication of the unit with the root by the positive sign of the exponent, division by the negative, since the double affirmation confirms itself, but the double denial annuls, and annihilates itself, the synthetic and analytic composition however accurately observe, and must observe, the same type.

In a mixed expression of positive and negative terms it is unimportant whether one regards the algebraic sign as the form of the combination or as the positive and negative property of each term. In the first case the position of the terms between the combination symbol may not be changed; in the second, the combination symbol proper is completely missing, since the algebraic sign belongs to the term. One then always understands by the whole an aggregate, i.e. an additive conjunction, and the order of the terms is completely arbitrary. In arithmetic one sometimes chooses the one, sometimes the other view, according as it is the aim to demand of an expression and the law which one will represent by it. Likewise it is unimportant whether one regards the minus sign before a parenthesis as a combination symbol or as its algebraic sign, since the application of the subtraction rule and the prefixing of the minus sign ahead of another sign of a magnitude are of equal effect.[15]

[15] Many modem textbooks now devote a section to the use of parentheses in formulas. But such a section seems to me to be completely superfluous, since everything that properly belongs in them can be expressed in a simple rule that I give here, since I do not recall having found it anywhere else. It reads as follows: *The conjunction of a higher type of calculation is tighter than that of a lower, and always precedes it.* Should the conjunction of a lower type of calculation nevertheless come first, then the magnitudes to be conjoined must be enclosed in parentheses. The content of a parenthesis is in all circumstances valid as a single magnitude; the division sign and the extended root symbol have the effect of a parenthesis. The synthetic and analytic conjunction of the same order are of equal rank, and it is unimportant in which order one applies them.

Proof that there can be no conjunction higher than exponentiation

In several textbooks it is a common assumption, and is also asserted by philosophers,[16] that exponentiation is not necessarily the highest order of numerical conjunction, but rather that one sets the root and exponent equal, and then can express by a number how many times one has applied this fundamental conjunction. I know of no serious attempt really to introduce such a higher conjunction into arithmetic, much less any analytic expression that would seem to require it; but an investigation of its admissibility cannot be without interest, and is necessary here in order to define the extent of the pure theory of number. If this new order should have its characteristic significance, and can give us new information about the combination of numbers, then a new field for number theory would thereby be attained; if on the other hand it is shown that such a higher conjunction in the sense of the others can no longer exist instead, then number theory remains at a fixed limit, the endeavor to externalize it is terminated, and it is turned inward so much the more, which likewise can be regarded as an important achievement.

What was said above about the orders of enumeration applies here; that, if there is such a higher conjunction, it results from a new higher enumeration, whereby the root and the exponent are made equal, and in consequence there must be a passive and an active number, a basis and a quantifier. We will now denote the alleged higher conjunction by setting the quantifying number to the left of the basis, after the fashion of power-exponentiation. Thus ${}^{3}4$ denotes $4^{4^{4}}$, i.e. 4 to the 4th power raised to the 4th, or else 4 to the power 4 raised to the 4th, according as one interprets it as $(4^4)^4 = 416$, or as $4(4^4) = 4^{256}$. From expressions of this type, which are considered by several

---

To clarify this rule I will provide an example. Consider $a+\frac{bc}{d}-ef^3\sqrt{g^4}$ Here it is not important whether I first add the second term to the first, and subtract the third from the sum, or let the subtraction of the third term from the second precede the addition, and add the first term to the difference of the second and third, since addition and subtraction are of the same order. Likewise in the second term it is unimportant whether it is taken as $bc : d$ or $\frac{b}{d} \cdot c$, and in the third, whether one takes $a\sqrt{g^4}$ *as* $\sqrt{(g^4)}$ *or* $\left(\sqrt{g^4}\right)$. On the other hand the multiplication of *b* by *c* and division by *d* must precede addition to *a*,since the chemical conjunction is tighter than the mechanical. In the third term *f* must be raised to the third power, and the root of g^4 be taken, before one combines the magnitudes by multiplication, and this multiplication must occur before one can subtract – The usual notation of many mathematicians, in which $\sqrt{ab}$ denotes, not $\left(\sqrt{a}\right) \cdot b$, but rather $\sqrt{(ab)}$, should but for custom be set aside, since it is the only exception from an otherwise completely uniform rule. That by this law of conjunction the application of parentheses is unambiguously defined is self-evident.

[16] Thus, *Fries* in his *Mathematischen Naturphilosophie*, Sect. 23, P. 77, says that the arithmetic operations previously developed do not themselves set any limit, but on the contrary one can proceed ever further in the operations of combination, and then indicates the next higher order as is shown by analogy with the previous ones.

mathematicians, such as *Condorcet* and *Euler,* the situation can already appear doubtful, since it can have a different value according as one relates an exponent either to the one just previous, or to the entire preceding expression. One can indeed now assume a definite succession of numbers, but such an assumption would be arbitrary even so, and by no means one that proceeds from the nature of the subject itself, and is demanded by the science. What is decisive here however is the circumstance that the root can in no way be transformed into an exponent, as the multiplier can be transformed into a multiplicand. But without such a transformation the root and exponent, as numerical magnitudes of completely different types, absolutely cannot be enumerated as definite magnitudes. Their equality as absolute numbers is therefore of much less significance, since they are different by nature, and even if equal in the sequence of numbers, they are not equal in this combination. The above expression gives neither three roots nor three exponents, but one root and two exponents; the quantifying number in the combination $^{3}4$ is therefore without any sense, and it is the same way with all similar expressions. In the dynamic composition the unit was the multiplicand, the root the multiplier, and the exponent indicated how many times it was applied as such. Now if one wishes to transform the multiplier into an exponent, which must necessarily happen if one is to enumerate the elements of this combination, and this may perhaps be accomplished by adding the unit as multiplier, then one always has a power of the number 1, whose value is not in the least changed by all the appended exponents. But such a transformation is also erroneous, since $1 \cdot 4$ is by no means equivalent to $1 \cdot 1^4$ any longer. The demand for a higher order of numerical conjunction than exponentiation, analogous to the earlier orders, is therefore completely inadmissible, and being inconsistent must be rejected, not only from the pure theory of number but from arithmetic generally.

The theory of progressions also leads to the same result. The nature of the derivation of one magnitude from another amounts in essence to a relation, if one takes this word in its broader sense, in which it is applied to the new. The equal derivation of two relations gives the ratio equation, or proportion; a conjunction of equal relations, with which one proceeds at once, without further intervening members, to the next one derived, yields the progression. It becomes clear from this that there are properly as many types of relations, proportions, and progressions as orders of numerical conjunction; mechanical, usually called arithmetic, chemical or geometric, and dynamic. Only the first two are treated in textbooks; for the last, one does not even have a symbol, it also seems to be of little use. We will call the relation $a * x$ dynamic in as much as we see how the next sequential term results from the preceding one by exponentiation. Retaining the symbol chosen, then, $a * a^e = b * b^e$ is the general scheme of a dynamic proportion, and

$$a, a^e, (a^e)^e, ((a^e)^e)^e, (((a^e)^e)^e)^e, \ldots,$$

or

$$a, a^e, a^{e^2}, a^{e^3}, \ldots, a^{e^n}$$

represents the scheme of a dynamic progression for which the logarithm of the terms stand in geometric progression.

Now the transition from one order of numerical conjunction to the next is by means of each order of progression. The general scheme of an arithmetic progression is

$$\begin{matrix} 0 & 1 & 2 & & n \\ a, & a+d, & a+2d, & \ldots, & a+nd. \end{matrix}$$

If we mechanically remove the first, here as the term denoted the 0th, as arbitrary, which has no influence on the law of the series, then the series is transformed into the following:

$$0, d, 2d, 3d, \ldots, nd,$$

where all terms appear as chemical conjunctions, the *n*th as the general scheme of a two-term chemical conjunction. The general scheme of a chemical or geometric progression is

$$a, ea, e^2a, e^3a, \ldots, e^na.$$

If as before we chemically remove the foremost from all the terms, then we have

$$1, 1 \cdot e, 1 * e^2, 1 \cdot e^3, \ldots, 1 \cdot e^n,$$

a series of powers of the multiplier e, and the general term e^n is the general scheme of a dynamic conjunction. The transition from the mechanical to the chemical conjunction is therefore by means of the arithmetic progression, the transition from the chemical to the dynamic conjunction by the geometric. But if we add the unit to the terms in the scheme presented above of the dynamic conjunction (which belongs to it, as we have seen above, and as was also shown from the geometric progression), namely

$$1 \cdot a, 1 \cdot a^e, 1 \cdot a^{e^2}, \ldots, 1 \cdot a^{e^n},$$

then remove the leading term, as root, from all of them, then we obtain

$$1, 1^e, 1^{e^2}, 1^{e^3}, \ldots, 1^{e^n}.$$

Here the exponent refers, incorrectly to be sure, to the unit, as root, for the unit is the multiplicand, the root the multiplier. Letting this pass, there remains

$$1, e, e^2, e^3, \ldots, e^n.$$

This is the same series that we found above for the geometric progression, but it must be taken in another sense, for here a is removed, not as factor but as root, and e^n remains, not as a factor of the unit, but as the exponent left over, which either has no root at all, and therefore no assignable significance, or refers to the unit, making the value of every term $= 1$, which is the same result we found above. From the dynamic series $a, a^e, a^{e^2}, \ldots$ follows the geometric progression

$$\log a, e \cdot \log a, e^2 \cdot \log a, \ldots, e^n \cdot \log a.$$

and for $a = 1$, i.e. for the dynamic series $1, e, e^2, \ldots,$

$$\log 1, e \cdot \log 1, e^2 \cdot \log 1, \ldots, e^n \cdot \log 1.$$

But $\log 1 = 0$, whence all the terms $= 0$, and since they are the logarithms of the terms of the principal series these terms are all $= 1$.

The preceding silhouette of a higher conjunction thus completely vanishes, arithmetic attains its limit with exponentiation, and its field is defined; but within it remains open to an infinite development.[17]

Concluding remarks

The pure theory of number represents for us the innermost property of arithmetic, the field to which it belongs in every relation, and which it shares with no other science. In its length it is defined by the three orders of enumeration, in its breadth by the pure unit, as the notion of a simple given, which admits no further specification, whereby it is precisely an arithmetic unit. Its depth is unlimited, as every organic entity, which is

[17] On the relation of the pure theory of number to geometric combination theory, and how both proceed side by side in an orderly process of formation, but that undivided arithmetic must enter simultaneously with geometry as the theory of spatial magnitudes, see the Foreword to my *Raumlehre*, Part 2, Berlin 1824. However I observe in the mean time that the aggregate of theorems noted here by no means ought all to belong in beginning instruction, since for this end one has rather still to single out the elementary theorems. The length of this already too extensive article prevents more being said about this here.

indeed externally defined, and assumes an easily graspable form, but conceals within an infinite depth, whereby it is a subject of never-ending investigation.

The concept of the pure theory of number is not arbitrary, its separation from non-isolated arithmetic not indefinite and uncertain, but that concept is provided by the nature of the subject itself, this separation is demanded by it. In it the unit is nothing but the unit. It admits no relativity whereby the unit can be regarded as positive or negative, no divisibility whereby it can enter as a fractional unit, no irrationality, in which a neutralizing of the innermost characteristic of number, the property of being discrete, is demanded. The negative numbers, fractions, and irrational numbers (if one can use this expression, which properly contains a contradiction, in general) have exactly the same relation with respect to the pure theory of number as the impossible magnitudes have with respect to arithmetic, i.e. they are not realizable for it, and have no absolute, but only a combination value.

Now if the concept of the pure theory of number is a necessity, then its development is at the same time an interior understanding, and it is not just the kernel of arithmetic, to which all the rest are attached in an orderly way; rather, it must simultaneously represent the type of their development, indeed in a certain sense the type of the scholarly method in general. But every interior understanding is simultaneously an understanding with nature, whose law of exterior development cannot be more manifest to us than we have developed the law of the interior synthesis within ourselves. The extraordinary progress that the natural scientists have made recently, and still make daily, imperatively demand not just a broadened, but in addition a more orderly and, as it were, intensive development of mathematics; indeed it is to some extent a sign that it will break forth and must break forth. In particular it seems to me that the theory of fixed proportions in chemistry, as well as the relation of fundamental and derived structures in mineralogy, stand in a remarkable, if as yet not completely clear relation to the pure theory of number, and one appreciates this in general if one does not also take the different orders of oxidation as enumerations of nature, the combination of a basic factor with an acid not as multiplicative, – as indeed language already points to an analogous relation in many respects.

Remarks on illustrations

Illustration 1, page x

Conference venue at University of Potsdam. The opening of the Grassmann Bicentennial Conference (16–19 September 2009) took place in the reception hall of the main building of the University of Potsdam on Neues Palais campus. The Faculty of Mathematics and Natural Sciences of the University of Szczecin was connected to the event via video conferencing. The conference culminated in a day at the Faculty of Mathematics and Natural Sciences of the University of Szczecin (19 September 2009), which not only brought fruitful exchange, but also the inauguration of an exhibition of rediscovered documents of the Grassmann family and an organ concert with the music of Carl Loewe, a long-time friend of the Grassmann family. The concert took place in Saint Jacob's Cathedral of Szczecin, where Loewe was active for many years.

Illustration 2, page xv

Opening session of the Grassmann Bicentennial Conference in September 2009 in Potsdam. Seated on the right-hand side, first row: Prof. Dr. Andrzej Witkowski, Vice-Rector for Science and International Cooperation of the University of Szczecin (first from the left); Dr. Manfred Stolpe – Patron of the Conference, Premier of the State of Brandenburg, 1990–2002/German Federal Minister, 2002–2005 (second from the left); Birgit Müller, Vice-chairman of the Potsdam City Council Assembly (second from the right); Prof. Dr. Dieter Wagner, Vice-President for Knowledge and Technology Transfer of the University of Potsdam (first from the right). The entire conference was recorded on audio and video. (Photo taken by Fritze.)

Illustration 3, page xv

Hermann Grassmann's great-grandson Dietrich Grassmann, the conference's guest of honor, and the main organizer of the conference Hans-Joachim Petsche in front of a picture showing the Stettin Masonic lodge garden in the nineteenth century.

Illustration 4, page xvi

Inauguration of a Grassmann exhibition unveiling recently discovered documents in the Pomeranian Library (during the conference's stay in Szczecin).

During the inauguration of a Grassmann-exhibition at the Pomeranian library, Jolanta Liskowacka, director of the department of autographs, presented recently discovered documents of the Grassmann family from family chronicles dating back to the sixteenth century.

Illustration 5, page xvi

The participants of the Grassmann Conference in front of the former Marienstiftsgymnaisum in Szczecin. The members of the Bicentennial Conference did not pass up on the occasion to have their picture taken on the impressive stairs leading up to the former Marienstiftsgymnasium, the secondary-school where Hermann Grassmann worked as a teacher for over a quarter of a century. The building still houses a secondary school today.

Illustration 6, page xx

Pestalozzi in Stanz, lithography from 1846, most likely published to commemorate the 100th anniversary of Pestalozzi's birth(1746–1827), accompanied by a text by Adolph

Diesterweg (1790–1866). Among the followers, pupils and friends of Pestalozzi we also find school councilor Friedrich Heinrich Gotthilf Grassmann (1784–1866), Hermann Grassmann's uncle and Robert Grassmann's foster-father. The two brothers Friedrich Heinrich Gotthilf and Justus Günther Grassmann (1779–1852), as well as school councilor Georg Wilhelm Bartholdy (1765–1815) were known to be fervent followers of Pestalozzi, see [Dilthey 1921, 478 ff.]. The mathematical work of Justus Grassmann took its point of departure in his Geometry for elementary schools, written in 1817 in the spirit of Pestalozzi's pedagogical principles [J. Grassmann 1817]. For more information on Pestalozzi, his pupils and followers, see [Diesterweg, Massmann and Kalisch 1845; Dejung 1980]. Concerning the importance of Pestalozzi's pedagogy for Justus and Hermann Grassmann, see also [Petsche 2010].

Illustration 7, page 2

First published description of the life of Hermann Grassmann by his son Justus, presumably from the year 1877. The German text runs as follows:[1]

Hermann Günther Graßmann wurde am 13ten April 1809 zu Stettin geboren. Den ersten Unterricht erhält er durch seine Mutter Johanne geborene Medenwald; später besuchte er das vereinigte Königl[iche] und Stadtgymnasium zu Stettin, an welchem sein Vater, Justus Günther Graßmann, Lehrer der Mathematik u. Physik war. Nach bestandenem Maturitätsexamen bezog er zusammen mit seinem älteren Bruder Gustav (geb. 2/7 07) die Universität Berlin, um wie dieser Theologie zu studiren. (Auch der Vater hatte ursprünglich Theologie studirt) Er hörte daselbst während der drei Studienjahre Vorlesungen bei Boekh, Hengstenberg, Marheineke, Neander, Schleiermacher, v. Raumer, Ritter, und scheint sich besonders eng // an Schleiermacher angeschlossen zu haben, von dessen Vorlesungen und Werken sich mehrfache Bearbeitungen unter seinen hinterlassenen Papieren finden. Mathematische Vorlesungen hat er nicht gehört; doch hat er die Beschäftigung mit der Mathematik, in der er schon auf dem Gymnasium hervorragendes leistete nicht fallen lassen, und hat besonders *als* seines Vaters bedeutendstes Werk "Zur physischen Krystallonomie und geometr. Combinationslehre 1829" erschien, dasselbe genau studirt und durchgearbeitet. Auch mit den anderen Werken seines Vaters, so besonders mit dessen

[1] Editorial notes:
- Page breaks in the notes of Justus Grassmann are indicated by "//".
- Text struck out is taken over as text struck out.
- Insertions in the margin or above the lines are written in italics.
- Proper completions are inserted between brackets "[,]".

Abhandlung "Über die reine Zahlenlehre" (Programm des Stettiner Gymnasiums 1827) // dessen Raumlehre und Trigonometrie hat er sich eingehend beschäftigt, und also die diesen Werken eigenthümliche mathematische Anschauungsweise von nicht geringem Einfluße auf seine späteren math[ematische] Arbeiten gewesen. Ostern 1831 trat er in das mit dem Stettiner Gymnasium verbundene Königl. Seminar für gelehrte und höhere Bürgerschulen ein, und absolvirte November desselben Jahres sein philologisches Staatsexamen (bes[onders] für die Fächer Lat[ein], Griech[isch] u[nd] Mathematik.) Trotzdem gab er seinen Lieblingsgedanken, Prediger zu werden, vorläufig noch nicht auf, sondern trieb fleißig Theologie und bestand 1834 22/5 sein erstes, 1839 12/7 sein zweites theologisches Examen – ersteres mit d. Prädikate "gut," letzteres mit d. Prädikate "sehr gut." 1840 1/5 endlich vervollstän- // digte er durch eine Nachprüfung sein wissenschaftliches Prüfungszeugniß von Nov[ember] 31 und erwarb sich eine unbedingte Facultas in Math[ematik], Physik, Mineralogie und Chemie, für welche Fächer das Zeugniß ihn als vorzugsweise befähigt empfiehlt. (Examensarbeit: Ueber die Theorie der Ebbe und Fluth.) Mittlerweile war er 1834 als Lehrer an die Gewerbeschule in Berlin gegangen, wo er besonders im Umgange mit Jacob Steiner neue Anregung *zu math[ematischen] Arbeiten* erhielt. 1836 1/1 wurde er Lehrer an der Ottoschule (Beschäftigung mit Chemie) 1842 Mich[aelis] am Gymnasium zu Stettin. In letzteres Jahr fallen seine Vorlesungen über sein wichtigstes math[ematisches] Werk, die Ausdehnungslehre, vor dessen // Veröffentlichung gehalten im Privatkreise vor seinem Bruder Robert Gr[aßmann], von Kamecke (jetzt Kriegsminister) Jungklass (jetzt Schulrath in Bernburg) Scheibert (jetzt Geh. Regierungs-u[nd] Schulrath a.D. in Schlesien.) Ostern 1843 wurde er Lehrer (1847 5/5 Oberlehrer) an der Friedr[ich-]Wilh[elms] Schule (Real[schule] I.O[rdnung]) zu Stettin; (seiner Lehrthätigkeit an dieser Schule verdanken mehrere kleine Lehrbücher ihre Entstehung, so: Leitfaden für den ersten Unterricht der lateinischen Sprache (1843), Leitfaden der deutschen Sprache 1848 (zusammen mit seinem Bruder Robert) Deutsches Lesebuch (zusammen mit W. Langbein) – Das Jahr 1848 nahm seine Thätigkeit hauptsächlich auf politischem Gebiete in Anspruch. In treuer Anhänglichkeit an das Königshaus erhob er, einer der ersten, öffentlich seine Stimme gegen // das revolutionäre Treiben in Berlin (z.B. Zeitungsartikel in der Vossischen Zeitung "Die Früchte des Berliner Barrikadenkampfes" und gründete zusammen mit seinem Bruder Robert eine Zeitschrift (Deutsche Wochenschrift) in der die brennendsten Tagesfragen, hauptsächlich die neu zu gestaltenden Verfassungen des deutschen Reiches, des preußischen Staates und der Kirche in Preußen besprochen werden. – In demselben Jahre (10/8) verlobte er sich mit Marie Therese Knappe, Tochter der verwittweten Frau Rittergutsbesitzer Knappe auf Alt-Storkow bei Nörenberg i[n] Pommern. Am 12ten April folgenden Jahres fand die Hochzeit statt, und bald sammelte sich um die Eltern ein glücklicher Kreis von Kindern (im Ganzen 11, noch lebend 8), für deren große und kleine Erlebnisse der Vater trotz // seiner amtlichen und wissenschaftlichen Thätigkeit stets ein gleich wachsames und liebevolles Auge hatte. Im Jahre 1852 1/7 erhielt er, nachdem sein Vater am 9/9 gestorben, die Stelle des ersten Mathematikers am vereinigten Königl[ichen] u[nd] Stadtgymnasium (dem späteren Marienstifts-Gymnasium) und bald darauf den Professoren-Titel. In dieser Stellung ist er bis an sein Lebensende thätig gewesen. *Er starb an den Folgen eines Herzfehlers, der ihm früher unbekannt gewesen war und sich zuerst 2 Jahre vor seinem Tode bemerkbar machte und ihn unter allmählicher Abnahme der Kräfte (Zuletzt trat Wassersucht in Folge der Blutverdünnung hinzu) sanft entschlafen ließ.*

Im Jahre 1844 erschien der erste Theil der Ausdehnungslehre bei Wigand in Leipzig (in Folge davon Verkehr mit Professor Moebius) 1847 seine geometrische Analayse, eine Bearbeitung der von der Fürstlich Jablonowskischen Gesellschaft gestellten Preisaufgabe, die Wiederherstellung und weitere Ausbildung des von Leibnitz erfundenen geometrischen Kalküls oder die Aufstellung eines ihm ähnlichen Kalküls betreffend (gekrönt 1/7 46) und nun folgen // zahlreiche Abhandlungen, meist geometrischen Inhalts in Crelles Journal; einige physikalische in Poggendorfs Annalen. Indessen hatten diese Arbeiten im Ganzen nicht den erwünschten Erfolg. Zwar fehlte es nicht an einzelner Anerkennung (Moebius, Bellavitis, Cremona), und gaben besonders manche der kleineren geometrischen Abhandlungen zu verwandten Bestrebungen Anlaß. Der eigentliche Kern aber der Arbeiten und besonders der Ausdehnungslehre blieb fremd und nirgends geschah dieses Werk eine Erwähnung. Da entschloß er sich, in dem er diesen Mißerfolg der Form der Darstellung zuschreiben zu müssen glaubte, mit der Veröffentlichung des zweiten Theils zugleich auch den ersten einer völligen Umarbeitung zu unterwerfen. Als aber auch das neue Werk (*erschienen 1862*) nicht mehr Beachtung fand // als das erste und zugleich alle Versuche, einen academischen Lehrstuhl zu erlangen und so einen Kreis von Schülern ~~zu~~ um sich zu sammeln und für seine Ideen zu gewinnen, fehlschlugen, da wandte er sich, wenn auch schweren Herzens von der Mathematik ab und den sprachlichen Studien zu, die er schon länger neben seinen math[ematischen] Arbeiten zur Erholung getrieben hatte. Schon im Jahre 1852 hatte er sich mit Sanskrit beschäftigt (1857 gemeinschaftlich mit Dr. Bressler) später hatte er Griechisch, Littauisch, Zend etc. getrieben, und als seine ersten sprachl[ichen] Arbeiten in Kuhns Zeitschrift eine günstige Aufnahme fanden, entschloß er sich zur Veröffentlichung eines größeren Werkes, zunächst eines Rig-Veda-Wörterbuchs, dann einer Rig-Veda- // Übersetzung, welchen beiden noch eine Grammatik folgen sollte. In diese Zeit fällt auch sein kleines Werk über "Deutsche Pflanzennamen" (1870) durch welches er eine deutsche Pflanzenbenennung in die Botanik einzuführen versucht. – Endlich nahm er 1872 nach mehr als zehnjähriger Pause seine math[ematischen] Arbeiten wieder auf, nachdem besonders Herm[ann] Hankel, Alfred Clebsch, Victor Schlegel auf die Bedeutung seiner Ausdehnungslehre aufmerksam gemacht hatten.

Die überaus reiche Anerkennung, die ihm für seine sprachlichen Werke zu theil wurde und die Freude auch seine mathematischen Arbeiten nicht mehr müßig liegen zu sehn, sondern, empfohlen durch bedeuten- // de Mathematiker, in die verschiedensten Disziplinen eingreifen zu sehn, haben ihm den Abend seines Lebens in hohem Maße verschönt. Noch wenige Tage vor seinem Tode theilte ihm Prof. Burmester in Dresden mit, daß er im kommenden Winter Vorles. über die Ausdehnungslehre zu halten gedenke; und mit Freuden konnte er nun Andern das überlassen, was ihm selbst ein ungünstiges Geschick versagt hatte. – Zeichen ~~der~~ *äußerer* Anerkennung waren die Ernennung zum correspondirenden Mitglied der *Königl[ichen]* Gesellschaft der Wissensch[aften] zu Goettingen (2/2 71), die Ernennung zum Ehren-Doctor der Philos[ophischen] Falcultät der Universität Tübingen (1876), die Ernennung zum Mitgliede der American Oriental // Society (1876) etc.

Auch auf andere Gebiete war seine Thätigkeit gerichtet. So hat ~~hat~~ er in den Jahren 61–72 zahlreiche Volkslieder gesammelt, ihre Melodien nach dem Gehör aufgeschrieben und zum Singen in der Familie dreistimmig gesetzt (Vier derselben sind zufällig im Festgruß an Ludwig Erk, Heilbronn 1876 veröffentlicht worden) – Ferner leitete er einen Männer*gesang*verein von Stettiner Schülern. – Er war Mitglied der Loge, in die er Weihnachten 1835 eintrat. Seit dem Jahre 1856 ist er als Redner in derselben thätig gewesen. – Seit 1857 war er Mitglied des Vorstandes des chinesischen Missionsvereins in Pommern, seit 1859 Sekretär und später Vorsitzen der. – Seit 1852 leitete er die von seinem Vater in Stettin gegründete physikalische Gesellschaft. – u. s. w. // Noch zu erwähnen ist eine kleine theologische Schrift: "Ueber den Abfall vom Glauben," die er nicht lange vor seinem Tode verfaßte und welche in Kürze im Druck erscheinen wird.

(Transcribed from the original by G. Keßler and H.-J. Petsche, August 2009)

Illustration 8, page 8

Cover sheet of Robert Grassmann's teacher's examination file from 1840. Unfortunately, the actual texts submitted by Robert Grassmann are no longer among the documents.

Illustration 9, page 16

First page of Hermann Grassmann's final examination essay from secondary school, written in 1827 on the topic: "What is genuine scientific spirit and how does it reveal itself as such?" The essay topic originated with great probability from his father, who at that time taught not only mathematics, but also German.

Transcription

Hermann Graßmann

studirt Theologie und Philosophie.

Thema.

Was ist wahrhaft wissenschaftlicher Sinn und wodurch erweist sich ein solcher

Wenn wir eine neue Lebenslaufbahn beginnen, so ist wohl nichts natürlicher und ~~w~~ zugleich wichtiger, als daß wir nachdenken über den neuen Beruf, damit wir die uns vorgezeichnete Bahn mit dem wahren ~~Feu~~ Sinn, und mit echter Treue und Liebe betreten mögen. Wenn gleich wir nun ~~gleich~~ auch schon früher uns mit wissenschaftlichen Gegenständen beschäftigten, so treten wir doch jetzt in einen ganz neuen Kreis ~~unser~~ der ~~Wirksamkeit~~ wissenschaftlichen Thätigkeit. Denn während früher unsere Bestrebungen durch die treue Hand der Aeltern und Lehrer geleitet wurden, so sind wir jetzt uns selbst überlassen, und sollen durch eigne Kraft, ohne fremde Hülfe, ~~die f~~ auf der Bahn der Wissenschaften fortschreiten. Auf welche Irrwege würde uns nicht unser Leichtsinn führen, wenn wir nicht ~~ge~~ jetzt ~~mit Ernst~~ über den Weg ~~nachdächten~~, den wir einschlagen wollen und über den Geist ~~und den Sinn~~, der uns begleiten und beleben soll in unserem neuen Berufe, mit Ernst nachdächten? Wir sehen ja so viele umherirren auf dem weitem Gebiete der ~~Wissenschaf~~ Wissenschaften, ohne daß sie den wahren // wissenschaftlichen Sinn in sich tragen. Viele treiben die Wissenschaften aus selbstsüchtigen Absichten, ~~um sie~~ und betrachten sie als ein Mittel, um sich ~~dadurch sie sich~~ für die Zukunft einen festen Lebensunterhalt zu erwerben, die ~~W~~ Wissenschaften sind ihm, was dem Handwerker sein Handwerk, er hegt keine Liebe für die Wissenschaften sondern nur für die äußer~~lichen~~ (en) Vortheile die sie mit sich führen. Daß dies nicht der rechte Weg sei sieht wohl jeder ein. – Andere ~~hinge~~ streben-nach dem Ruhme ~~des Gelehrten; der äußere Gl Glanz die und~~ und ~~nach~~ der ~~äußeren~~ Ehre, die dann ~~dem~~

Translation

Hermann Graßmann

studied theology and philosophy.

Theme.

What is genuine scientific spirit and how does it reveal itself as such

When we begin a new course of life there is nothing more natural and ~~w~~ at the same time more important than that we think over the new occupation, in order that we may enter the way sketched for us with genuine ~~Feu~~ spirit, and with genuine faith and love. Even if we ~~just~~ already concerned ourselves with scientific subjects before, now we enter into a completely new sphere ~~of our~~ of ~~effectiveness~~ scientific activity. For although earlier our endeavors were guided by the faithful hands of parents and teachers, now we are left to ourselves, and will by our own power, without outside help, ~~that f~~ proceed along the road of science. On what torturous path would our foolishness not lead us if we did not ~~ge~~ now ~~with seriousness think~~ think seriously about the path we would follow and the spirit ~~and the spirit~~ that will accompany and enliven us in our new occupation? Indeed, we see so many wandering about in the broad domain of the ~~sciene~~ sciences without bearing genuine // scientific spirit. Many such a one pursues the sciences with egotistical intentions, ~~in order that~~ and considers it ~~thereby itself se~~ as a means to acquire for himself a solid living, the ~~s~~ sciences are to him what his craft is to the craftsman, he feels no love for sciences but only for the external advantage to which they lead him. That this is not the correct path anyone can see. – Others ~~sacri~~ strive after glory ~~of scholars; the external gl gleam and~~ and ~~after external~~ for the honor, which the true scholars will then assign

~~wahrhaft~~ Gelehrten zu theil wird, ~~sie~~ diese beharren schon eifriger auf der Bahn der Wissenschaften; allein das, wonach sie streben, ist nur ein äußerer Schein; und diesen vermögen sie auch nur zu erlangen, ~~in~~ das wahre Heiligthum der ~~Wi~~ Wissenschaft bleibt ihnen verschlossen. Andere endlich, wiewohl die wenigsten, wurden nicht durch Liebe zu der Sache, sondern vielleicht durch den Wunsch derer die ihnen theuer sind oder durch andere an ~~fü~~ und für sich nicht tadelswerthe Gründe, bewogen, den Wissenschaften ihren Fleiß zu weihen. Sie suchen die Sache mit Eifer zu treiben, sind treu und fleißig in ihrem Berufe; aber ihr Fleiß ~~hat~~ trägt nicht ~~dann~~ die erwünschten ~~Fort~~ Früchte; denn die Liebe für den Beruf, der, wie der Sonnenschein, // die jungen Keime entfaltet und die Früchte zur Reife bringt, mangelt ih~~m~~nen. Also die Liebe zu den Wissenschaften ist es, die den wahrhaft wissenschaftlichen Sinn in uns erzeugt und nährt, jene Liebe, die hervorgeht aus dem Streben nach der Vollkommenheit und ~~nach~~ der Ausbildung des Geistes. Dieses Streben aber regt sich in uns allen, und es ist wohl kaum irgend einer dem dieses Streben ganz fremd sei. Ja die todte Natur strebt enger zu einer gewissen Stufe der Vollkommenheit; dieses Streben vereinigt gleichsam die ganze Natur und Alles ~~g~~ wandelt hin nach einem hohen Zweck. ~~Dieses Streben trieb~~ Es ist das Ebenbild Gottes, was in dem Menschen lebt, es hebt ihn enger zu seinem Urbilde. Aber leider ist in der sündhaften Natur des Menschen dieses Ebenbild durch unreine Stoffe befleckt; der Mensch hält oft den äußeren Schein des Guten und Vortheilhaften für das Gute und Nützliche selbst und wird so ~~irre~~ geleitet von dem Wege zur Vollkommenheit; seine Selbstliebe artet in Eigenliebe aus. Es möchte vielleicht scheinen als stehe das eben gesagte in keinem Zusammenhange mit dem, worüber ich sprechen wollte; allein diese Entwicklung schien mir nöthig zu sein, um daraus den wahren wissenschaftlichen Sinn herleiten zu können; denn dieser Sinn ~~steht im~~ kann nur hervorgehen aus dem eifrigen // Streben nach Vollkommenheit. Dieses Streben ist es ~~der,~~ welches unsern wissenschaftlichen Bestrebungen innere Stärke, ~~und~~ Ausdauer, und ein stets sich selbst erzeugendes Leben verleiht; – welches sie bewahrt vor jeder Einmischung eines fremden, eigennützigen Zweckes, welches ~~sie rein erhält vor~~ in unsern Bemühungen Ausbildung des Geistes und des Herzens vereint. Denn wie kann jemand danach wissenschaftlichen Sinn hegen, wenn er nicht zugleich nach moralischer Veredlung strebt; ~~ich ka~~ und

them, for they will persist ever more eagerly on the road of science; only that after which they strive is only an external appearance; this only are they able to achieve, ~~in~~ the genuine shrine of ~~sei~~ science remains closed to them. Finally, yet others, although the smallest number, not from love of the subject, but perhaps by the wish of those dear to them or through other ~~fo~~ grounds not unworthy in themselves, are moved to dedicate their diligence to the sciences. They seek to pursue the subject with enthusiasm, are faithful and diligent in their profession; but their diligence ~~has~~ does not ~~then~~ bear the desired ~~fort~~ fruits; for the love for the profession which, like sunshine, // unfolds the young shoots and brings the fruit to ripeness, is lacking in them. Thus the love for the sciences is that which genuine scientific spirit produces and draws us closer, that love that comes from striving toward perfection and the development ~~after~~ of the spirit. ~~Of these strivings pursue~~ These strivings however move themselves in all of us, and indeed there is scarcely anyone to whom this striving is completely alien. Indeed, dead Nature strives more closely to a certain level of perfection; this striving unites all of nature as it were and turns everything ~~g~~ toward a high goal. It is the image of God that lives in man, it raises him closer to his archetype. But unfortunately in the sinful nature of man this image is stained by impure matter; man often holds the external appearance of the good and the advantageous for the good useful itself and is thus ~~wand~~ led away from the path to perfection; his self-love degenerates into egotism. It might perhaps seem that what was just said bears no relation to that about which I would speak; only this development seems to me necessary, in order to be able to derive thereby genuine scientific spirit; for this spirit ~~stands in the~~ can only come from the enthusiastic striving for perfection. This striving it is ~~which~~ that lends our scientific endeavors inner strength, and staying power, and an ever self-generating life; – which protects it from mixture with alien, self-interested goals, which ~~he pursues purely for~~ in our endeavors unites development of the mind and heart. For how can someone give attention to scientific spirit if

wenn wir gleich viele sehen, die sich den Ruf der Gelehrsamkeit erworben haben; die aber in moralischer Hinsicht ~~aber~~ weit ~~hinter andern~~ zurück sind; so kann ich doch nicht glauben, daß diese~~n~~ von dem wahrhaft wissenschaftlichen Sinn innig durchdrungen ~~sind~~ sein sollten. Denn dieser Sinn strebt enger nach der Vollkommenheit, ~~der welchen dieser Sinn belebt,er sucht sich selbst auszubilden~~ sucht sich selbst zu veredlen und so seinem himmlichen Vorbilde immer ähnlicher zu werden; sein~~e~~ Wissen~~schaften~~ ist ihm nicht ein todter Schatz, sondern eine stets strömende Quelle der Veredlung. ~~Sie~~ Je höher er engerstrebt, desto mehr belebt sich seine Thätigkeit, je näher er seinem Ziele kommt desto mehr entzündet sich seine~~m~~ Liebe zu den Wissenschaften. – Bei dem hingegen welcher // aus andern Absichten die Wissenschaften treibt, zeigt sich bald eine Erschlaffung der Kräfte, die Triebfedern, die ihn bewegten reichten nicht hin, um ihn für immer mit Eifer zu beleben. Sobald daher diese Zwecke ~~d~~ verschwinden, so verläßt er die Wissenschaften, welche er nicht liebte, ohne etwas anderes gewonnen zu haben, als ein eitles nutzloses Gut. Wie edel erscheint dagegen derjenige der seine Geistesgaben durch wahre ~~wi~~ wissenschaftliche Bildung veredelt, der durch sein Wissen dem Menschen-Geschlechte wahrhaften Nutzen erweiset, und so ~~der~~ jenes Streben in der ~~Nä~~ Nächstenliebe seine Anwendung finden läßt! –

Ja groß ist dieses Ziel, was ich mir vorgestellt; aber ich fühle es, es ist das wahre Ziel aller wahrhaft wissenschaftlichen Bestrebungen, und diesen Zweck möge der Allmächtige stets in mir beleben und wenn ich ihn aus dem Auge~~n~~ verliere von neuem in mir hervorrufen, damit ich auf der neuen gefährlichen Bahn fest beharre und mich durch keine äußere Lockungen von diesem Wege abschrecken lasse.

(Transcribed from the original by G. Keßler and H.-J. Petsche, March 2010)

he does not at the same time strive after moral improvement; ~~I ca~~ and if we see many, who have attracted to themselves reputation of learning; who however in the moral respect ~~however~~ are far ~~behind other~~ behind; thus I still cannot believe that this would be deeply imbued by genuine scientific spirit. For this spirit strives more closely to perfection, ~~that which this spirit enlivens, he seeks to shape himself~~ seeks to improve itself and thus to become ever more like its heavenly example; its ~~seien~~ knowledge is to him not a dead treasure, but an ever-flowing source of improvement. ~~He~~ The higher he strives, the more enlivened his activity, the nearer he comes to his goal, the more inflamed his love for the sciences. – On the other hand he // who pursues the sciences for other purposes, soon shows a flagging of power, the motivating force that moved him does not hold out, for him forever to be stimulated to enthusiasm about it. As soon therefore as this goal vanishes, he abandons the sciences, which he did not love, without having achieved anything but a vain and useless possession. How noble on the other hand appears one who improves his intellectual gifts by genuine ~~sci~~ scientific education, who by his knowledge proves truly useful to mankind, and thus ~~the~~ that endeavor finds its application in ~~bro~~ brotherly love!

Yes, great is the goal I have set myself; but I feel that it is the genuine goal of all scientific endeavors, and may the Almighty ever enliven this purpose in me, and if I lose sight of it call forth in me anew, wherewith I firmly persevere on the perilous new road and let myself by no external temptations be deterred from this path.

(Translation by Lloyd Kannenberg, April 2010)

One finds several strikethroughs and underlinings of the examining teachers.

Illustration 10, page 84

Hermann Grassmann's secondary school certificate from 1827. The German text was first printed in [Engel 1911. p. 15]:

Transcription
Gymnasium zu Stettin.
Entlassungs-Zeugnis
No. I.

Hermann Günther Grassmann aus Stettin, Sohn des Gymn. Professor Grassmann zu Stettin.

Zeit des Schulbesuchs: im Ganzen: 7 1/2 Jahre; in Prima 2 Jahre.

Aufführung:

a) gegen Mitschüler: Durch sanftmütigen Sinn, Friedfertigkeit und Gefälligkeit erwarb er sich die allgemeine Liebe seiner Kommilitonen und gab nie Veranlassung zu irgend einem Mißverhältnis.
b) Gegen Vorgesetzte: Bescheidenheit und Regelmäßigkeit zeichneten ihn vorteilhaft aus und gewannen ihm die allgemeine Zufriedenheit seiner Lehrer in vorzüglichem Grade.

Fleiß:

Durch ununterbrochenen, regelmäßigen Besuch der Lehrstunden, durch gespannte Aufmerksamkeit in den so wie durch erfreuliche Beweise eines eifrigen, gründlichen und glücklichen Fleißes hat er seinen Lehrern in allen Objekten des Unterrichts Befriedigung gewährt.

Kenntnisse:

In Sprachen und Wissenschaften hat er so befriedigende Fortschritte gemacht, daß ihm das Zeugnis der unbedingten Tüchtigkeit zur Universität nicht hat versagt werden können.

Stettin den 17ten September 1827

(Transcribed from the original by G. Keßler and H.-J. Petsche, März 2010)

Translation
Gymnasium at Stettin.
Discharge-Certificate
No. 1.

Hermann Günther Graßmann of Stettin, son of the Gymn. Professor Graßmann in Stettin.

Time of school attendance: in all 7 1/2 years; in Prima 2 years.

Performance:

a) With respect to schoolmates: By kind sensibility, peaceability, and helpfulness he acquired the general love of his fellow students and gave no occasion for any inadequacy.
b) With respect to superiors: Modesty and orderliness marked him out to advantage and won him the general satisfaction of his teachers to an excellent degree.

Diligence:

By uninterrupted, regular attendance at classes, by keen attention in them as well as by gratifying evidence of an eager, thorough and fortunate diligence he has given his teachers satisfaction in all objectives of instruction.

Knowledge:

In language and science he has made such satisfactory progress that the Certificate of Unconditional Competence at the University cannot be denied him.

Stettin, the 17th September 1827

(Translation by Lloyd Kannenberg, April 2010)

Illustration 11, page 128

Friedrich Engel does not mention this text in his listing of Hermann Grassmann's writings [HGW32, 356–367]. It is a tract on the didactics of experimentation in chemistry class, written in 1850 [H. Grassmann 1850]. Its approaches are extremely modern. Grassmann's experience with teaching chemistry dated back to his days at Friedrich-Wilhelms-Schule, a secondary school below the level of "Gymnasium" run by Carl Scheibert. When Grassmann began to teach at "Gymnasium" in 1852, he stopped teaching chemistry since it was not part of the curriculum of higher secondary schools like "Gymnasium."

Illustration 12, page 240

Hermann Grassmann's petition from 2 January 1869 to minister von Mühler, in which he applied for the second professorship of mathematics at the University of Greifswald (with handwritten comments by the ministry and a draft of the negative reply).[2]

Transcription	**Translation**
Stettin den 2. Januar 1869	Stettin the 2. January 1869
Gehorsamste Bitte des Professors Graßmann um Verleihung der zweiten ordentlichen Professur der Mathematik an der Universität Greifswald.	Most humble request of Professor Graßmann to be granted the second Ordinary Professorship of mathematics at the University of Greifswald
Euer Excellenz	Your Excellency
bitte ich ganz gehorsamst, mir die zu Ostern vacant werdende zweite ordentliche Professur der Mathematik an der Universität Greifswald hochgeneigtest verleihen zu wollen. Obgleich mein Einkommen, welches sich gegenwärtig auf 1420 Reichsthaler beläuft, sich bei diesem Uebergange bedeutend verringern würde, und ich diese Verringerung bei meiner starken Familie (von 9 Kindern) nicht ganz unbeachtet lassen darf, so glaube ich doch, mich mit dieser meiner gehorsamsten Bitte an Euer Excellenz wenden zu müssen, da ich, seit mir meine mathematischen Studien neue, fruchtverheißende Gebiete dieser Wissenschaft aufschlossen, es als die Aufgabe meines Lebens angesehen habe und noch heute ansehe, diese Gebiete zu bearbeiten und in das gesammte Gebiet des mathematischen Wissens einzufügen, und es mir noch heute als das Ziel meiner Bestrebungen vor Augen steht, die akademische Jugend in diese Gebiete des Wissens einzuführen, wie	I most humbly request your most gracious favor that I be granted the second Ordinary Professorship in mathematics at the University of Greifswald, which is to become vacant at Easter. Although my income, which at present amounts to 1420 Reichsthaler, would upon this transfer be significantly reduced, a reduction to which, with my large family (of 9 children), I cannot remain completely inattentive, I still believe I must turn to Your Excellency with my most humble request, for, as my mathematical studies opened to me new potentially fruitful areas of this science, I saw it as my life's task, and see it so even today, to cultivate this area and insert it into the entire domain of mathematical knowledge, and to this very day there stands as the object of my endeavors to intro -duce academic youths to this area of knowledge, as I have expressed myself repeatedly

[2] Editorial notes see footnote 1.

ich dies wiederholt auch öffentlich ausgesprochen habe, namentlich in der Ausdehnungslehre von 1844 Vorrede p. XVII und in der Ausdehnungslehre von 1862 Vorrede p. IX. Ich habe mir daher erlaubt, diese beiden Werke gehorsamst beizufügen. Ebenso habe ich bei dem Erscheinen des zweiten dieser Werke // Seiner Excellenz dem damaligen Minister v. Bethmann-Hollweg den gehorsamsten Wunsch ausge-sprochen, als Professor der Mathematik an eine Universität berufen zu werden und habe von Seiner Excellenz unter dem 12. Februar 1862 die Antwort erhalten, daß er diesen Wunsch, wenn sich eine geeignete Gelegenheit darbiete, gern in Erwägung nehmen werde, er mich jedoch schon im Voraus darauf aufmerksam mache, daß die meisten Besoldungen derartiger Universitätsstellen hinter dem Betrage meiner jetzigen Einnahmen zurückstehe. Wenn gleich nun diese Antwort keine Zusicherung irgend welcher Art enthält, so hielt ich es doch für meine Pflicht, derselben auch gegen Euer Excellenz Erwähnung thun zu müssen.

In der Hoffnung der Gewährung meiner
gehorsamsten Bitte unterzeichne ich
Euer Excellenz

ganz gehorsamster
H. Graßmann
Professor am Gymnasium

(Zahlreiche Bearbeitungsvermerke sowie handschriftlicher Vermerk auf dem linken Rand der ersten Seite. Dort auch der nachfolgend transkribierte Entwurf einer Antwort des Ministers)

Berlin, den 10ten März 1869.
An den Herrn Professor H. Graßmann Wohlgeboren

in Stettin

Euer Gesuch um Verleihung der zweiten Professur der Mathematik an der Universität zu Greifswald habe ich mit Rücksicht auf den Vorschlag der dortigen Philosophischen Facultät wegen Wiederbesetzung dieses Lehrstuhles ~~zu meinem Bedauern~~ nicht willfahren können; wovon ich Sie hirdurch bei Rücksendung (2 Bücher) der Anlagen Ihrer Eingabe vom 2ten Januar d.J. in Kenntniß setze.

Der Minister
i.V.
(Transcribed from the original by G. Keßler and H.-J. Petsche, März 2010)

and publicly, namely in the Ausdehnungslehre of 1844 Foreword p. XVII and in the Ausdehnungslehre of 1862 Foreword p. IX. I have therefore ventured most humbly to enclose these two Works. Likewise I had, upon the appearance of the second of these Works, // expressed to His Excellency the then Minister v. Bethmann-Hollweg the most humble wish to be appointed as Professor of Mathematics at a University and have from His Excellency, under the date 12.February 1862, received the answer that he would be pleased to take this wish into consideration if a suitable opportunity presented itself, however he already pointed out to me in advance that the pay of such a University position usually takes second place behind the amount of my present income. Now even though this answer contains no promise of any sort, I still regard as my duty that I must make mention of it to Your Excellency as well.

In the hope of the granting of my
Most humble request I sign myself,
Your Excellency's

Most obedient
H. Graßmann
Professor at the Gymnasium

(Numerous signatures and handwritten remarks on the left-hand margin of the page. Here we also find the following draft of the minister's reply to Grassmann)

Berlin, the 10th March 1869.
To Herr Professor H Graßmann Sir

in Stettin

With your request for the bestowal of the second Professor of Mathematics at the University of Greifswald I cannot ~~to my regret~~ comply, in consideration of the proposal of the Philosophical Faculty there for the refilling of this Chair; whereby I inform you herewith, along with the return (2 books) the enclosures of your petition of 2nd January of this year.

The Minister
p.p.
(Translation by Lloyd Kannenberg, April 2010)

Comments:

Friedrich Engel, in his biography of Graßmann [Engel 1911], quotes the draft of the letter of Hermann Grassmann. The most essential difference is the time reference [Engel 1911, 281/282]:

In Engel, instead of "12. February 1862," probably as erroneously written by Grassmann: "2. November 1862." The date is very relevant due to the fact that a change at the head of the ministry took place in the meantime.

Further differences:

In Engel [ibid., 282], instead of "potentially fruitful areas" ("fruchtverheißende Gebiete") we find "richly fruitful areas" ("fruchtreiche Gebiete");

in Engel, instead of "however he already pointed out to me in advance" ("er mich jedoch schon im voraus") originally "that however he already pointed out to me in advance" ("daß er mich jedoch schon im voraus");

in Engel, instead of "takes second place" ("zurückstehe") originally "take second place" ("zurückstehen");

in Engel, instead of "must make mention" ("Erwähnung thun zu müssen") originally "make mention" ("Erwähnung zu thun").

Note on Minister of Culture v. Mühler:

Heinrich v. Mühler (1813–1874) was a Prussian "Minister of Spiritual, Educational, and Medical Affairs" from 18 March 1862 until 12 January 1872. Mühler's appointment was controversial from the beginning. His Ministry was very soon embroiled in criticism on account of his reproachful prevention of necessary reforms and all-too-meager budget, and especially of concern was the payment of teachers and the relevant pension funds.

Illustration 13, page 342

Reproduction of a recently discovered negative reply to a petition by Hermann Grassmann, dated February 1873. He had asked for a paid sabbatical in order to complete scientific works. It is likely that he had aimed to use the time to publish his Rigveda dictionary rapidly (it appeared in six volumes between 1873 and 1875), see also [Engel 1911, 302 f.].

Transcription	**Translation**
Abschrift	**Copy**
Auf das von Euer Wohlgeboren unter dem 2. Februar d. J. an den Herrn Minister der geistlichen, Unterrichts- und Medicinal-Angelegenheiten gerichtete Gesuch ertheilen wir Ihnen unter Rückgabe der Anlage im Auftrage des Herrn Ministers hierdurch den Bescheid, daß die thatsächlichen Verhältnisse des Marienstifts-Gymnasiums gegenwärtig nicht gestatten, Sie zum Zweck wissenschaftlicher Arbeiten mit Belassung Ihres Gehalts auf ein Jahr zu beurlauben. Es fehlt sowohl an einem geeigneten Vertreter für Sie als auch an Mitteln, einen solchen zu remuneriren.	On the request of Your Honor of 2 February of this year, directed to the Herr Minister for Spiritual, Educational and Medical Affairs, we herewith inform you under return of the enclosure of the Herr Minister, that the actual conditions of the Marienstifts-Gymnasium at present does not permit granting you for the purpose of scientific work with retention of your income for one year. In addition there is lacking a suitable substitute, as well as a means to remunerate such a person.
Königliches Provinzial-Schul-Collegium von Pommern v. Jentze.	Royal Provincial-School-Collegium of Pomerania v. Jentze.
An den Herrn Professor Graßmann Wohlgeboren *hier*	To Herr Professor Graßmann His Honor *here*
Das Original der Verfügung erhalten, Abschrift genommen und das Original befördert an den Prof. Graßmann am 24. April 1873.	The original held at disposal, copy taken and the original posted to Prof. Grassmann on 24 April.
Heydemann. *(Transcribed from the original by G. Keßler and H.-J. Petsche, März 2010)*	Heydemann. *(Translation by Lloyd Kannenberg, April 2010)*

Illustration 14, page 388

Hermann Grassmann was a collector of Pomeranian folksongs. For "singing in the family," he arranged them – exclusively by ear – for three voices. The song reproduced here, "Es steht ein Schloss...," printed in [Birlinger and Crecelius 1876, 5–9], was among Grassmann's favorites. Unfortunately, Grassmann's large collection of folk songs has disappeared.

Illustration 15, page 452

Hermann Grassmann's son Karl Justus (1851–1909) became his father's successor at the Stettin Marienstiftsgymnasium.

Transcription	**Translation**
Marienstifts-Curatorium. Stettin, den 19. December 1877. J.-No. 649	Marienstifts-Curatorium. Stettin, the 19. December 1877. J.-No. 649
Empf 20./12 Ko	Empf 20./12 Ko
Der Direction theilen wir die Verfügung des Königlichen Provinzial-Schul-Collegiums vom 13.d. M.S. 3949, welche das in Folge des Todes des Professor Dr. Graßmann beschlossene Aufrücken der Lehrer des Marienstifts-Gymnasiums und die Berufung des Dr. Justus Graßmann aus Pyritz betrifft, in Abschrift zur Nachricht hierbei ergebenst mit. Die Lehrer sind von ihrer Beförderung sämmtlich durch uns benachrichtigt und dem pp. Wienke ist mitgetheilt, was der Herr Minister in Bezug auf ihn bemerkt hat. Zugleich fügen wir den Lebenslauf, die Prüfungszeugnisse pro maturitate und pro facultate docendi und das Zeugnis über das Probejahr des Dr. Justus Graßmann zur Kenntnisnahme bei.	We inform the Administration of the order of the Royal Provincial-School-Collegium of the 13th of this month, S. 3949, which, in consequence of the death of Professor Dr. Graßmann, concerns the decision of the promotion of the teacher of the Marienstifts-Gymnasium and the appointment of Dr. Justus Graßmann of Pyritz, herewithin the copy of the Notice announcing it. The teacher is by your dispatch fully informed and the pp. Wienke as to what the Herr Minister has remarked to him in this regard. At the same time we enclose the life history, the examination certificates pro maturitate and facultate docendi and the certificate of the probationary year of Dr. Justus Graßmann for your attention.
v Gronefeld Wehrmann	v. Gronfeld Wehrmann
An die Direction des Marienstifts-Gymnasiums hier, z.H. des Herrn Professor Dr. Kolbe Wohlgeboren hier	To the Administration of the Marienstifts-Gymnasium here, attn. of Herr Professor Dr. Kolbe here
(Transcribed from the original by G. Keßler and H.-J. Petsche, März 2010)	*(Translation by Lloyd Kannenberg, April 2010)*

Illustration 16, page 454

Title page of the school publication series in which Justus Grassmann's work on the theory of number was published in 1827. This tract features many approaches which Hermann Grassmann would continue in his Extension Theory of 1844, see [Petsche 2009, 101–116].

Notes on contributors

Biryukov, Boris Vladimirovitch; Prof. Dr.

Educational background: Study of philosophy, logic, and history. Doctor of Philosophy.
Current position: Head of an Intercollege Center of research of reading and information culture, Professor at Moscow State Linguistic University.
Main research interests: Logic and history of logic, methodology of science and philosophical questions of computer science.
Selected publications: *Gottlob Frege: a modern view* (2000). // *Grassmann Brothers: career landmarks* (2008) with L.G. Biryukova. // *Hard times of philosophy*, vol. 1–6 (2005–2009).
Address: 119034 Moscow, ul. Ostozhenka, 38, Russia.
E-mail: gerus2003@rambler.ru.

Biryukova, Lyubov Gavrilovna; Prof. Dr.

Educational background: Study of mathematics, history of mathematics and logic, Ph.D.
Current position: Professor at Moscow State Linguistic University.

Main research interests: Mathematics, history of mathematics and logic, philosophy of mathematics
Selected publications: *Grassmann Brothers: career landmarks* (2008) with B.V. Biryukov.
Address: 119034 Moscow, ul. Ostozhenka, 38, Russia.
E-mail: gerus2003@rambler.ru.

Browne, John M.; Dr.

Educational background: Bachelor of Engineering, Bachelor of Science (Mathematics), Master of Applied Science, Doctor of Philosophy (Engineering Mathematics).
Current position: Mathematica consultant.
Main research interests: Computer algebra systems, invariant algebras.
Selected publications: *Graßmann Algebra: Exploring extended vector algebra with Mathematica – draft book and Mathematica package.* Published online at http://sites.google.com/site/grassmannalgebra.
Address: 180 Laughing Water Road, Eltham, VIC, 3095, Australia.
E-mail: john.browne@alumni.unimelb.edu.au.

Cantù, Paola; Dr.

Educational background: Study of philosophy (Milano), Master (DES) in History and Philosophy of Logic (Genève), Ph.D. in Philosophy of Science (Genova).
Current position: Researcher CNRS at the University of Provence, France.
Main research interests: Philosophy and history of logic and mathematics, argumentation theory.
Selected publications: *Logic and Pragmatism. Selected Writings of Giovanni Vailati,* (2009) ed. with C. Arrighi, M. De Zan, and P. Suppes. // *Theories of argumentation* (Italian, 2006) with I. Testa. // Mathematik als Größenlehre. In *Wolffiana, eds.* J. Stolzenberg and O.-P. Rudolph (2008).
Address: CEPERC – BP 68, Université de Provence, 3, pl. Victor Hugo, 13331 Marseille cedex 03, France.
E-mail: paola.cantu@univ-provence.fr.

Crapo, Henry; Prof. Dr.

Educational background: Ph.D. in Mathematics, M.I.T., Professor of Mathematics at the University of Waterloo, Ontario, Canada.
Current position: Founding member of the research group Topologie structurale, Université de Montréal, then at the Institut d'informatique et en automatique, Rocquencourt, France, and at the École des hautes études en sciences sociales. Now retired, at the Centre de Recherche "Les Moutons matheux", 34520 La Vacquerie, France.
Main research interests: Combinatorial geometry and matroid theory, geometric lattices; Whitney algebra and Grassmann's regressive product; applications of projective geometry to scene analysis and rigidity of structures; homological approaches to geometric invariant theory.
Selected publications: See the arXiv, http://arxiv.org/archive/math.
Address: Grand' rue, 34520 La Vacquerie, France.
E-mail: crapo@ehess.fr.

Demonte-Barbera, Filippo; Dr. Ing.

Educational background: Dr. Ing. Industrial Electrical Engineering (Politecnico di Torino), 1960; M.Sc. Secure Electronic Commerce (Royal Holloway, University of London), 2000.
Past positions: R&D Engineering, Systems Engineering, Industrial Management, General Management, [Cryogenics (Italian National Research Council, 1960), Information Technology (IBM, 1960–1962; Olivetti, 1962–1994)], Technology and Market Consulting & Management (Information Communication Technologies), 1994–2007.
Current position: Freelance, scientific and business translations; research in the History of Mathematics.
Main research interests: Past, present, and future of Geometric Algebras and related fields.
Selected publications: *Apparatus for Recognizing Graphic Symbols* (British Patent, 1973). // m-parking, an application of m-commerce (2000). // Editor of the new edition of *Calcolo geometrico secondo l'Ausdehnungslehre di H. Grassmann preceduto dalle operazioni della logica deduttiva* by G. Peano 2010.
Address: via XXV Aprile 10, 10013 Borgofranco d'Ivrea, TO, Italy.
E-mail: fdb@iol.it.

Freguglia, Paolo; Prof. Dr.

Educational background: Study of mathematics and physics (Laurea in Matematica (1971), Sapienza University of Rome); Study of philosophy at Pontificia Universitas Gregoriana of Roma.
Current position: Full Professor, Dept. of Pure and Applied Mathematics, University of L'Aquila, Italy.
Main research interests: History of mathematics (algebra and geometry in the XVI and XVII century and in the XIX and XX century [geometric calculus]); Logic and foundations (modern aspects of syllogistics, inductive statistics, analysis of Peano's school contributions to foundations of mathematics); Mathematical models in physics (geometrical optics and beam dynamics) and in biology (geometrical and dynamics models in evolutionary theories).
Selected publications: Dall'eredità grassmanniana alla teoria delle omografie nella scuola di Peano. *La Matematica nella società e nella cultura. Journal of Unione Matematica Italiana.* 1 (2008), with C. Bocci. // Calcolo geometrico e numeri ipercomplessi: origini e primi sviluppi ottocenteschi. *Bollettino dell'unione matematica italiana.* A, 8, 7–A (2004). // *Geometria e numeri. Storia, teoria elementare ed applicazioni del calcolo geometrico* (book, 2006), with a contribution by A. Bazzani.
Address: Via Panoramica, 233, I – 58019 Porto S.Stefano GR, Italy.
E-mail: paolo.freguglia@technet.it or fregugli@univaq.it.

Gotō, Toshifumi; Prof. Dr.

Educational background: Study of Indology, Indo-European linguistics, Indo-Iranian philology, Dr.phil. in Indo-European linguistics (Erlangen).
Current position: Professor in Indology and History of Indian Buddhism at Tōhoku University, Sendai, Japan.
Main research interests: Indo-European studies, the extension of Indo-European speaking people; historical grammar of Old Indo-Iranian, especially in the verb; religion and Weltanschauung in the Veda, Avesta, and Buddhism.
Selected publications: Die "I. Präsensklasse" im Vedischen (1987). Vasiṣṭha und Varuṇa in RV VII 88 (2000). // *"Purūravas und Urvaśī" aus dem neuentdeckten Vādhūla-Anvākhyāna* (Ed. Ikari) (2000).
Address: Department of Indology and History of Indian Buddhism, Graduate School of Arts and Letters, Tōhoku University, Kawauchi 27–1, Aobaku, Sendai, 980–8576, Japan.
E-mail: gotop@sal.tohoku.ac.jp.

Grattan-Guinness, Ivor; Prof. Dr.

Educational background: M.A. in Mathematics at Oxford University in 1967, then an M.Sc (Econ.) in Mathematical Logic and the Philosophy of Science at the London School of Economics in 1966. Ph.D. and D.Sc. in the History of Science at the University of London.
Current position: Emeritus Professor of the History of Mathematics and Logic at Middlesex University, England; Visiting Research Associate in the Center for Philosophy of Natural and Social Science at the London School of Economics.
Main research interests: History and philosophy of mathematics and logics.
Selected publications: *Convolutions in French mathematics, 1800–1840.* From the calculus and mechanics to mathematical analysis and mathematical physics (three volumes, 1990). // *The Fontana/Norton history of the mathematical sciences. The rainbow of mathematics* (1997, 1998). // *The search for mathematical roots, 1870–1940.* Logics, set theories and the foundations of mathematics from Cantor through Russell to Gödel (2000). // *Routes of learning.* Highways, pathways and byways in the history of mathematics (2009).
Address: 43, St. Leonard's Road, Bengeo, Herts. SG14 3JW, UK.
E-mail: eb7io6gg@waitrose.com.

Hafner, Johann Ev.; Prof. Dr.

Educational background: Study of catholic theology and philosophy, Ph.D. in Philosophy, habilitation in Systematic Theology.
Current position: Full Professor in Religious Studies at University of Potsdam; Dean of the Faculty of Arts (Philosophische Fakultät) University of Potsdam.
Main research interest: Theory of religion, angelology, systems theory, inter-religious dialogue.
Selected publications: *Binnendifferenzierung und Verbindlichkeit der Konfessionen* (2010) ed. with M. Hailer. // *Angelologie* (2009). // *Parallelwelten* (2009) ed. with J. Valentin. // *Katholische Theologie an der Universität* (2009) ed. with J. Schmiedl.
Address: Universität Potsdam, Institut für Religionswissenschaft, 14469 Potsdam, Am Neuen Palais 10, Germany.
E-mail: hafner@uni-potsdam.de.

Hartimo, Mirja; Dr.

Educational background: Ph.D. in Philosophy (Boston University 2005) with a dissertation: *Edmund Husserl's Phenomenology and the Development of Mathematics in the Late Nineteenth Century*.
Current position: Post-doc researcher (Academy of Finland) at the University of Helsinki.
Main research interests: Husserl, history and philosophy of mathematics.
Selected publications: Phenomenology and Mathematics (ed.) (2010). // From Geometry to Phenomenology. Synthese. 162.2 (2008). // Towards completeness: Husserl on theories of manifolds 1890–1901. Synthese. 157.1 (2007).
Address: Department of Philosophy, PO Box 24, 00014 University of Helsinki, Finland.
E-mail: mirja.hartimo@helsinki.fi.

Hestenes, David; Prof. Dr.

Educational background: BA in philosophy and speech (Pacific Lutheran University); Ph.D. in physics (UCLA).
Current position: Emeritus Professor of Physics at Arizona State University.
Main research interests: Geometric calculus, relativistic electron physics, cognitive science and physics education research.
Selected publications: Differential Forms in Geometric Calculus. In *Clifford Algebras and their Applications in Mathematical Physics*, eds. F. Brackx et al. (1993). // [Hestenes 2003a]. //
[Hestenes 2009]. // Notes for a Modeling Theory of Science Cognition and Physics Education, In *Modelling in Physics and Physics Education*, eds. E. van den Berg, A. Ellermeijer, and O. Slooten (2008).
Address: Department of Physics, Arizona State University, USA.
E-mail: hestenes@asu.edu.

Heuser, Marie-Luise; Dr.

Educational background: Study of philosophy, history, physics, and mathematics; Ph.D. in philosophy.

Current position: Assistant Professor and Lecturer in Philosophy at Braunschweig University, project leader of the European Project *Space and Culture*.
Main research interests: Philosophy of mathematics, Naturphilosophie and philosophy of culture and technology.
Selected publications: Dynamisierung des Raumes und Geometrisierung der Kräfte (1997). In *Fessellos durch die Systeme*, eds. W. Ch. Zimmerli et al. (1997). // Mathematik und Zeit im 19. Jahrhundert. In *Die Wiederentdeckung der Zeit*, eds. A. Gimmler et al. (1997). // Die Anfänge der Topologie in Mathematik und Naturphilosophie. In *Topologie – Zur Raumbeschreibung in den Kultur- und Medienwissenschaften*, ed. S. Günzel (2007).
Address: Technische Universität Braunschweig, Seminar für Philosophie, 38106 Braunschweig, Bienroder Weg 80, Germany (www.space-culture.eu).
E-mail: m-l.heuser@tu-bs.de.

Hildenbrand, Dietmar; Dr.

Educational background: Study and Ph.D. in Computer Science.
Current position: Senior scientist at TU Darmstadt.
Main research interests: The application of geometric algebra in computer graphics, computer vision and robotics, the efficient implementation of geometric algebra algorithms.
Selected publications: *Geometric computing in computer graphics using conformal geometric algebra* (2005). // Efficient inverse kinematics algorithm based on conformal geometric algebra using reconfigurable hardware. Chapter in *Grapp conference book* (2008) with Holger Lange, Florian Stock, and Andreas Koch. // Gaalop – High performance parallel computing based on conformal geometric algebra. In *Geometric algebra computing for engineering and computer science* (2010) with Joachim Pitt and Andreas Koch.
Address: TU Darmstadt, Hochschulstr. 10, 64289 Darmstadt, Germany
E-mail: Hildenbrand@informatik.tu-darmstadt.de.

Hitzer, Eckhard; Dr.

Educational background: Ph.D. in Theoretical Physics.
Current position: Part-time Lecturer at The University of Fukui.

Main research interests: Theory and applications of geometric calculus, crystallography, visualization, neural computation.
Selected publications: *Real Clifford Algebra Cl(n, 0), n = 2, 3(mod4) Wavelet Transform* (2009). // *Geometric Roots of -1 in Clifford Algebras Cl(p, q) with p + q <= 4* (2009), with R. Abşamowicz. // *Interactive 3D Space Group Visualization with CLUCalc and the Clifford Geometric Algebra Description of Space Groups* (2008), with C. Perwass.
Address: Department of Applied Physics, University of Fukui, 3–9–1 Bunkyo, 910–8507 Fukui, Japan.
E-mail: hitzer@mech.u-fukui.ac.jp.

Horn, Martin Erik; Dr.

Educational background: Study of physics and chemistry, State Examination as Teacher of Physics and Mathematics, Ph.D. in Physics Education.
Current position: Visiting Professorship (Vertretungsprofessur) for Physics Education at Johann Wolfgang Goethe University Frankfurt on the Main and Lecturer of Physics at Beuth University of Applied Science Berlin.
Main research interests: Physics education (main focus: geometric algebra and space-time algebra), foundations of mathematics (main focus: bilateral hypergeometric functions).
Selected publications: Quaternionen und Geometrische Algebra. In *Didaktik der Physik. Beiträge zur Frühjahrstagung in Kassel.* (2006): Beitrag 28.2. // Vom Raum zur Raumzeit. In *Chemie- und Physikdidaktik für die Lehramtsausbildung,* ed. D. Höttecke (2009). // *Bilateral Binomial Theorem* (2003), SIAM Problem, see reference at Bruce C. Berndt, Wenchang Chu: Two entries on bilateral hypergeometric series in Ramanujan's lost notebook, *Proceedings of the American Mathematical Society* 135 (2007).
Address: Johann Wolfgang Goethe University Frankfurt, Physics Education Department, Faculty of Physics, Max-von-Laue-Str. 1, 60438 Frankfurt on the Main, Germany.
E-mail: m.horn@physik.uni-frankfurt.de. // martinhorn@beuth-hochschule.de.

Kannenberg, Lloyd; Prof. Dr.

Educational background: S.B., MIT 1961; M.S. in Physics, minor in Mathematics, University of Florida 1963; Ph.D. in Physics, Northeastern University, 1967.

Current position: Professor Emeritus of Physics, University of Massachusetts, Lowell.
Main research interests: Quantum field theory, foundations of quantum mechanics.
Selected publications: Quantum formalism via signal analysis. *Found. Phys.* 19 (1989). // Uniqueness of solutions to Helmholtz's equation with linear boundary conditions. *Am. J. Phys.* 57 (1989). // Killing vectors in gauge supersymmetry. *J. Math, Phys.* 18 (1978).
Address: Department of Physics and Applied Physics, University of Massachusetts Lowell, Lowell MA 01854, USA.
E-mail: lloyd_Kannenberg@uml.edu.

Katsikadeli, Christina; Dr.

Educational background: Studies in German language and philology and general linguistics; Ph.D. in General Linguistics
Current position: Project leader, University of Salzburg.
Main research interests: Vedic lexicography and lexicology.
Selected publications: Articles in RIVELEX (2006). // Poetische Lizenzen im Rgveda? In *Poetische Lizenzen, eds.* W. Dressler and O. Panagl (2007) together with Stefan Niederreiter.
Address: Universtität Salzburg, Fachbereich Linguistik, Mühlbacherhofweg 6, 5020 Salzburg, Austria.
E-mail: Christina.Katsikadeli@sbg.ac.at.

Kozianka, Maria; Dr.

Educational background: Study of English, German, Sanskrit, and Comparative Indo-European Linguistics, Ph.D. thesis in Comparative Indo-European Linguistics.
Current position: Team leader of the project *Etymologisches Wörterbuch des Althochdeutschen (Etymological Dictionary of Old High German)*, SAW Leipzig.
Main research interests: Etymology of German and Indo-European languages, word formation, history of linguistics.
Selected publications: *Etymologisches Wörterbuch des Althochdeutschen*, vol. 3 (fadum – fûstslag), vol. 4 (gâba – hylare), ed. by A.L. Lloyd and R. Lühr, arranged by G. Kohlrusch, M. Kozianka, K.K. Purdy, and R. Schuhmann (2007), (2009).

Address: Etymologisches Wörterbuch des Althochdeutschen, Sächsische Akademie der Wissenschaften zu Leipzig, Außendienststelle: Lehrstuhl für Indogermanistik, FSU Jena, Zwätzengasse 12, 07743 Jena, Germany.
E-mail: Maria.Kozianka@uni-jena.de.

Krisch, Thomas; Ao. Univ.-Prof. Dr.

Educational background: Studies in German language and philology, in general and historical linguistics, and in music.
Current position: Associate Professor in Linguistics and project leader, University of Salzburg.
Main research interests: Vedic lexicography; historical and comparative syntax.
Selected publications: RIVELEX (2006). // *Zur Genese und Funktion der altindischen Perfekta mit langem Reduplikationsvokal. Mit kommentierter Materialsammlung.* (1996). // On the "Syntax of Silence" in Proto-Indo-European. In *New Approaches to Word order Variation and Change in Germanic,* eds. R. Hinterhölzl and S. Petrova (2009).
Address: Universtität Salzburg, Fachbereich Linguistik, Mühlbacherhofweg 6, 5020 Salzburg, Austria.
E-mail: Thomas.Krisch@sbg.ac.at.

Kuzicheva, Zinaida Andreyevna; Dr.

Educational background: Study of mathematics, history of mathematics and philosophy, Candidate's degree in mathematics and *physics.*
Current position: Senior research assistant at MSU, Department of Mathematics and Mechanics.
Main research interests: History of foundations of mathematics and mathematical logic.
Selected publications: *Leibniz's logical program and its role in the history of logic and cybernetics* (1982). // *Formation and development of mathematical logic* (1997). // *From the history of formation of logical-mathematical constructivism (XIX – beginning of XX)* (2009).
Address: 119991 Leninskye Gory, MSU, Russia.
E-mail: zakuzicheva@mail.ru.

Lewis, Albert C.; Dr.

Educational background: Mathematics, classical Greek, computer science; Doctorate in History of Mathematics.
Current position: Trustee, Educational Advancement Foundation, Austin, Texas, USA.
Main research interests: Early twentieth-century mathematics in the US, especially the topology school of R.L. Moore. Grassmann.
Selected publications: *The Collected Papers of Bertrand Russell*, Vol. 2: *Philosophical Papers, 1896–99* (1990) edited with N. Griffin. // The beginnings of the R.L. Moore school of topology, *Historia Mathematica* 31 (2004). // Kenneth O. May and information retrieval in mathematics, *Historia Mathematica* 31 (2004).
Address: 2303 Rio Grande, Austin, Texas, 78701–5131, USA.
E-mail: alewis@edu-adv-foundation.org.

Li, Hongbo; Prof. Dr.

Educational background: Ph.D. from the Department of Mathematics, Beijing University in 1994; post-doctoral researcher on machine proving at the Institute of Systems Science, Chinese Academy of Sciences; post-doctoral researcher on Geometric Algebra at Arizona State University, USA.
Current position: Full Professor of mathematics at the Academy of Mathematics and Systems Science of the Chinese Academy of Sciences, Deputy Director of the Key Laboratory of Mathematics Mechanization of the Chinese Academy of Sciences.
Main research interests: Automated geometric reasoning, symbolic computation, geometric algebra, invariant theory, robotics, computer numerical control, computational geometry, computer vision, differential geometry, and computational topology.
Selected publications: *Invariant Algebras and Geometric Reasoning* (2008). // *Symbolic Computational Geometry with Advanced Invariant Algebras* (2010). // *Computer Algebra and Geometric Algebra with Applications* (2005), ed. with P.-J. Olver and G. Sommer.
Address: Key Laboratory of Mathematics Mechanization, Academy of Mathematics and Systems Science, Chinese Academy of Sciences, Beijing 100190, P.R. China.
E-mail: hli@mmrc.iss.ac.cn.

Liesen, Jörg; Prof. Dr.

Educational background: Ph.D. (1999) and habilitation (2004) in Mathematics.
Current position: Professor of Mathematics at TU Berlin, Germany.
Main research interests: Numerical mathematics and scientific computing.
Selected publications: Numerical solution of saddle point problems, *Acta Numerica* (2005) with M. Benzi and G.H. Golub. // On optimal short recurrences for generating orthogonal Krylov subspace bases, with Zdeněk Strakoš, *SIAM Review* (2008).
Address: TU Berlin, Institute of Mathematics, MA 4–5, Straße des 17. Juni 136, 10623 Berlin, Germany.
E-mail: liesen@math.tu-berlin.de.

Marchuk, Nikolay; Prof. Dr.

Educational background: Study of mathematics, Ph.D. in Mathematics;
Current position: Associate Professor at Moscow State University; Senior scientific staff of Steklov Mathematical Institute of the Russian Academy of Sciences.
Main research interests: Mathematics: Grassmann algebra, Clifford algebra, Quantum field theory.
Selected publications: Model Dirac-Maxwell equations with pseudounitary symmetry, *Theor. and Math. Phys.*, 157.3 (2008). // *Field theory equations and Clifford algebras* (2009).
Address: Steklov Mathematical Institute, Gubkina str. 8, Moscow 119991, Russia.
E-mail: nmarchuk2005@yandex.ru, nmarchuk@mi.ras.ru.

Mazzola, Guerino; Prof. Dr.

Educational background: Study of mathematics, theoretical physics, and crystallography at the University of Zurich, Switzerland.
Current position: Professor at the School of Music of the University of Minnesota and Lecturer at the Computer Science Department of the University of Zurich.
Main research interests: Mathematical music theory, philosophy of mathematics, mathematical gesture theory, semiotics, theoretical computer science, Topos theory.
Selected publications: *The Topos of Music* (2002). // *La vérité du beau dans la musique* (2007). // *Flow, Gesture, and Spaces in Free Jazz* (2009).

Address: School of Music, University of Minnesota, 2106 Fourth Street South, Minneapolis MN 55455, USA.
E-mail: mazzola@umn.edu.

Niederreiter, Stefan; Dr.

Educational background: Studies in historical linguistics and in German language and philology.
Current position: Project leader, University of Salzburg.
Main research interests: Vedic lexicography and lexicology.
Selected publications: *Articles in RIVELEX* (2006). // Poetische Lizenzen im Ṛgveda? In *Poetische Lizenzen, eds.* W. Dressler and O. Panagl (2007) together with Ch. Katsikadeli. // *Morphologische Varianz und semantische Konkurrenz. Verbalabstrakta im Rig-Veda* (2001).
Address: Universtität Salzburg, Fachbereich Linguistik, Mühlbacherhofweg 6, 5020 Salzburg, Austria.
E-Mail: Stefan.Niederreiter@sbg.ac.at.

Ohala, John; Prof. Dr.

Educational background: Study of the history of English, phonetics, phonology; Ph.D., Linguistics, 1969, UCLA. [= University of California, Los Angeles].
Current position: Professor Emeritus, Department of Linguistics, University of California, Berkeley.
Main research interests: Phonetics, phonology, ethological aspects of animal communication.
Selected publications: The listener as a source of sound change. In *Papers from the Parasession on Language and Behavior,* eds. C.S. Masek, R.A. Hendrick, and M.F. Miller (1981). // The origin of sound patterns in vocal tract constraints. In *The production of speech,* ed. P.F. MacNeilage (1983). // An ethological perspective on common cross – language utilization of F0 of voice. *Phonetica* 41 (1984).
Address: 1149 Hillview Road Berkeley, CA 94708, USA.
E-mail: ohala@berkeley.edu.

Otte, Michael; Prof. Dr.

Educational background: Study of mathematics, physics, and German literature (Literaturwissenschaft)
Current position: Prof. Emeritus University of Bielefeld, Germany.
Main research interests: History and philosophy of mathematics.
Selected publications: The Ideas of H. Grassmann in the Context of the Mathematical and Phil. Tradition since Leibniz. *Historia Mathematica 19* (1989). // Does Mathematics have Objects? *Synthese* 134 (2003). // Proof-Analysis and Continuity, *Foundations of Science* 11 (2006).
Address: Neuenkirchenerstr. 6, 49324 Melle, Germany.
E-mail: michaelontra@aol.com.

Peckhaus, Volker; Prof. Dr.

Educational background: Studied mechanical engineering, history, German language and philosophy; Ph.D. in philosophy (history and philosophy of mathematics), habilitation in philosophy (history of logic).
Current position: Professor for Philosophy of Science and Technology; Dean of the Faculty for Arts and Humanities.
Main research interests: History of modern logic, history, and philosophy of mathematics, philosophy of science, epistemology.
Selected publications: Hilbertprogramm und Kritische Philosophie (1990). // *Logik, Mathesis universalis und Allgemeine Wissenschaft* (1997). // *Oskar Becker und die Philosophie der Mathematik*, ed. (2005).
Address: Universität Paderborn, Institut für Humanwissenschaften: Philosophie, Warburger Str. 100, 33100 Paderborn, Germany.
E-mail: volker.peckhaus@upb.de.

Petsche, Hans-Joachim; Prof. Dr.

Educational background: Studied mathematics, physics and philosophy, Ph.D. in history and philosophy of mathematics, habilitation in the field of philosophy of science and technology.

Current Position: Professor and Senior Lecturer in Philosophy at the University of Potsdam, Deputy Director of the Institute of Philosophy.
Main research interests: Philosophy of science and technology, history and philosophy of mathematics, the philosophical aspects of computer science, the internet and the new media.
Selected publications: *LOGO in FORTH* (1989), with H. Schachtzabel and H.-J. Sprengel. // *Knowledge Management and Philosophy* (2003), ed. with K. Freyberg and H.-J. Klein. // *Hermann Graßmann – Biography* (2009).
Address: Universität Potsdam, Institut für Philosophie, Am Neuen Palais 10, 14469 Potsdam, Germany.
E-mail: petsche@uni-potsdam.de.

Pfalzgraf, Jochen; Prof. Dr.

Educational background: Study of mathematics, physics, and some basic computer science, Diploma, Ph.D., and Habilitation in Mathematics.
Current position: Full Professor (Ordinarius) at the Department of Computer Sciences, University of Salzburg.
Main research interests: Mathematical informatics, i.e. mathematical methods applied to modeling problems in computer science and AI. Applied and computational category theory, multi-agent systems, geometric spaces, neural network structures, logical fiberings.
Selected publications: Computer-aided construction of finite geometric spaces: Automated verification of geometric constraints. *Journal of Automated Reasoning* 26 (2001), together with W. Gehrke. // On a category of geometric spaces and geometries induced by group actions. *Ukrainian Jour. Physics,* 43, 7 (1998). // On logical fiberings and automated deduction in many-valued logics using Gröbner bases. *RACSAM, Rev. Real. Acad. Ciencias, Ser. A. Mat.*, 98.1 (2004).
Address: University of Salzburg, Department of Computer Sciences, Jakob Haringer Str. 2, A–5020 Salzburg, Austria.
E-mail: jochen.pfalzgraf@sbg.ac.at.

Radu, Mircea; Dr.

Educational background: Mathematics, mathematics education; Doctorate in History of Mathematics.

Current position: Research worker, teaches mathematics, mathematics education, and epistemology at the Oberstufen-Kolleg des Landes Nordrhein-Westfalen an der Universität Bielefeld, Germany.
Main research interests: History of axiomatics; semiotics; philosophy of mathematics education.
Selected publications: Justus Grassmann's contributions to the foundations of mathematics: mathematical and philosophical aspects. *Historia Mathematica* 27.1 (2000). // A debate about the axiomatization of arithmetic: Otto Hölder against Robert Graßmann. *Historia Mathematica* 30.3 (2003). // Symbolic language versus understanding in mathematics education: A brief archaeologic investigation of mathematics education discourse. In *Semiotics in Mathematics Education*, eds. L. Radford, G. Schubring, and F. Seeger (2008).
Address: Oberstufen-Kolleg des Landes Nordrhein-Westfalen an der Universität Bielefeld, Postfach 100131, 33501 Bielefeld, Germany.
E-mail: mircea.radu@uni-bielefeld.de.

Reich, Karin; Prof. Dr.

Educational background: Study of mathematics, physics and astronomy, Ph.D. and habilitation in History of Mathematics.
Current position: Since 1995 Professor for History of Science at the University of Hamburg, retired in 2007.
Main research interests: History of mathematics.
Selected publications: Publications etc., see homepage http://www.math.uni-hamburg.de/home/reich/
Address: Universität Hamburg, Dep. Mathematik, 20146 Hamburg, Bundesstr. 55, Germany.
E-mail: reich@math.uni-hamburg.de.

Riche, Jacques M.; Dr.

Educational background: Studied philosophy, linguistics, and computer science. Ph.D. in Logic. Australian National University.
Current position: Teaches history of science and technology at the University of Louvain (Louvain-la-Neuve); researcher in logic and computer science at the University of Louvain (Leuven).

Main research interests: Non-classical logics, history of logic and mathematics, logic in computer sciences, Whitehead.
Selected publications: From Universal Algebra to Universal Logic. In *Perspectives on Universal Logic*, eds. J.-Y. Béziau et al (2007). // *A.N. Whitehead: mathematician, logician and natural philosopher* (forthc.).
Address: Katholieke Universiteit Leuven, Department Computer Wetenschappen, 200A, Celestijnenlaan, B–3001 Heverlee, Belgium.
E-mail: jacques.riche@cs.kuleuven.be.

Russ, Steve; Dr.

Educational background: Study of mathematics, theology, and philosophy, Ph.D. on The Mathematical Works of Bernard Bolzano.
Current position: Associate Professor of Computer Science at University of Warwick, UK.
Main research interests: Empirical modeling, history and philosophy of mathematics, science and computing.
Selected publications: *The Mathematical Works of Bernard Bolzano* (2004, reprinted 2006). // Experimenting with Computing. *Journal of Applied Logic* 6 (2008) with W.M. Beynon. // Human Computing: Modelling with Meaning. *Literary and Linguistic Computing* 21.2 (2006) with W.M. Beynon and W. McCarty.
Address: Department of Computer Science, University of Warwick, Coventry CV4 7AL, UK.
E-mail: steve.russ@warwick.ac.uk.

Sühring, Peter; Dr.

Educational background: Study of musicology, comparative literary science, and philosophy.
Current position: Scientific Collaborator at the Institute of Musicology, Faculty of Music, University of Arts Berlin.
Main research interests: History of music, theory of music, and the cultural environment of musicology.
Selected publications: *Der Rhythmus der Trobadors* (2003). // *Die Rekonstruktion der Sammlung Spitta* (2005). // *Die frühesten Opern Mozarts* (2006).

Address: Universität der Künste, Fakultät Musik, Fasanenstr. 1B, 10623 Berlin, Germany.
E-mail: Peter.Suehring@gmx.de.

Swimmer, Alvin; Prof. Dr.

Educational background: B.A. 1952; M.S. 1954 (Master's Thesis, Grassmann Algebra); Ph.D. University of California, Berkeley, 1963 (Ph.D. Dissertation, Resolution of Singularities in Polynomial Rings).
Current position: Emeritus and Associate; Arizona State University.
Main research interests: Projective Geometry, Spirals, Grassmann Algebra.
Selected publications: *The Completion of Grassmann's Natur-Wissenschaftliche Methode*. In [Schubring 1996a] (1996). // *The Key To Arithmetic* (1990).
Address: Arizona State University, PO BOX 871804, Tempe, AZ 85287–1804, USA.
E-mail: aswimmer@asu.edu.

Szukała, Maciej; Dr.

Educational background: Study of history und ethnology. Ph.D. in History.
Current position: Collaborator at the State Archive in Szczecin.
Main research interests: Prussian archives in the nineteenth and twentieth centuries; ideology and politics.
Selected publications: *Powstanie i działalność Towarzystwa Historii i Starożytności Pomorza w Szczecinie w latach 1824–1918* (Szczecin 2000). // *Stettiner Archivare und die "deutsche Ostforschung"*. Die Korrespondenz zwischen Albert Brackmann und Adolf Diestelkamp in den Jahren 1935 bis 1941 (2002). // *Staatsarchiv Stettin – Wegweiser durch die Bestaende bis zum Jahr 1945* (2004) with R. Gazinski and P. Gut.
Address: Archiwum Panstwowe w Szczecinie ul. sw. Wojciecha 13, 70–410 Szczecin, Poland.
E-mail: alakusz@wp.pl.

Ziche, Paul; Prof. Dr.

Educational background: Study of philosophy, physics, and psychology in Munich and Oxford, Ph.D. in Philosophy, Assistant Professor for History of Science at Jena University, habilitation in München on the Interaction between Philosophy and the Sciences around 1900.
Current position: Professor (chair) for the History of Modern Philosophy, University of Utrecht, Netherlands.
Main research interests: German idealism; philosophy and the sciences; history of science, including the history of psychology.
Selected publications: *Mathematische und naturwissenschaftliche Modelle in der Philosophie Schellings und Hegels*. (1996). // *Introspektion.* Texte zur Selbstwahrnehmung des Ichs. (1999). // *Wissenschaftslandschaften um 1900.* Philosophie, die Wissenschaften und der nicht-reduktive Szientismus (2008).
Address: University of Utrecht, Department of Philosophy, Heidelberglaan 8, Postbus 80126, 3508 TC Utrecht, Netherlands.
E-mail: paul.ziche@phil.uu.nl.

References

A1 = Grassmann, Hermann. 1844. *Die Wissenschaft der extensiven Grösse oder die Ausdehnungslehre, eine neue mathematische Disciplin. Erster Theil: Die lineale Ausdehnungslehre ein neuer Zweig der Mathematik, dargestellt und durch Anwendungen auf die übrigen Zweige der Mathematik, wie auch auf die Statik, Mechanik, die Lehre vom Magnetismus und die Krystallonomie erläutert.* 1. Aufl. Leipzig: Wigand (2. Aufl. Leipzig 1878). [Reprinted in GW11, 4–312].

$A1_K$ = Grassmann, Hermann. 1995b. *Linear extension theory* [Die Lineale Ausdehnungslehre, 1844], transl. Lloyd C. Kannenberg. In [H. Grassmann 1995a, 1–312].

A2 = Grassmann, Hermann. 1862. *Die Ausdehnungslehre. Vollständig und in strenger Form begründet.* Berlin: Enslin. [Reprinted in GW12, 1–383].

$A2_K$ = Grassmann, Hermann. 2000. *Extension theory* [Die Ausdehnungslehre, 1862], transl. Lloyd C. Kannenberg. American Mathematical Society, London Mathematical Society.

Abbe, Ernst. 1986. *Briefe an seine Jugend- und Studienfreunde Carl Martin und Harald Schütz.* Eds. Volker Wahl and Joachim Wittig. Berlin: Akademie-Verlag.

Abłamowicz, Rafał and Bertfried Fauser. 2009. *CLIFFORD – A Maple package for Clifford algebra computations with Bigebra, SchurFkt, GfG – Groebner for Grassmann, Cliplus, Define, GTP, Octonion, SP, SymGroupAlgebra, and code_support* (Software, © Rafal Ablamowicz and Bertfried Fauser). http://math.tntech.edu/rafal/ (accessed March 3, 2010).

Ahlfors, Lars V. 1986. Möbius transformations in $\mathbb{R}^n$ expressed through 2×2 matrices of Clifford numbers. *Complex Variables* 5: 215–224.

Altenburg, Otto. 1942. Ludwig Giesebrecht: Zur Wiederkehr seines Geburtstages. *Monatsblätter der Gesellschaft für Pommersche Geschichte und Altertumskunde* 56: 1–6.

Altenburg, Otto. 1966. Ludwig Giesebrecht (1792–1873). In *Pommersche Lebensbilder*, vol. 4, ed. Walter Menn, 292–314. Köln-Graz: Böhlau.

André, Johannes. 1974. On finite noncommutative affine spaces. In *Combinatorics*, eds. Marshall Hall Jr. and Jacobus H. van Lint. 2nd ed., 65–113. Amsterdam: Mathematisch Centrum. (Mathematical Centre Tracts 56).

André, Johannes. 1983. Coherent configurations and noncommutative spaces. *Geometriae Dedicata* 13: 351–360.

André, Johannes. 1988. Endliche nichtkommutative Geometrie. (Lecture Notes). *Annales Universitatis Saraviensis (Series Mathematicae)* 2: 1–136.

André, Johannes. 1992. Configurational conditions and digraphs. *Journal of Geometry* 43: 22–29.

André, Johannes and Hans Ney. 1991. On Anshel-Clay-Nearrings. *Proceedings of the Conference on Nearrings and Nearfields*. Oberwolfach: Math. Forschungsinstitut.

Anon. 1798. Review of *Cours-Tabelle, mit beygefügter Gebrauchs-Anweisung und Erklärung* von Abraham Crailsheim. 1798. *Allgemeinen Literatur-Zeitung*, No. 219 (19.7.1798). Jena and Leipzig: 133–134.

Asinovskiy, Erik I., Vladimir P. Vizgin, Vitaliy L. Ginzburg, Yu. A. Ivanov, T.D. Ilina, V.S. Kirsanov, Yuriy A. Lebedev et al. 2002. In memory of Lev Solomonovich Polak. *Physics Uspekhi* 45 (8): 903–905.

Aster, Ernst von. 1969. *Historia filozofii* [Die Geschichte der Philosophie]. Warszawa: PWN.

Atkins, B.T. Sue. 2008. Theoretical lexicography and its relation to dictionary-making. In *Practical Lexicography. A Reader,* ed. Thierry Fontenelle. 31–50. Oxford: University Press.

Atkins, B.T. Sue and Michael Rundell. 2008. *The Oxford guide to practical lexicography.* Oxford: University Press.

Auerbach, Felix. 1918. *Ernst Abbe: sein Leben, sein Wirken, seine Persönlichkeit. Nach den Quellen und aus eigener Erfahrung geschildert,* Leipzig: Akademische Verlags-Gesellschaft.

Aufrecht, Theodor, ed. 1861, 1863. *Die Hymnen des Ṛigveda.* 2 vols. Berlin: Dümmler. [Indische Studien, ed. Albrecht Weber, vol. 6 and 7].

Aufrecht, Theodor, ed. 1877. *Die Hymnen des Rigveda.* 2nd ed., 2 vols. Bonn: Adolph Marcus.

Ayer, Jules. 1964. The a priori. In *Philosophy of mathematics: selected readings,* eds. Paul Benacerraf and Hilary Putnam. 315–328. Cambridge: Cambridge Univ. Press.

Ball, Robert Stawell. 1901. Report on the mathematical work of Sir A.N. Whitehead, Fellow of Trinity College, Cambridge. Bulletin de la Société Physico-Mathématique de Kasan, 2ème Série, Tome X, n° 4, 19–22.

Ballauff, Ludwig. 1878. Review of R. Grassmann, *Die Wissenschaftslehre oder Philosophie* (1875–1876), *Jahrbuch des Vereins für Wissenschaftliche Pädagogik* 10: 172–194.

Barnabei, Marilena, Andrea Brini and Gian-Carlo Rota. 1985. On the exterior calculus of invariant theory. *Journal of Algebra* 96: 120–160.

Bartholomae, Christian. 1904. *Altiranisches Wörterbuch.* Strassburg: Trübner.

Baylis, William Eric. 1999. *Electrodynamics. A modern geometric approach.* Boston: Birkhäuser.

Baylis, William Eric. 2002. *Electrodynamics. A modern geometric approach.* 2nd ed. Boston: Birkhäuser.

Bayro-Corrochano, Eduardo Jose and Gerik Scheuermann, eds. 2009. *Geometric algebra computing for engineering and computer science.* London: Springer.

Bellavitis, Giusto. 1854. Sposizione del metodo delle equipollenze. *Memorie della Società Italiana delle Scienze* 25: 1–85.

Benfey, Theodor. 1848. *Die Hymnen des Sâma-Veda. Herausgegeben, übersetzt und mit Glossar versehen.* Leipzig: Brockhaus. [Reprinted 1978. Hildesheim: Olms].

Berg, Jan. 1962. *Bolzano's logic.* Stockholm: Almquist & Wiksell.

Bergaigne, Abel. 1886. La Saṃhitā primitive du Ṛig-Veda. *Journal asiatique,* Septembre-Octobre 1886: 193–271.

Bergaigne, Abel. 1887. Nouvelles recherches sur l'histoire de la Saṃhitā du Ṛig-Veda. *Journal asiatique,* Février-Mars 1887: 191–287.

Beyvers, Gottfried and Elvira Krusch. 2009. *Kleines 1 x 1 der Relativitätstheorie. Einsteins Physik mit Mathematik der Mittelstufe.* 2nd ed. Berlin: Springer.

Biener, Klaus. 2005. Hermann Minkowski – Mathematiklehrer Einsteins. *cms-Journal* 27: 77–78.

Birkhoff, Garrett. 1940. *Lattice theory.* 3rd ed. New York: American Mathematical Society. (Colloquium Publications XXV).

Birlinger, Anton and Wilhelm Crecelius. 1876. *Deutsche Lieder. Festgruss an Ludwig Erk zum fünfzigjährigen Dienstjubiläum, Berlin, 10. Juni 1876.* Heilbronn: Gebr. Henninger.

Biryukov, Boris Vladimirovitch and Lyubov Gavrilovna Biryukova. 1982. "Uchenie o formach (velichinakh)" German i Roberta Grasmanov kak predvoskhishchenie konstruktivnogo napravleniya v matematike, I. [Hermann and Robert Grassmann's theory of forms (theory of magnitude) as an anticipation of the constructive direction in mathematics, I. Questions of Cybernetics. Cybernetics and logical formalization. Historical and methodological aspects. Moscow]. In *Voprosy kibernetiki.* Kibernetika i logicheskaya formalizaciya. Istoricheskie i metodologicheskie aspekty. 36–92. Moskva: Akademiya nauk.

Biryukova, Lyubov Gavrilovna. 1984. "Uchenie o formach (velichinakh)" German i Roberta Grasmanov kak predvoskhishchenie konstruktivnogo napravleniya v matematike, II. [Hermann and Robert Grassmann's theory of forms (theory of magnitude) as an anticipation of the constructive direction in mathematics, II. Questions of cybernetics. Cybernetics and mathematical logic in historic-mathematical aspect. Moscow]. In *Voprosy kibernetiki*. Kibernetika i matematicheskaya logika v istoriko-metodologicheskom aspekte. 45–111. Moskva: Akademiya nauk.

Biryukova, Lyubov Gavrilovna and Boris Vladimirovitch Biryukov. 1997. On the axiomatic sources of fundamental algebraic structures: the achievements of Hermann Grassmann and Robert Grassmann. *Modern Logic* 7: 131–159.

Blake, Stephen. 2005. A.N. Whitehead's geometric algebra. http://www.stebla.pwp.blueyonder.co.uk/papers/Euclid.pdf (accessed January 12, 2010).

Bobynin, Viktor Viktorovitch. 1886a. *Biografii znamenitykh matematikov XIX stoletiya*. Vyp. I. R. Sturm, E. Shreder, L. Zonke. Germann Grassman. [Biographies of important mathematicians of the 19th century: R. Sturm, E. Schröder, L. Sohncke: Hermann Grassmann]. Moskva: Mamontov.

Bobynin, Viktor Viktorovitch. 1886b. *Opyty matematicheskogo izlozheniya logiki*. [Attempts towards a mathematical account of logic]. Moskva.

Bohlen, Julius von. 1865. *Die Erwerbung Pommerns durch die Hohenzollern: Zur Erinnerung an die vor fünfzig Jahren erfolgte Wiedervereinigung des ganzen Pommerns unter die Herrschaft seines erlauchten Königshauses*. Berlin: Decker.

Bohm, David. 2006. Physics and perception. In *The special theory of relativity*, ed. David Bohm. Appendix A. 219–275. London: Routledge Classics.

Böhtlingk, Otto and Rudolph Roth. 1855–1875. *Sanskrit-Wörterbuch*. 7 vols. St. Petersburg: Kaiserl. Akad. der Wiss. [PW].

Böhtlingk, Otto and Rudolph Roth. 1856–1858. *Sanskrit-Wörterbuch*. St. Petersburg: Buchdruckerei der Kaiserlichen Akademie der Wissenschaften.

Böhtlingk, Otto and Rudolph Roth. 1881. *Sanskrit-Wörterbuch in kürzerer Fassung*. 2. part. St. Petersburg: Buchdruckerei der Kaiserlichen Akademie der Wissenschaften.

Bokowski, Jürgen and Bernd Sturmfels. 1989. *Computational synthetic geometry*. Berlin and Heidelberg: Springer. (Lecture Notes in Mathematics, 1355).

Bolzano, Bernard. 1804. *Betrachtungen über einige Gegenstände der Elementargeometrie*. Prague: Karl Barth. [Translation 'Considerations on some Objects of Elementary Geometry' appears in [Russ 2004, 24–81].].

Bolzano, Bernard. 1810. *Beyträge zu einer begründeteren Darstellung der Mathematik. Erste Lieferung*. Prague: Caspar Widtmann. [Translation 'Contributions to a Better-Grounded Presentation of Mathematics' appears in [Russ 2004, 82–137].].

Bolzano, Bernard. 1837. *Wissenschaftslehre. Versuch einer ausführlichen grossentheils neuen Darstellung der Logik mit steter Rücksicht auf deren bisherigen Bearbeiter*. vol. 1–4. Sulzbach: Seidel. [Reprinted 1929. Ed. Wolfgang Schultz, vol.1–4. Leipzig: Meiner].

Bondi, Hermann. 1967. *Assumption and myth in physical theory. The Tarner lectures, delivered November 1965*. Cambridge: Cambridge University Press.

Boole, George. 1847. *The mathematical analysis of logic. Being an essay towards a calculus of deductive reasoning*. Cambridge: Macmillan, Barclay, and Macmillan; London: George Bell. [Reprinted 1951. Oxford: Blackwell].

Boole, George. 1854. *An investigation of the laws of thought: on which are founded the mathematical theories of logic and probabilities*. London: Macmillan.

Borga, Marco, Paolo Freguglia and Dario Palladino. 1985. *I contributi fondazionali della scuola di Peano*. Milano: F. Angeli.

Boring, Edwin G. 1941. An operational restatement of G.E. Müller's psychophysical axioms. *The Psychological Review* 48: 457–464.

Borkowski, Hans. 1898. Schleiermacher als Mathematiker. *Archiv der Mathematik und Physik* 16: 337–346.

Bottazzini, Umberto 1986. *The higher calculus: A history of real and complex analysis from Euler to Weierstrass*. Transl. Warren Van Egmond. New York: Springer.

Böttcher, Karl-Heinz and Bertram Maurer. 2008. *Stuttgarter Mathematiker. Geschichte der Mathematik an der Universität Stuttgart von 1829 bis 1945 in Biographien.* Stuttgart: Universität Stuttgart.

Branig, Hans. 1959. Die Oberpräsidenten der Provinz Pommern. *Baltische Studien*, NF 46: 99–107.

Branig, Hans. 1979. Wesen und Geist der höheren Verwaltungsbeamten in Preussen in der Zeit des Vormärz. In *Neue Forschungen zur Brandenburg-Preußischen Geschichte 1*, eds. Friedrich Benninghoven and Cécile Lowenthal-Hensel, 161–171. Köln, Wien: Böhlau.

Bravi, Paolo and Adrea Brini. 2001. Remarks on invariant geometric calculus, Cayley-Grassmann algebras and geometric Clifford algebras. In *Algebraic combinatorics and computer science, a tribute to Gian-Carlo Rota*, eds. Henry Crapo and Domenico Senato. 129–150. Milano and New York: Springer.

Brini, Andrea and Antonio G. B. Teolis. 1996. Grassmann progressive and regressive products and CG-algebras. In [Schubring 1996a, 231–242].

Brini, Andrea, Francesco Regonati and Antonio G. B. Teolis. 2001. Grassmann geometric calculus, invariant theory and superalgebras. In *Algebraic combinatorics and computer science, a tribute to Gian-Carlo Rota*, eds. Henry Crapo and Domenico Senato. 151–196. Milano and New York: Springer.

Browne, John M. 2009. *Grassmann algebra: Draft book and Mathematica package.* http://sites.google.com/site/grassmannalgebra. (accessed March 3, 2010).

Buchheim, Arthur. 1884. On the theory of matrices. *Proceedings of the London Mathematical Society* 16: 63–82.

Bühler, Karl. 1930. *Die geistige Entwicklung des Kindes.* 6th. ed. Jena: Gustav Fischer.

Bülow, Gottfried von. 1882. Robert Klempin. In *Allgemeine Deutsche Bibliographie,* vol. 16: 144–154. Leipzig: Duncker & Humblot.

Burali-Forti, Cesare. 1938. Elementi di calcolo vettoriale. In *Enciclopedia delle matematiche elementari e complementi*, vol. 2.2., eds. Luigi Berzolari, Giulio Vivanti and Duilio Gigli. 105–140. Milano: U. Hoepli.

Burali-Forti, Cesare. 1986. *Geometria analitico proiettiva.* Torino: Ed. Petrini.

Burali-Forti, Cesare and Roberto Marcolongo. 1907a. Per l'unificazione delle notazioni vettoriali, (Nota I). *Rendiconti del Circolo Matematico di Palermo*: 23, 324–328.

Burali-Forti, Cesare and Roberto Marcolongo. 1907b. Per l'unificazione delle notazioni vettoriali, (Nota II). *Rendiconti del Circolo Matematico di Palermo*: 24, 65–80.

Burali-Forti, Cesare and Roberto Marcolongo. 1907c. Per l'unificazione delle notazioni vettoriali, (Nota III). *Rendiconti del Circolo Matematico di Palermo*: 24, 318–332.

Burali-Forti, Cesare and Roberto Marcolongo. 1908a. Per l'unificazione delle notazioni vettoriali, (Nota IV). *Rendiconti del Circolo Matematico di Palermo*: 25, 352–375.

Burali-Forti, Cesare and Roberto Marcolongo. 1908b. Per l'unificazione delle notazioni vettoriali, (Nota V). *Rendiconti del Circolo Matematico di Palermo*: 26, 369–377.

Burali-Forti, Cesare and Roberto Marcolongo. 1909a. *Elementi di calcolo vettoriale con numerose applicazioni alla geometria, alla meccanica e alla fisica-matematica.* (2d ed., 1921). Bologna: Nicola Zanichelli.

Burali-Forti, Cesare and Roberto Marcolongo. 1909b. *Omografie vettoriali con applicazioni alle derivate rispetto ad un punto e alla fisica-matematica.* Torino: Petrini.

Burali-Forti, Cesare and Roberto Marcolongo. 1912, 1913. *Analyse vectorielle générale* (I and II vol.). Pavia: Mattei & C. Editeurs.

Burali-Forti, Cesare and Tommaso Boggio. 1924. *Espaces courbes. Critique de la relativité.* Torino: Sten Editrice.

Calvet, Ramon González. 2001. *Treatise of plane geometry through geometric algebra.* http://www.xtec.cat/~rgonzal1/treatise2.htm (accessed March 3, 2010).

Cantù, Paola. 2003. *The definition of mathematics: From a theory of quantities to a theory of forms. Grassmann's extension theory* (Italian). PhD thesis, Università di Genova.

Cantù, Paola. 2010. Aristotle's prohibition rule on kind-crossing and the definition of mathematics as a science of quantities. *Synthese* 174: 225–235.

Cartan, Élie. 1907. Nombres complexes. Exposé, d'après l'article allemand de E. Study. *Encyclopédie des sciences mathématiques pures et appliqués*, volume I.1. 329–468. Paris: Gauthier-Villars.

Cartan, Élie. 1922. *Leçons sur les invariants intégraux*. Paris: Hermann.
Cartan, Élie. 1937. *Leçons sur la théorie des spineurs*. Paris: Hermann. (Engl. transl. 1966. Cambridge: MIT Press).
Cartan, Élie. 1981. *The theory of spinors*. (First ed. in French 1937 as "Leçons sur la théorie des spineurs", gathered and arranged by André Mercier). New York: Dover Publications.
Carus, Paul. 1889. The old and the new mathematics. *Open Court* 2: 1468–1472.
Carus, Paul. 1903. The philosophical foundations of mathematics. *The Monist* 13: 273–294.
Cassirer, Ernst. 1907. Kant und die moderne Mathematik. (Mit Bezug auf Bertrand Russells und Louis Couturats Werke über die Prinzipien der Mathematik.) *Kant Studien* XII: 1–49.
Cassirer, Ernst. 1910. *Substanzbegriff und Funktionsbegriff. Untersuchungen über die Grundfragen der Erkenntniskritik*. Berlin: Bruno Cassirer.
Cassirer, Ernst. 1913. Erkenntnistheorie nebst den Grundfragen der Logik. *Jahrbücher der Philosophie* 1: 1–59.
Cassirer, Ernst. 1923. *Substance and function and Einstein's theory of relativity*. Chicago: Open Court.
Cassirer, Ernst. 1927. Erkenntnistheorie nebst den Grundfragen der Logik und Denkpsychologie. *Jahrbücher der Philosophie* 3: 31–92.
Cassirer, Ernst. 1953. *Substance and function*. New York: Dover.
Cassirer, Ernst. 1995. *Zur Metapysik der symbolischen Formen*. Ed. John Michael Krois. Hamburg: Felix Meiner (= *Nachgelassene Manuskripte und Texte*. Vol. 1).
Cauchy, Augustin Louis. 1822. 'Sur le développement des functions en séries ...', *Bulletin des sciences, par la Société Philomatique de Paris*, 49–54. [Reprinted in *Oeuvres complètes*, ser. 2, vol. 2, Paris, Gauthier-Villars, 1958, 276–282].
Cauchy, Augustin Louis. 1829. Sur l'équation à l'aide de laquelle on détermine les inégalités séculaires des mouvements des planètes. *Exercises de Mathématiques* 4: 174–195.
Cayley, Arthur. 1858. A memoir on the theory of matrices. *Philosophical Transactions of the Royal Society of London* 148: 17–37.
Chapman, Barbara, Gabriele Jost and Ruud van der Pas. 2008. *Using OpenMP: portable shared memory parallel programming*. Cambridge: MIT Press.
Chisholm, Monty. 2002. *Such silver currents. The story of William and Lucy Clifford 1845–1929*. Cambridge: The Lutterworth Press.
Chisholm, Roy J. S. and Alan. K. Common, eds. 1986. *Clifford algebras and their applications in mathematical physics*. Dordrecht and Boston: Reidel.
Church, Alonzo. 1936. Bibliography of symbolic logic. *Journal of symbolic logic*, 1: 121–218.
Clifford, William Kingdon. 1878. Applications of Grassmann's extensive algebra. *American Journal of Mathematics*: 1 (4): 350–358.
Clifford, William Kingdon. 1882a. *Mathematical papers*. Ed. Robert Tucker. London: Macmillian.
Clifford, William Kingdon. 1882b. *Applications of Grassmann's extensive algebra*. In [Clifford 1882a, 266–276].
Clifford, William Kingdon. 1882c. *On the classification of geometric algebras*. In [Clifford 1882a, 397–401].
Cooke, Roger. 1993. Zbl 0780.01004. Review of *William Hamilton. 1805–1865*, (Uil'yam Gamil'ton. 1805–1865.) (Russian) by Lev Solomonovich Polak. ZMATH – The database Zentralblatt MATH, http://www.zentralblatt-math.org/zmath/en/.
Corry, Leo. 1997. Hermann Minkowski and the postulate of relativity. *Archive for History of Exact Sciences* 51 (4): 273–314.
Cox, Homersham. 1882. Homogeneous coordinates in imaginary geometry and their application to systems of forces. *Quarterly journal of pure and applied mathematics* 18: 178–215.
Cox, Homersham. 1883a. On systems of circles and bicircular quartics. *Quarterly journal of pure and applied mathematics* 19: 74–124.
Cox, Homersham. 1883b. On the application of quaternions and Grassmann's Ausdehnungslehre to different kinds of uniform space. *Transactions of the Cambridge Philosophical Society* 13: 69–143.

Coxeter, Harold Scott MacDonald and William Oscar Jules Moser. 1980. *Generators and relations for discrete groups*. 4th ed. New York: Springer.

Crapo, Henry and Jürgen Richter-Gebert. 1994. Automatic proving of geometric theorems. In *Invariant methods in discrete and computational geometry*, ed. Neil L. White. 107–139. Dordrecht and Boston: Kluwer Academic Publishers.

Crapo, Henry and William Schmitt. 2000. The Whitney algebra of a matroid. *Journal of Combinatorial Theory (A)* 91: 215–263.

Creuzer, Friedrich. 1812–1821. *Symbolik und Mythologie der alten Völker, besonders der Griechen*. 4 Vol. Leipzig and Darmstadt: Leske.

Crowe, Michael J. 1967. *A history of vector analysis. The evolution of the idea of a vectorial system*. Notre Dame: University of Notre Dame University Press. (Dover 1994)

Dahlhaus, Carl. 1966. Über den Begriff der tonalen Funktion. In *Beiträge zur Musiktheorie des 19. Jahrhunderts*, ed. Martin Vogel. 96–97. Regensburg: Bosse.

Darrigol, Olivier. 2003. Number and measure: Hermann von Helmholtz at the crossroads of mathematics, physics, and psychology. *Studies in History and Philosophy of Science* 34: 515–573.

Darwin, Erasmus. 1803. *The temple of nature; or the origin of society: a poem with philosophical notes*. London: Printed for J. Johnson by T. Bensley.

Dathe, Uwe. 1993. Theoretische Quellen des frühen Frege. In *Philosophie und Logik*. Frege-Kolloquien Jena 1989/91. Ed. Werner Stelzner. Berlin and New York: de Gruyter. 39–44.

Dathe, Uwe. 2005. Frege in Jena: academic contacts and intellectual influences. In *Gottlob Frege: Critical assessments of leading philosophers*, Vol. I: Frege's philosophy in context. Eds. Michael Beaney and Erich H. Reck. New York and London: Routledge. 40–53.

Dawson, Jerry. F. 1966. *Friedrich Schleiermacher. The making of a nationalist*. Austin, Texas: University of Texas Press.

Dejung, Emanuel. 1980. *"Pestalozzi in Stanz"; ein Bild und seine Aussagen*. Ed. Christoph Dejung. http://www.heinrich-pestalozzi.de/fr/documentation/einzelbeitraege/dejung_christoph/level_3/emanuel_dejung_pestalozzi_in_stanz/index.htm. (accessed March 3, 2010).

Delbrück, Bertold. 1888. *Altindische Syntax*. (Syntaktische Forschungen, V.) Halle a. S.: Waisenhaus.

Descartes, René. 1661. Principia matheseos universalis, seu introduction ad geometriae methodum Renati Des Cartes. Sec. ed. Amsterdam: Elzevier.

Diderot, Denis. 1754. *Gedanken zur Interpretation der Natur*. [Reprinted in *Philosophische Schriften*. Vol 1, 1984, ed. and transl. Theodor Lücke. Berlin: Das Europäische Buch, 415–472].

Diesterweg, Friedrich Adolph Wilhelm. 1956. *Sämtliche Werke*. 1st section: Zeitschriftenbeiträge. Vol. 1. Berlin: Volk und Wissen. 195–197, 264, 525.

Diesterweg, Friedrich Adolph Wilhelm. 1959. *Sämtliche Werke*. 1st section: Zeitschriftenbeiträge. Vol. 3. Berlin: Volk und Wissen. 224.

Diesterweg, Friedrich Adolph Wilhelm, Hans Ferdinand Massmann and Ernst Wilhelm Kalisch. 1845. *Die Feier des 100sten Geburtstages Heinrich Pestalozzi's in Berlin am 12. Januar 1845*. Vorfeier des hundertjährigen Jubiläums seiner Geburt am 12. Januar 1846. Berlin: Vossische Buchhandlung.

Dieudonne, Jean. 1971. *La geometrie des groupes classiques*. Berlin etc.: Springer.

Dieudonne, Jean. 1979. The tragedy of Grassmann. *Linear and Multilinear Algebra* 8: 1–14.

Dilthey, Wilhelm. 1921. *Die Jugendgeschichte Hegels und andere Abhandlungen zur Geschichte des deutschen Idealismus*. Leipzig: Teubner (= Wihelm Diltheys gesammelte Schriften, Vol. 4).

Dirac, Paul Adrien Maurice. 1928. The quantum theory of the electron. *Proceedings of the Royal Society A* 117: 610–624.

Dittmer, Johannes M. 2001. Schleiermachers Wissenschaftslehre als Entwurf einer prozessualen Metaphysik in semiotischer Perspektive. Berlin: de Gruyter.

Doran, Chris J.L. 1994. *Geometric algebra and its application to mathematical physics*. Dissertation, University of Cambridge.

Doran, Chris J.L. and Anthony Lasenby. 2003. *Geometric algebra for physicists*. Cambridge: Cambridge University Press.

Doran, Chris J.L., David Hestenes, Franciscus Sommen and Nadine van Acker. 1993. Lie groups as spin groups. *Journal of Mathematical Physics* 34: 3642–3669.

Dorst, Leo, Daniel Fontijne and Stephen Mann. 2007. *Geometric algebra for computer science*. San Francisco: Morgan Kaufmann.

Dugac, Pierre. 1973. Eléments d'analyse de Karl Weierstrass. *Archive for History of Exact Sciences* 10: 42–176.

Echeverria, Javier. 1979. L'analyse geometrique de Grassmann et ses rapports avec La characteristique geometrique de Leibniz. *Studio Leibnitiana* 11: 223–273.

Eggert, Oskar. 1965. *Stände und Staat in Pommern im Anfang des 19. Jahrhunderts*, Köln-Graz: Böhlau.

Einstein, Albert. 1988. *The meaning of relativity*. 5th ed. Princeton, New Jersey: Princeton University Press.

Eisenbud, David and Joe Harris. 2000. *The geometry of schemes*. New York: Springer. (Graduate texts in mathematics, Vol. 197).

Elizarenkova, Tatjana Jakovlevna. 1999. *Rigveda*. Izdanie vtoroe, ispravlennoe. 3 vols. (Mandaly I–IV, Mandaly V–VIII, Mandaly IX–X). Moskva: Nauka.

Emmet, Dorothy. 1948. A.N. Whitehead: The last phase. *Mind, New Series* 57: 265–274.

Engel, Friedrich. 1911. *Grassmanns Leben. Nebst einem Verzeichnisse der von Grassmann veröffentlichten Schriften und einer Übersicht des handschriftlichen Nachlasses*. In [HGW32, 1–400].

Engel, Friedrich. 1936. Hermann Graßmann. In *Pommersche Lebensbilder*, vol. 2, ed. Martin Wehrmann, Adolf Hofmeister, Wilhelm Braun, 74–84. Stettin: Saunier.

Fearnley-Sander, Desmond. 1979. Hermann Grassmann and the creation of linear algebra. *American Mathematical Monthly* 86: 809–817.

Fermi, Enrico. 1995. *Notes on quantum mechanics. A Course given by Enrico Fermi at the University of Chicago. 2nd ed., with problems compiled by Robert A. Schluter*. Chicago: The University of Chicago Press.

Ferrein, Antoine. 1741. De la formation de la voix de l'homme. *Mémoires de l'académie royale*. 409–432. Paris: Imprimerie Royale.

Feynman, Richard. 1987. The reason for antiparticles. In Richard Feynman and Steven Weinberg. *Elementary particles and the laws of physics*. The 1986 Dirac memorial lectures. Cambridge: Cambridge University Press.

Fisher, Nicholas. 1990. Classification of the sciences. In *Companion to the history of modern science*, eds. Geoffrey Cantor et al. 853–867. London: Routledge.

Forder, Henry George. 1941. *The calculus of extension*. Cambridge: University Press. (2nd ed. 1960).

Forsyth, Andrew Russell. 1912. *Lehrbuch der Differential-Gleichungen: mit den Auflösungen der Aufgaben von Hermann Maser*. 2., autoris. Aufl./nach der 3. Aufl. des engl. Orig. besorgt und mit einem Anh. von Zusätzen vers. von Walther Jacobsthal. Braunschweig: Vieweg.

Frank, Manfred. 2005. Metaphysical foundations: a look at Schleiermacher's *Dialectic*. In *Friederich Schleiermacher*, ed. Jacqueline Mariña. 15–34. Cambridge: Cambridge University Press.

Frank, Philip. 1911. Das Verhalten der elektromagnetischen Feldgleichungen gegenüber linearen Transformationen der Raumzeitkoordinaten. *Annalen der Physik*, 4. Folge, 35: 599–607.

Franz, Walter. 1935. *Zur Methodik der Dirac-Gleichung. Sitzungsberichte der Bayerischen Akademie der Wissenschaften, Mathematisch-naturwissenschaftliche Abteilung*. München: C.H. Beck.

Frege, Gottlob. 1884. *Die Grundlagen der Arithmetik. Eine logisch mathematische Untersuchung über den Begriff der Zahl*. Breslau: Wilhelm Koebner.

Freguglia, Paolo. 1992. *Dalle equipollenze ai sistemi lineari. Il contributo italiano al calcolo geometrico*. Urbino: QuattroVenti ed.

Freguglia, Paolo. 2006. *Geometria e numeri. Storia, teoria elementare ed applicazioni del calcolo geometrico*. Torino: Bollati Boringhieri.

Freguglia, Paolo and Cristiano Bocci. 2008. Dall'eredità grassmanniana alla teoria delle omografie nella scuola di Peano. *La Matematica nella Società e nella Cultura: Rivista dell' Unione Matematica Italiana* 1: 131–164.

Fries, Jakob Friedrich. 1822. *Die mathematische Naturphilosophie nach philosophischer Methode bearbeitet.* Heidelberg: Christian Friedrich Winter.

Fritsch, Rudolf. 1990. Mehmke, Rudolf. *Neue Deutsche Biographie* 16: 621–623.

Frobenius, Ferdinand Georg. 1878. Ueber lineare Substitutionen und bilineare Formen. *Journal für die reine und angewandte Mathematik (Crelle's Journal)* 84: 1–63.

Frobenius, Ferdinand Georg. 1879. Ueber homogene totale Differentialgleichungen. *Journal für die reine und angewandte Mathematik (Crelle's Journal)* 86: 1–19.

Fulton, William and Joe Harris. 1991. *Representation theory: A first course.* New York: Springer. (Graduate Texts in Mathematics, Vol. 129).

Gaidenko, Piama. 1981. Ontologic foundation of scientific knowledge in 17th and l8th century rationalism. In [Jahnke and Otte 1981, 55–63].

Gehrke, Wolfgang and Jochen Pfalzgraf. 2001. Computer-aided construction of finite geometric spaces: Automated verification of geometric constraints. *Journal of Automated Reasoning* 26: 139–160.

Geldner, Karl Friedrich. 1907. *Der Rigveda in Auswahl.* First part. Glossar. Stuttgart: W. Kohlhammer.

Geldner, Karl Friedrich, transl. 1951. *Der Rig-Veda. Aus dem Sanskrit ins Deutsche übersetzt und mit einem laufenden Kommentar versehen.* 3 vols. Cambridge, Mass., London: Harvard University Press.

Gerhardt, Volker, Reinhard Mehring and Jana Rindert. 1999. *Berliner Geist: eine Geschichte der Berliner Universitätsphilosophie bis 1946. Mit einem Ausblick auf die Gegenwart der Humboldt-Universität.* Berlin: Akademie Verlag.

Gibbs, Josiah Willard. 1961. *The scientific papers of J.W. Gibbs*, volume 1–2, New York: Dover.

Goethe, Johann Wolfgang. 1970. Tonlehre. In *Die Schriften zur Naturwissenschaft/Goethe.* Ed. Dorothea Kuhn and Wolf von Engelhardt. 1. section, vol.11. 134–138. Weimar: Böhlau.

Goethe, Johann Wolfgang. 1991. Briefwechsel zwischen Goethe und Zelter in den Jahren 1799 bis 1832. Eds. Hans-Günter Ottenberg and Edith Zehm. In *Johann Wolfgang Goethe: Sämtliche Werke nach Epochen seines Schaffens.* Münchner Edition. Ed. Karl Richter. Vol. 20.1. München: Hanser.

Gollmert, Luis. 1856. Die Preußischen Staats-Archive. *Archiv für Landeskunde der Preußischen Monarchie* 4: 113–163.

Gotō, Toshifumi. 1987. *Die "I. Präsensklasse" im Vedischen. Untersuchung der vollstufigen thematischen Wurzelpräsentia.* Wien: Österreichische Akademie der Wissenschaften.

Gotō, Toshifumi. 2005. Ai. *ádbhuta-*, *ádabdha-*, jav. *abda-*, *dapta-*, und ai. *addhā*, aav. ap. *azdā*. In *Indogermanica: Festschrift Gert Klingenschmitt; indische, iranische und indogermanische Studien dem verehrten Jubilar dargebracht zu seinem fünfundsechzigsten Geburtstag*, ed. Günter Schweiger. 45–64. Taimering: Schweiger.

Grassmann, Hermann. 1840. *Theorie der Ebbe und Flut. Prüfungsarbeit von 1840.* [Reprinted in HGW31, 8–203].

Grassmann, Hermann. 1844. *Die lineale Ausdehnungslehre ein neuer Zweig der Mathematik dargestellt und durch Anwendungen auf die übrigen Zweige der Mathematik, wie auch auf die Statik, Mechanik, die Lehre vom Magnetismus und die Krystallonomie erläutert*, Leipzig: Otto Wigand. [A1], [2nd ed. H. Grassmann 1878. Reprinted in HGW11, 4–312].

Grassmann, Hermann. 1845. Kurze Uebersicht über das Wesen der Ausdehnungslehre. *Archiv der Mathematik und Physik* 6 (1845): 337–350. [Reprinted in HGW11, 297–312].

Grassmann, Hermann. 1847. *Geometrische Analyse geknüpft an die von Leibniz erfundene geometrische Charakteristik.* Gekrönte Preisschrift. Leipzig: Weidmann, Brockhaus. [Reprinted in HGW11, 321–398].

Grassmann, Hermann. 1850. Der chemische Unterricht auf der höheren Bürgerschule. *Pädagogische Revue* 11 (1850) vol. 25, Abt. 1, H. 7: 45–54.

Grassmann, Hermann. 1851. Besprechung zweier Religionslehrbücher: 1. Lehrbuch der christlichen Religion für die oberen Classen der Stadt- und Landschulen [...] von E. Thierbach, 3. Aufl. Freiberg 1849. 2. Lehrbuch der Religion für die oberen Classen protestantischer hoher Schulen von L.A. Petri, 3. Aufl. Hannover 1850. *Pädagogische Revue*. 1. Abt. a, (27): 341–348.

Grassmann, Hermann. 1853. Zur Theorie der Farbenmischung. *Poggendorffs Annalen der Physik und Chemie* 89: 69–84 [Reprinted in HGW22, 161–173].

Grassmann, Hermann 1854. *Uebersicht der Akustik und der niedern Optik*. Programm des Könglichen und Stadtgymnasiums zu Stettin. Stettin. [Reprinted in HGW22, 174–202].

Grassmann, Hermann. 1855. Sur les différents genres de multiplication. *Journal für die reine und angewandte Mathematik (Crelle's Journal)* 49 (1855): 123–141. [Reprinted in HGW21, 199–217].

Grassmann, Hermann. 1861. *Lehrbuch der Mathematik für höhere Lehranstalten. Erster Teil: Arithmetik*. Berlin: Enslin. [LA].

Grassmann, Hermann. 1862. *Die Ausdehnungslehre. Vollständig und in strenger Form bearbeitet*. Berlin: Th. Chr. Fr. Enslin. [A2], [Reprinted in HGW12, 1–383].

Grassmann, Hermann. 1863. Über die Aspiraten und ihr gleichzeitigen Vorhandensein im An- und Auslaute der Wurzeln. *Zeitschrift für vergleichende Sprachforschung auf dem Gebiete des Deutschen, Griechischen und Lateinischen* 12 (2): 81–138.

Grassmann, Hermann. 1865. *Lehrbuch der Mathematik für höhere Lehranstalten*. Zweiter Teil: Trigonometrie, Berlin: Enslin.

Grassmann, Hermann. 1873–1875. *Wörterbuch zum Rig-Veda*. Leipzig: Brockhaus. [Reprinted 1976. 5th ed. Wiesbaden: Otto Harrassowitz].

Grassmann, Hermann. 1874. Die älteste Religion der Inder. *Allgemeine Missions-Zeitschrift. Monatshefte für geschichtliche und theoretische Missionskunde* 1: 473–482.

Grassmann, Hermann. 1876. *Rig-Veda. Übersetzt und mit kritischen und erläuternden Anmerkungen versehen von Hermann Grassmann. Erster Theil. Die Familien-Bücher des Rig-Veda: (zweites bis achtes Buch)*. Leipzig: F.A. Brockhaus. [RV1]. [Reprinted 1990. Frankfurt: Minerva Verlag].

Grassmann, Hermann. 1877a. *Rig-Veda. Übersetzt und mit kritischen und erläuternden Anmerkungen versehen von Hermann Grassmann. Zweiter Theil. Sammelbücher des Rig-Veda: (erstes, neuntes, zehntes Buch)*. Leipzig: Brockhaus. [RV2]. [Reprinted 1990. Frankfurt: Minerva Verlag].

Grassmann, Hermann. 1877b. Bemerkungen zur Theorie der Farbenempfindungen. Briefliche Mittheilung an den Verfasser. In [Preyer 1877, 629–637]. [Reprinted in HGW22, 213–221].

Grassmann, Hermann. 1877c. Der Ort der Hamiltonschen Quaternionen in der Ausdehnungslehre. *Mathematische Annalen* 12 (1877): 375–386. [Reprinted in HGW31, 268–282].

Grassmann, Hermann. 1877d. Über die physikalische Natur der Sprachlaute. *Wiedemanns Annalen der Physik und Chemie* 1 (4): 606–629 [Reprinted in HGW22, 222–240].

Grassmann, Hermann. 1878. *Ueber den Abfall vom Glauben: Mahnungen an die wissenschaftlich Gebildeten der Neuzeit*. Stettin: Otto Brandner.

Grassmann, Hermann. 1894–1911. *Gesammelte mathematische und physikalische Werke*, 3 Bände in 6 Teilen, herausgegeben von F. Engel unter Mitwirkung von E. Study (Band 1.1), H. Grassmann d. J. (Band 1.2), E. Study und G. Scheffers (Band 2.1), J. Lüroth (Band 2.2), J. Grassmann (Band 3.1), Teubner Verlag, Leipzig, 1894–1911. [HGW11–HGW32].

Grassmann, Hermann. 1947. *Teoría de la extensión. Nueva disciplina matemática expuesta y aclarada mediante aplicaciones*. Transl. Emilio Oscar Roxin. Buenos Aires and Mexico: Espasa-Calpe.

Grassmann, Hermann. 1994. *La science de la grandeur extensive*. French transl. Dominique Flament and Bernd Bekemeier, Blanchard, Paris.

Grassmann, Hermann. 1995a. *A new branch of mathematics: The "Ausdehnungslehre" of 1844 and other works*. Engl. transl. Lloyd C. Kannenberg. Chicago, Ill.: Open Court.

Grassmann, Hermann. 1995b. *Linear extension theory (Die lineale Ausdehnungslehre, 1844)*, Engl. transl. Lloyd C. Kannenberg. In [H. Grassmann 1995a, 1–312]. [$A1_K$].

Grassmann, Hermann. 1995c. *Geometric analysis (Geometrische Analyse, 1847)*, Engl. transl. Lloyd C. Kannenberg. In [H. Grassmann 1995a, 314–414].
Grassmann, Hermann. 1996. *Wörterbuch zum Rig-Veda*. 6., revised and complemented edition by Maria Kozianka. Wiesbaden: Harrassowitz.
Grassmann, Hermann. 2000. *Extension theory* (Die Ausdehnungslehre, 1862), Engl. transl. Lloyd C. Kannenberg. American Mathematical Society, London Mathematical Society. [$A2_K$].
Grassmann, Hermann and Robert Grassmann. 2008. Logika i filosofiya matematiki. Izbrannoe. (Logic and philosophy of mathematics. Selected works – Russian), eds. Lyubov Gavrilovna Biryukova and Zinaida Andreevna Kuzicheva, transl. Boris Vladimirovitch Biryukov. Moskva: Nauka.
Graßmann, Justus Günther. 1817. *Raumlehre*, 1. Teil. Berlin: Realschulbuchhandlung.
Graßmann, Justus Günther. 1824, *Raumlehre*, 2. Teil. Berlin: Reimer.
Graßmann, Justus Günther. 1827. *Ueber den Begriff und Umfang der reinen Zahlenlehre*. Programmabhandlung des Stettiner Gymnasiums. Stettin.
Graßmann, Justus Günther. 1829. *Zur physischen Krystallonomie und geometrischen Combinationslehre*. Erstes Heft. Stettin: Morin.
Graßmann, Justus Günther. 1833. Combinatorische Entwicklung der Krystallgestalten. *Annalen der Physik* 106 (1): 1–43.
Grassmann, Robert. 1872a. *Die Formenlehre oder Mathematik*. Stettin: R. Grassmann. [Reprinted in R. Grassmann 1966].
Grassmann, Robert. 1872b. *Die Grösenlehre. Erstes Buch der Formenlehre oder Mathematik*. Stettin: R. Grassmann. [Reprinted in R. Grassmann 1966].
Grassmann, Robert. 1872c. *Die Begriffslehre oder Logik. Zweites Buch der Formenlehre oder Mathematik*. Stettin: R. Grassmann. [Reprinted in R. Grassmann 1966].
Grassmann, Robert. 1872d. *Die Bindelehre oder Combinationslehre. Drittes Buch der Formenlehre oder Mathematik*: Stettin: R. Grassmann. [Reprinted in R. Grassmann 1966].
Grassmann, Robert. 1872e. *Die Zahlenlehre oder Arithmetik. Viertes Buch der Formenlehre oder Mathematik*. Stettin: R. Grassmann. [Reprinted in R. Grassmann 1966].
Grassmann, Robert. 1872f. *Die Ausenlehre oder Ausdehnungslehre. Fünftes Buch der Formenlehre oder Mathematik*. Stettin: R. Grassmann. [Reprinted in R. Grassmann 1966].
Grassmann, Robert. 1890a. *Die Geschichte der Philosophie und erste Einleitung in die Philosophie*. Stettin: R. Grassmann.
Grassmann, Robert. 1890b. *Die Denklehre oder die Lehre von den Arten der wissenschaftlichen Denkakte und Verknüpfungen*. Stettin: R. Grassmann.
Grassmann, Robert. 1890c. *Die Logik und die andern logischen Wissenschaften*. Stettin: R. Grassmann.
Grassmann, Robert. 1890d. *Formelbuch der logischen Wissenschaften*. Stettin: R. Grassmann.
Grassmann, Robert. 1890e. *Die Erkenntnislehre*. Stettin: R. Grassmann.
Grassmann, Robert. 1895. *Die Formenlehre oder Mathematik in strenger Formelentwicklung. Gebäude des Wissens*, vol. 23. Stettin: R. Grassmann.
Grassmann, Robert. 1966. *Die Formenlehre oder Mathematik*. Introduction by Joseph Ehrenfried Hofmann, Hildesheim: Georg Olms.
Grassmann, Robert. 2005. *The theory of forms [Formenlehre] or mathematics. An English reading* (trans. Lloyd C. Kannenberg), unpublished.
Grattan-Guinness, Ivor. 1996. Where does Grassmann fit in the history of logic. In [Schubring 1996a, 211–216].
Graumann, Helmut. 2000. *50 berühmte Deutsche aus Pommern: Ein biographischer Überblick*. Berlin: Frieling & Huffmann.
Gregory, Duncan Farquharson. 1840. On the real nature of symbolical algebra. *Transactions of the Royal Society of Edinburgh* 14: 208–216.
Grehn, Joachim, ed. 1991. *Metzler Physik*. 2nd ed. Stuttgart: J.B. Metzlersche Verlagsbuchhandlung.
Grehn, Joachim, ed. 1993. *Metzler Physik*. Lehrerband. 2nd ed. Hannover: Schroedel.

Grimm, Jacob. 1822. *Deutsche Grammatik*. Vol. 1. 2nd ed. Göttingen: Dieterichschen Buchhandlung.

Gull, Stephen, Anthony Lasenby and Chris J.L. Doran. 1993. Imaginary numbers are not real – the geometric algebra of spacetime. *Foundations of Physics* 23 (9): 1175–196.

Gwiazdowska, Ewa. 2001. *Widoki Szczecina: Źródła ikonograficzne do dziejów miasta od XVI wieku do 1945 r.* [Stettin Prospects. Iconographic sources for the history of the city from the 16th century to 1945]. Szczecin: Muzeum Narodowe.

Hahn, Theo. 2005. *International tables of crystallography*. Vol. A. Dordrecht: Springer. (Also online at: it.iucr.org).

Hamilton, William Rowan. 1853. *Lectures on quaternions*. Dublin: Hodges and Smith.

Hamilton, William Rowan. 1882. *Elemente der Quaternionen*. Transl. Paul Glan. Barth: Leipzig.

Hankel, Hermann. 1867. *Vorlesung über die complexen Zahlen und ihre Functionen*. 1. Theil: Theorie der complexen Zahlensysteme insbesondere der gemeinen imaginären Zahlen und der Hamiltonschen Quaternionen nebst ihrer geometrischen Darstellung. Leipzig: Leopold Voss.

Hankel, Hermann. 1871. "Grenze". In *Allgemeine Encyklopädie der Wissenschaften und Künste*. Vol. 90. Eds. Johann Samuel Ersch and Johann Gottfried Gruber. 185–211, Leipzig: Brockhaus.

Harris, Joe. 1992. *Algebraic geometry: A first course*. New York: Springer. (Graduate Texts in Mathematics, Vol. 133).

Hart, John C., George K. Francis and Louis H. Kauffman. 1994. Visualizing quaternion rotation. *ACM Transactions on Graphics* 13 (3): 256–276.

Hartimo, Mirja Helena. 2007. Towards completeness: Husserl on theories of manifolds 1890–1901. *Synthese* 156: 281–310.

Hartung, Fritz. 1981. Zur Geschichte der preußischen Verwaltung im 19. und 20. Jahrhundert (1948). In *Moderne Preußische Geschichte, vol. 2*, eds. Otto Büsch and Wolfgang Neugebauer, 680–738. Berlin, New York: de Gruyter.

Haüy, René Just. 1804a. *Lehrbuch der Mineralogie ausgearbeitet vom Bürger Haüy*. Aus dem Französischen übersetzt und mit Anmerkungen versehen von Dietrich Ludwig Gustav Karsten. Vol. 1 u. 2. Paris and Leipzig: Reclam.

Haüy, René Just. 1804b. *Handbuch der Physik, für den Elementarunterricht in den französischen National-Lyceen ausgearbeitet von R.J. Haüy*. Aus dem Französischen übersetzt und mit Anmerkungen und Zusätzen vermehrt von Christian Samuel Weiß. Vol. 1u. 2. Paris and Leipzig: Reclam.

Hawking, Stephen. 1998. Dirac memorial address. In *Paul Dirac – The man and his work*, ed. Peter Goddard. XIII–XV. Cambridge: Cambridge University Press.

Hawking, Stephen. 2003. *Das Universum in der Nußschale*. München: dtv.

Hawkins, Thomas. 2008. Frobenius and the symbolical algebra of matrices. *Archive for History of Exact Sciences* 62: 23–57.

Hawkins, Thomas. 1975. Cauchy and the spectral theory of matrices. *Historia Mathematica* 2: 1–29.

Hawkins, Thomas. 1977a. Another look at Cayley and the theory of matrices. *Archives Internationales d'histoire des Sciences* 27: 82–112.

Hawkins, Thomas. 1977b. Weierstrass and the theory of matrices. *Archive for History of Exact Sciences* 17: 119–163.

Hawkins, Thomas. 2000. *Emergence of the theory of Lie groups: an essay in the history of mathematics 1869–1926*. New York: Springer.

Hay, William H. 1956. Paul Carus: A case-study of philosophy on the frontier. *Journal of the History of Ideas*. 17: 498–510.

Heijenoort, Jean van. 1967. Logic as a calculus and logic as language. *Synthese* 17: 324–330.

Helmholtz, Hermann. 1863. *Die Lehre von den Tonempfindungen als physiologische Grundlage für die Theorie der Musik*. Braunschweig: Friedrich Vieweg u. Sohn.

Henderson, Harold. 1993. *Catalyst for controversy: Paul Carus of Open Court*. Carbondale and Edwardsville: Southern Illinois University Press.

Hestenes, David. 1966. *Space-time algebra*. New York: Gordon & Breach.

Hestenes, David. 1967. Real spinor fields. *Journal of Mathematical Physics* 8: 798–808.
Hestenes, David. 1985. *New foundations for classical mechanics*. Dordrecht and Boston: Kluwer.
Hestenes, David. 1990. *New foundations for classical mechanics*. (Reprint with corrections). Dordrecht: Kluwer.
Hestenes, David. 1991. The design of linear algebra. *Acta Applicandae Mathematicae* 23: 65–93.
Hestenes, David. 1996. Grassmann's vision. In [Schubring 1996a, 243–54].
Hestenes, David. 2001. Old wine in new bottles: A new algebraic framework for computational geometry. In *Advances in geometric algebra with applications in science and engineering*, eds. Eduardo Bayro-Corrochano and Garret Sobczyk. 3–17. Boston: Birkhauser.
Hestenes, David. 2003a. Oersted medal lecture 2002: Reforming the mathematical language of physics. *American Journal of Physics* 71(2): 104–121.
Hestenes, David. 2003b. Spacetime physics with geometric algebra. *American Journal of Physics* 71 (7): 691–714.
Hestenes, David. 2009. Zitterbewegung in quantum mechanics. *Foundations of Physics* DOI 10.1007/s10701-009-9360-3.
Hestenes, David. 2010. New tools for computational geometry and rejuvenation of screw theory. In *Geometric algebra computing for engineering and computer science*, eds. Eduardo Bayro-Corrochano and Gerik Scheuermann. London: Springer.
Hestenes, David and Garret Sobczyk. 1984. *Clifford algebra to geometric calculus*. A unified language for mathematics and physics. Dordrecht and Boston: Reidel.
Hestenes, David and Jeremy W. Holt. 2007. The crystallographic space groups in geometric algebra. *Journal of Mathematical Physics* 48 (2): 023514–023514–22.
Hestenes, David and Renatus Ziegler. 1991. Projective geometry with Clifford algebra. *Acta Applicandae Mathematicae* 23: 25–63.
Heuser, Marie-Luise. 1991. Georg Cantors transfinite Zahlen und Giordano Brunos Unendlichkeitsidee. In *Selbstorganisation. Jahrbuch für Komplexität in den Natur-, Sozial- und Geisteswissenschaften*. Bd. 2: Der Mensch in Ordnung und Chaos, ed. Uwe Niedersen, 222–244. Berlin: Duncker & Humblot.
Heuser, Marie-Luise. 1992. Schelling's Concept of Self-Organization. In *Evolution of Dynamical Structures in Complex Systems*, eds. Rudolf Friedrich and Arne Wunderlin, 395–415. Berlin/Heidelberg/New York: Springer.
Heuser, Marie-Luise. 1996. Geometrical product – exponentiation – evolution. Justus Günther Grassmann and dynamist Naturphilosophie. In [Schubring 1996a, 47–58].
Heuser, Marie-Luise. 2007. Die Anfänge der Topologie in Mathematik und Naturphilosophie. In *Topologie. Zur Raumbeschreibung in den Kultur- und Medienwissenschaften*, ed. Stephan Günzel, 183–200. Bielefeld: transcript Verlag.
HGW11 = Grassmann, Hermann. 1894. *Gesammelte mathematische und physikalische Werke*. Vol. 1.1. Ed. Friedrich Engel assisted by Eduard Study. Leipzig: Teubner. [Reprinted 1972. New York].
HGW12 = Grassmann, Hermann. 1896. *Gesammelte mathematische und physikalische Werke*. Vol. 1.2. Ed. Friedrich Engel, assisted by Hermann Grassmann (Jun.). Leipzig: Teubner. [Reprinted 1972. New York].
HGW21 = Grassmann, Hermann. 1904. *Gesammelte mathematische und physikalische Werke*. Vol. 2.1. Eds. Eduard Study, Georg Scheffers and Friedrich Engel. Leipzig: Teubner. [Reprinted 1972. New York].
HGW22 = Grassmann, Hermann. 1902. *Gesammelte mathematische und physikalische Werke*. Vol. 2.2. Eds. Jakob Lüroth and Friedrich Engel. Leipzig: Teubner. [Reprinted 1972. New York].
HGW31 = Grassmann, Hermann. 1911. *Gesammelte mathematische und physikalische Werke*. Vol. 3.1. Eds. Justus Grassmann and Friedrich Engel. Leipzig: Teubner. [Reprinted 1972. New York].
HGW32 = Grassmann, Hermann. 1911. *Gesammelte mathematische und physikalische Werke*. Vol. 3.2. Ed. Friedrich Engel. Leipzig: Teubner. [Reprinted 1972. New York].
Hilbert, David. 1899. *Grundlagen der Geometrie*. Leipzig: Teubner. [Transl. *Foundations of Geometry*, transl. Leo Unger. LaSalle, Illinois: Open Court 1971].

Hilbert, David. 1918. Axiomatisches Denken. *Mathematische Annalen* 78: 405–415. English translation: [Hilbert 1996].
Hilbert, David. 1996. Axiomatic thought. In *From Kant to Hilbert. A source book in the foundations of mathematics*. 2 Vols. Ed. William Ewald, Oxford: Clarendon Press 1996, 1105–1115.
Hildenbrand, Dietmar. 2005. Geometric computing in computer graphics using conformal geometric algebra. *Computers & Graphics* 29(5): 802–810.
Hildenbrand, Dietmar, Daniel Fontijne, Christian Perwass and Leo Dorst. 2004. Tutorial geometric algebra and its application to computer graphics. In *Eurographics conference Grenoble*.
Hildenbrand, Dietmar, Daniel Fontijne, Yusheng Wang, Marc Alexa and Leo Dorst. 2006. Competitive runtime performance for inverse kinematics algorithms using conformal geometric algebra. In *Eurographics conference Vienna*.
Hildenbrand, Dietmar, Holger Lange, Florian Stock and Andreas Koch. 2008. Efficient inverse kinematics algorithm based on conformal geometric algebra using reconfigurable hardware. In *GRAPP conference Madeira*.
Hill, Claire Ortiz. 2000. Husserl's Mannigfaltigkeitslehre. In *Husserl or Frege? Meaning, objectivity, and mathematics*, eds. Claire Ortiz Hill and Guillermo E. Rosado Haddock, 109–135. New York and La Salle, Ill.: Open Court.
Hindenburg, Carl Friedrich. 1776. *Beschreibung einer ganz neuen Art, nach einem bekannten Gesetze fortgehende Zahlen durch Abzählen oder Abmessen bequem und sicher zu finden: nebst Anwendung der Methode auf verschiedene Zahlen, besonders auf eine darnach zu fertigende Factorentafel, mit eingestreueten, die Zahlenberechnung überhaupt betreffenden Anmerkungen*. Leipzig: Crusius.
Hitzer, Eckhard and Christian Perwass. 2004. Crystal cells in geometric algebra. *Proceedings of the International Symposium on Advanced Mechanical Engineering* (ISAME), Fukui, 290–295.
Hitzer, Eckhard and Christian Perwass. 2005a. Full geometric description of all symmetry elements of crystal space groups by the suitable choice of only three vectors for each Bravais cell or crystal family. In *Proceedings of the International Symposium on Advanced Mechanical Engineering*. 19–25. Busan: Pukyong National University.
Hitzer, Eckhard and Christian Perwass. 2005b. Crystal cell and space lattice symmetries in Clifford geometric algebra. In *International conference on numerical analysis and applied mathematics*, eds. Theodore E Simos, George Psihoyios and Charalampos Tsitouras. 937–941. Weinheim: Wiley-VCH.
Hitzer, Eckhard and Christian Perwass. 2006. Three vector generation of crystal space groups in geometric algebra. *Bulletin of the Society for Science on Form* 21: 55–56.
Hitzer, Eckhard and Christian Perwass. 2010. *Interactive 3D space group visualization with CLUCalc and the Clifford geometric algebra description of space groups*. Accepted for Jayme Vaz and Waldyr A. Rodrigues (eds.) [Proceedings of ICCA8, 26–30 May 2008, Las Campinas, Brazil].
Hodge, William Vallance Douglas and Daniel Pedoe. 1954. *Methods of algebraic geometry*, in 3 volumes. Cambridge: University Press.
Hoenigswald, Henry M. 1978. The annus mirabilis 1876 and posterity. *Transactions of the Philological Society* 76: 17–35.
Hoffmann, Karl. 1955. Ein grundsprachliches Possessivsuffix. *Münchener Studien zur Sprachwissenschaft* 6: 35–40 (= *Aufsätze zur Indoiranistik II*, ed. by Johanna Narten. Wiesbaden: Reichert 1976: 378–383).
Hoffmann, Karl. 1967. *Der Injunktiv im Veda*. Heidelberg: Carl Winter.
Homeyer, Carl Gustav. 1854. Sechs und zwanzigster Jahresbericht, betreffend die Jahre 1850. 51. 51. 52. 52. In *Baltische Studien*, AF 15, no. 2: 3–47.
Horn, Martin Erik. 2009a. Vom Raum zur Raumzeit. In *Chemie- und Physikdidaktik für die Lehramtsausbildung*, ed. Dietmar Höttecke (GDCP Band 29). 455–457. Berlin: LIT-Verlag Dr. W. Hopf.
Horn, Martin Erik. 2009b. Die Spezielle Relativitätstheorie im Kontext der Raumzeit-Algebra. In *Didaktik der Physik*. Tagungs-CD zur Frühjahrstagung der DPG in Bochum, ed. Volkhard Nordmeier. Paper 30.40. Berlin: Lehmanns Media.

Hültenschmidt, Erika. 1996. Hermann Grassmann's contribution to the construction of a German "Kulturnation". In [Schubring 1996a, 87–113].

Husemoller, Dale. 1966. *Fibre bundles*. 2nd ed., New York: Springer. (Graduate texts in mathematics, Vol. 20).

Husserl, Edmund. 1970. *Philosophie der Arithmetik. Mit Ergänzenden Texten (1890–1901).* Husserliana Band XII. The Hague: Martinus Nijhoff.

Husserl, Edmund. 1982. *Studien zur Arithmetik und Geometrie. Texte aus dem Nachlass (1886–1901).* Husserliana Band XXI. Ed. Ingeborg Strohmeyer. The Hague/Boston/Lancaster: Martinus Nijhoff.

Husserl, Edmund. 1994a. *Early writings in the philosophy of logic and mathematics*. Transl. Dallas Willard. Dordrecht/Boston/London: Kluwer.

Husserl, Edmund. 1994b. *Briefwechsel, Band I. Die Brentanoschule*. Ed. Elisabeth Schuhmann. Dordrecht/Boston/London: Kluwer.

Husserl, Edmund. 1994c. *Briefwechsel, Band V. Die Neukantianer*. Ed. Elisabeth Schuhmann. Dordrecht/Boston/London: Kluwer.

Husserl, Edmund. 1994d. *Briefwechsel, Band VI. Philosophenbriefe*. Ed. Elisabeth Schuhmann. Dordrecht/Boston/London: Kluwer.

Husserl, Edmund. 1994e. *Briefwechsel, Band IX. Familienbriefe*. Ed. Elisabeth Schuhmann. Dordrecht/Boston/London: Kluwer.

Husserl, Edmund. 2001. *Logical investigations*. Transl. John N. Findlay. London and New York: Routledge.

Husserl, Edmund. 2003. Philosophy of arithmetic: Psychological and logical investigations with supplementary texts from 1887–1901. In *Edmund Husserl collected works*, vol. 10, transl. Dallas Willard. Dordrecht: Kluwer.

Hyde, Edward W. 1890. *The directional calculus*. Boston: Ginn.

Hyrkkänen, Markku. 2009. All history is, more or less, intellectual history: R. G. Collingwood's contribution to the theory and methodology of intellectual history. *Intellectual History Review* 19 (2): 251–263.

Irtel, Hans. 1987. A conjoint Grassmann structure for testing the additivity of binocular color mixtures. *Journal of Mathematical Psychology* 31: 192–202.

Israel, Giorgio. 1981. "Rigor" and "axiomatics" in modern mathematics. *Fundamenta Scientiae* 2 (2): 205–219.

Jacobsthal, Erwin. 1945–46. *Wechselspiel eines Lebens. Lebenserinnerungen (bis 1909).* Written in Guatemala (personal photocopy of Jacobsthal's typed manuscript).

Jacobsthal, Erwin. 1946–47. *Aus dem unbedeutenden Leben eines nicht bedeutenden Arztes. Fragment einer für die Öffentlichkeit bestimmten Autobiografie.* Written in Guatemala (personal photocopy of Jacobsthal's typed manuscript).

Jacobsthal, Gustav. 1870. *Die Mensuralnotation des 12. und 13. Jahrhunderts*. Dissertation. Berlin: Humboldt University Archive.

Jacobsthal, Gustav. 1888. *Vorlesungsskizze Die Opern Mozarts*. Sommer 1888. In Musikalischer Nachlass Gustav Jacobsthal in der Musikabteilung der Staatsbibliothek zu Berlin, B13.

Jacobsthal, Gustav. 2002. Vorläufige Gedanken zur Verbesserung der musikalischen Zustände an den preußischen Universitäten. Memorandum an das preußische Kultusministerium 1883. Ed. Peter Sühring. In *Jahrbuch des Staatlichen Instituts für Musikforschung*, ed. by Günther Wagner. 295–322. Stuttgart and Weimar: J.B. Metzler.

Jacobsthal, Gustav. 2010. *Übergänge und Umwege in der Musikgeschichte.* Aus Straßburger Studien und Vorlesungen Gustav Jacobsthals. Codex Montpellier · Palestrina · Monteverdi · Haydn · Emanuel Bach · Mozart, ed. Peter Sühring. Hildesheim: Olms.

Jacobsthal, Walther. 1899. *Ueber die asymptotische Darstellung der Integrale einer gewissen linearen Differentialgleichung zweiter Ordnung.* Dissertation. Universität Strassburg.

Jacobsthal, Walther. 1917. *Mondphasen, Osterrechnung und Ewiger Kalender*. Berlin: Springer.

Jahnke, Hans Niels. 1978. *Zum Verhältnis von Wissensentwicklung und Begründung in der Mathematik – Beweisen als didaktisches Problem.* Materialien und Studien des Instituts für Didaktik der Mathematik der Universität Bielefeld, vol. 10. Bielefeld: IDM.

Jahnke, Hans Niels. 1990. *Mathematik und Bildung in der Humboldtschen Reform.* Goettingen: Vandenhoeck + Ruprecht.

Jahnke, Hans Niels and Michael Otte. eds. 1981. *Epistemological and social problems of the sciences in the early nineteenth century.* Dordrecht: Reidel.

Jevons, William Stanley. 1864. *Pure logic or the logic of quality apart from quantity with remarks on Boole's system and the relation of logic and mathematics.* London: E. Stanford. Also in [Jevons 1890].

Jevons, William Stanley. 1890. *Pure logic and other minor works.* Eds. Robert Adamson and Harriet A. Jevons. London and New York: Macmillan and Co. [Reprinted 1991. Bristol: Thoemmes Press].

Johannsen, Rolf Hermann. 2007. *Friedrich Wilhelm IV. von Preußen. Von Borneo nach Rom. Sanssouci und Residenzprojekte 1815–1848.* Kiel: Ludwig.

Jordan, Camille. 1870. *Traité des substitutions et des équations algébriques.* Paris: Gauthier-Villars.

Kagan, Veniamin Fedorovitch. 1913. Theoretical foundations of mathematics. In *Granat Encyclopedic Dictionary*, vol. 41, part VII, 7th ed., columns 327–468. St.Petersberg-Odessa.

Kanitscheider, Bernulf. 1971.*Geometrie und Wirklichkeit.* Berlin: Duncker & Humblot.

Kant, Immanuel. 1763. *Versuch, den Begriff der negativen Größen in die Weltweisheit einzuführen.* Königsberg: Kanter. [Reprinted in Immanuel Kant, *Vorkritische Schriften, 1757–1777.* Kant, *Gesammelte Schriften*, Akademie-Ausgabe, Vol.2, 1968, 165–204. Berlin: de Gruyter] [Transl. *Attempt to introduce the concept of negative magnitudes into philosophy*, transl. and ed. David Walford and Ralf Meerbote. In The Cambridge edition of the works of Immanuel Kant. Cambridge: Univ. Press 2003, 203–241]

Kant, Immanuel. 1764. *Untersuchung über die Deutlichkeit der Grundsätze der natürlichen Theologie und der Moral.* [Reprinted in Immanuel Kant, *Vorkritische Schriften*, 1757–1777. Kant, *Gesammelte Schriften*, Akademie-Ausgabe, Vol.2, 1968, 273–301. Berlin: de Gruyter].

Kant, Immanuel. 1781. *Kritik der reinen Vernunft.* Riga: J.F. Hartknoch. [Reprinted in Immanuel Kant, *Werke*, ed. Wilhelm Weischedel, Vol. 4. 1956. Frankfurt: Suhrkamp].

Kant, Immanuel. 1786. *Metaphysische Anfangsgründe der Naturwissenschaft.* Riga: J. F. Hartknoch. [Reprinted in Immaniel Kant, *Gesammelte Schriften.* Akademie-Ausgabe. Vol. 4. 1968, 465–566. Berlin: de Gruyter], [Transl. 2004. Metaphysical foundations of natural science, transl. and ed. Michael Friedmann. Cambridge: University Press]

Kant, Immanuel. 1788. *Über den Gebrauch teleologischer Principien in der Philosophie.* [Reprinted in Immanuel Kant, *Abhandlungen nach 1781.* Kant, *Gesammelte Schriften*, Akademie-Ausgabe, Vol.8, 1968, 157–184. Berlin: de Gruyter].

Kelchner, Ernst. 1879. Johann Christian Ludwig Haken. In *Allgemeine Deutsche Biographie*, vol. 10: 396–7. Leipzig: Duncker & Humblot.

Kempelen, Wolfgang von. 1791. *Mechanismus der menschlichen Sprache.* Wien: J. B Degen.

Kennedy, Hubert C. 1974. Peano, Giuseppe. *Dictionary of scientific biography* 10: 441–444.

Kennedy, Hubert C. 1980. *Life and works of Giuseppe Peano.* Dordrecht: Reidel.

Kennedy, Hubert C. 2002. *Giuseppe Peano.* German transl. Ruth Amsler. San Francisco: Peremptory Publications, http://home.pacbell.net/hubertk/Peano-German.pdf. (accessed March 3, 2010).

Kern, Franz. 1875. *Ludwig Giesebrecht als Dichter, Gelehrter und Schulmann.* Stettin: Nahmer.

Kern, Franz. 1879. *Ludwig Theodor Giesebrecht.* In *Allgemeine Deutsche Biographie*, vol. 9: 159–69. Leipzig: Duncker & Humblot.

Kern, Reinold. 1914. *Ludwig Giesebrecht als Dichter.* Berlin: Weidmann.

Killing, Wilhelm. 1893–1898. *Einführung in die Grundlagen der Geometrie.* Bd. 1,2. Paderborn: Schöningh.

Kircher, Ernst, Raimund Girwidz and Peter Häußler. 2007. *Physikdidaktik – Theorie und Praxis.* Berlin: Springer.

Kitcher, Philip. 1975. Bolzano's ideal of algebraic analysis. *Studies in History and Philosophy of Science* 6: 229–269.

Klein, Felix. 1872. *Vergleichende Betrachtungenüber neuere geometrische Forschungen.* Erlangen: Andreas Deichert.

Klein, Felix. 1895. Über Arithmetisierung der Mathematik. *Zeitschrift für den mathematischen und naturwissenschaftlichen Unterricht* 27: 143–149.

Klein, Felix. 1926. *Vorlesungen über die Entwicklung der Mathematik im 19. Jahrhundert.* Vol. 1 eds. R. Courant and O. Neugebauer: Berlin: Springer. [Reprinted 1956. New York: Chelsea Publishing].

Klein, Felix. 1939. *Elementary mathematics from an advanced standpoint. Part II: Geometry.* English translation of the 1925 edition. New York: Dover.

Klein, Felix. 1979. *Vorlesungen über die Entwicklung der Mathematik im 19. Jahrhundert.* Two volumes reprinted as one. Foreword by Richard Courant and Otto Neugebauer. Berlin etc.: Springer.

Koch, Friedrich. 1821. Der Fürst und die Schule oder über die Verdienste Pommerscher Fürsten und die Schulen ihres Landes. Womit zu der Säcular-Feier der Vereinigung Pommerns mit dem Preussischen Staate, welche am 4. August Nachmittag um 3 Uhr mit einer öffentlichen Redeübung in dem großen Hörsale des Gimnasiums begangen werden soll alle Beschützer, Gönner und Freunde unserer Schulanstalt ehrerbietigst und ergebenst einlädt. In *Programm des Gymnasiums zu Stettin.* Stettin.

Koselleck, Reinhart. 1967. *Preußen zwischen Reform und Revolution: Allgemeines Landrecht, Verwaltung und soziale Bewegung von 1791 bis 1848.* Stuttgart: Klett.

Krantz, David H. 1975. Color measurement and color theory: I. Representation theorem for Grassmann structures. *Journal of Mathematical Psychology* 12: 203–303.

Kratzenstein, C. G. 1781. *Tentamen resolvendi problema ab academia scientiarum imperiali petropolitana...* Petropoli: Typis Academiae Scientiarum.

Kremer, Richard L.: Innovation through synthesis. Helmholtz and color research. In *Hermann von Helmholtz and the foundations of nineteenth-century science,* ed. David Cahan, 205–228. Berkeley/Los Angeles/London: University of California Press.

Krisch, Thomas. 1996. *Zur Genese und Funktion der altindischen Perfekta mit langem Reduplikationsvokal. Mit kommentierter Materialsammlung.* Innsbruck: Institut für Sprachwissenschaft.

Krisch, Thomas. 2006. *RIVELEX. Rigveda-Lexikon.* Band 1/Vol. 1. *Wörter beginnend mit "a"* [Words incipient with "a"]. Mitarbeiter/Collaborators: Christina Katsikadeli, Stefan Niederreiter, Thomas Kaltenbacher. Graz: Leykam.

Kroll, Frank-Lothar, ed. 1997. *Die Königin von Borneo: ein Roman/Friedrich Wilhelm IV.* 1817. Berlin: Nicolai.

Kryłowicz, Jerzy. 2004. *The young Kryłowicz.* A collection of articles translated by Axel Holvoet and ed. Wojciech Smoczyński. Kraków: Poligrafix. [Articles from 1927–1928].

Kryłowicz, Jerzy. 1935. *Études indoeuropéennes I.* Kraków: Polska Akademja Umiejętności.

Külpe, Oswald. 1904. Versuche über Abstraktion. *Bericht über den ersten Kongreß für experimentelle Psychologie* 56–68.

Külpe, Oswald. 1908. *Immanuel Kant. Darstellung und Würdigung.* (1st ed. 1907) Leipzig: Teubner.

Külpe, Oswald. 1910. *Erkenntnistheorie und Naturwissenschaft. Vortrag gehalten am 19. September 1910 auf der 82. Versammlung Deutscher Naturforscher und Ärzte in Königsberg.* Leipzig: Hirzel.

Külpe, Oswald. 1912–1923. *Die Realisierung. Ein Beitrag zur Grundlegung der Realwissenschaften.* 3 vols. Leipzig: Hirzel.

Kuntze, Friedrich. 1908. Die Bedeutung der Ausdehnungslehre Hermann Grassmanns für die Transzendentalphilosophie. *Bericht über den III. internationalen Kongress für Philosophie zu Heidelberg.* Theodor Elsenhans, 429–437. Heidelberg: Hörning.

Kusch, Martin. 1999. *Psychological knowledge. A social history and philosophy.* London: Routledge.

LA = Grassmann, Hermann. 1861. *Lehrbuch der Mathematik für höhere Lehranstalten. Erster Teil: Arithmetik.* Berlin: Enslin.

Lagrange, Joseph-Louis. 1797. *Théorie des fonctions analytiques,* 1st ed. Imprimerie Impériale, Paris. [2nd ed. (1813) as *Oeuvres,* vol. 9].

Lapointe, Sandra. 2010. Bolzano's philosophy of mathematical knowledge. *Internet Encyclopedia of Philosophy* at http://www.iep.utm.edu/bol-math/#H7 (accessed 12th April 2010).
Lasenby, Anthony N., Chris J. L. Doran and Stephen F. Gull. 1998. Gravity, gauge theories and geometric algebra. *Philosophical Transactions of the Royal Society, London* A 356: 487–582.
Lawvere, F. William. 1996. Grassmann's dialectics and category theory. In [Schubring 1996a, 255–64].
Lehman, W. P. 1967. *A reader in nineteenth-century historical Indo-European linguistics*. Bloomington and London: Indiana University Press.
Leibniz, Gottfried Wilhelm. 1666a. Disputatio arithmetica de complexionibus. Lipsiae: literis Spörelianis.
Leibniz, Gottfried Wilhelm. 1666b. *Brief an Christian Daum, 26. März (5. April) 1666*. In [Leibniz, Gottfried Wilhelm 2006, 5–6].
Leibniz, Gottfried Wilhelm. 1666c. *Dissertatio de Arte Combinatoria*, Lipsiae: Fick et Seubold.
Leibniz, Gottfried Wilhelm. 1956. *The Leibniz-Clarke correspondence*. Ed. Henry Gavin Alexander. Manchester: Manchester Univ. Press.
Leibniz, Gottfried Wilhelm. 2006. *Sämtliche Schriften und Briefe*. 2. Reihe. 1. Bd. Berlin: Akademie.
Lemcke, Hugo. 1892. Zur Erinnerung an Ludwig Giesebrecht. *Monatsblätter der Gesellschaft für Pommersche Geschichte und Altertumskunde* 6: 113–14.
Lepenies, Wolf. 1978. *Das Ende der Naturgeschichte*. Frankfurt: Suhrkamp.
Lewis, Albert C. 1977. H. Grassmann's 1844 Ausdehnungslehre and Schleiermacher's Dialektik. *Annals of Science* 34: 103–162.
Lewis, Albert C. 1981. Justus Grassmann's school programs as antecedents of Hermann Grassmann's 1844 "Ausdehnungslehre". In [Jahnke and Otte 1981, 255–268].
Lewis, Albert C. 1996. The influence of Grassmann's theory of tides on the 'Ausdehnungslehre'. In [Schubring 1996a, 29–35].
Lewis, Albert C. 2004. The unity of logic, pedagogy and foundations in Grassmann's mathematical work. *History and Philosophy of Logic* 25: 15–36.
Li, Hongbo. 2008. *Invariant algebras and geometric reasoning*. Singapore: World Scientific.
Li, Hongbo and Yihong Wu. 2003. Automated short proof generation in projective geometry with Cayley and Bracket algebras I, II. *Journal of Symbolic Computation* 36 (5): 717–762, 763–809.
Lie, Sophus. 1871. *Over en classe geometriske transformationer*. (On a class of geometric transformations) Dissertation. Christiania (now Oslo) University.
Lindeman, Frederik Otto. 1965. La loi de Sievers et le début du mot en indo-européen. *Norsk Tidsskrift for Sprogvidenskap* 20: 38–108.
Loria, Gino. 1888. JFM 20.0689.04. Review of *Calcolo geometrico secondo l'Ausdehnungslehre di H. Grassmann preceduto dalle operazioni della logica deduttiva* by Giuseppe Peano. ZMATH – The database Zentralblatt MATH, http://www.zentralblatt-math.org/zmath/en/.
Lottner, Carl. 1862. Ausnahmen der ersten Lautverschiebung. *Z. f. Vergleichende Sprachforschung auf dem Gebiete des Deutschen, Griechischen und Lateinischen* 11 (3): 161–205.
Lotze, Hermann. 1852. *Medicinische Psychologie oder Psychologie der Seele*. Leipzig: Weidmann.
Lounesto, Pertti. 1997. *Clifford algebras and spinors*. Cambridge: Cambridge University Press.
Lubotsky, Alexander M. 1997. *A Rigvedic word concordance*. 2 vols. New Haven, Connecticut: American Oriental Society (= American Oriental Series Vol. 82–83).
Luciano, Erika and Clara Silvia Roero. 2008. Cronologia della vita e delle opere di Giuseppe Peano. http://www.peano2008.unito.it/crono.php.
Ludwig, Alfred, transl. 1876–1888. *Der Rigveda oder Hymnen der Brâhmana. Zum ersten Male vollständig ins Deutsche übersetzt mit Commentar und Einleitung*. 6 vols. Prag: Tempsky.
Luhmann, Niklas. 2002a. *Die Wissenschaft der Gesellschaft*. Frankfurt a.M.: Suhrkamp.
Luhmann, Niklas, ed. 2002b. *Theories of distinction: redescribing the descriptions of modernity*. Stanford, Calif.: Stanford Univ. Press.
Lützen, Jesper, ed. 2001. *Around Caspar Wessel and the geometric representation of complex numbers*. Copenhagen: C. A. Reitzels Forlag.

Mädler, Inken. 1997. *Kirche und bildende Kunst der Moderne: ein an F. D. E. Schleiermacher orientierter Beitrag zur theologischen Urteilsbildung.* Tübingen: Mohr Siebeck.

Mancosu, Paolo. 1999. Bolzano and Cournot on mathematical explanation. *Revue d'histoire des sciences* 52: 429–455.

MAP 2009. The homepage of Maple. http://www.maplesoft.com/products/maple. (accessed March 3, 2010).

Marchuk, Nikolay Gurjevitch. 2009. *Uravneniya teorii polya i algebry Klifforda.* [Field theory equations and Clifford algebras]. Moscow: RCD.

Marcolongo, Roberto. 1905. *Meccanica razionale* (2 vol.). Milano: Manuali Hoepli.

Marcolongo, Roberto. 1921. *Relatività.* Messina: Casa Editrice Giuseppe Principato.

Marx, Carl Michael. 1825. *Geschichte der Crystallkunde.* Carlsruhe und Baden: D. R. Marx'sche Buchhandlung.

Mausfeld, Rainer. 1998. Color perception: From Grassmann codes to a dual code for object and illumination colors. In *Color vision. Perspectives from different disciplines*, eds. Werner G.K. Backhaus, Reinhold Kliegl and John S. Werner, 219–250. Berlin and New York: De Gruyter.

Mayrhofer, Manfred. 1956–1980. *Kurzgefaßtes etymologisches Wörterbuch des Altindischen: a concise etymological Sanskrit Dictionary.* Vol. 1–4. Heidelberg: Winter.

Mayrhofer, Manfred. 1981. *Nach hundert Jahren. Ferdinand de Saussures Frühwerk und seine Rezeption durch die heutige Indogermanistik. Mit einem Beitrag von Roland Zwanziger.* Heidelberg: Carl Winter.

Mayrhofer, Manfred. 1986. *Indogermanische Grammatik*, 1. vol., 2. part. Heidelberg: Carl Winter.

Mayrhofer, Manfred. 1986–2001. *Etymologisches Wörterbuch des Altindoarischen.* 3 vols. Heidelberg: Carl Winter. [EWAia].

Mayrhofer, Manfred. 2005. *Die Fortsetzung der indogermanischen Laryngale im Indo-Iranischen.* Wien: Verlag der Österreichischen Akademie der Wissenschaften (= Philosophisch-Historische Klasse, Sitzungsberichte 730. Band).

Mayrhofer, Manfred. EWAia. See [Mayrhofer 1986–2001].

Mazzola, Guerino. 1985. *Gruppen und Kategorien in der Musik.* Berlin: Heldermann.

Mazzola, Guerino. 2002. *The topos of music.* Basel: Birkhäuser.

Medem, Friedrich Ludwig Baron von. 1835. Das königliche Provinzial — Archiv zu Stettin. *Zeitschrift für Archivkunde, Diplomatik und Geschichte* 2: 1–92.

Mehmke, Rudolf. 1880. *Anwendung der Grassmann'schen Ausdehnungslehre auf die Geometrie der Kreise in der Ebene.* Stuttgart: Greiner & Pfeiffer.

Mehmke, Rudolf. 1904. Vergleich zwischen der Vektoranalysis amerikanischer Richtung und derjenigen deutsch italienischer Richtung. *Jahresbericht der Deutschen Mathematiker-Vereinigung* 13: 217–228.

Mehmke, Rudolf. 1913. *Vorlesungen über Punkt- und Vektorenrechnung in zwei Bänden.* Erster Band: Punktrechnung. Leipzig und Berlin: Teubner.

Mehrtens, Herbert. 1979. *Die Entstehung der Verbandstheorie.* Hildesheim: Gerstenberg (*arbor scientiarum*; A.VI).

Meinong, Alexius. 1903. Bemerkungen über den Farbenkörper und das Mischungsgesetz. In *Gesamtausgabe*, vol. 1, eds. Rudolf Haller and Rudolf Kindinger, 1969, 495–576. Graz: Akademische Druck- und Verlagsanstalt.

Meinong, Alexius. 1904. Über Gegenstandstheorie. In *Gesamtausgabe*, vol. 2, ed. Rudolf Haller, 1971, 481–535. Graz: Akademische Druck- und Verlagsanstalt.

Meyer, Donald Harvey. 1962. Paul Carus and the religion of science. *American Quarterly* 14: 597–607.

Miller, William H. 1868. *On the crystallographic method of Grassmann, and its employment in the investigation of the general geometric properties of crystals.* Cambridge: University Press.

Milne, E. A. 1948. *Vectorial mechanics.* London: Methuen.

Milnor, John W. and James D. Stasheff. 1974. *Characteristic classes.* Princeton: University Press.

Minkowski, Hermann. 1909. *Raum und Zeit.* Vortrag gehalten auf der 80. Naturforscher-Versammlung zu Köln am 21. September 1908. (Sonderabdruck). Leipzig: Teubner.

Misner, Charles W., Kip S. Thorne and John Archibald Wheeler. 1995. *Gravitation*. New York: W. H. Freeman.

Möbius, August Ferdinand. 1827. *Der barycentrisch Calcul, ein neues Hülfsmittel zur analytischen Behandlung der Geometrie etc.* Leipzig: Barth.

Möbius, August Ferdinand. 1885. *Gesammelte Werke*. Eds. Felix Klein and Wilhelm Scheibner. Vol. 1. Ed. Richard Baltzer. Leipzig: Hirzel.

Moigno, Abbé Francois Napoleon Marie. 1840. *Leçons de calcul différentiel et de calcul intégral,* vol. 1, Paris: Bachelier.

Monna, Antonie Frans. 1973. Hermann Hankel. *Nieuw Archief voor Wiskunde* 21 (3): 64–87.

Mourrain, Bernard. 1991. *Approche effective de la théorie des invariants des groupes classiques*. Dissertation. Paris: École Polytechnique.

Müller, Gerhard. 1934. Von Brüggemann zu Berghaus. *Monatsblätter der Gesellschaft für Pommersche Geschichte und Altertumskunde* 48: 17–24.

Narten, Johanna. 1960. Das vedische Verbum *math*. *Indo-Iranian Journal* 4: 121–135 (= *Kleine Schriften* I, eds. Marcos Albino and Matthias Fritz, Wiesbaden: Reichert 1995, 11–25).

Niedzielska, Magdalena. 1997. 1997th Tożsamość elit politycznych w Prusach Wschodnich w pierwszej połowie XIX w. [The identity of political elites in East Prussia of the first half of the nineteenth century]. *Zapiski historyczne*, vol. 62, no. 1: 53–80.

Nooten, Barend A. van and Gary B. Holland. 1994. *Rig Veda. A metrically restored text with an introduction and notes*. Cambridge, Mass. and London: Harvard University Press.

NVIDIA. 2009. The CUDA home page. http://www.nvidia.com/object/cuda home.html. (accessed March 3, 2010).

Ohala, John J. 1993. The phonetics of sound change. In *Historical linguistics: Problems and perspectives*, ed. Charles Jones. 237–278. London: Longman.

Oldenberg, Hermann, 1888. *Die Hymnen des Ṛigveda*. Vol. 1. *Metrische und textgeschichtliche Prolegomena*. Berlin: Wilhelm Hertz.

Oldenberg, Hermann, 1909. *Ṛgveda. Textkritische und exegetische Noten*. 1.–6. book. Berlin: Weidmann.

Oldenberg, Hermann, 1912. *Ṛgveda. Textkritische und exegetische Noten*. 7.–10. book. Berlin: Weidmann.

Otte, Michael. 1989. The ideas of Hermann Grassmann in the context of the mathematical and philosophical tradition since Leibniz. *Historia Mathematica* 16: 1–35.

Otte, Michael. 1992. Gegenstand und Methode in der Geschichte der Mathematik. *Philosophia Naturalis* 29: 31–68.

Otte, Michael. 1997. Analysis and synthesis in mathematics from the perspective of Peirce's philosophy. In *Analysis and synthesis in mathematics*, eds. Michael Otte and Marco Panza, 327–364. Dordrecht: Kluwer.

Pais, Abraham. 2002. *Inward bound. Of matter and forces in the physical world.* (Reprint with corrections). New York: Clarendon Press/Oxford & Oxford University Press.

Parra Serra, Josep Manel and Emma Sallent del Colombo. 2002. Covariància i Invariància a l'Espaces Courbes de Cesare Burali-Forti i Tommaso Boggio. In *Actes de la VI Trobada d'Història de la Ciència i de la Tècnica*, eds. Josep Batlló Ortiz, Pasqual Beernat López and Roser Puig Aguilar. 437–442. Barcelona: SCHCT.

Pashler, Hal and Steven Yantis, eds. 2002. *Steven's handbook of experimental psychology*. Vol. 1: *Sensation and perception*. 3rd ed. New York: Wiley & Sons.

Pauli, Wolfgang. 1927. Zur Quantenmechanik des magnetischen Elektrons. *Zeitschrift für Physik* 43 (9/10): 601–623.

Peano, Giuseppe. 1887. *Applicazioni geometriche del calcolo infinitesimale*. Torino: Fratelli Bocca.

Peano, Giuseppe. 1888. *Calcolo geometrico secondo l'Ausdehnungslehre di H. Grassmann preceduto dalle operazioni della logica deduttiva*. Torino: Fratelli Bocca.

Peano, Giuseppe. 1891a. *Gli elementi di calcolo geometrico*. Torino: Candeletti.

Peano, Giuseppe. 1891b. *Die Grundzüge des geometrischen Calculs von G. Peano*. Transl. Adolf Schepp. Leipzig: Teubner.

Peano, Giuseppe. 1894. Hermann Grassmann's Gesammelte mathematische und physikalische Werke, vol. I, part I. *Rivista di Matematica* 4: 167–169.

Peano, Giuseppe. 1896. Saggio di calcolo geometrico. *Atti della Reale Accademia delle Scienze di Torino* 31: 952–975.

Peano, Giuseppe. 1897a. *Zarys Rachunku geometrycznego*. Transl. Samuel Dickstein. Warszawa: Prae matematyczno-fizycnych.

Peano, Giuseppe. 1897b. *Entwicklung der Grundbegriffe des geometrischen Calculs*, Transl. Alois Lanner. Salzburg: Zaunrithsche.

Peano, Giuseppe. 1898. Analisi della teoria dei vettori. *Atti della Reale Accademia delle Scienze di Torino* 33: 513–534.

Peano, Giuseppe. 2000. *Geometric calculus according to the Ausdehnungslehre of H. Grassmann*. Transl. Lloyd C. Kannenberg. Boston: Birkhäuser.

Peano, Giuseppe. 2010. *Calcolo geometrico secondo l'Ausdehnungslehre di H. Grassmann preceduto dalle operazioni della logica deduttiva*. Ed. Filippo Demonte-Barbera. Torino: Aragno.

Peckhaus, Volker. 1996. The influence of Hermann Günther Grassmann and Robert Grassmann on Ernst Schröder's algebra of logic. In [Schubring 1996a, 217–227].

Peckhaus, Volker. 1997. *Logik, Mathesis universalis und allgemeine Wissenschaft. Leibniz und die Wiederentdeckung der formalen Logik im 19. Jahrhundert*. Berlin: Akademie.

Peckhaus, Volker. 2004. Schröder's logic. In *Handbook of the history of logic*. Vol. 3: *The rise of modern logic: From Leibniz to Frege*. Eds. Dov M. Gabbay and John Woods. Amsterdam etc.: Elsevier North Holland, 557–609.

Peirce, Charles Sanders. 1877. Note on Grassmann's calculus of extension. *Proceedings of the American Academy of Arts and Sciences* 13: 115–116.

Penrose, Roger. 2005. *The road to reality. A complete guide to the laws of the universe*. London: Vintage Books.

Perwass, Christian. 2001. *CLUCalc – a visual calculator*. http://www.clucalc.info/. (accessed March 3, 2010).

Perwass, Christian. 2005. The CLU home page. http://www.clucalc.info. (accessed March 3, 2010).

Perwass, Christian and Eckhard Hitzer. 2005a. Space group visualizer. http://www.spacegroup.info/. (accessed March 3, 2010).

Perwass, Christian and Eckhard Hitzer. 2005b. Interactive visualization of full geometric description of crystal space groups. In *Proceedings of the international symposium on advanced mechanical engineering*. 276–282. Busan: Pukyong National University.

Petrich, Hermann. 1887. Johann August Sack. In *Pommersche Lebens- und Landesbilder*, part 2.2, ed. Hermann Petrich, 257–339. Stettin: Saunier.

Petrich, Hermann. 1890. Johann August Sack. In *Allgemeine Deutsche Biographie*, vol. 30, 152–153. Leipzig: Duncker & Humblot.

Petsche, Hans-Joachim. 1979. *Leben und Wirken Hermann Günther Graßmanns*. Dissertation. Potsdam: Päd. Hochschule Potsdam.

Petsche, Hans-Joachim. 2006. *Graßmann*. Basel: Birkhäuser. (*Vita Mathematica*; 13).

Petsche, Hans-Joachim. 2009. *Hermann Graßmann. Biography*. Basel: Birkhäuser.

Petsche, Hans-Joachim. 2010. It began with Pestalozzi and Schleiermacher – Reflections on the polymath Hermann Graßmann (1809–1877). *EMS Newsletter*, 76 (June 2010): 39–46.

Petsche, Hans-Joachim, Lloyd Kannenberg, Gottfried Keßler and Jolanta Liskowacka, eds. 2009. *Hermann Graßmann – roots and traces. Autographs and unknown documents*. Basel: Birkhäuser.

Petsche, Hans-Joachim and Peter Lenke. 2010. *International Grassmann Conference. Hermann Grassmann Bicentennial: Potsdam and Szczecin, 16 - 19 September 2009. Video Recording of the Conference*. Potsdam: Universitätsverlag (ISBN 978-3-86956-093-9).

Pfalzgraf, Jochen. 1985a. On a model for noncommutative geometric spaces. *Journal of Geometry* 25: 147–163.

Pfalzgraf, Jochen. 1985b. *Sobre la existencia de espacios semiafines y finitos*. Archivos de Investigación, 3 (1), (Proceedings). Chillán: Instituto Profesional de Chillán.

Pfalzgraf, Jochen. 1986. On geometries associated with group operations. *Geometriae Dedicata* 21: 193–203.

Pfalzgraf, Jochen. 1987a. A note on simplices as geometric configurations. *Archiv der Mathematik* 49: 134–140.

Pfalzgraf, Jochen. 1987b. Representation of geometric spaces as fibered structures (in German). *Results in Math.* 12: 172–190.

Pfalzgraf, Jochen. 1991. Logical fiberings and polycontextural systems. In *Fundamentals of artificial intelligence research*, eds. Philippe Jorrand and Jozef Kelemen. 170–184. Berlin: Springer. (FAIR'91).

Pfalzgraf, Jochen. 1996. An application of Grassmann geometry to a problem of robotics. In [Schubring 1996a, 337–343].

Pfalzgraf, Jochen. 1998. On a category of geometric spaces and geometries induced by group actions. *Ukrainian Journal of Physics* 43 (7): 847–856.

Pfalzgraf, Jochen. 2004. On logical fiberings and automated deduction in many-valued logics using Gröbner bases. *RACSAM, Revista de la Real Academia de Ciencias Exactas, Serie A, Matematicas* 98 (1): 213–227.

Plummer, L. Gordon. 1970. *The Mathematics of the cosmic mind.* 2nd revised ed. Wheaton, Ill.: The Theosophical Publishing House.

Polak, Lev Solomonovich. 1993. *Uilyam Gamilton. 1805–1865.* (William Hamilton, 1805–1865) Moskva: Nauka.

Polaschegg, Andrea. 2005. *Der andere Orientalismus. Regeln deutsch-morgenländischer Imaginationen im 19. Jahrhundert.* Berlin u.a.: de Gruyter.

Poretzky, Platon Sergejewitsch. 1881. *Izlozhenie osnovnykh nachal matematicheskoy logiki v vozmozhno bolee naglyadnoy i obshchedostupnoy forme.* [Explanation of the foundations of mathematical logic in an as clear and understandable form as possible]. Sobranie protokolov zasedaniy sekciy fiziko-metematicheskikh nauk obshchestva estestvoispytateley pri Kazanskom universitete. Vol. 1. 2–31. Kazan.

Poretzky, Platon Sergejewitsch. 1884. *O sposobakh resheniya logicheskikh ravenstv i ob obratnom sposobe matematicheskoy logiki.* [On methods of solving logical equalities and on the reverse method of mathematical logic]. Sobranie protokolov zasedaniy sekciy fiziko-metematicheskikh nauk obshchestva estestvoispytateley pri Kazanskom universitete. Vol. 2. Kazan.

Prandtl, Ludwig. 1904. Über eine einheitliche Bezeichnungsweise der Vektorenrechnung im technischen und physikalischen Unterricht. *Jahresbericht der Deutschen Mathematiker-Vereinigung* 13: 36–40.

Preyer. Wilhelm Thiery. 1876a. *Über die Aufgabe der Naturwissenschaft. Ein Vortrag.* Jena: Dufft.

Preyer. Wilhelm Thiery. 1876b. *Die Grenzen der Tonwahrnehmung.* Jena: Dufft.

Preyer. Wilhelm Thiery. 1877. Elemente der reinen Empfindungslehre. In *Sammlung physiologischer Abhandlungen*, ed. Wilhelm Thiery Preyer, vol. 1, 537–637. Jena: Dufft.

Preyer. Wilhelm Thiery. 1989. *Die Seele des Kindes*, ed. Georg Eckardt. Berlin et al.: Springer.

Radu, Mircea. 2000a. Justus Grassmann's contributions to the foundations of mathematics – mathematical and philosophical aspects. *Historia Mathematica* 27: 4–35.

Radu, Mircea. 2000b. *Nineteenth century contributions to the axiomatization of arithmetic – a historical reconstruction and comparison of the mathematical and philosophical ideas of Justus Grassmann, Hermann and Robert Grassmann, and Otto Hölder.* Ph.D. Thesis. University Bielefeld. Germany.

Radu, Mircea. 2005. Did Hermann and Robert Grassmann contribute to the emergence of formal axiomatics? *Activity and sign: Grounding mathematics education*, eds. Michael H. G. Hoffmann, Johannes Lenhard and Falk Seeger, 263–274. New York: Springer.

Rajagopalan, Mohan and Anwar Ghuloum. 2007. Ct: C for throughput computing home page. Intel White Paper. http://techresearch.intel.com/articles/Tera-Scale/1514.htm. (accessed March 3, 2010).

Randt, Erich. 1934. *Robert Klempin*. In *Pommersche Lebensbilder*, vol. 1, eds. Adolf Hofmeister, Erich Randt and Martin Wehrmann, 175–189. Stettin: Saunier.

Rask, Rasmus Kristian. 1811. *Bejledning til det Islandiske eller gamle Nordiske Sprog.* Kjøbenhavn: Schubothes Forlag.

Rask, Rasmus Kristian. 1818. *Undersögelse om det gamle Nordiske eller Islandske Sprogs Oprindelse.* Kjöbenhavn: Gyldendalske Boghandlings Forlag.

Rasmussen, Jens Elmegård. 1985. Der Prospektiv – eine verkannte indogermanische Verbalkategorie? In *Grammatische Kategorien. Funktion und Geschichte. Akten der VII. Fachtagung der Indogermanischen Gesellschaft. Berlin 20.–25. Februar 1983*, ed. Bernfried Schlerath assisted by Veronica Rittner. 384–399. Wiesbaden: Reichert Verlag.

Raumer, Rudolf von. 1837. *Die Aspiration und die Lautverschiebung. Eine sprachgeschichtliche Untersuchung.* Leipzig: Brockhaus [Reprinted in Rudolf von Raumer. 1863. *Gesammelte sprachwissenschaftliche Schriften*. 1–21. Frankfurt a. M. & Erlangen: von Heyder & Zimmer].

Reich, Karin. 1993. Der Mathematiker Rudolf Mehmke: Bausteine zu Leben und Werk. In *Naturwissenschaft und Technik in der Geschichte.* 25 Jahre Lehrstuhl für Geschichte der Naturwissenschaften und Technik am Historischen Institut der Universität Stuttgart, ed. Helmuth Albrecht. 263–285. Stuttgart: Verl. für Geschichte der Naturwiss. und der Technik.

Reich, Karin. 1994. *Die Entwicklung des Tensorkalküls. Vom absoluten Differentialkalkül zur Relativitätstheorie.* Basel et al.: Birkhäuser.

Reich, Karin. 1996a. Die Rolle Arnold Sommerfelds bei der Diskussion um die Vektorrechnung, dargestellt anhand der Quellen im Nachlaß des Mathematikers Rudolf Mehmke. In *History of Mathematics. States of the Art*, eds. Joseph W. Dauben, Menso Folkerts, Eberhard Knobloch and Hans Wußing. 319–341. San Diego: Academic Press.

Reich, Karin. 1996b. The Emergence of vector calculus in physics: the early decades. In [Schubring 1996a, 197–210].

Reich, Karin. 2004. Der Briefwechsel Rudolf Mehmkes. In *Mathematik im Fluss der Zeit. Tagung zur Geschichte der Mathematik in Attendorn/Neu-Listernohl*, eds. Wolfgang Hein and Peter Ullrich. 398–406. Augsburg: Rauner.

Renou, Louis. 1955–1969. *Études Védiques et Paninéennes.* (Tomes I–XVII) Paris: Éditions E. de Bocard.

Reti, Rudolph. 1951. *The thematic process in music.* New York: MacMillan. [2nd ed. 1978].

Richardson, Alan W. 1992. Logical idealism and Carnap's construction of the world. *Synthese* 93: 59–92.

Riche, Jacques. 2007. From universal algebra to universal logic. In *Perspectives on universal logic*, eds. Jean-Yves Beziau and Alexandre Costa-Leite, 3–39. Monza: Polimetrica.

Richter-Gebert, Jürgen. 1995. Mechanical theorem proving in projective geometry. *Annals of Math. and Artificial Intelligence* 13: 159–171.

Riemann, Bernhard. 1854. *Ueber die Hypothesen, welche der Geometrie zu Grunde liegen.* (Habilitationsschrift). [Reprinted 1876 in *Bernhard Riemann's gesammelte mathematische Werke und wissenschaftlicher Nachlaß*. Ed. Heinrich Weber with assistance from Richard Dedekind. 254–269. Leipzig: Teubner].

Riemann, Bernhard. 1873. On the hypotheses which lie at the base of geometry, transl. William Kingdom Clifford, *Nature* 8: 14–17, 36–37.

Riesz, Marcel. 1958. See [Riesz 1993].

Riesz, Marcel. 1993. *Clifford numbers and spinors.* Dordrecht and Boston: Kluwer. (Originally distributed as lecture notes at the University of Maryland, 1958.).

Ringel, Claus Michael. 2008. Whitehead's theory of extension in 'Process and Reality'. *Analysis and Metaphysics* 7: 36–63.

RIVELEX = Krisch, Thomas. 2006. *RIVELEX. Rigveda-Lexikon* Band 1/Vol. 1. *Wörter beginnend mit "a"* [Words incipient with "a"]. Mitarbeiter/Collaborators: Christina Katrsikadeli, Stefan Niederreiter, Thomas Kaltenbacher. Graz: Leykam.

Rix, Helmut, Martin Kümmel, Thomas Zehnder, Reiner Lipp and Brigitte Schirmer. 2001. *Lexikon der indogermanischen Verben*. 2nd edition. Wiesbaden: Reichert.

Roero, Clara Silvia, ed. 2002. *L'Opera Omnia di Giuseppe Peano.* Torino: Università.

Rota, Gian-Carlo and Joel Stein. 1973. Applications of Cayley algebras. In *Colloquio Internazionale sulle Teorie Combinatoire*, vol. 2. 71–97. Roma: Accad. Naz. Lincei.

Rowe, David E. 1996. On the reception of Grassmann's work in Germany during the 1870's. In [Schubring 1996a, 131–145].

Runze, Maximilian. 1894. *Ludwig Giesebrecht und Carl Loewe*. Berlin: Duncker.

Runze, Maximilian. 1899. *Notizen zu den bisher unveröffentlichten und vergessenen Liedern, Gesängen, Romanzen und Balladen*. In *Carl Loewes Werke. Gesamtausgabe der Balladen, Legenden, Lieder und Gesänge für eine Singstimme, im Auftrag der Loeweschen Familie,* ed. Maximilian Runze, vol. 2. VI–XVIII. Leipzig: Breitkopf and Härtel.

Rusnock, Paul. 2000. *Bolzano's philosophy and the emergence of modern mathematics*. Amsterdam: Rodopi.

Russ, Steve. 2004. *The mathematical works of Bernard Bolzano.* Oxford: Oxford University Press. [Reprinted with corrections 2006].

Russell, Bertrand. 1983. *Cambridge essays, 1888–99*. London; Boston: G. Allen & Unwin.

Russell, Bertrand. 1990. *Philosophical papers, 1896–99*. London; Boston: Unwin Hyman.

RV1 = Grassmann, Hermann. 1876. *Rig-Veda. Übersetzt und mit kritischen und erläuternden Anmerkungen versehen von Hermann Grassmann. Erster Theil. Die Familien-Bücher des Rig-Veda: (zweites bis achtes Buch).* Leipzig: F. A. Brockhaus.

RV2 = Grassmann, Hermann. 1877a. *Rig-Veda. Übersetzt und mit kritischen und erläuternden Anmerkungen versehen von Hermann Grassmann. Zweiter Theil. Sammelbücher des Rig-Veda: (erstes, neuntes, zehntes Buch).* Leipzig: Brockhaus.

Salmonowicz, Stanisław. 1987. *Prusy: Dzieje państwa i społeczeństwa.* [Prussia. History of State and Society]. Poznań: Wydawnictwo Poznańskie.

Salmonowicz, Stanisław. 1993. Podziały terytorialne i ustrój ziem pomorskich w latach 1815–1850. [Territorial division and the system of Pommeranian areas in the year 1815–1850]. In *Historia Pomorza*, vol. 3 (1815–1850), part 1, ed. Gerard Labuda, 55–142. Poznań: Poznańskie TPN.

Saussure, Ferdinand de. 1879. *Mémoire sur le système primitif des voyelles dans les langues indo-européennes*. Leipzig: Teubner.

Schelling, Friedrich Wilhelm Josef. 1799. *Einleitung zu seinem Entwurf eines Systems der Naturphilosophie*. [Reprinted 1988. Stuttgart: Reclam].

Schelling, Friedrich Wilhelm Josef. 1802/03. Über die Konstruktion in der Philosophie. In *Kritisches Journal der Philosophie*, eds. Friedrich Wilhelm Josef Schelling and Georg Wilhelm Friedrich Hegel. [Reprinted 1985. Berlin: Das Europäische Buch. 201–224].

Schindler, Jochem. 1977. Notizen zum Sieversschen Gesetz. *Die Sprache* 23: 56–65.

Schlegel, Victor. 1872. *System der Raumlehre. Nach den Prinzipen der Grassmann'schen Ausdehnungslehre und als Einleitung in dieselbe*. Erster Theil. Leipzig: Teubner.

Schlegel, Victor. 1878. *Hermann Grassmann: Sein Leben und seine Werke*. Leipzig: Brockhaus.

Schleiermacher, Friedrich. 1803. *Grundlinien einer Kritik der bisherigen Sittenlehre*. Berlin: Realschulbuchhandlung. [Reprinted 1834. Berlin: Reimer].

Schleiermacher, Friedrich. 1839. *Dialektik. Aus Schleiermacher's handschriftlichen Nachlasse herausgeg. von L. Jonas*. In Schleiermacher, Friedrich. *Sämmtliche Werke*. 3rd secton. Vol. 4.2. Berlin: Reimer.

Schleiermacher, Friedrich. 1913. *Über die Religion. Reden an die Gebildeten unter ihren Verächtern (1799)*. In ursprüngl. Gestalt neu herausgeg. von Rudolf Otto. Göttingen: Vandenhoeck & Ruprecht. [Transl. *On Religion: Speeches to its Cultured Despisers*].

Schleiermacher, Friedrich. 1984. *Der christliche Glaube (1830/31)*, ed. Martin Redeker. Berlin: de Gruyter. [English translation 1928: *The Christian Faith*, ed. Hugh Ross Mackintosh and James Stuart Stewart. Edinburgh: T. & T. Clark].

Schmid, Wolfgang P. 1956. Vedisch *īmahe* und Verwandtes. *Indogermanische Forschungen* 62: 219–239.

Schmitt, Rüdiger. 2009. *Aus Karl Brugmanns Jugenderinnerungen*. Wien: Österr. Akad. der Wiss.

Scholz, Erhard. 1989. *Symmetrie – Gruppe – Dualität. Zur Beziehung zwischen theoretischer Mathematik und Anwendungen in Kristallographie und Baustatik des 19. Jahrhunderts*. Berlin: Deutscher Verlag der Wissenschaften.

Scholz, Erhard. 1994. Schelling und die dynamistische Kristallographie. In *Selbstorganisation. Jahrbuch für Komplexität in den Natur-, Sozial-, und Geisteswissenschaften*. Vol. 5, eds. Marie-Luise Heuser and Wilhelm G. Jacobs, 219–230. Berlin: Duncker & Humblot.

Scholz, Erhard. 1996. The influence of Justus Grassmann's crystallographic works on Hermann Grassmann. In [Schubring 1996a, 37–45].

Schreiber, Peter, ed. 1995. *Hermann Graßmann – Werk und Wirkung. Internationale Fachtagung anläßlich des 150. Jahrestages des ersten Erscheinens der "linealen Ausdehnungslehre" (Lieschow/Rügen, 23.–28.5.1994)*. Greifswald: Ernst-Moritz-Arndt-Universität.

Schröder, Ernst. 1873. *Lehrbuch der Arithmetik und Algebra für Lehrer und Studirende*. Vol. 1 [no further volumes published]: *Die sieben algebraischen Operationen*. Leipzig: B.G. Teubner.

Schröder, Ernst. 1874. *Über die formalen Elemente der absoluten Algebra. Beilage zum Programm des Pro- und Real-Gymnasiums in Baden-Baden für 1873/74*. Stuttgart: Schweizerbart'sche Buchhandlung.

Schröder, Ernst. 1877. *Der Operationskreis des Logikkalkuls*. Leipzig: Teubner. (Reprinted as "Special issue", Sonderausgabe Darmstadt: Wissenschaftliche Buchgesellschaft).

Schröder, Ernst. 1895. *Vorlesungen über die Algebra der Logik (exakte Logik)*. Vol. 3, pt. 1: *Algebra und Logik der Relative*. Leipzig: Teubner.

Schubring, Gert, ed. 1996a. *Hermann Günther Graßmann (1809–1877): Visionary mathematician, scientist and neohumanist scholar. Papers from a sesquicentennial conference*. Dordrecht, Boston and London: Kluwer.

Schubring, Gert, ed. 1996b. Introduction – Reflections on the complex history of Grassmann's reception. In [Schubring 1996a, ix–xxix].

Schubring, Gert, ed. 1996c. The cooperation between Hermann and Robert Grassmann on the foundations of mathematics. In [Schubring 1996a, 59–70].

Schubring, Gert, ed. 2008. Petsche, H.-J.: *Graßmann*. Basel 2006 (Review). *The Mathematical Intelligencer* 30 (4): 62–63.

Schubring, Gert, ed. 2009. Hermann Günther Graßmann (1809–1877). Ein vielseitiger Innovator. *Mitteilungen der Deutschen Mathematiker-Vereinigung* 17: 177–185.

Schuhmann, Karl. 1977. *Husserl-Chronik. Denk- und Lebensweg Edmund Husserls*. The Hague: Martinus Nijhoff.

Schulze, Friedrich, ed. 1906. *Briefe eines alten Schulmannes*. Leipzig: Voigtländer.

Schwartze, Heinz. 1996. On Hermann Grassmann's life and his work as a mathematics teacher. In [Schubring 1996a, 7–18].

Schweitzer, Arthur Richard. 1915. *Les idées directrices de la logique génétique des mathématiques*. Dissertation. Chicago: University of Chicago.

Sebestik, Jan. 1992. Logique et Mathématique chez Bernard Bolzano. Paris: Vrin.

Semler, Christian August. 1811. *Versuch über die combinatorische Methode, ein Beytrag zur angewandten Logik und allgemeinen Methodik*. Dresden: Walther.

Sheremetyevskiy, Vladimir Petrovitch. 1940. *Ocherki po istorii matematiki*. [Outline on history of mathematics]. Moskva: Uchpedgiz.

Silberstein, Ludwig. 1912. Quaternionic form of relativity. *The London, Edinburgh, and Dublin Philosophical Magazine and Journal of Science*, 6th series, 23: 790–809.

Silberstein, Ludwig. 1913. Second memoir on quaternionic relativity. *The London, Edinburgh, and Dublin Philosophical Magazine and Journal of Science*, 6th series, 25:135–144.

Skidelsky, Edward. 2008. *Ernst Cassirer: the last philosopher of culture*. Princeton: University Press.

Skolem, Thoralf. 1923. Begründung der elementaren Arithmetik durch die rekurrierende Denkweise ohne Anwendung scheinbarer Veränderlichen mit unendlichem Ausdehnungsbereich. *Skrifter utgiv av Videnskapsselskapet i Kristiania, I. Matematisk-naturvidenskabelig klasse*, no. 6: 1–38 [Reprinted in

English in *From Frege to Gödel*, ed. van Heijenoort. Cambridge, Mass.: Harvard University Press, 1967, 302–333].
Smirnov, Vladimir Alekseevich. 1962. Geneticheskiy metod postroeniya nauchnoy teorii. [Genetic method of construction of scientific theory]. In *Filosofskie voprosy sovremennoy formalnoy logiki*. 263–284. Moskva: Nauka.
Sommerfeld, Arnold. 1910a. Zur Relativitätstheorie. I. Vierdimensionale Vektoralgebra. *Annalen der Physik*, 4. Folge, 32: 749–776.
Sommerfeld, Arnold. 1910b. Zur Relativitätstheorie. II. Vierdimensionale Vektoralgebra. *Annalen der Physik*, 4. Folge, 33: 649–689.
Sossinsky, Alexei. 2002. *Knots – Mathematics with a twist*. Cambridge, Mass.: Harvard University Press.
Spranger, Eduard. 1928. *Wilhelm v. Humboldt und die Humanitätsidee*. Berlin: Reuther & Reichard.
Stang, Christian Schweigaard. 1965. Indo-européen * *Gʷ ŌM*, * *D(I)I̯ĒM*. *Symbolae linguisticae in honorem Georgii Kuryłowicz*. Wrocław: 292–296 (= *Opscula linguistica. Ausgewählte Aufsätze und Abhandlungen*. Oslo: Universitetsforlaget 1970: 40–44).
Steffens, Wilhelm, ed. 1931a. *Briefwechsel Sacks mit Stein und Gneisenau*. Stettin: Saunier.
Steffens, Wilhelm, ed. 1931b. Johann August Sack. *Pommersche Heimatpflege* 2, no. 2: 95–102.
Steffens, Wilhelm, ed. 1932. Johann August Sack und Gottlieb Mohnicke. *Monatsblätter der Gesellschaft für Pommersche Geschichte und Altertumskunde* 46: 17–23.
Steffens, Wilhelm, ed. 1938a. Aus dem Briefwechsel Wilhelm Meinhold mit dem Oberpräsidenten JA Sack. *Monatsblätter der Gesellschaft für Pommersche Geschichte und Altertumskunde* 52: 45–54.
Steffens, Wilhelm, ed. 1938b. Die Geschichte der "Pommerschen Provinzialblätter" und die Entstehung der "Baltischen Studien", *Baltische Studien*, NF 40: 232–259.
Steinitz, Ernst. 1910. Algebraische Theorie der Körper. *Journal für die reine und angewandte Mathematik (Crelle's Journal)* 137: 167–309.
Stępiński, Włodzimierz. 1983. Pruska biurokracja prowincjalna wobec gospodarczej działąlności Szczecina w latach 1815–1870. [The Prussian provincial bureaucracy in relation to the economic activity of Stettin 1815–1870]. *Zeszyty Historyczne* no. 1–2: 131–163.
Stępiński, Włodzimierz. 1993. Szlachta na Pomorzu Zachodnim wobec pruskich reform uwłaszczeniowych w poczatkach XIX w. [West Pomeranian nobles in view of the Prussian reforms in the vesting of ownership at the beginning of the nineteenth century.]. *Zapiski Historyczne* 58, no. 4: 29–53.
Stępiński, Włodzimierz. 1994. Szczecin w latach 1806–1870: Na drodze do kapitalizmu. [Stettin in the years 1806–1870. On the course to capitalism.] In *Dzieje Szczecina 1806–1945* [History of Stettin 1806–1945], vol. 3, ed. Bogdan Wachowiak, 9–278. Szczecin: Wydawnictwo "13 Muz".
Stępiński, Włodzimierz. 1998. Szczecińscy filolodzy między państwem a gimnazjum. W kwestii stosunków profesorów gimnazjalnych Gimnazjum Mariackiego do państwowej polityki edukacyjnej w dobie restauracji. [Szczecin philologists between State and school. On the question of the relationship of teachers of the Marienstift-Gymnasium to the state education policy in the time of the Restoration]. *Szczecińskie Studia Historyczne* 11: 49–68.
Stolfi, Jorge. 1991. *Oriented projective geometry*. Boston: Academic Press.
Strohmeier, Katharina, Denis Knauer and Michael Stas, eds. 2007. *Potsdamer Ge(h)schichte 04. Exotisches Potsdam*. Berlin-Brandenburg: be.bra.
Stuloff, Nikolai Nikolajevich. 1968. Über den Wissenschaftsbegriff der Mathematik in der 1. Hälfte des 19. Jahrhunderts. In *Beiträge zur Entwicklung der Wissenschaftstheorie im 19. Jahrhundert*, ed. Alwin Diemer, 71–89. Meisenheim am Glan: Hain.
Sturm, Charles François. 1829. Analyse d'un Mémoire sur la résolution des équations numériques. *Bulletin des Sciences Mathématiques* 11: 419–422.
Sturm, Rudolf, Ernst Schröder and Leonhard Sohncke. 1878. Hermann Grassmann. Sein Leben und seine mathematisch-physikalischen Arbeiten. *Mathematische Annalen* 14: 1–45.
Sühring, Peter. 2005. *Die frühesten Opern Mozarts. Untersuchungen im Anschluß an Jacobsthals Straßburger Vorlesungen*. Kassel: Bärenreiter.

Sühring, Peter. Forthcoming. *Musik inmitten von Natur, Geschichte und Sprache. Gustav Jacobsthal – ein Musikologe im deutschen Kaiserreich. Eine ideen- und kulturgeschichtliche Biografie mit Briefen und Dokumenten.*

Sulloway, Frank J. 1997. *Born to rebel: birth order, family dynamics, and creative lives,* New York: Pantheon.

Sveistrup, Hans. 1929. Friedrich Kuntze. *Kant-Studien*. 34(1–4): 291–299.

Swimmer, Alvin. 1990. *Ceva, Menelaus and the associative law*. (Unpublished manuscript, available from Author).

Swimmer, Alvin. 1996. The completion of Grassmann's Natur-Wissenschaftliche Methode. In [Schubring 1996a, 265–280].

Sylvester, James Joseph. 1850. Additions to the articles in the September number of this journal, "On a new Class of Theorems," and on Pascal's theorem. *Philosophical Magazine* 37: 363–370.

Sylvester, James Joseph. 1852. A demonstration of the theorem that every homogeneous quadratic polynomial is reducible by real orthogonal substitutions to the form of a sum of positive and negative squares. *Philosophical Magazine* 4: 138–142.

Sylvester, James Joseph. 1869. Presidential Address to section 'A' of the British Association. In *The collected mathematical papers of James Joseph Sylvester*, vol. 2, 1908, 650-661; 714–731. Cambridge: University Press.

Sylvester, James Joseph. 1888. The late Arthur Buchheim. *Nature* 38: 515–516.

Tacke, Otto. 1923. Ludwig Giesebrecht als religiöser Erzieher. *Unser Pommerland* 8, no. 4: 107–10.

Tait, Peter Guthrie. 1890. *An elementary treatise on quaternions*. 3rd ed. Oxford: Clarendon Press.

Tappenden, Jamie. 1995. Geometry and generality in Frege's philosophy of arithmetic. *Synthese* 102: 319–361.

The Holy Bible, New International Version. 1984. Colorado Springs: International Bible Society.

Tichy, Eva. 1986. Vedisch *r̥tā́van-* und avestisch *aṣ̌auuan-*. *Die Sprache* 32: 91–105.

Tom Dieck, Tammo. 2000. Topologie. 2nd ed. Berlin and New York: de Gruyter. (De Gruyter Lehrbuch). 264.

Töpfer, Heinrich August. 1793. *Combinatorische Analytik und Theorie der Dimensionszeichen in Parallele gestellt*. Leipzig: Crusius.

Tsabary, Guy and Aviv Censor. 2004. An alternative mathematical model for special relativity. *arXiv:math-ph/0402054v1*. http://arxiv.org/abs/math-ph/0402054 (accessed March 3, 2010).

Turek-Kwiatkowska, Lucyna. 1986. *Obraz przeszłości regionu w świadomości historycznej społeczeństwa pomorskiego w pierwszej połowie XIX w.* [The image of the regional past in the view of the Pomeranian society of the first half of the nineteenth century]. Szczecin: Wydawnictwa Nauk.

Turek-Kwiatkowska, Lucyna. 1997. Wybitni szczecinianie XVIII i XIX w. [Outstanding Stettiners of the eighteenth and nineteenth centuries]. In *Regiony w dziejach Polski. Pomorze Zachodnie, z. 5, Szczecinianie* [Regions in the history of Poland. Part 5, Stettin], ed. Kazimierz Kozłowski, 49–59. Szczecin: PTH.

Turner, R. Steven. 1987. Paradigms and productivity: The case of physiological optics, 1840–1894. *Social Studies of Science* 17: 35–68.

Turner, R. Steven. 1996. The origins of colorimetry: what *did* Helmholtz and Maxwell learn from Grassmann? In [Schubring 1996a, 71–86].

Varèse, Edgard. 1960. Erinnerungen und Gedanken. In *Darmstädter Beiträge zur Neuen Musik*, vol. 3, ed. Wolfgang Steinecker. 65–72. Mainz: Schott.

Varrentrapp, Georg. 1880. Nekrolog. Dr. Alexander Crailsheim. *Jahresbericht über die Verwaltung des Medizinalwesens*. Frankfurt, M.: 252–257.

Vasiliev, Alexander Vasilievitch, ed. 1913. *Novye idei v matematike*. [New ideas in mathematics]. Neperiodich. izd. Sankt Peterburg: Obrazovanie.

Venn, John. 1894. *Symbolic logic*. 2nd ed., "revised and rewritten". London: Macmillan & Co.

VIA = Werba, Chlodwig H. 1997. *VIA Verba Indoarica. Die primären und sekundären Wurzeln der Sanskrit Sprache.* Pars I:*Radices Primariae.* Wien: Verlag der Österreichischen Akademie der Wissenschaften.

Vogel, Barbara. 1983. Beamtenkonservatismus: Sozial- und verfassungsgeschichtliche Voraussetzungen der Parteien in Preussen im frühen 19. Jahrhundert. In *Deutscher Konservatismus im 19. und 20. Jahrhundert*, ed. Bernd-Jürgen Wendt, 1–31. Bonn: Neue Gesellschaft.

Wackernagel, Jacob. 1942. Indoiranica. Aus dem Nachlaß von Jacob Wackernagel, herausgegeben von Albert Debrunner. *Zeitschrift für vergleichende Sprachforschung auf dem Gebiete der indogermanischen Sprachen* (Kuhns Zeitschrift) 67: 154–182 (= *Kleine Schriften I*, Göttingen: Vandenhoeck 1955: 370–398).

Walras, Léon. 1896. *Études d'économie sociale. Théorie de la repartition de la richesse sociale,* Lausanne: Corbaz.

Wang, Hao. 1957. The axiomatization of arithmetic. *Journal of Symbolic Logic* 22: 145–157.

Wang, Hao. 1970. *Logic, computers and sets.* New York: Chelsea.

Warner, Frank W. 1983. *Foundations of differentiable manifolds and Lie groups.* New York: Springer. (Graduate Texts in Mathematics, Vol. 94).

Webb, Judson C. 1980. *Mechanism, mentalism, and metamathematics.* Dordrecht: D. Reidel Pub. Co.

Weber, Heinrich and Josef Wellstein. 1905. *Encyklopädie der Elementar-Mathematik. Ein Handbuch für Lehrer und Studierende.* Vol. 2. *Elementare Geometrie.* Eds. Heinrich Weber, Josef Wellstein and Walther Jacobsthal. Leipzig: Teubner.

Webster's. 1981. *Third new international dictionary of the English language.* Vol. 1 (A to G). Chicago etc.: Encyclopædia Britannica.

Wehrmann, Martin. 1894. *Geschichte des Königlichen Marienstifts-Gymnasiums in Stettin 1544–1894,* Stettin: Herrcke & Lebeling.

Wehrmann, Martin. 1921. *Geschichte von Pommern.* Vol. 2. 2. Gotha: Perthes.

Weierstraß, Karl. 1868. Zur Theorie der bilinearen und quadratischen Formen. *Monatsberichte der Königlichen Preußischen Akademie der Wissenschaften zu Berlin* (1868): 311–338.

Weiss, Christian Samuel. 1803. *Lectures on philosophy at the university of Leipzig: Kants Metaphysische Anfangsgründe der Naturwissenschaft.* Unpublished manuscripts.

Weiss, Christian Samuel. 1804a. Dynamische Ansicht der Krystallisation. In [Haüy 1804a, vol. 1, 365–389].

Weiss, Christian Samuel. 1804b. Ueber die Krystallisation des Feldspathes. In [Haüy 1804a, vol. 2, 711–723].

Weiss, Christian Samuel. 1804c. Entwickelung der hauptsächlichsten Resultate neuerer philosophischer Nachforschungen über das Wesen und die allgemeinsten Verhältnisse der Materie. In [Haüy 1804b, vol. 1, 641–667].

Weiss, Christian Samuel. 1804d. Ueber die Zustände von Festigkeit und Flüssigkeit und ihre verschiedenen Arten. In [Haüy 1804b, vol. 1, 668–699].

Weiss, Christian Samuel. 1804e. Über Licht, Wärme, Electricität und Magnetismus. In [Haüy 1804, vol. 2, 601–647].

Weiss, Christian Samuel. 1810–1817. *Lectures on philosophy at the university of Berlin.* Unpublished manuscripts.

Werba, Chlodwig H. 1997. *VIA Verba Indoarica. Die primären und sekundären Wurzeln der Sanskrit Sprache.* Pars I: *Radices Primariae.* Wien: Verlag der Österreichischen Akademie der Wissenschaften.

Wettersten, John. 1990. Integrating psychology and methodology. How can psychology and methodology be integrated? *Journal for General Philosophy of Science* 21: 293–308.

Wheeler, Lynde Phelps. 1951. *Josiah Willard Gibbs, the history of a great mind,* Yale University Press. [Reprinted 1998. Woodbridge: Ox Bow Press].

White, Neil L. 1975. The Bracket ring of combinatorial geometry I. *Transactions of the American Mathematical Society* 202: 79–103.

White, Neil L. 2005. Grassmann-Cayley algebra and robotics applications. In *Handbook of geometric computing*, ed. Eduardo Bayro-Corrochano. 629–656. Heidelberg: Springer.

Whitehead, Alfred North. 1897–98. The geodesic geometry of surfaces in non-Euclidean space. *Proceedings of the London Mathematical Society* 29: 275–234.

Whitehead, Alfred North. 1898. *A treatise on universal algebra with applications*. Cambridge: University Press.

Whitehead, Alfred North. 1929. *Process and reality. An essay in cosmology*. Cambridge: University Press.

Whitehead, Alfred North. 1978. *Process and reality, Corrected edition*. New York: The Free Press.

Wielopolski, Alfred. 1965. *Ustrój polityczny Pomorza Zachodniego w XIX wieku*. [The political system of West Pomerania in the nineteenth century], Poznań: Wydawnictwo Poznańskie.

Wienand, Sabine. 2009. Der Vater der Vektoren, *Frankfurter Allgemeine Sonntagszeitung*, 13. Sept. 2009. http://www.faz.net/ (accessed March 3, 2010).

Willis, Robert. 1830. On the vowel sounds, and on reed organ-pipes. *Transactions of the Cambridge Philosophical Society* 3: 231–252 [+ 3 plates].

Windisch, Ernst. 1920. *Geschichte der Sanskrit-Philologie und indischen Altertumskunde*. Vol. 2. Berlin: Walter de Gruyter.

Witzel, Michael and Toshifumi Gotō, eds. 2007. *Rig-Veda. Das Heilige Wissen: Erster und zweiter Liederkreis*. Transl. of Rig-Veda by Michael Witzel and Toshifumi Gotō, with Eijirō Dōyama and Mislav Ježić. Frankfurt and Leipzig: Verlag der Weltreligionen im Insel Verlag.

Wu, Wen-Tsun. 1984. *Basic principles of mechanical theorem proving in geometries*. Vol. I: *Part of elementary geometries*. Beijing: Science Press. (Springer, 1994).

Wundt, Wilhelm. 1910. *Grundzüge der physiologischen Psychologie*. Vol. 2. Leipzig: Engelmann.

Yanovskaya, Sofya Aleksandrovna. 1958. Iz istorii aksiomatiki. [On the history of axiomatics]. *Istoriko-matematicheskie issledovaniya*, Vyp. 11. 63–96. Moskva: Fizmatgiz.

Zaddach, Arno. 1988. *Algebra de Grassmann y Geometrı a Proyectiva*. Universidad de Tarapacá: Facultad de Ciencias (Chile).

Zaddach, Arno. 1994. *Grassmanns Algebra in der Geometrie mit Seitenblicken auf verwandte Strukturen*. Mannheim et al.: Bibliographisches Institut.

Zaddach, Arno. 1996. Regressive products and Bourbaki. In [Schubring 1996a, 285–295].

Ziche, Paul. ed. 1999. *Introspektion. Texte zur Selbstwahrnehmung des Ichs*. Wien and New York: Springer.

Ziche, Paul. ed. 2002. Wilhelm Thiery Preyers Psychomathematik: Die Mathematisierung der reinen Empfindungen. *Psychologie und Geschichte* 10: 283–295 (= *Empirische Psychologie und philosophischer Empirismus*. Schlaglichter auf 250 Jahre Psychologiegeschichte. Festschrift Georg Eckardt, eds. Kitty Dumont, Matthias John and Temilo van Zantwijk).

Ziche, Paul. ed.. 2008. *Wissenschaftslandschaften um 1900. Philosophie, die Wissenschaften und der nicht-reduktive Szientismus*. Zürich: Chronos.

Ziche, Paul. ed.. 2009. Wilhelm Ostwald als Begründer der modernen Logik. Logik und künstliche Sprachen bei Ostwald und Louis Couturat. In *Ein Netz der Wissenschaften? Wilhelm Ostwalds "Annalen der Naturphilosophie" und die Durchsetzung wissenschaftlicher Paradigmen*, eds. Pirmin Stekeler-Weithofer, Heiner Kaden and Niko Psarros. 46–66. Leipzig: Sächsische Akademie der Wissenschaften.

Index of names and citations

The following index offers dates of birth and death as well as brief biographical notes on figures of historical significance. Page numbers after such entries indicate where a person is mentioned in the book. Name entries in italic type refer to authors quoted in this book. The respective year of publication is given in square brackets, followed by the page numbers indicating where these publications are cited.

E

F

S

MIX
Papier aus verantwortungsvollen Quellen
Paper from responsible sources
FSC® C105338

If you have any concerns about our products,
you can contact us on
ProductSafety@springernature.com

In case Publisher is established outside the EU,
the EU authorized representative is:
Springer Nature Customer Service Center GmbH
Europaplatz 3, 69115 Heidelberg, Germany

Printed by Libri Plureos GmbH
in Hamburg, Germany